Bivalve Molluscs

Biology, Ecology and Culture

Bivalve Molluscs

Biology, Ecology and Culture

Elizabeth Gosling

Fishing News Books
An imprint of Blackwell Science

© 2003 by Fishing News Books, a division of Blackwell Publishing
Editorial offices:
Blackwell Publishing Ltd, 9600 Garsington Road, Oxford OX4 2DQ, UK
  Tel: +44 (0)1865 776868
Blackwell Publishing, Inc., 350 Main Street, Malden, MA 02148-5020, USA
  Tel: +1 781 388 8250
Blackwell Science Asia Pty, 550 Swanston Street, Carlton, Victoria 3053, Australia
  Tel: +61 (0)3 8359 1011

First published 2003
Reprinted 2004

Library of Congress Cataloging-in-Publication Data

Gosling, E.M.
  Bivalve molluscs / Elizabeth Gosling.
     p.   cm.
Includes bibliographical references.
  ISBN 0-85238-234-0 (alk. paper)
  1. Bivalvia. I. Title.
  QL430.6 .G67 2002
  594'.4–dc21   2002010263

ISBN 0-85238-234-0

A catalogue record for this title is available from the British Library

Set in 10.5/12pt Bembo
by SNP Best-set Typesetter Ltd., Hong Kong
Printed and bound in Great Britain
by MPG Books Ltd, Bodmin, Cornwall

The publisher's policy is to use permanent paper from mills that operate a sustainable
forestry policy, and which has been manufactured from pulp processed using acid-free
and elementary chlorine-free practices. Furthermore, the publisher ensures that the text
paper and cover board used have met acceptable environmental accreditation standards.

For further information on Blackwell Publishing, visit our website:
www.blackwellpublishing.com

Contents

Preface

The phylum Mollusca is one of the largest, most diverse and important groups in the animal kingdom. There are over 50 000 described species and about 30 000 of these are found in the sea. The class Bivalvia is one of six classes of molluscs and comprises animals enclosed in two shell valves. Examples are mussels, oysters, scallops and clams. Although the class contains a relatively small number of species, about 7500, it elicits substantial interest chiefly because so many of its members are eaten by people in large amounts. In 1999, production of bivalves from fisheries and aquaculture was over 10.6 million metric tonnes worldwide with a monetary value of over $9.3 billion (FAO, 2001).

Some years ago while teaching a course on Bivalve Biology to aquaculture students I realized that although the students could access information from several texts, a single book covering all aspects of the biology, ecology and culture of bivalve molluscs did not exist. Thus the idea to write such a book was conceived. What started off as a small undertaking very quickly snowballed into a substantial task, which took several years to complete, primarily due to my having to read and digest the wealth of published information. The focus of the book is on marine bivalves of commercial importance and it has been written primarily for undergraduate students. However, the book with its 1200 references will serve also as a useful starting point for postgraduate students and professionals working on reproductive physiology, ecology, genetics, disease, public health, aquaculture, and fisheries management of commercial bivalves.

Chapter 1 is a brief introduction to the class Bivalvia, and describes the evolution of the group from a creeping ancestral mollusc to the burrowers and attached surface dwellers of today. Chapter 2 provides a detailed description of external and internal anatomy, while Chapter 3 describes global and local distribution patterns, and the physical and biological factors influencing distribution and abundance. Filter feeding and controlling factors are covered in Chapter 4. Chapter 5 deals with reproduction, larval development and settlement, while Chapter 6 covers methods of measuring growth, and the various factors influencing growth. The processes of circulation, respiration, excretion and osmoregulation are described in Chapter 7. The fishery assessment and management methods that are used in commercial fisheries of mussels, oysters, scallops and clams are covered in Chapter 8. The fundamentals of bivalve aquaculture, and the application of genetic methods are treated in Chapters 9 and 10, respectively. Chapter 11 deals with diseases and parasites, and the diversity of defense mechanisms utilized by bivalves. Finally, Chapter 12 deals with the role of bivalves in disease transmission to humans.

Acknowledgements

I would like to express my gratitude to the many colleagues and friends who have helped and given advice during the preparation of this book. In particular I wish to thank my partner Jim, and son Daniel, for their continual support and patience. My sincere thanks to the following people who generously took time to review chapters, or parts of chapters: Andy Beaumont, Peter Beninger, Andy Brand, Norbert Dankers, Antonio Figueras, Susan Ford, Jim Gosling, Philippe Goulletquer, Pauline King, Per Kristensen, Dave McGrath, Mark Norman, Kevin O Kelly, Brian Ottway, Maarten Ruth, Matthias Seaman, Ray Seed, Sandra Shumway and James Weinberg.

The following people gave invaluable advice and information, particularly for the Bivalve Culture, and Fisheries and Management chapters: Gavin Burnell, Ryan Carnegie, Jonathan Clarke, Noel Coleman, Eric Edwards, Luca Garibaldi, Ximing Guo, David Gwyther, Wolfgang Hagena, Marc Herlyn, Bob Hickman, Jaap Holstein, C. Barker Jørgensen, Yoshinobu Kosaka, Ian Laing, John Lucas, Clyde MacKenzie Jr., K.S. Naidu, John Nell, Maurizio Perotti, Bettina Reineking, David Roberts, Luisa Salaris, Aad Smaal, Sue Utting and J. Evan Ward.

My gratitude to all those who supplied slides or artwork: Andy Beaumont, Peter Beninger, Susan Bower, Craig Burton, John Costello, Caroline Cusack, Per Dolmer, Antonio Figueras, Susan Ford, Philippe Goulletquer, Niall Herriott, Akira Komaru, Maria Lyons Alcantara, Dave McGrath, Paul Moran, Kevin O Kelly, Brian Ottway, Christine Paillard, Linda Park, Chris Richardson, Maarten Ruth, Ray Seed, Gudrun Thorarinsdóttir, Michelle (Paraso) Tomlinson, James Weinberg and Michael Wilkins.

To my colleagues in the School of Science, Galway–Mayo Institute of Technology for their steadfast support: Patricia Dineen, Imelda Divelly-McCann, Pauline King, Kathleen Lough, Declan Maher, Dave McGrath, Íde Ní Fhaolain, Brian Ottway, Colin Pybus, Gerry Quinn and Malachy Thompson. A very special thanks to my office mate, Miriam Pybus, who saw this book from conception to delivery. Thanks also to the library staff of the GMIT, the National University of Ireland, Galway (NUI Galway), Amanda Mahon, librarian, at the Marine Institute, Angela Gallagher (NUI Galway) who gave me helpful advice on artwork, and Noel Wilkins (NUI Galway) for his constant encouragement during the writing of this book.

Finally, thanks to the team at Blackwell Science: Nigel Balmforth, Fionnguala Sherry-Brennan, Steve Lockwood and Josie Severn, for their efforts in keeping me on course from start to finish.

1 An Introduction to Bivalves

The phylum Mollusca is one of the largest, most diverse and important groups in the animal kingdom. There are over 50 000 described species in the phylum and about 30 000 of these are found in the sea. Molluscs are soft-bodied animals but most are protected by a hard protective shell. Inside the shell is a heavy fold of tissue called the mantle. The mantle encloses the internal organs of the animal. Another feature of the phylum is a large muscular foot that is generally used for locomotion.

Although most molluscs share this basic body plan the group is characterised by a great diversity of form and habit. As Morton (1967) aptly puts it:

> 'Molluscs range from limpets clinging to rocks, to snails which crawl or dig or swim, to bivalves which anchor or burrow or bore, to cephalopods which torpedo through the water or lurk watchfully on the bottom. They penetrate all habitats: the abysses of the sea, coral reefs, mudflats, deserts, and forests, rivers, lakes and under ground. They may become hidden as parasites in the interior of other animals. They feed on every possible food and vary in size from giant squids and clams to little snails a millimetre long.'

There are six major classes of molluscs. The Class Gastropoda is the largest (40 000 species) and most diverse, containing spirally coiled snails, flat-shelled limpets, shell-less sea slugs and terrestrial snails and slugs (Fig. 1.1). The Class Bivalvia with about 7500 species, includes animals with two shell valves such as mussels, oysters, scallops and clams. Octopus, squid and cuttlefish are in the Class Cephalopoda. There are about 650 species in this class and they represent the most organised and specialised of all the molluscs. The Classes Polyplacophora (chitons) and Scaphopoda (tusk shells) together contain about 1000 species.

The primitive Class Monoplacophora, known only before 1952 as limpet-like fossils from the Devonian and Cambrian eras (300–500 mya) contains the single Genus *Neopilina*. In many respects *Neopilina* resembles the hypothetical form that zoologists have invented as the ancestral mollusc. This is envisioned as a small, shelled animal that lived in shallow pre-Cambrian seas, and crept over the substrate on a large foot, scraping algae off the rocks with its specialised mouth parts (Fig. 1.2). At the posterior of the animal was a pair of ciliated filamentous gills, which functioned solely as respiratory organs. It is believed that all present-day molluscan classes share a common ancestor that closely resembled one such archetype mollusc. As this book is concerned with bivalve molluscs, attention will now be fully focused on the Class Bivalvia, also known as Lamellibranchia.

The bivalves are in some ways the most highly modified of all the molluscs. Over evolutionary time they have become flattened side to side (Fig. 1.3). Two mantle lobes that secrete two shell valves, hinged dorsally, cover the body organs. Two muscles, the anterior and posterior adductors, control the opening and closing of the shell valves. Lateral compression has resulted in

1

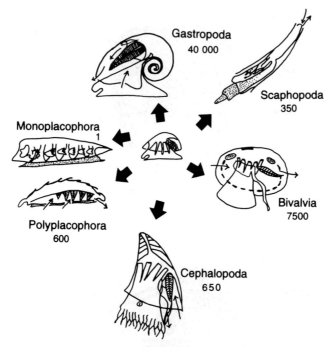

Fig. 1.1. The radiation of the six classes of molluscs from a hypothetical ancestor. Approximate numbers of species in each class are included. Adapted from Morton (1967).

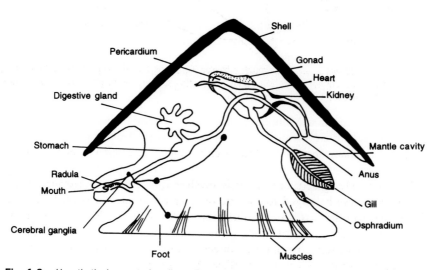

Fig. 1.2. Hypothetical ancestral mollusc. From Willmer (1990).

the raising of the mouth off the substrate, and the role of catching food has shifted from the mouth to the two gills. As a result, these have become enormously enlarged, and function as one of the most efficient systems of ciliary feeding in the Animal Kingdom. The foot has lost its flat creeping sole, is wedge-shaped and must extend out between the valves in order for the animal

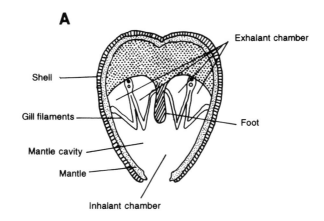

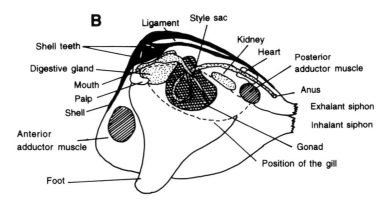

Fig. 1.3. (A) Transverse section through a bivalve illustrating lateral compression and the position of the mantle, foot and gills. (B) Longitudinal section showing the major organs; gill omitted for clarity. Adapted from Barnes *et al.* (1993).

to move. In many species the foot is either lost entirely, or is greatly reduced in size.

It is generally agreed that the early bivalves were shallow burrowers in soft substrates. These belonged to the Protobranchia and are represented by fossil forms that date back to the Cambrian era (500 mya), and also by some living forms such as the little nut shells, genus *Nucula*. These lie just barely covered in muddy sand, with the anterior directed downward and the posterior end directed toward the soil–water interface. *Nucula* is a typical isomyarian bivalve, i.e. anterior and posterior adductor muscles are about the same size. Unlike most other bivalves, the flow of water into the animal is from anterior and posterior directions (see Fig. 1.4A). *Nucula* feeds on surface deposits by means of palps, long fleshy extensions of the mouth. Therefore, the gills are primarily respiratory organs. The development of labial palps was perhaps a necessary stage in the evolution of filter feeding, making it possible for the mouth to be lifted off the substrate. There is no doubt that study of this group makes

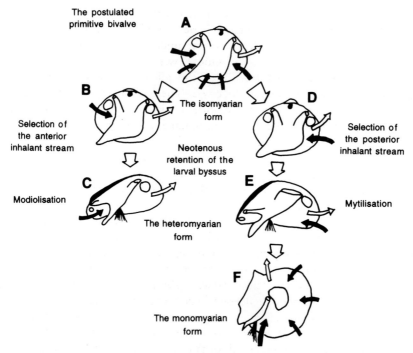

Fig. 1.4. The evolution of the heteromyarian form, and ultimately monomyarian form, from an isomyarian ancestor. (A) Postulated primitive isomyarian bivalve such as *Nucula* or *Glycymeris*, with water capable of entering the mantle from anterior and posterior directions. (B) Selection of the anterior inhalant stream by representatives of such groups as the Lucinoidea (shallow burrowers in tropical mud) can only result in the process of heteromyarianisation leading to (C), a modioliform shell found in ark shells, *Arca*. (D) Selection of the posterior inhalant stream can result in full expression of the heteromyarian form (E) e.g. in the mussel, *Mytilus*, and ultimately, the monomyarian form (F), in oyster and scallop species. Reprinted from Morton (1992) with permission from Elsevier Science.

it easier to understand the transition from the primitive mollusc to more modern bivalves.

One of the most important developments in the evolution of modern bivalves was moving the site of water intake to the posterior of the animal (Fig. 1.4D). This made it possible for bivalves to penetrate sand or mud 'head first' with the posterior end in free communication with the water above. Extensions of the mantle to form siphons at the posterior enabled the animals to live deeper and deeper under the surface. As bivalves evolved, plankton in the incoming current was increasingly adopted as a source of food, the gills replacing the palp processes as the feeding organs.

The chief modification of the gills for filtering was the lengthening and folding of individual gill filaments. In addition, many extra filaments were added so that they extended as far forward as the labial palps. Both of these modifications greatly increased the surface area of the gills. It is believed that the triangular-shaped filaments of the primitive bivalve gill progressively changed over evolutionary time to the W-shaped filaments of the modern bivalve gill (Barnes, 1980). There is a notch at the bottom of each side of the W and these line up with similar notches on adjacent filaments to form a food groove that extends the length of the underside of the gill. Yonge (1941) has suggested that since the food groove was necessary for nutrition these notches probably preceded folding of the gill filaments. Changes in both ciliation and water circulation followed. The exploitation of filter feeding made it possible for bivalves to colonise a wide variety of habitats that had hitherto been inaccessible to their protobranch ancestors.

Burrowing into the substrate is the habit most extensively exploited by bivalves. They burrow into the protection offered by sand, mud or gravel using the foot. Contact is maintained with the surface by way of siphons that extend from the posterior end of the animal (Fig. 1.3B). Water flows in through the inhalant siphon, through the gills, where filtering of suspended food particles takes place, and exits through the exhalant siphon. Some bivalves burrow only deep enough to cover the shell and have correspondingly short siphons and others burrow very deeply and have long siphons. Cockles (e.g. *Cardium* spp.) are shallow burrowers, while many clam species, e.g. razor clams (*Siliqua*, *Ensis*), burrow as deep as 60 cm. The geoducks (*Panopea*) on the west coast of the United States are among the deepest burrowers, digging down to a depth of over a metre. Their siphons are so large that they can no longer be retracted into the shell. Other adaptations to a deep burrowing lifestyle include the development of a streamlined shell for fast burrowing and fusion of the mantle edges (apart from a small gape for the large muscular foot) to prevent entry of sediment into the mantle cavity.

Many bivalves that burrow deeply (>30 cm) live in permanent burrows, moving deeper as they grow larger. When dug out, large individuals cannot re-burrow. This lifestyle is brought to an extreme by bivalves that bore into hard substrates such as shell, coral, wood and rock. Sharp-edged ridges on the anterior face of the shell valves facilitate the task of boring. In shell and coral-borers there is evidence that an acid-mucous secretion of the mantle edges helps to dissolve the calcareous substrate. Boring bivalves are permanently locked in their burrows and are, therefore, inevitably dependent on outside sources of food. However, in wood-boring bivalves excavated 'sawdust' is the principal food source and phytoplankton is only used to supply the nitrogen and vitamins missing from an all-wood diet.

Although most bivalves have adopted the burrowing lifestyle there is a large and successful group that lives permanently attached on the surface. Attachment is provided either by byssus threads, or by a cement-like substance that fixes one valve to the substrate. The byssal apparatus is seen as a persistent post-larval structure that evolved for temporary attachment of the animal to the substrate during the vulnerable stage of metamorphosis. In most species of oysters, clams and scallops the byssal apparatus is subsequently lost. However, in mussels it persists into adult life (Fig. 1.4C & E). This return to the epifaunistic life of their ancestors is believed to have occurred early on in the evolution of bivalves, about 400 mya (Yonge & Thompson, 1976). In byssally attached forms there has been a tendency for the anterior (head) end of the animal to become smaller with a corresponding enlargement of the posterior end. Accompanying this change there has also been a reduction of the anterior adductor muscle and an increase in the size of the posterior adductor muscle. The evolution of this heteromyarian form led to the development of a pronounced triangular shape. This is very marked in mussels in the Family Mytilidae (see Fig. 1.4E) and is believed to be an adaptation to living in clusters; expansion of the posterior shell allowing free access, posteriorly, to the water above (Morton, 1992).

The heteromyarian condition has been seen as a stepping-stone towards the monomyarian form and the adoption of a horizontal posture (Fig. 1.4F). Monomyarian bivalves have largely circular shells, all trace of the anterior adductor muscle is lost, and the body has been reorganised around the enlarged and more or less centrally-placed posterior muscle. Water enters around two-thirds or more of the rounded margins of the shell. In oysters the shell is fixed to the substrate by either the right or left valve, depending on the species. Shell attachment has led to varying degrees of inequality in the size of the two shell valves.

Scallops also display the same monomyarian condition as oysters. The shell valves are circular but both can be concave and similar, or the left (uppermost) valve may be flat. Like oysters, they also lie in a horizontal position on the substrate. However, scallops far from being fixed are active, swimming bivalves. In early life they use byssus threads for attachment to algae, but before they attain a size of 15 mm the majority of species have detached themselves to take up a free-living existence on the seabed.

The number of species of bivalve molluscs is only about 20% of that documented for gastropods. Yet, there is substantial interest in this group chiefly because so many of its members are eaten by humans in large amounts. In the following chapters attention will be focused only on bivalves of commercial importance: mussels, oysters, scallops and clams. While the general term 'shellfish' will sometimes be used to refer to this group, the author is well aware that for many people the term has a wider meaning and incorporates many other non-bivalve molluscs not dealt with in this book, such as abalone, periwinkles, whelks and also crustaceans such as crabs, prawns and shrimp.

References

Barnes, R.D. (1980) *Invertebrate Zoology*, 4th edn. W.B. Saunders Co., Philadelphia.

Barnes, R.S.K., Callow, P. & Olive, P.J.W. (1993) *The Invertebrates: a New Synthesis*, 2nd edn. Blackwell Scientific Publications, Oxford.

Morton, B. (1992) The evolution and success of the heteromyarian form in the Mytiloida. In: *The Mussel Mytilus: Ecology, Physiology, Genetics and Culture* (ed. E.M. Gosling), pp. 21–52. Elsevier Science Publishers B.V., Amsterdam.

Morton, J.E. (1967) *Molluscs*. Hutchinson University Library, London.

Willmer, P. (1990) *Invertebrate Relationships: Patterns in Animal Evolution*. Cambridge University Press, Cambridge.

Yonge, C.M. (1941) The Protobranchiata Mollusca: a functional interpretation of their structure and evolution. *Philos. Trans. R. Soc. Lond.*, Ser. B., **230**, 79–147.

Yonge, C.M. & Thompson, T.E. (1976) *Living Marine Molluscs*. William Collins, Glasgow.

2 Morphology of Bivalves

Introduction

In this chapter the approach is to use much studied, representative species from each of the groups: mussels, oysters, scallops and clams rather than to attempt an exhaustive description of the morphology of all bivalve species. Instead, for mussels, species from the genus *Mytilus* have been chosen; for oysters *Ostrea* and *Crassostrea* species; for scallops *Pecten maximus* and for clams *Mercenaria mercenaria*.

The general morphology and functions of the shell, mantle, foot, gill, alimentary canal, gonad, heart, kidney and nervous tissue are described in the following sections. Additional information on their particular roles in feeding, reproduction, circulation, excretion and osmoregulation, is presented in Chapters 4, 5 and 7.

The shell

The mollusc shell has several functions: it acts as a skeleton for the attachment of muscles, it protects against predators, and in burrowing species it helps to keep mud and sand out of the mantle cavity.

Its main component is calcium carbonate and is formed by the deposition of crystals of this salt in an organic matrix of the protein, conchiolin. Three layers make up the shell: (1) a thin outer periostracum of horny conchiolin, often much reduced due to mechanical abrasion, fouling organisms, parasites or disease, (2) a middle prismatic layer of aragonite or calcite, a crystalline form of calcium carbonate, and (3) an inner calcareous (nacreous) layer, that is either of dull texture or iridescent mother-of-pearl, depending on the species.

Very early in larval development an area of ectodermal cells in the dorsal region of the developing embryo secretes the first larval shell. The secretion of a second larval shell by the mantle, rather than the shell gland, follows soon after. Following metamorphosis the secretion of the adult shell begins. This is more heavily calcified, has different pigmentation and more conspicuous sculpturing than the larval shell. The outer mantle fold secretes the periostracum and prismatic layers, while the inner nacreous layer is secreted by the general mantle surface (see below). The shell grows in circumference by the addition of material from the edge of the mantle, and grows in thickness by deposition from the general mantle surface. Calcium for shell growth is obtained from the diet, or taken up from seawater. Carbonate is derived from the CO_2/bicarbonate pool in the animal's tissues. The energy required for shell growth is not an insignificant portion of a bivalve's total energy budget (Hawkins & Bayne, 1992).

The colour, shape and markings on the shell vary considerably between the different groups of bivalves. Not surprisingly therefore, shell characters are consistently used in species identification (Table 2.1).

Table 2.1. The major shell characters used in species identification.

Character	Variations
Shell shape	Oval, circular, triangular, elongate, quadrate
Shell valves	Similar (equivalve), or dissimilar (inequivalve)
Colour	Shell exterior: background/surface; patterns
	Shell interior: white, pearly etc.
Ribs	Number, width, prominence (distinct, flattened)
Sculpturing	Concentric lines, ridges, grooves
Ligament	Position (internal, external)
Umbo	Position (anterior, terminal, subterminal)
Adductor scars	Number, size, position
Hinge line	Straight or curved, presence of 'ears' (size, shape)
Hinge teeth	Number, type
Pallial line	With or without a sinus
Pallial sinus	Size

Mussels

In mussels the two shell valves are similar in size, and are roughly triangular in shape (Fig. 2.1A & B). The valves are hinged together at the anterior by means of a ligament. This area of the shell is called the umbo. When empty shells are examined the interior of the shell is white with a broad border of purple or dark blue. This is called the pallial line and is the part of the shell along which the mantle is attached. On the inside of each valve are two muscle scars, the attachment points for the large posterior adductor muscle and the much reduced anterior adductor muscle. The anterior adductor scar is absent in *Perna* species. Anterior and posterior retractor muscles are also attached to the shell, these control the movement of the foot (see below). The foot in turn secretes a byssus, a bundle of tough threads of tanned protein. These threads emerge through the ventral part of the shell and serve as mooring lines for attachment of the mussel to the substrate or other mussels.

The colour of mussel shells is controlled by several genes (Innes & Haley, 1977; Newkirk, 1980) and also varies depending on the age and location of the animal (Mitton, 1977). In the intertidal zone the blue mussel *Mytilus edulis* has a blue-black and heavy shell, while in the sublittoral region, where mussels are continuously submerged, the shell is thin and brown with dark brown to purple radial markings. In the rock mussel *Perna perna* the shell is red-maroon with irregular patches of light brown and green. Juvenile green mussels, *Perna viridis*, have bright green or blue-green shells, but older individuals tend to have more brown in the shell (Siddall, 1980).

The presence of concentric rings on bivalve shells has been extensively used to age them. In many species of scallops and clams these rings have been shown to be annual in origin and therefore can be used as a reliable estimate of age. However, in mussels there are few geographic locations where the rings provide an accurate estimate of age (Lutz, 1976). One must, therefore, resort to shell sectioning for age determination. When longitudinal sections are examined microscopically, distinct growth bands in the inner nacreous layer

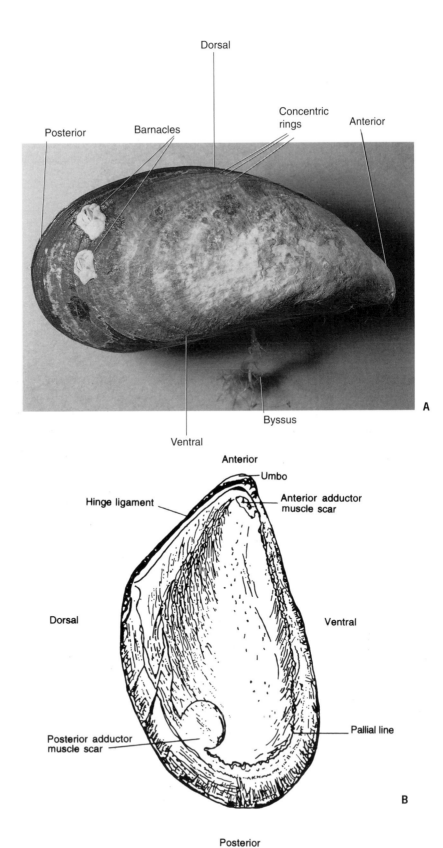

Dorsal

Posterior   Barnacles   Concentric
                          rings   Anterior

A

Byssus

Ventral

Anterior

Umbo

Hinge ligament   Anterior adductor
                  muscle scar

Dorsal   Ventral

Posterior adductor
muscle scar   Pallial line

B

Posterior

Fig. 2.1. The external (A), internal (B) shell and (C) internal features of the mussel *Mytilus edulis*.
Photos © Craig Burton 2001.

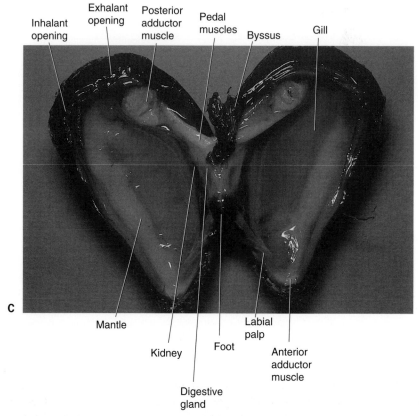

Fig. 2.1. *Continued*

are clearly seen (see Chapter 6 Figs. 6.6 & 6.7). These are formed at a rate of one per year during spring (Lutz, 1976). In addition, in the middle prismatic layer there are micro-growth bands which have a tidal periodicity (Richardson, 1989). These bands, together with the annual bands, can be used to track individual short-term and long-term variations in growth rates, respectively.

The convention used for taking the principal shell measurements is shown in Fig. 2.2. Height is the distance from the hinge line to the shell margin. Length is the widest part across the shell at 90 degrees to the height. The width is measured at the thickest part of the two shell valves (Dore, 1991).

Under optimal conditions, such as in the sublittoral zone, *M. edulis* and the Mediterranean mussel, *M. galloprovincialis*, attain a shell length of 100–130 mm, whereas in marginal conditions, e.g. the high intertidal zone on an exposed shore, mussels may measure as little as 20–30 mm, even after 15–20 years (Seed, 1976). Shell shape is also very variable in these two mussel species. The shells of densely packed mussels have higher length to height ratios than those from

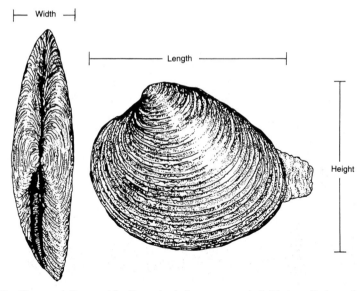

Fig. 2.2. The convention used for the main shell measurements in bivalves. Redrawn from Dore (1991).

less crowded conditions. This is most extreme in older mussels and ensures that they can more readily exploit posterior feeding currents, since they are effectively elevated above younger mussels in the same clump (Seed & Suchanek, 1992).

Mussels have been extensively used to assess environmental contamination. Radionucleotides and metals such as uranium, vanadium and lead are highly concentrated in the outer periostracal layer of the shell (Widdows & Donkin, 1992; Livingstone & Pipe, 1992). Some anti-fouling agents, e.g. tributyl tin (TBT), cause shell deformities, characterised by the production of cavities within the shell, in both mussels and oysters.

Oysters

In the European flat oyster, *Ostrea edulis*, the shell valves are approximately circular and are hinged together on the dorsal side by a horny ligament (Fig. 2.3Ai). The right valve is flat while the left is cupped (Fig. 2.3Aii). At rest on the sea-bed the flat valve is uppermost and the cupped valve is cemented to the substrate. The American eastern oyster, *Crassostrea virginica*, also has dissimilar valves, but the general shell shape is more elongated and the left valve is more deeply cupped than in *Ostrea* (Fig. 2.3Bi & ii). In both oyster species the shell colour is off-white, yellowish or cream but often with purple or brown radial markings in *C. virginica*. The inside of the shell valves is pearly-white and there is a single large adductor scar. The shell in both oyster groups

A i

A ii

Fig. 2.3. External shell of (A) *Ostrea edulis* (i) & (ii) and (B) *Crassostrea gigas* (i) & (ii). (C) Internal features of *C. gigas*. © Craig Burton 2001.

is thick and solid and both valves have distinct concentric sculpturing, with the surface of the cupped valve more raised and frilled in *Crassostrea*. The concentric markings cannot be used to age oysters, and one must resort to sectioning of the shell or ligament for an accurate estimate. In general, *Ostrea edulis* has a maximum shell height of 100 mm, while *C. virginica* grows as large as 350 mm. For a very comprehensive account of larval and adult shell structure in *C. virginica* see Carriker (1996).

B i

B ii

Fig. 2.3. *Continued*

Scallops

Scallops more than any other group of bivalves have attracted the interest of naturalists and collectors for centuries. 'In appearance no other molluscan shells have so pleasing a design and range of colours as pecten shells . . .' (Cox, 1957). In the king scallop, *Pecten maximus*, the left valve is flat and is slightly over-lapped by the right one, which is convex (Fig. 2.4A). Adults recess in the substrate with the flat valve uppermost. The two valves, which are roughly circular, are held together along the hinge line by a rubbery internal ligament. Typically, there are 15–16 ribs radiating from the hinge. These alternate with

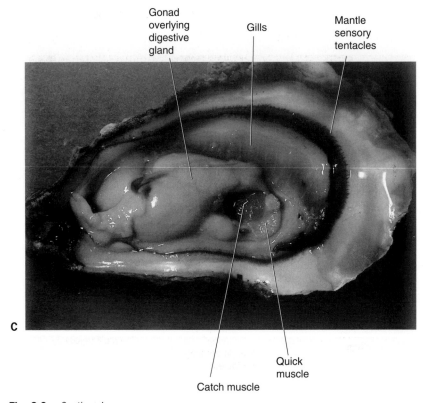

Gonad overlying digestive gland

Gills

Mantle sensory tentacles

c

Quick muscle

Catch muscle

Fig. 2.3. *Continued*

grooves and give the scallop its distinct comb-like appearance. There are two projecting 'ears' or auricles on either side of the umbo; these vary in size and shape and are used, along with other shell characters, to differentiate one species of scallop from the next (see Table 2.1 for shell characters used in species identification). There is a large, centrally placed, adductor muscle, a standard seafood commodity that is widely traded and universally available (Dore, 1991). Distinct annual rings on the shell, make ageing of scallops a relatively easy task, compared to mussels and oysters. Shell size in scallops varies quite a bit depending on the species: *P. maximus* can be up to 150 mm in length, while the sea scallop, *Placopecten magellanicus*, and the yesso scallop, *Patinopecten yessoensis*, can reach a size of 200–230 mm. Other species such as the queen scallop, *Aequipecten* (*Chlamys*) *opercularis*, the Icelandic scallop, *Chlamys islandica* and the bay scallop, *Argopecten irradians*, seldom grow larger than 100 mm.

The beautiful colours that are a feature of scallop shells are laid down when the shell is being formed. In *P. maximus* the colours range from off-white to yellow to light brown, often overlaid with bands or spots of darker pigment. In *A. irradians*, background colour of the shell and overlying pigment distribution appear to be coded by at least two separate genes (Adamkewicz & Castagna, 1988).

Fig. 2.4. (A) External shell and (B) internal anatomy of the scallop, *Pecten maximus*. © Craig Burton 2001.

Clams

Clams are a very diverse group of bivalves in that there is notable variation in the shape, size, thickness, colour and degree of sculpturing of the shell from one species to the next. The one feature that all clams have in common is that they burrow into the sea-bed. Consequently, both shell and body (see below) display modifications necessary for this type of existence. The quahog clam, *Mercenaria mercenaria*, has a thick, triangular shell (Fig. 2.2). It is grey or brown with a sculpturing of numerous shallow concentric rings, which run around the shell, parallel to the hinge. Annual rings are clearly visible on the shell exterior and thus ageing in this species, and indeed in many of the other commercially important clam species, is an easy task. The inside of the shell is glossy white, often with bluish-purple tints. It was this feature that made them valued as currency in earlier times (Dore, 1991). There are three conspicuous teeth on each valve and each tooth fits into a corresponding socket on the opposing valve (Fig. 2.5). This ensures an intimate fit when the valves are closed. The shell interior is marked by an anterior and posterior adductor muscle scar, a distinct pallial line and a short pallial sinus – the indentation indicating the position of the retracted siphons in the closed shell (see below). The depth of the pallial sinus is a very reliable indicator of the length of the siphons, and thus the burrowing depth of a particular clam species.

Mercenaria mercenaria and the softshell clam, *Mya arenaria*, can reach a shell length of 150 mm, while the surf clam, *Spisula solidissima*, grows as large as 220 mm. The palourde, *Ruditapes decussata* and the Manila clam, *Ruditapes philippinarum* (Fig. 2.6A) are much smaller, with a maximum shell length of

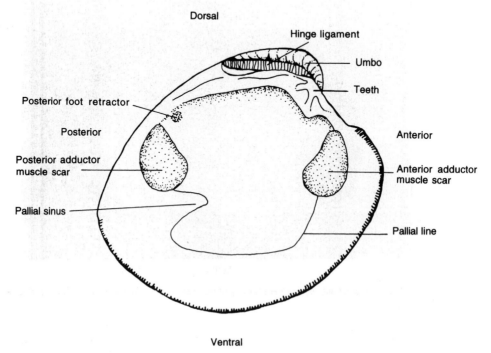

Fig. 2.5. Inner surface of the left shell valve of the clam, *Mercenaria mercenaria*. From Wallace & Taylor (1997). Reprinted with permission from Pearson Education.

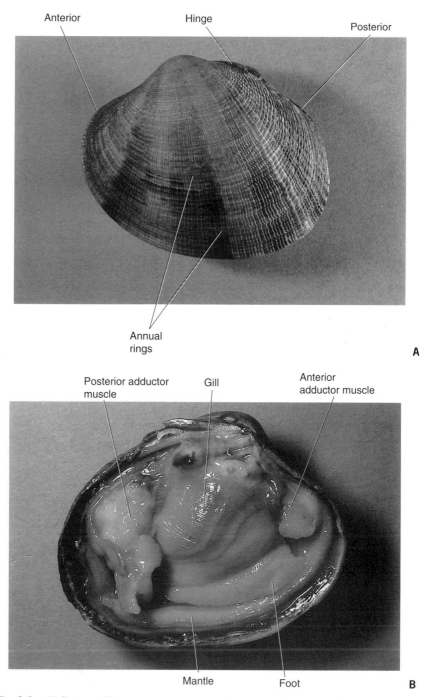

Anterior   Hinge   Posterior

Annual
rings

A

Posterior adductor   Gill   Anterior
muscle   adductor muscle

Mantle   Foot   **B**

Fig. 2.6. (A) Shell and (B) internal anatomy of the Manila clam, *Ruditapes philippinarum*. © Craig Burton 2001.

about 75 mm. The European cockle, *Cerastoderma edule*, is even smaller with a maximum shell length of 65 mm. Cockles and arkshells are close relatives of clams and for most practical purposes can be considered as clams (Dore, 1991).

In bivalves where the two shell valves are the same shape, such as mussels and clams, the valves are drawn together by an anterior and posterior adductor muscle. When these are relaxed the shell is opened by the elasticity of the ligament. Contraction of the adductor muscles closes the shell. When a bivalve dies these muscles can no longer contract and the ligament forces the shell open. A dead bivalve always has a gaping shell.

In bivalves with dissimilar valves, e.g. oysters and scallops, there is a single centrally placed adductor muscle. This performs the same function as the two adductors in mussels and clams. The muscle is divided into two visually distinct parts: the catch, made up of smooth muscle fibres is responsible for sustained closure of the valves with little energy expenditure, and the quick, composed of striated muscle fibres, is concerned with fast repetitive movements such as for swimming. The catch and quick portions can be clearly seen in Fig. 2.3C. In mussels and clams these two types of fibres are intermingled so that the adductor muscles appear homogeneous.

In a 'swimming' scallop, repeated rapid closure of the widely open valves by contraction of the quick muscle, and backward ejection of water on either side of the hinge, causes the scallop to move forward in a series of jerks (Fig. 2.7A). By suitably adjusting the edges of the mantle to direct a stream of water downwards the scallop can move both forwards and upward in an erratic movement, looking like it is taking a series of bites out of the water. In some situations water is jetted through only one side of the shell causing the animal to rotate (Fig. 2.7B). An approaching predator such as a starfish can provoke a different type of movement; forceful expulsion of water from the margins of the shell drives the animal backward in what looks like a very definite 'escape' movement (Fig. 2.7C). Using a slight variation of jet propulsion the scallop can execute a somersault if it finds itself in an overturned position (Fig. 2.7D). By suitably adjusting the mantle margins it rights itself by suddenly expelling water downwards all around the free margins of the shell (Cox, 1957). Details on the dynamics of swimming from high-speed film analysis are in Carsen *et al.* (1996) and Cheng & Demont (1996 a,b).

Scallops in the process of burying themselves on the seabed (recessing) utilise similar mechanisms. A series of powerful contractions of the adductor muscle eject water from the mantle cavity in the region of the posterior auricle (one of the shell projections at the hinge line). The water jets are directed downwards by muscular control of the velum (inner mantle fold); this lifts the shell at an angle to the seabed and subsequent water jets blow a hollow in the sediment (Brand, 1991). Once the hollow is of sufficient depth an extra powerful contraction lifts the scallop and it lands precisely in the recess. Sediment settles on the shell and after a few days this may be sufficiently thick to make detection difficult. Surprisingly, the main benefit is feeding related as recessing lowers the inhalant water current to the level of the sediment surface and detritus can be more easily drawn into the mantle cavity (see Chapter 4).

When bivalves are submerged and feeding the shell valves are always open. In burrowing bivalves the siphons are extended above the substrate through a

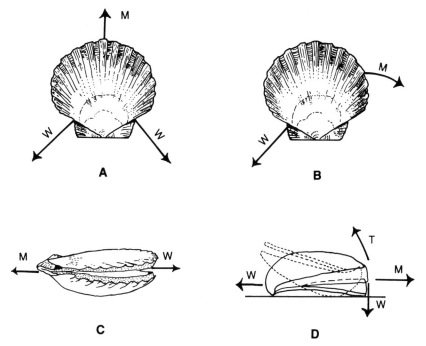

Fig. 2.7. Swimming and escape movements in scallops. Arrows indicate direction of movement (M), direction of jet stream (W) and turning (T). (A) Swimming movement, (B) twisting movement, (C) escape movement and (D) somersaulting movement. (A), (B) & (C) redrawn from Cox (1957) with permission from Shell Transport and Trading Co. UK; (D) redrawn from Wilkens (1991), with permission from the author.

gap in the shell. Closure of the shell occurs when the animal is out of the water, or when predators or adverse environmental conditions, e.g. sudden salinity changes or high levels of toxicants, threaten it.

Figs. 2.1C, 2.3C, 2.4B & 2.6B, which accompany the shell photos of mussels, oysters, scallops and clams, show the various organ systems described in the following sections of this chapter.

Mantle

In bivalves the mantle consists of two lobes of tissue which completely enclose the animal within the shell. Between the mantle and the internal organs is a capacious mantle cavity. With the exception of mussels (see below) the mantle is thin and transparent but the edges of the mantle and siphons are usually darkly pigmented, which probably gives protection from the harmful effects of solar radiation (Seed, 1971). The mantle consists of connective tissue with haemolymph ('blood') vessels, nerves and muscles that are particularly well developed near the mantle margins. Cilia on the inner surface of the mantle play an important role in directing particles onto the gills and in deflecting heavier material along rejection tracts towards the inhalant opening, the entry point on the mantle for incoming water. Periodically, the rejected material is discharged by sudden and forceful closure of the shell valves; this is sufficient to blow the rejected material out of the mantle cavity through the inhalant opening.

In mussels the mantle contains most of the gonad (Fig. 2.1C). Gametes proliferate within the mantle and are carried along ciliated channels to paired gonoducts that discharge into the mantle cavity (see below). After mussels have released their gametes the mantle is thin and transparent. The mantle is not only the site of gametogenesis but is also the main site for the storage of nutrient reserves, especially glycogen. In *M. edulis* reserves are laid down in summer and are utilised in autumn and winter in the formation of gametes. For a full discussion of energy metabolism in the mantle and other tissues see de Zwann & Mathieu (1992).

The mantle plays a role in the bioaccumulation of metal and organic contaminants in mussels, although the gills, kidney and digestive gland are considered as more important sites. Accumulated organic contaminants are actively metabolised and eliminated through the kidney, while heavy metals are sequestered by a specialised group of proteins called metallothioneins in gill, mantle and digestive gland tissue, or by lysosomes in digestive gland and kidney cells. For a full discussion on the bioaccumulation, toxicity and elimination of environmental contaminants in bivalves see reviews by Widdows & Donkin (1992), Livingstone & Pipe (1992) and Roesijadi (1996).

Pearl formation

Sometimes a foreign object like a sand grain or a parasitic larva lodges between the mantle and the shell. The bivalve encapsulates the object with layers of nacreous shell and so a pearl is formed. Although all bivalves are capable of forming pearls it is only those with an inner mother-of-pearl layer that can produce pearls of commercial importance. The best quality natural pearls are produced by pearl oysters, *Pinctada* spp. which live throughout the tropical Indo-Pacific. Nowadays, most oyster pearls are produced artificially by inserting a small piece of freshwater mussel shell surrounded in mantle tissue from a donor oyster, into the gonad at the base of the foot. The implanted tissue develops and covers the nucleus with layers of nacre. A useable pearl may take 3–5 years to form in this way but the industry is a lucrative one; the total world production value for 1994 was more than $1.3 billion (Fassler, 1995). Mussels (*Mytilus* sp.) produce pearls but this is in response to infection by the larva of a small parasitic flatworm. Unfortunately, the high incidence of pearls in bottom-grown mussels on the Atlantic coast of North America, Denmark, England and France is an impediment to marketing these mussels. The problem can be eliminated if mussels are grown on ropes and brought to market before the pearls reach a detectable size (Lutz, 1980).

Mantle margins

The mantle margins are thrown into three folds: the outer one, next to the shell, is concerned with shell secretion (see above); the middle one has a sensory function (see below), and the inner one is muscular and controls water flow in the mantle cavity. A minute space that contains pallial fluid separates the mantle from the shell, except in the regions of muscle attachment. The calcareous and organic materials for shell formation are deposited into this space.

The mantle is attached to the shell by muscle fibres in the inner fold; the line of attachment, the pallial line, runs in a semicircle a short distance from the edge of the shell (Fig. 2.5). In most bivalves the mantle margins are fused between the inhalant and exhalant openings. In mussels the exhalant opening is small, smooth and conical, and the inhalant aperture is wider and fringed by sensory papillae (Fig. 2.8). In oysters the inhalant opening is very large (Fig. 2.3C), mirroring the gills and the exhalant opening is again small. There is no mantle fusion in scallops so water enters around the entire mantle edge and exits on either side of the hinge line.

Additional modification of the mantle edge occurs in deep-burrowing forms such as clams. To minimise fouling of the mantle cavity there is complete fusion of the mantle edges except for three openings, the inhalant and exhalant apertures at the posterior end, and an opening at the anterior end through which the foot protrudes. In addition, the margins of the exhalant and inhalant openings are elongated to form siphons that extend out of the substrate for feeding, but which can be retracted into the shell. Siphon extension is mediated by haemolymph pressure or by water pressure in the mantle cavity when the shell is closed. Muscles in the inner mantle fold control siphon retraction. Some siphonate forms, e.g. the sand gaper, *Mya arenaria*, are not capable of totally retracting the siphons. Consequently, their siphons have a covering of periostracum to protect against predation. A novel strategy to avoid predation

Fig. 2.8. Exhalant (white and smooth) and inhalant (fringed with tentacles) openings in the mantle of the mussel *Mytilus edulis*. Photo courtesy of John Costello, Aquafact International Services Ltd., Galway, Ireland.

is used by the butter clam, *Saxidomus gigantus*. This species sequesters diet-derived paralytic algal toxins, highly potent neurotoxins, in its siphons as a defence against siphon–nipping fish (Kvitek, 1991).

As mentioned already, the outer mantle fold secretes the shell. The middle fold is primarily sensory having assumed this role in the evolution of the bivalve form from the ancestral mollusc – a change that involved the loss of the head and associated sense organs. The middle fold is frequently drawn out into short tentacles that contain tactile and chemoreceptor cells. Both of these cell types play an important role in predator detection and avoidance. Ocelli, which are sensitive to sudden changes in light intensity, may also be present on the middle fold. These 'eyes' can be simple invaginations lined with pigment cells and filled with a mucoid substance or 'lens'. However, in *Pecten maximus* the eyes, unquestionably better developed than in other bivalves, consist of a cornea, a lens and a retina. It is not clear whether they can perceive images clearly but any moving object that can cast a shadow, or that creates a near-field disturbance to which the scallop is not accustomed, will cause the shell to close briefly (Wilkens, 1991). While sensory receptors are found at the edge of the open mantle in sessile or surface dwelling bivalves, in burrowing forms they are concentrated at the tips of the siphons. Further information on the nervous system in bivalves is given below.

The inner mantle fold, or velum, is the largest of the three mantle folds and is particularly conspicuous in scallops. Small sensory tentacles or papillae (Fig. 2.9) usually fringe the fold and there is a large muscular component, especially on the inhalant opening. The velum plays an important role in controlling the flow of water into and out of the mantle cavity. It also plays a very important role in the so-called 'escape' response and swimming movement of scallops, as described above.

Fig. 2.9. The scallop *Pecten maximus* resting on the seabed with sensory tentacles and muscular velum of the mantle clearly visible. Photo courtesy of John Costello, Aquafact International Services Ltd., Galway, Ireland.

Pathogens and symbionts

The mantle in mussels is also host to various non-pathogenic viruses, potentially pathogenic protozoans, commensal cnidarians and parasitic flatworms (Bower, 1992). The parasitic flatworm *Proctoeces maculatus* seriously reduces glycogen energy reserves in heavily infected mussels. This can lead to disturbances of gametogenesis, and possible sterilisation and death. In the scallop, *Aequipecten opercularis*, the mantle eyes can be infected with the ciliate ectoparasite *Licnophora auerbachi*, thus leading to epidermal damage and pigment loss. This is especially serious in young developing scallops where it interferes with their escape response.

Many algae are seen as symbionts of marine bivalves. For example, in the giant tropical clam, *Tridacna gigas*, the edges of the mantle are packed with symbiotic zooxanthellae that presumably utilise carbon dioxide, phosphates and nitrates supplied by the clam. In return the algae provide the clam with a supplementary source of nutrition. This explains why these clams grow as large as 100 cm in length in impoverished coral-reef waters. Not all algae found in bivalves are so benign. *Zoochlorella* sp. produce lesions of the mantle, tentacles and eyes in the scallops *Argopecten irradians* and *Placopecten magellanicus*, and *Coccomyxa parasitica* infect the mantle of *P. magellanicus*, causing gross deformities of the shell valves (Getchell, 1991). A full account of diseases, parasites and pests of bivalves is presented in Chapter 11.

Gills

Filter feeding is believed to have evolved in some group of early protobranch molluscs, giving rise to the lamellibranchs, the dominant class of modern bivalves. Lamellibranchs feed by using the incoming current as a source of food, the gills having replaced the palps as the feeding organs. One important development in the evolution of filter feeding was movement of the site of water intake to the posterior of the animal (see Chapter 1).

Structure

The lamellibranch gills, or ctenidia, are two large, curtain-like structures that are suspended from the ctenidial axis that is fused along the dorsal margin of the mantle (Fig. 1.3A & 2.10A). Within the ctenidial axis are the branchial nerve and afferent and efferent branchial haemolymph vessels. Generally, the gills follow the curvature of the shell margin with the maximum possible surface exposed to the inhalant water flow (Figs. 2.3C & 2.4B). Each gill is made up of numerous W-shaped (or double V) filaments, and an internal skeletal rod rich in collagen strengthens each filament. Each V is known as a demibranch and each arm is called a lamella, giving an inner descending and outer ascending lamella (Fig. 2.10Bi). In the space between the descending and ascending lamellae is the exhalant chamber, connected to the exhalant area of the mantle edge; the space ventral to the filaments is the inhalant chamber connected to the inhalant area of the mantle edge (Fig. 2.10A).

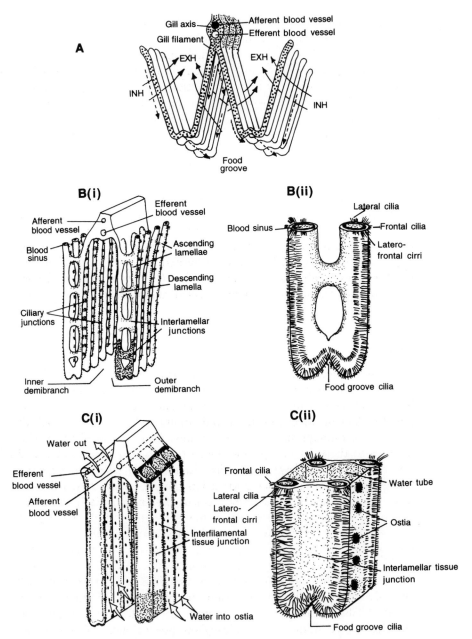

Fig. 2.10. (A) Section of a lamellibranch gill showing the ctenidial axis and four W-shaped filaments. For greater clarity the descending and ascending lamellae of each demibranch have been separated. Solid arrows indicate direction of water flow through the filaments from inhalant (INH) to exhalant (EXH) chambers and broken arrows indicate path of particle transport to the food grooves. (B) (i) Section of a fillibranch gill in the mussel, *Mytilus edulis*. Adjacent filaments are joined together by ciliary junctions. (ii) Transverse section through one fillibranch gill filament, shaded in B (i), showing pattern of ciliation. (C) (i) Section of a eulamellibranch gill in the clam *Mercenaria mercenaria*. Adjacent filaments are joined by tissue connections, called interfilamental junctions. Water enters the gill through perforations (ostia). (ii) Transverse section through a demibranch [shaded in C (i)] showing pattern of ciliation. (A) Redrawn from Barnes *et al.* (1993); (B) & (C) redrawn from Pechenik (1991) with permission from the McGraw-Hill Companies.

In more primitive lamellibranchs neighbouring gill filaments are attached to one another simply through interlocking clumps of cilia (Fig. 2.10Bi). This rather delicate gill type is termed fillibranch, and is seen in mussels and scallops. In more advanced bivalves neighbouring filaments are joined to each other at regular intervals by tissue connections (interfilament junctions), leaving narrow openings or ostia between them (Fig. 2.10Ci). This gill type, termed eulamellibranch, is a solid structure and is found in the majority of bivalves. In oysters interfilament junctions are less extensive than in most eulamellibranch species, so the gills are often referred to as pseudo-eulamellibranch.

In some bivalves, e.g. cockles, razor clams, oysters and scallops, the surface area of the gills has been greatly increased by folding or plica. This is believed to be an adaptation to living on coarse substrate; the plicate surface allows coarse filtration of large particles, while cilia on the filaments deal with finer filtration. Filaments are often structurally differentiated into principal and ordinary filaments, in which case the gill is termed heterorhabdic (Chapter 4 Fig. 4.8). The principal filaments are located in the troughs of plicae, separated from each other by a variable number (10–20) of ordinary filaments. A gill with undifferentiated filaments is called homorhabdic.

Functions

Cilia on the gill filaments have specific arrangements and functions (Figs. 2.10Bii & Cii). Lateral cilia are set along the sides of the filaments in fillibranch gills and in the ostia of eulamellibranch gills. These cilia are responsible for drawing water into the mantle cavity and passing it through the gill filaments or through the ostia, and then upwards to the exhalant chamber and onwards to the exhalant opening. Lying between the lateral and frontal cilia (see below) are the large feather-like latero-frontal cilia that are unique to bivalves. When the incoming current hits the gill surface these cilia flick particles from the water and convey them to the frontal cilia. The frontal cilia, which are abundantly distributed on the free outer surface of the gill facing the incoming current, convey particles aggregated in mucous downwards towards the ciliated food grooves on the ventral side of each lamella. In some bivalves there are actually two sets of frontal cilia, large coarse ones that carry larger particles for eventual rejection as pseudofaeces, and rows of small fine cilia that convey small particles towards the labial palps and mouth. The movement of cilia is under nervous control. Each gill axis is supplied with a branchial nerve from a visceral ganglion (see below), which subdivides to innervate individual groups of filaments. A more detailed description of the structure and function of the gills in filtration and feeding is presented in Chapter 4.

In bivalves the gills have a respiratory as well as a feeding role. Their large surface area and rich haemolymph supply make them well suited for gas exchange. Deoxygenated haemolymph is carried from the kidneys to the gills by way of the afferent gill vein. Each filament receives a small branch of this vein. The filaments are essentially hollow tubes within which the haemolymph circulates. Gas exchange takes place across the thin walls of the filaments. The oxygenated haemolymph from each filament is collected into the efferent gill vein that goes to the kidney and on to the heart (Fig. 2.15). It is probable that gas exchange also occurs over the general mantle surface.

The gills are also involved in the bioaccumulation of pesticides, soluble heavy metals and hydrocarbons. Exposure to high concentrations of zinc, copper and lead causes varying types of gill damage, such as degeneration of mucous secretory cells, loss of cilia and fusion of filaments (Livingstone & Pipe, 1992). Metallothioneins, specialised proteins that bind heavy metals, and free radical scavengers such as glutathione and vitamin C, are widely distributed in tissues such as gill and digestive gland.

The foot

The primitive mollusc had a broad ciliated flat foot, well supplied with mucous gland cells, and the animal is believed to have moved over the lubricated substrate in a gliding motion, using a combination of ciliary action and muscular contractions. In the evolution of bivalves the body became laterally compressed. Consequently, the foot lost its flat creeping sole and became blade-like and directed in an anterior direction as an adaptation for burrowing. Early on in this evolutionary process several groups of bivalves returned to the epifaunal life-style of their ancestors. This was made possible, among other factors, by the retention of the byssal apparatus of the foot into adult life to facilitate attachment of the animal to a hard substrate (see Chapter 1). In some epifaunal bivalves, such as mussels, the foot has retained some locomotory capacity, especially in the juvenile stage.

The foot first appears when bivalve larvae are about 200 µm in length, and it becomes functional in crawling and attachment at ~260 µm shell length (Bayne, 1971). This is the pediveliger stage of development, which immediately precedes settlement and metamorphosis (see Chapter 5). The foot is proportionately very large and sock shaped, and is made up of layers of circular and longitudinal muscles surrounding a capacious haemolymph space. The ventral surface or 'sole' of the foot is covered in cilia and in the mussel, *Mytilus edulis*, there are as many as nine different kinds of gland, each of which plays it own specific role in crawling and attachment (Lane & Nott, 1975). A byssal duct opens at the 'heel' of the foot and a byssal groove extends forward along the 'sole' from this opening (Hodgson & Burke, 1988).

At the onset of settlement, when the pediveliger larva is between 250–300 µm shell length, it slowly descends from the plankton to the seabed. Then follows a period of swimming and crawling behaviour before attachment and metamorphosis occurs.

Attachment in mussels

While swimming, the foot is fully extended, and periodically the velum (larval swimming organ) is withdrawn, and the larva sinks to the bottom and begins to crawl. If the substrate is unsuitable, i.e. one that does not stimulate the secretion of byssus, the foot is withdrawn and the larva once again swims off

(Lutz & Kennish, 1992). This cycle can be repeated many times over a period of a few days. In *Mytilus*, when a suitable substrate is found the larva continues to crawl for some time, gradually ceases movement, protrudes the foot and quickly secretes a single byssal thread. In the newly attached mussel larva this thread can be repeatedly broken and reformed before final settlement takes place. As the mussel grows in length more and more attachment threads are secreted. To resist dislodgement mussels cluster their threads in the direction of applied forces, e.g. facing ebb and flow of tide.

There are four distinct regions in the mussel byssus: root, stem, thread and plaque (Fig. 2.11). The root is embedded in the muscular tissue at the base of the foot; the stem is divided into sections and each has a thread attached; each thread is tipped by a plaque at which attachment to the substrate takes place. In adult *Mytilus*, threads average 2–4 cm in length, have a diameter of 0.2–0.3 mm and are attached to plaques 2–3 mm in diameter. The core of each thread consists of a collagen/elastin protein gradient with mainly collagen at the plaque end and elastic fibres at the stem end. A tough durable varnish mostly composed of polyphenolic protein covers the thread. Waite (1992) gives more detail on structure and formation of byssus in mussels, and also highlights the increasing interest that industry is showing in byssal components.

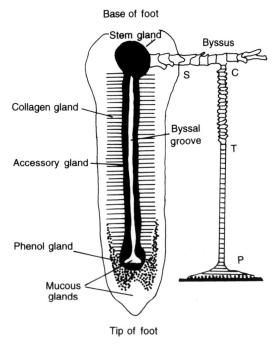

Fig. 2.11. Byssus-secreting glands in the foot (ventral surface) of the mussel, *Mytilus edulis*, including the byssus with stem (S), cuff (C), thread (T) and plaque (P). Root not shown. The mucous glands secrete acidic glycosaminoglycans and glycoproteins; the phenol gland contains proteins rich in 3,4 dihydroxyphenylalanine (DOPA); the collagen gland contains collagenous and elastic proteins, the accessory gland has catecholoxidase activity, and the stem gland secretes the lamellar proteins of the stem. Redrawn from Waite (1992).

One adhesive protein deposited into the byssal attachment plaques has now been characterised (Waite & Qin, 2001).

Attachment threads are not to be confused with the long drifting threads secreted by young post-larval mussels that prolong the planktonic phase (see Chapter 5). Although similar in diameter and structure to attachment threads, they differ in that they are long, single filament structures that are secreted by special glands that atrophy and disappear later in settled mussels (Lane *et al.*, 1985).

Attachment in other species

Oyster larvae use cement for attachment, squeezing this out from the byssal gland and then applying the left shell valve to the cement. The cement is a highly organised complex of micro-fibres and is similar in composition to the byssal threads of mussels (Cranfield, 1975). Once settled, oysters are not capable of detaching and reattaching (see Chapter 5). The attached larva undergoes metamorphosis, a series of morphological changes that herald the transition from a pelagic to a sessile existence. In mussels, scallops and clams the larval foot is retained into adult life. However, in oysters such as *Ostrea* and *Crassostrea* spp., which attach by cement, there is complete loss of the foot at metamorphosis. In scallops, such as *Pecten maximus* and *Aequipecten opercularis*, the byssus serves only for temporary attachment, and the animals break away on reaching a certain size. Consequently, the foot is a very small degenerate structure in adults of these species. Some scallops however, e.g. *Chlamys varia*, spend their whole life attached by byssus, as do all mussels. The foot is, therefore, a prominent feature in such species. In young mussels the foot also has a role in locomotion and removal of debris from the surface of the shell.

Digging

Clams initially use a byssus for attachment to particles of sediment but they soon abandon this and use their large foot to burrow into the substrate. Burrowing speed in clams (*Mya arenaria*) is routinely used as a bioassay for contaminated sediments (Phelps, 1989). The burrowing movement is brought about by a combination of haemolymph pressure and muscle action in the foot. When a bivalve is placed on a suitable substrate it proceeds to burrow into it in a series of steps or digging cycles, that continue until the animal is beneath the surface. The activity from initiation to completion of burrowing is called the digging period. Digging cycles are repeated many times during a digging period.

A digging cycle consists of a series of co-ordinated activities, similar in all burrowing bivalves investigated to date; Fig 2.12A illustrates the various shell and pedal muscles involved in a digging cycle:

- The shell adductor muscles relax and the two valves open a little and press against the substratum because of the elastic nature of the ligament. The projecting foot probes and pushes into the surrounding sand. Rhythmi-

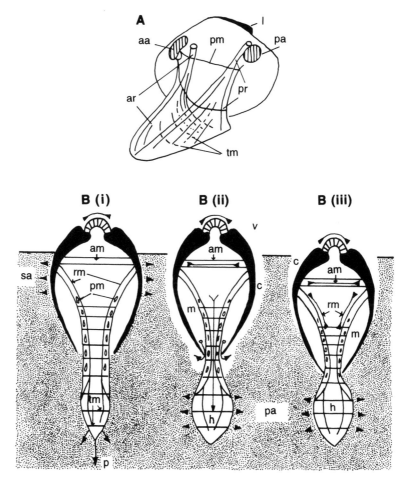

Fig. 2.12. (A) The principal muscles used in burrowing. aa, Anterior adductor; l, ligament; pa, posterior adductor; pm, protractors, circular muscles around the upper part of the foot; pr, posterior retractor; ar, anterior retractor; tm, transverse pedal muscles. (B) Successive stages in the burrowing of a generalized bivalve showing shell (sa) and pedal (pa) anchorages (arrowheads). (i) Valves press against the substrate by means of the opening thrust of the ligament and the foot extends by probing (p). (ii) Contraction of the adductor muscles (am) ejects water from the mantle cavity (m), thus loosening the substrate (c) around the valves (v); high pressure simultaneously produced in the haemocoele (h) gives rise to pedal dilation. (iii) Contraction of retractor muscles (rm) pulls the shell down into the loosened substrate. tm, Transverse pedal muscles; pm, pedal protractor muscles; ▶———◀ tension in ligament, adductor or retractor muscles. Redrawn from Trueman (1968).

cally repeated contraction and relaxation of the pedal protractor and transverse muscles initiate protraction of the foot (Fig. 2.12Bi).
- As the foot protrudes the shell valves begin to close by contraction of the adductor muscles (Fig. 2.10Bii). Some water is expelled from the mantle cavity, which helps to loosen the sand and facilitates the movement of the foot. The remaining water in the mantle cavity and the haemolymph act as a hydrostatic skeleton. Closure of the shell valves serves to increase the

pressure of these two fluids. Haemolymph from the body is forced down into the foot, causing the foot to swell and anchor into the substrate.

- Once the foot is anchored pedal retractor muscles contract, pulling the shell downwards into the loosened sand (Fig. 2.10Biii). These muscles insert on the shell in the region of the adductor muscles. Contraction of the anterior retractor occurs before that of the posterior retractor muscle. This has the effect of rocking the shell, which in turn facilitates its movement through the sand. Ridges or projections on the valves often aid anchorage, especially during the rocking movements.
- The shell adductors relax, the valves open and pedal dilation and anchorage is lost.
- There is now a static period during which the foot repeatedly probes the surrounding substrate.

As the animal penetrates further into the substrate the digging period shortens. This is probably not due to fatigue since animals will repeatedly burrow when they are removed from the substrate after completing a digging period (Trueman, 1968). To ascend towards the surface they back out, pushing against the anchored end of the foot, although some actually turn around and burrow upwards.

Burrowers such as the razor shells, *Ensis* or *Solen*, have long, thin shells and a large foot that takes up half of the mantle cavity. Such clams can burrow as fast as one can dig for them. On the other hand, *Mercenaria mercenaria* and *Mya arenaria* with their smaller foot and ovoid shells are not well suited for fast burrowing.

The ciliate *Peniculistoma mytili* inhabits the foot and adjacent surfaces of almost 100% of *Mytilus edulis* in the North and Baltic Seas, but there is no evidence that it has deleterious effects. However, metacercaria of the flatworm *Himasthla elongata* can penetrate the foot and heavy infestations can seriously affect byssal thread secretion and shell cleaning behaviour in *Mytilus edulis*, and burrowing activity in the cockle, *Cerastoderma edule* (Lauckner, 1983).

Labial palps and alimentary canal

Each gill terminates within a pair of triangular shaped palps that are situated on either side of the mouth. The inner surface of each palp faces the gill, and is folded into numerous ridges and grooves that carry a complicated series of ciliary tracts. The outer surfaces of the palps are smooth and between the inner and outer surfaces there is muscular connective tissue (see Fig. 4.10, Chapter 4).

The main function of the labial palps is continually to remove material from the food tracts on the gills in order to prevent gill saturation. In dense suspensions, sorting and rejection tracts on the palps channel most of the filtered material away from the mouth and deposit it as pseudofaeces so that the animal can continue to filter and ingest at an optimum rate (Bayne *et al.*, 1976a). The pseudofaeces is carried along rejectory tracts on the mantle to the inhalant opening. Periodically, the pseudofaeces is forcefully ejected through this opening. When the ingestive capacity is not exceeded particles from the gill move along acceptance tracts on the labial palps towards the mouth (see Chapter 4).

Stomach and style

The mouth is ciliated and leads into a narrow ciliated oesophagus. Ciliary movement helps to propel material towards the stomach. Indeed, this method of moving material is found throughout the length of the alimentary canal, primarily because it lacks a muscular wall. The stomach is large and oval-shaped and lies completely embedded in the digestive gland, which opens into it by several ducts (Fig. 2.13). A semi-transparent gelatinous rod about 3 cm long, the crystalline style, originates in the style sac at the posterior end of the stomach and projects forward and dorsally across the cavity of the stomach to rest against the gastric shield, a thickened area of the stomach wall. The projecting anterior end of the style is rotated against the gastric shield by the style sac cilia, and in the process the style end is abraded and dissolved, releasing carbohydrate-splitting enzymes in the process. This loss is made good by continual additions by the style sac to the base of the style (see Purchon, 1957 for a detailed description of the bivalve stomach). This remarkable structure is possibly the only example of a rotating part in an animal. In intertidal bivalves, such as oysters, the style is not a permanent structure; it dissolves at low tide when the animal is not feeding and is reformed when the tide comes in (Langton & Gabbott, 1974).

The style has additional functions in the digestive process. As it revolves the food-laden mucous string becomes wrapped round the head of the style, which acts like a winch, winding the mucous string around it (Fig. 2.13B). The low pH of the stomach facilitates the dislodgement of particles from the mucous string. These particles are then mixed with the other contents of the stomach, including the liberated enzymes from the style. The rotation of the style helps the mixing process to take place. While all the mixing and extra-cellular digestion is taking place, the stomach contents come under the influence of ciliary tracts that cover all areas of the stomach except those occupied by the gastric shield (Yonge & Thompson, 1976). These ciliated tracts have fine ridges and grooves and act as sorting areas in much the same way as the labial palps. Finer particles and digested matter are kept in suspension by cilia at the crests of the ridges, and this material is continually swept towards the digestive gland duct openings. Larger particles segregate out and are channelled into the intestine along a deep rejectory groove on the floor of the stomach.

Digestion

The digestive gland, which is brown or black and consists of blind-ending tubules that connect to the stomach by several ciliated ducts, is the major site of intracellular digestion. Within these ducts there is a continuous two-way flow: materials enter the gland for intracellular digestion and absorption and wastes leave *en route* to the stomach and intestine (Fig. 2.14). The tubules are

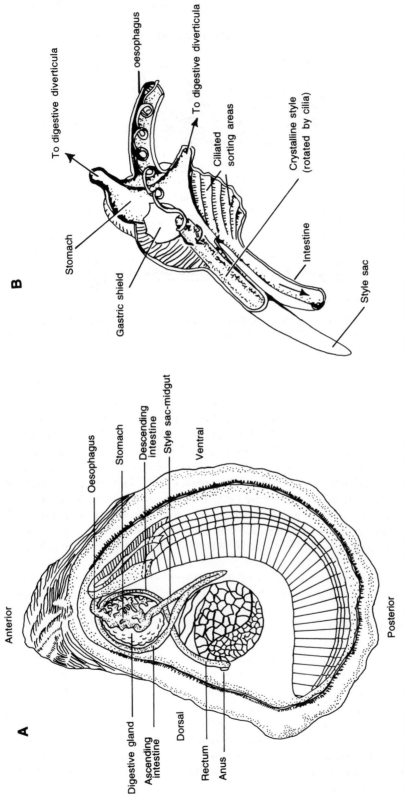

Fig. 2.13. (A) The digestive system of the oyster *Crassostrea virginica*. Redrawn from Langdon & Newell (1996), after Galtsoff (1964). (B) Bivalve stomach showing rotation of crystalline style and winding of food string. Rejectory groove on floor of stomach not shown. Redrawn from Pechenik (1991) with permission from the McGraw-Hill Companies.

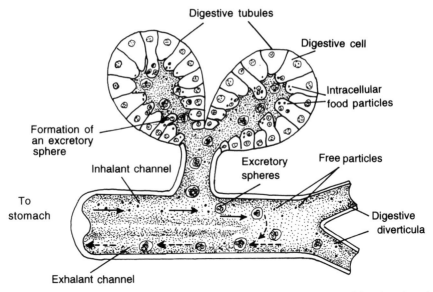

Fig. 2.14. A section of the digestive gland showing absorption and intracellular digestion of material coming from the stomach (solid arrows) and outward movement of wastes (broken arrows). Redrawn from Owen (1955).

composed of two cell types, digestive cells and basophil (secretory) cells (Weinstein 1995 and see Chapter 4). Digestive cells take up material by pino-cytosis (uptake of extracellular fluid into a cell) and digest it within vacuoles. The end products of digestion are released directly into the system (Mathers, 1972), and the waste is retained in membrane-bound residual bodies that are later released into the lumen of the tubules. The digestive cells eventually rupture and are replaced by new cells. The released waste enclosed in excre-tory spheres, is swept towards the stomach and ultimately to the intestine. This waste may also contain digestive enzymes which could be utilised by the stomach for extra-cellular digestion (Mathers, 1972). The secretory cells carry out extensive protein synthesis but their exact role is not yet known. The cellular structure and function of the digestive gland has been reviewed for scallops by Beninger & Le Pennec (1991) and for oysters by Langdon & Newell (1996).

The stomach, digestive gland and, sometimes the intestine, produces a wide variety of digestive enzymes. Carbohydrate-splitting enzymes, e.g. amylases have been found in high concentration in the style, while cellulases, laminar-inase, β-galactosidase and others were mostly confined to the digestive gland. Fat digesting enzymes e.g. esterases and acid and alkaline phosphatases have been reported from the stomach, digestive gland and intestine of the oysters, *Ostrea edulis* and *Crassostrea angulata* (Mathers, 1973; Langdon & Newell, 1996 and references therein). Proteolytic activity is characteristically low in bivalves but endopeptidases such as chymotrypsin and cathepsin have been reported from the digestive gland, and trypsin activity in the stomach and intestine of some bivalves (Reid, 1968).

The digestive gland also plays an important role in the storage of metabolic reserves, which are used as an energy source during the process of gameto-genesis (see Chapter 5) and during periods of physiological stress (Bayne *et al.*, 1976a).

Rejected particles from the stomach as well as waste material from the digestive gland pass into the long coiled intestine (Fig. 2.13A). The waste is formed into faecal pellets that are voided through the anus and are swept away through the exhalant opening.

Pollutants

Because of the pronounced endocytotic nature of the digestive gland it is a major site for the uptake of metal and organic contaminants. A wide range of enzymes which deal with the bio-transformation of organic contaminants has been detected in the digestive gland of numerous bivalves (see Livingstone & Pipe, 1992 for review). Also, metallothioneins, antioxidant enzymes and oxyradical scavengers, such as glutathione, have high activities in digestive gland tissue. The response of this tissue to contaminant exposure has been well documented for molluscs (Moore, 1991; Lowe & Pipe, 1994; Cajaraville *et al.*, 1995; Livingstone *et al.*, 2000). When challenged with a wide variety of contaminants the tubules show some degree of tissue breakdown, mediated through the release of hydrolytic enzymes from absorptive cells of the tubules. The formation of granulocytomas (inflammatory lesions) in digestive gland tissue and in the mantle has also been reported (Lowe & Moore, 1979).

Gonads

The reproductive system in bivalves is exceedingly simple. The gonads are paired but are usually so close together that the pair is difficult to detect. Each gonad is little more than a system of branching tubules, and gametes are budded off the epithelial lining of these tubules. The tubules unite to form ducts that lead into larger ducts and eventually terminate in a short gonoduct. In primitive bivalves, e.g. the nut shell, *Nucula*, the gonoducts open into the kidneys, and eggs and sperm exit through the kidney opening (nephridiopore) into the mantle cavity. In most bivalves the gonoducts open through independent pores into the mantle cavity, close to the nephridiopore. With the exception of oysters (*Ostrea* spp.), fertilisation is external and the gametes are shed through the exhalant opening.

Mussels are dioecious, i.e. the sexes are separate, and the gonad develops within the mantle tissue (Fig. 2.1C). In ripe mussels (*Mytilus edulis*) the mantle containing the gametes is typically orange in females and creamy-white in males. In oysters the gonad covers the outer surface of the digestive gland (Fig. 2.3C). When *Ostrea edulis* is in a ripe condition the gonad forms a layer 2–3 mm thick and its creamy colour obscures the brown colour of the digestive gland. In ripe *Crassostrea gigas* the gonad is often 6–8 mm thick and may comprise a third of the total body weight, exclusive of the shell (Walne, 1974). In clams the gonad is situated at the base of the foot. In oysters and clams the sexes are separate but they do have the ability to change sex (see Chapter 5). Most scallop species are hermaphrodites. In *Pecten maximus* the mature gonad is divided into a dorsal testis which is white in colour and a ventral ovary which is orange-red, and both of these curve around the large central adductor muscle (Fig. 2.4B). The sperm are shed before the eggs, thus minimising the chances of self-fertilisation. In dioecious scallop species such as *Chlamys varia* and *Placopecten magellanicus*, the female gonad is pink while the male gonad is white.

In *Mytilus* gamete development can be seriously affected when the gonad is infected with protozoans such as *Steinhausia mytilovum* or *Marteilia* sp. (Bower, 1992). The larval stages of trematodes (*Prosorhynchus squamatus*, *Proctoeces maculatus*, *Bucephalus cuculus*, *Cercaria pectinata*) inhabit the gonad, as well as other tissues of mussel, clam and oyster species. One of the common effects of these parasites is castration. In oysters this has been seen as actually beneficial from the gastronomic point of view in that *Bucephalus*-infected animals do not spawn but remain fat and retain an excellent flavour all summer (Menzel & Hopkins, 1955)!

Heart and haemolymph vessels

The heart lies in the mid-dorsal region of the body, close to the hinge line of the shell. It lies in a space called the pericardium, which surrounds the heart dorsally and a portion of the intestine ventrally. The heart consists of a single, muscular ventricle and two thin-walled auricles. Haemolymph flows from the auricles into the ventricle, which contracts to drive the haemolymph into a single vessel, the anterior aorta. A posterior aorta is also present in pseudo- and eulamellibranch bivalves. The anterior aorta divides into many arteries, the most important of these are the pallial arteries that supply the mantle, and the visceral arteries (gastro-intestinal, hepatic and terminal) that supply the stomach, intestine, shell muscles and foot with haemolymph (Fig. 2.15). The arteries break up into a network of vessels in all tissues and these then join to form veins which empty into three extensive spaces, the pallial, pedal and median ventral sinuses. The circulatory system is therefore an open system with the haemolymph in the sinuses bathing the tissues directly. From the sinuses the haemolymph is carried to the kidneys for purification.

In *Mytilus* some of the haemolymph from the kidney network enters the gills, discharging into the afferent gill vein, which gives off a branch to each gill filament, descending on one side and ascending on the other (Fig. 2.15). The ascending vessels join to form an efferent gill vein that passes back to the kidneys. The haemolymph from the kidneys returns to the auricles of the heart. In other bivalves haemolymph from the gills does not return to the kidney but flows directly from the gills to the heart. In all bivalves there are well-developed circulatory pathways through the mantle, which therefore serves as an additional site of oxygenation.

Haemolymph plays a number of important roles in bivalve physiology. These include gas exchange, osmoregulation (Chapter 7), nutrient distribution, waste elimination and internal defence (Chapter 11). It also serves as a fluid skeleton, giving temporary rigidity to such organs as the labial palps, foot (see

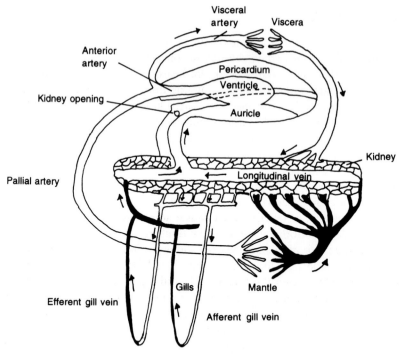

Fig. 2.15. Circulatory system of the mussel *Mytilus edulis*. Redrawn from Borradaile *et al.* (1961).

above) and mantle edges. The haemolymph contains cells called haemocytes that float in colourless plasma. There is no respiratory pigment because the haemolymph oxygen concentration is similar to seawater (Bayne *et al.*, 1976b). Also the sedentary lifestyle, together with large surfaces for oxygen uptake, precludes the need for circulating respiratory pigments in bivalves.

Several different categories of haemocytes have been described in mussels (Moore & Lowe, 1977), scallops (Auffret, 1985), clams (Foley & Cheng, 1974) and oysters (Cheng, 1996). Two broad categories are recognised, granulocytes (10–20 μm diameter), and agranulocytes (4–6 μm diameter). The majority of haemocytes belong to the former group and are characterised by granules of various sizes and shapes in the cytoplasm. Granulocytes phagocytose bacteria, algae, cellular debris and protozoan parasites. Agranular haemocytes have few, if any, cytoplasmic granules, and appear to be less phagocytic then granulocytes. There is evidence of at least three different types of agranulocytes in bivalves and perhaps even two types of granulocytes (reviewed by Hine, 1999).

Haemocytes are not confined to the haemolymph system but move freely out of the sinuses into surrounding connective tissue, the mantle cavity and gut lumen. Therefore, it is not surprising that these cells play an important role in physiological processes such as nutrient digestion and transport, excretion, tissue repair, and internal defence (see Chapter 7).

Excretory organs

There are two types of excretory organs in bivalves, the pericardial glands and the paired kidneys. In *Mytilus* the reddish-brown U-shaped kidneys lie ventral to the pericardial cavity surrounding the heart, and dorsal to the gill axis, and extend the complete length of the gill axis from the labial palps to the posterior adductor muscle. In scallops the paired kidneys are attached to the anterior margin of the central adductor muscle, partially hidden by the gonad. One arm of each kidney is glandular and opens into the pericardium, and the other end is a thin-walled bladder that opens through a nephridiopore into the mantle cavity (Fig. 2.15).

The brown-coloured pericardial glands, sometimes referred to as Keber's organs, develop from the epithelial lining of the pericardium and come to lie over the auricular walls of the heart. Waste accumulates in certain cells of the pericardial glands and this is periodically discharged into the pericardial cavity and from there it is eliminated via the kidneys. Other cells of the pericardial glands are involved in filtering the haemolymph, the first stage of urine formation. The filtrate then flows to the glandular part of the kidney where the processes of secretion and re-absorption of ions occurs. The end-result is urine that has a high concentration of ammonia, and smaller amounts of amino acids and creatine (see Table 7.6, Chapter 7). Most aquatic invertebrates excrete ammonia as the end product of protein metabolism. Ammonia is highly toxic but its small molecular size and high solubility in water ensure that it diffuses extremely rapidly away from the animal.

While the kidneys and pericardial glands are the major excretory organs, excretory products are probably also lost across the general body surface and particularly across the gills (see Chapter 7).

In mussels and scallops the kidney plays a very important role in the storage and elimination of hydrocarbons and heavy metals such as zinc and cadmium. Metals are compartmentalised into kidney lysosomes and are eventually excreted in the urine. One of the effects of metal sequestration is an increased incidence of renal cysts (references in Livingstone and Pipe, 1992).

The kidney is the site of infection for *Haplosporidium* and larval stages of the trematode, *Proctoeces maculatus*, in *Mytilus*, and for *Rickettsia* bacteria and protozoan coccidians in *Argopecten irradians* (Lauckner, 1983; Bower, 1992).

Nerves and sensory receptors

The nervous system of bivalves is fundamentally simple. It is bilaterally symmetrical and consists of three pairs of ganglia and several pairs of nerves (Fig. 2.16). The cerebral ganglia are joined by a short commissure dorsal to the oesophagus. From each cerebral ganglion two pair of nerve cords extend to the posterior of the animal. One pair extend directly back to the visceral ganglion, which is located on the surface of the posterior adductor muscle. The second pair extend posteriorly and ventrally to the pedal ganglia in the foot. The cerebral ganglia innervate the palps, anterior adductor muscle, and part of the mantle, as well as the statotocysts and osphradia (see below). The pedal ganglia control the foot. The visceral ganglia control a large area: gills,

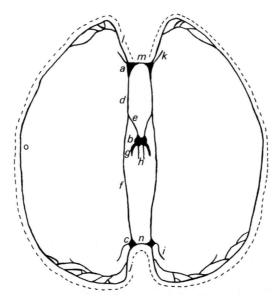

Fig. 2.16. Schematic representation of the nervous system in the mussel *Mytilus edulis*. Dashed line indicates the outline of the two shell valves. a: cerebral ganglia; b: pedal ganglia; c: visceral ganglia; d: cerebrovisceral pedal connective; e: cerebropedal connective; f: cerebrovisceral connective; g: pedal nerve; h: byssal retractor nerve; i: branchial nerve; j: posterior pallial nerve; k: anterior pallial nerve; l: buccal nerve; m: cerebral commissure; n: visceral commissure; o: circumpallial nerve. Reprinted from de Zwann & Mathieu (1992), with permission from Elsevier Science.

heart, pericardium, kidney, digestive tract, gonad, posterior adductor muscle, part or all of the mantle, siphons and pallial sense organs. In scallops the paired visceral ganglia are completely fused, thus forming an elementary brain with distinct optic lobes that innervate the eyes on the edge of the mantle. This is the largest and most complex nervous structure found in the Bivalvia (details in Beninger & Le Pennec, 1991).

The ganglia also have a major neurosecretory role in bivalves. Several different types of neurosecretory cells have been identified in mussels and most of these are located in the cerebral ganglia (Illanes–Bucher, 1979). These cells (neurones) produce peptides that are released into the circulatory system. Some of these peptides have been characterised in *Mytilus* and have been shown to be important in the regulation of growth, gametogenesis and in glycogen metabolism (see de Zwann & Mathieu, 1992 for review and references). More details on the role of neurohormones in gametogenesis, and growth are in Chapters 5 and 6, respectively.

During the evolution of bivalves, with loss of a distinct head, most of the sense organs withdrew from the anterior end and have come to lie at the edge of the mantle. In burrowing forms they are concentrated at the tips of the siphons. Most sensory receptors are located on the middle fold of the mantle. This fold is thick and bears a large number of pallial tentacles, their length and number varying with the species. The tentacles are covered in epithelial sensory cells that are sensitive to touch. A slight tactile stimulus elicits local contraction of the mantle or siphon musculature. This is a reflex action and is not under the control of the central nervous system. A strong stimulus produces a co-ordinated retraction of the whole animal into its shell. This more general and clearly adaptive type of contraction is under the control of the visceral ganglion. Chemoreceptor cells are probably also present on the tentacles. In scallops, for example, starfish extracts elicit tentacle withdrawal, contraction of the velum, and even the swimming response at higher concentrations (Wilkens, 1991).

Ocelli, which can detect sudden changes in light intensity, may also be present on the middle fold of the mantle or siphons. These may take the form of invaginated eyecups lined with pigmented sensory cells and filled with a mucoid substance that acts as a lens, or they may be very well developed structures as in scallops. As mentioned already it is not clear whether scallop eyes can perceive images clearly.

Sensory receptors called osphradia are well known in gastropods but in bivalves they are difficult to detect, either because of their small size or because they are absent in some species. These receptors consist of patches of a single layer of sensory epithelium lying close to the visceral ganglia. In scallops, e.g. *Placopecten magellanicus*, what are believed to be osphradia continue along the outermost side of the gill axis as a raised ridge of tissue innervated by the branchial nerve from the visceral ganglion (Beninger & Le Pennec, 1991). In gastropods, such as carnivorous whelks, the osphradium is an organ of chemoreception for testing incoming water into the mantle cavity. It is, therefore, believed to play an important role in the detection of prey but what its role is in bivalves is not at all clear. However, it is very likely that the osphradia of gastropods and bivalves are not homologous.

A pair of statoreceptors lies in the foot near the pedal ganglia and are enervated by the cerebro-pleural ganglia. There are two types, statocysts that are closed vesicles containing either single or small multiple endogenous concretions, or statocrypts that communicate with the exterior and contain many exogenous particles, e.g. sponge spicules cemented by mucous (Beninger & Le Pennec, 1991). There is evidence that the statoreceptors function in orientation when scallops are swimming (Barber, 1968), but their role in other bivalves is not yet clear.

References

Adamkewicz, L. & Castagna, M. (1988) Genetics of shell color and pattern in the Bay scallop *Argopecten irradians. J. Hered.*, **79**, 14–17.

Auffret, M. (1985) Morphologie comparative des types hémocytaires chez quelque Mollusques Bivalves d'intéret commercial. Thèse de Doctorat d'État. Université de Bretagne Occidentale, Brest, France.

Barber, V.C. (1968) The structure of mollusc statocysts, with particular reference to cephalopods. *Symp. Zool. Soc. Lond.*, **23**, 37–62.

Barnes, R.S.K., Callow, P. & Olive, P.J.W. (1993) *The Invertebrates: a New Synthesis*, 2nd edn. Blackwell Scientific Publications, Oxford.

Bayne, B.L. (1971) Some morphological changes that occur at the metamorphosis of the larvae of *Mytilus edulis*. In: *Fourth European Marine Biology Symposium* (ed. D.J. Crisp), pp. 259–80. Cambridge University Press, London.

Bayne, B.L., Thompson, R.J. & Widdows, J. (1976a) Physiology: I. In: *Marine Mussels: Their Ecology and Physiology* (ed. B.L. Bayne), pp. 121–206. Cambridge University Press, Cambridge.

Bayne, B.L., Widdows, J. & Thompson, R.J. (1976b) Physiology: II. In: *Marine Mussels: Their Ecology and Physiology* (ed. B.L. Bayne), pp. 207–60. Cambridge University Press, Cambridge.

Beninger, P.G. & Le Pennec, M. (1991) Functional anatomy of scallops. In: *Scallops: Biology, Ecology and Aquaculture* (ed. S.E. Shumway), pp. 133–223. Elsevier Science Publishers B.V., Amsterdam.

Borradaile, L.A., Potts, F.A., Eastham, L.E.S. & Saunders, J.T. (1961) *The Invertebrata*, 4th edn. Cambridge University Press, Cambridge.

Bower, S.M. (1992) Diseases and parasites of mussels. In: *The Mussel Mytilus: Ecology, Physiology, Genetics and Culture* (ed. E.M. Gosling), 543–63. Elsevier Science Publishers B.V., Amsterdam.

Brand, A.R. (1991) Scallop ecology: distribution and behaviour. In: *Scallops: Biology, Ecology and Aquaculture* (ed. S.E. Shumway), pp. 517–84. Elsevier Science Publishers B.V., Amsterdam.

Cajaraville, M.P., Robledo, Y., Etxeberria, M. & Marigómez, I. (1995) Cellular biomarkers as useful tools in the biological monitoring of environmental pollution: molluscan digestive lysosomes. In: *Cell Biology and Environmental Toxicology* (ed. M.P. Cajaraville), 29–55. University of the Basque Country Press Service, Bilbao.

Carriker, M.R. (1996) The shell and ligament. In: *The Eastern Oyster Crassostrea virginica* (eds V.S. Kennedy, R.I.E Newell & A.F. Eble), pp. 75–168. Maryland Sea Grant, College Park, Maryland.

Carsen, A.E., Hatcher, B.G. & Scheibling, R.E. (1996) Effect of flow velocity and body-size on swimming trajectories of sea scallops, *Placopecten magellanicus* (Gmelin): a comparison of laboratory and field measurements. *J. Exp. Mar. Biol. Ecol.*, **203**, 223–43.

Cheng, T. (1996) Hemocytes: forms and functions. In: *The Eastern Oyster Crassostrea virginica* (eds V.S. Kennedy, R.I.E. Newell & A.F. Eble), pp. 299–333. Maryland Sea Grant, College Park, Maryland.

Cheng, J.Y. & Demont, M.E. (1996a) Jet-propelled swimming in scallops: swimming mechanics and ontogenic scaling. *Can. J. Zool.*, **74**, 1734–48.

Cheng, J.Y. & Demont, M.E. (1996b) Hydrodynamics of scallop locomotion: unsteady fluid forces on clapping shells. *J. Fluid Mech.*, **317**, 73–90.

Cox, I. (Ed) (1957) *The Scallop: Studies of a Shell and its Influence on Humankind*. Shell Transport and Trading Co. Ltd., UK.

Cranfield, H.J. (1975) The ultrastructure and histochemistry of the larval cement gland of *Ostrea edulis* L. *J. mar. biol. Ass. U.K.*, **55**, 497–503.

Dore, I. (1991) *Shellfish: a Guide to Oysters, Mussels, Scallops, Clams and Similar Products for the Commercial User*. Van Nostrand Reinhold, New York.

Fassler, C.R. (1995) Farming jewels: new developments in pearl farming. *World Aquaculture*, **26**, 5–10.

Foley, D. A. & Cheng, T.C. (1974) Morphology, hematological parameters and behavior of hemolymph cells of the pelecypods *Crassostrea virginica* and *Mercenaria mercenaria*. *Biol. Bull.*, **146**, 343–56.

Galtsoff, P.S. (1964) The American oyster *Crassostrea virginica* Gmelin. *Fish. Bull.*, **64**, 1–480.

Getchell, R.G. (1991) Diseases and parasites of scallops. In: *Scallops: Biology, Ecology and Aquaculture* (ed. S.E. Shumway), pp. 471–94. Elsevier Science Publishers B.V., Amsterdam.

Hawkins, A.J.S. & Bayne, B.L. (1992) Physiological interrelations and the regulation of production. In: *The Mussel Mytilus: Ecology, Physiology, Genetics and Culture* (ed. E.M. Gosling), pp. 171–222. Elsevier Science Publishers B.V., Amsterdam.

Hine, P.M. (1999) The inter-relationships of bivalve haemocytes. *Fish & Shellfish Immunol.*, **9**, 367–85.

Hodgson, C.A. & Burke, R.D. (1988) Development and larval morphology of the spiny scallop, *Chlamys hasata. Biol. Bull.*, **174**, 303–18.

Illanes–Bucher, J. (1979) Recherches cytologiqueset expérimentales sur la neurosécrétion de la moule *Mytilus edulis* L. (*Mollusque, Lamellibranche*). Thèse de 3eme cycle, Université de Caen, France.

Innes, D.J. & Haley, L.E. (1977) Inheritance of a shell-color polymorphism in the mussel. *J. Hered.*, **68**, 203–04.

Kvitek, R.G. (1991) Paralytic shellfish toxins sequestered by bivalves as a defense against siphon-nipping fish. *Mar. Biol.*, **111**, 369–74.

Lane, D.J.W. & Nott, J.A. (1975) A study of the morphology, fine structure and histochemistry of the foot of the pediveliger of *Mytilus edulis. J. mar. biol. Ass. U.K.*, **55**, 477–95.

Lane, D.J.W., Beaumont, A.R. & Hunter, J.R. (1985) Byssus drifting and the drifting threads of the young post-larval mussel *Mytilus edulis. Mar. Biol.*, **84**, 301–08.

Langdon, C.J. & Newell, R.I.E. (1996) Digestion and nutrition in larvae and adults. In: *The Eastern Oyster Crassostrea virginica* (eds V.S. Kennedy, R.I.E. Newell & A.F. Eble), pp. 231–69. Maryland Sea Grant, College Park, Maryland.

Langton, R.W. & Gabbott, P.A. (1974) The tidal rhythm of extracellular digestion and the response to feeding in *Ostrea edulis. Mar. Biol.*, **24**, 181–87.

Lauckner, G., (1983) Diseases of Mollusca: Bivalvia. In: *Diseases of Marine Animals* (ed. O. Kinne), pp. 477–961. Biologische Anstalt Helgoland, Hamburg.

Livingstone, D.R. & Pipe, R.K. (1992) Mussels and environmental contaminants: molecular and cellular aspects. In: *The Mussel Mytilus: Ecology, Physiology, Genetics and Culture* (ed. E.M. Gosling), pp.425–64. Elsevier Science Publishers B.V., Amsterdam.

Livingstone, D.R., Chipman, J.K., Lowe, D.M. *et al.* (2000) Development of biomarkers to detect the effects of organic pollution on aquatic invertebrates: recent molecular, genotoxic, cellular and immunological studies on the common mussel (*Mytilus edulis* L.) and other mytilids. *Int. J. Environ. Pollut.*, **13**, 56–91.

Lowe, D.M. & Moore, M.N. (1979) The cytology and occurrence of granulocytomas in mussels. *Mar. Poll. Bull.*, **10**, 137–41.

Lowe, D.M. & Pipe, R.K. (1994) Contaminant induced lysosomal membrane damage in marine mussel digestive cells: an in-vitro study. *Aquat. Toxicol.*, **30**, 357–65.

Lutz, R.A. (1976) Annual growth patterns in the inner shell layer of *Mytilus edulis. J. mar. biol. Ass. U.K.*, **56**, 723–31.

Lutz, R.A. (1980) Pearl incidences: mussel culture and harvest implications. In: *Mussel Culture and Harvest: a North American Perspective* (ed. R.A. Lutz), pp. 193–222. Elsevier Science Publishing Co. Inc., New York.

Lutz, R.A. & Kennish, M.J. (1992) Ecology and morphology of larval and early post-larval mussels. In: *The Mussel Mytilus: Ecology, Physiology, Genetics and Culture* (ed. E.M. Gosling), pp. 53–85. Elsevier Science Publishers B.V., Amsterdam.

Mathers, N.F. (1972) The tracing of a natural algal food labeled with a carbon[14] isotope through the digestive tract of *Ostrea edulis. Proc. Malac. Soc. Lond.*, **40**, 115–24.

Mathers, N.F. (1973) A comparative histochemical survey of enzymes associated with the processes of digestion in *Ostrea edulis* and *Crassostrea angulata* (Mollusca: Bivalvia). *J. Zool., Lond.*, **169**, 169–79.

Menzel, R.W. & Hopkins, S.H. (1955) The growth of oysters parasitised by the fungus *Dermocystidium marinum* and by the trematode *Bucephalus cuculus*. *J. Parasitol.*, **41**, 333–42.

Mitton, J.B. (1977) Shell colour and pattern variation in *Mytilus edulis* and its adaptive significance. *Chesapeake Sci.*, **18**, 387–90.

Moore, M.N. (1991) Environmental distress signals: cellular reactions to marine pollution. In: *Progress in Histo- and Cytochemistry as a Tool in Environmental Toxicology* (eds W. Graumann & J. Drucker), pp. 1–19. Fischer Verlag, Stuttgart.

Moore, M. & Lowe, D. (1977) The cytology and cytochemistry of the hemocytes of *Mytilus edulis* and their responses to injected carbon particles. *J. Invert. Pathol.*, **29**, 18–30.

Newkirk, G.F. (1980) Genetics of shell colour in *Mytilus edulis* L. and the association of growth rate with shell colour. *J. Exp. Mar. Biol. Ecol.*, **47**, 89–94.

Owen, G. (1955) Observations on the stomach and digestive diverticula of the lamellibranchia. *Quart. J. Microsc. Sci.*, **96**, 517–37.

Pechenik, J.A. (1991) *Biology of the Invertebrates*, 2nd edn. Wm. C. Brown (WCB) Publishers, Dubuque, Iowa.

Phelps, H. (1989) Clam burrowing bioassay for estuarine sediment. *Bull. Environ. Contam. Toxicol.*, **43**, 838–45.

Purchon, R.D. (1957) The stomach in the Filibranchia and Pseudolamellibranchia. *Proc. Zool. Soc. Lond.*, **129**, 27–60.

Reid, R.G.B. (1968) The distribution of digestive tract enzymes in lamellibranchiate bivalves. *Comp. Biochem. Physiol.*, **24**, 727–44.

Richardson, C.A. (1989) An analysis of the microgrowth bands in the shell of the common mussel *Mytilus edulis*. *J. mar. biol. Ass. U.K.*, **69**, 477–91.

Roesijadi, G. (1996) Environmental factors: response to metals. In: *The Eastern oyster Crassostrea virginica* (eds V.S. Kennedy, R.I.E. Newell & A.F. Eble), pp. 515–37. Maryland Sea Grant, College Park, Maryland.

Seed, R. (1971) A physiological and biochemical approach to the taxonomy of *Mytilus edulis* L. and *M. galloprovincialis* Lmk. from S.W. England. *Cah. Biol. Mar.*, **12**, 291–322.

Seed, R. (1976) Ecology. In: *Marine Mussels: Their Ecology and Physiology* (ed. B.L. Bayne), pp. 13–65. Cambridge University Press, Cambridge.

Seed, R. & Suchanek, T.H. (1992) Population and community ecology of *Mytilus*. In: *The Mussel Mytilus: Ecology, Physiology, Genetics and Culture* (ed. E.M. Gosling), pp. 87–169. Elsevier Science Publishers B.V., Amsterdam.

Siddall, S.E. (1980) A classification of the genus *Perna* (Mytilidae). *Bull. Mar. Sci.*, **30**, 858–70.

Trueman, E.R. (1968) The burrowing activities of bivalves. *Symp. Zool. Soc. Lond.*, **22**, 167–86.

Waite, J.H. (1992) The formation of mussel byssus: anatomy of a natural manufacturing process. In: *Results and Problems in Cell Differentiation*, Vol. 19 (ed. S.T. Case), pp. 27–54. Springer-Verlag, Berlin.

Waite, J.H. & Qin, X.X. (2001) Polyphosphoprotein from the adhesive pads of *Mytilus edulis*. *Biochemistry*, **40**, 2887–93.

Wallace, R.L. & Taylor, W.K. (1997) *Invertebrate Zoology: a Laboratory Manual*, 5th edn. Prentice Hall., Inc. New Jersey.

Walne, P.R. (1974) *Culture of Bivalve Molluscs: 50 Years' Experience at Conwy*. Fishing News Books, Oxford.

Weinstein, J.E. (1995) Fine structure of the digestive tubule of the eastern oyster, *Crassostrea virginica* (Gmelin 1791). *J. Shellfish Res.*, **14**, 97–103.

Widdows, J. & Donkin, P. (1992) Mussels and environmental contaminants: bioaccumulation and physiological aspects. In: *The Mussel Mytilus: Ecology, Physiology, Genetics and Culture* (ed. E.M. Gosling), pp.383–424. Elsevier Science Publishers B.V., Amsterdam.

Wilkens, L.A. (1991) Neurobiology and behavior of the scallop. In: *Scallops: Biology, Ecology and Aquaculture* (ed. S.E. Shumway), pp. 428–69. Elsevier Science Publishers B.V., Amsterdam.

Yonge, C.M. & Thompson, T.E. (1976) *Living Marine Molluscs*. William Collins, Glasgow.

Zwann, A. de, & Mathieu, M. (1992) Cellular biochemistry and endocrinology. In: *The Mussel Mytilus: Ecology, Physiology, Genetics and Culture* (ed. E.M. Gosling), pp. 223–307. Elsevier Science Publishers B.V., Amsterdam.

3 Ecology of Bivalves

Introduction

Ecology is the study of interactions between organisms and their environment. Environment embodies everything outside the organism that impinges on it, e.g. physical factors such as temperature, salinity and light, and biological factors such as predators, competitors and parasites. An organism's response to these factors influences its distribution and abundance both on a local and regional scale.

In this chapter the effects of temperature and salinity, probably the two most important physical factors governing the distribution of marine organisms, will be described in detail. Their effects on other aspects of bivalve biology, such as feeding, reproduction, growth, and respiration and osmotic regulation, are covered in Chapters 4, 5, 6 & 7, respectively. Other limiting factors such as aerial exposure, oxygen concentration, currents and substrate type will be reviewed briefly. In addition, effects of biological factors such as predation and competition on bivalve distribution and abundance will be considered. Pathogens and parasites, which also influence the ecology of bivalve populations, are dealt with in Chapter 11.

Before describing the major factors that influence bivalve distribution patterns, some information on the global and local ranges of representative species is presented in the following section.

Global and local distribution patterns

Global maps of the geographic distribution of commercially important bivalve species, based on published information, are illustrated in Figs. 3.1–3.4. A few points are worth noting:

- Some parts of the world e.g. Asia, Africa and S. America are under represented simply because there are few if any comprehensive accounts on species distributions from these regions.
- Several species, in particular oysters and clams, have become established, either through planned or accidental introductions, in regions outside of what would be considered their native range.
- Until relatively recent times distributions have been mapped solely on the basis of external features of the shell. With the advent of molecular markers some distributions have had to be revised, either by extending or reducing the previously reported geographic range (see below). Sometimes species distributions are based solely on molecular markers. This highlights the fluidity of current understanding of species' distributions and boundaries. With sufficient time, expertise and money this situation could be rectified.

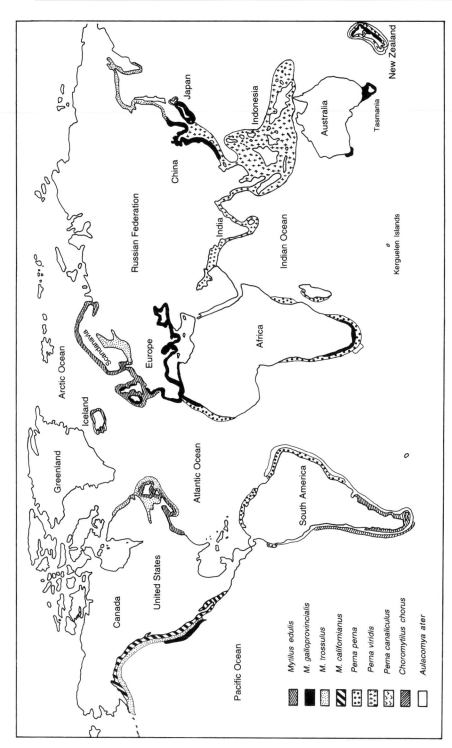

Fig. 3.1. Approximate global ranges of the main commercially important mussel species from Mason (1976); Siddall (1980); Vakily (1989); Gosling (1992) and references therein; Hickman (1992); Hockey & Schurink (1992); Sanjuan et al. (1994); Inoue et al. (1997); Suchanek et al. (1997); Toro (1998); Calvo-Ugarteburu & McQuaid (1998); Comesaña et al. (1998); (Rawson et al., 2001). In the Northern Hemisphere where the ranges of *M. edulis*, *M. trossulus* and *M. galloprovincialis* overlap variable amounts of hybridisation occurs between species pairs. In the Southern Hemisphere using shell morphology and protein markers mussels (*Mytilus*) were identified as either *M. edulis* or *M. galloprovincialis* (McDonald et al., 1991). However, recent results from mtDNA analysis indicate that mussels in this region are derived from two ancient migration events from the Northern Hemisphere (Hilbish et al. (2000); mussels are similar but not identical to Northern Hemisphere *M. edulis* and *M. galloprovincialis*.

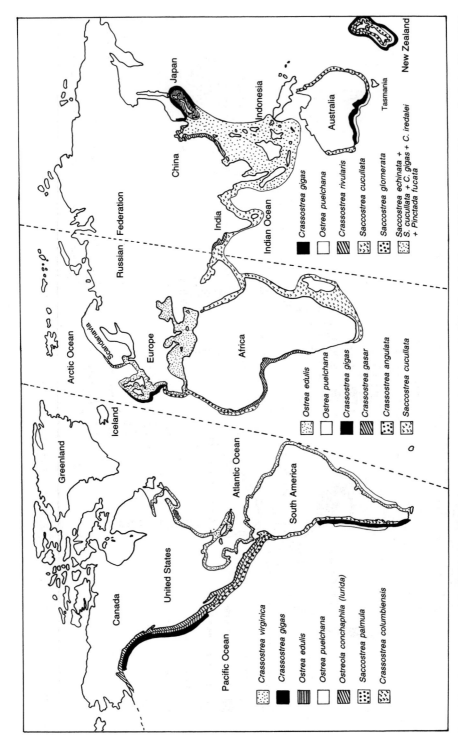

Fig. 3.2. Global ranges of the main commercially important oyster species from text of Carriker & Gaffney (1996). For clarity this map (and also Figs. 3.3 & 3.4) is divided into three sections because the number of species far exceeds the number of species identification patterns. The distribution of *Crassostrea virginica* may not be as extensive as shown in figure (see text).

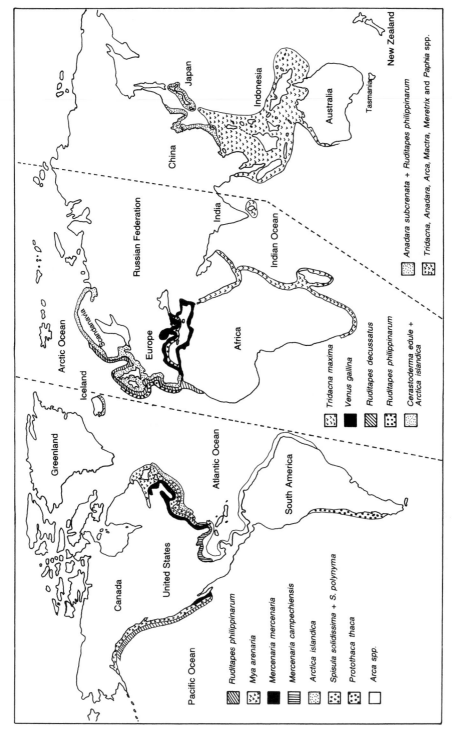

Fig. 3.3. Global ranges of the main commercially important clam species from Heslinga (1989); Manzi & Castagna (1989); Malouf & Bricelj (1989) and FAO (2001).

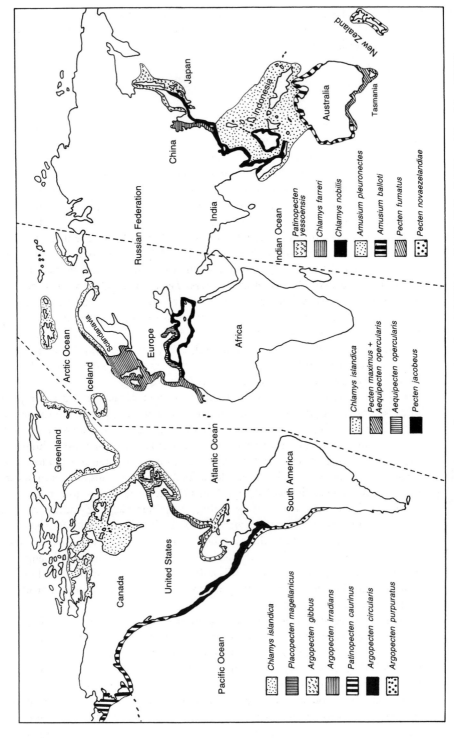

Fig. 3.4. Global ranges of the main commercially important scallop species from Brand (1991); Luo (1991); Waller (1991) and Guo & Luo (2002).

Instead of describing distribution patterns for all of the species illustrated in Figs. 3.1.–3.4. a few representative genera from each of the four bivalve groups will be considered.

Mussels

Mussels in the genus *Mytilus* are a dominant component of rocky shore communities in cooler waters of the northern and southern hemispheres. Of all of the species in the genus the blue mussel, *Mytilus edulis*, would appear to have the widest distribution, being found from mild subtropical to Arctic regions. Other *Mytilus* species have a more restricted range. For example, the Californian mussel, *M. calfornianus*, is confined to the Pacific coast of North America (Seed & Suchanek, 1992), while *Mytilus trossulus* is restricted to cool water regions of the Northern Hemisphere (Fig. 3.1). Incidentally, the distribution of *M. trossulus* has been mapped solely on the basis of molecular characters (see Chapter 10), and it remains to be seen whether with further study its range extends into Arctic waters. *Perna* has a subtropical to tropical distribution in the southern hemisphere. Of the three *Perna* species in Fig. 3.1 the green mussel, *Perna viridis*, has the most widespread distribution, extending from India eastwards to south-east Asia, the China coast, Indonesia, Philippines and Samoa. The rock mussel, *Perna perna*, is found on the coasts of South America and South Africa, while the green-lipped or New Zealand mussel, *Perna canaliculus* is found only in New Zealand.

On a local scale, mussels (*Mytilus*) dominate the intertidal to subtidal regions of rocky shores. *Mytilus edulis* has the widest distribution pattern in the genus, extending from high intertidal to subtidal regions, from estuarine to fully marine conditions, and from sheltered to extremely wave-exposed shores. At exposed sites the species prefers gently sloping, slow-draining platforms to steep rock faces. On such shores dense beds of small mussels, often several layers thick, blanket the rocks (Fig. 3.6). On sheltered shores and in estuaries fewer but larger mussels are found (Fig. 3.5). Subtidal populations on pier pilings, oil platforms and culture ropes enjoy a virtually predator-free environment and this, combined with continuous submersion, allows *Mytilus* to reach a large size in a relatively short period of time e.g. 50 mm shell length in 6–8 months (Page & Hubbard, 1987). *Perna* spp. are found both in brackish estuaries and in open, but not very exposed, waters in the lower intertidal to subtidal regions of the shore.

Mussels settle on a wide variety of substrates e.g. rock, stones, pebbles, shell, cement and wood, once the substrate is firm enough to provide a secure anchorage. In the case of *Perna*, mangrove mudflats represent a major habitat for tropical species. Early spat either attach to filamentous algae, from which they eventually migrate onto adult mussel beds, or else they settle directly onto adult beds (details in Chapter 5). Upper distribution limits for mussels are usually governed by physical factors, primarily temperature, while predators are mainly responsible for setting lower limits (Seed & Suchanek, 1992).

Fig. 3.5. Boulders covered with the mussel, *Mytilus edulis*, on a sheltered shore in Ireland. Photo courtesy of John Costello, Aquafact International Services Ltd., Galway, Ireland.

Oysters

In comparison to mussels there is a lot more information on the global distribution of oyster species, presumably because of their greater economic importance. Several of them have broad geographic ranges (Fig. 3.2). For example, the eastern oyster, *Crassostrea virginica*, extends from the Gulf of St. Lawrence in Canada (48°N) to the Gulf of Mexico and continues onto the coasts of Brazil and Argentina (50°S), a total distance of about 8000 km (but see below). Similarly, the range of the Indian rock oyster, *Saccostrea cucullata*, extends from the tropical coast of West Africa (10°W) eastward across the Indian Ocean as far as the Philippines (130°E). Other species with broad ranges are the Columbian oyster, *Crassostrea columbiensis*, the European flat oyster, *Ostrea edulis*, the black-bordered oyster, *Saccostrea echinata* and the Indian rock oyster (Fig. 3.2). The Pacific oyster *Crassostrea gigas* is native to the Indo-West Pacific region, but because of successful introductions onto the Pacific coast of North America, western Europe and Australia, it virtually has a global distribution. As mentioned above, distributions may need to be revised in the light of new evidence. For example, the Chilean flat oyster, *Ostrea puelchana*, shown to have a circum-global distribution between latitudes 35° and 50°S

Fig. 3.6. Mussels (*Mytilus*) on an exposed rocky shore in the west of Ireland. Individuals only attain a maximum shell length of 2.5 cm, and that after many years. Photo courtesy of D. McGrath, Galway-Mayo Institute of Technology, Galway, Ireland.

(Fig. 3.2), is probably not a single species. Results from DNA molecular phylogenetic analyses suggest that in Argentina the species is indeed *O. puelchana* but that in Chile and New Zealand it is *O. chiliensis*, in S. Africa, *O. algoensis* and in Australia *O. angasi* (Jozefowicz & Ó Foighil, 1998; Ó Foighil *et al.*, 1999). There is genetic evidence that the species on the coasts of Brazil and Argentina is not *C. virginica*, but the mangrove oyster, *C. rhizophorae* (Hedgecock & Okazaki, 1984), so that the former may not have so wide a distribution after all. Also, there is recent evidence that a second species, *C. brasiliana*, occurs sympatrically with *C. rhizophorae* (Ignacio *et al.*, 2000). It is clearly important that in mapping the geographic distribution of a particular species that we know whether it is genetically homogeneous or a mixture of two or more species.

Oysters are commonly found in the low intertidal to subtidal regions of shallow, sheltered estuaries (Table 3.1). Compared to mussels they settle on a more restricted range of substrates, mainly rock and shell, and roots and branches in the case of mangrove species.

Clams

Global distribution of the major clam species is presented in Fig. 3.3. The hard clam, *Mercenaria mercenaria* is distributed on the Atlantic coast of North America from the Gulf of St. Lawrence to Florida, and is particularly abundant from Maine to Virginia. In the southern part of its range it is sympatric with *M. campechiensis*, which extends into the Gulf of Mexico. *M. mercenaria* has been successfully introduced into California, but with limited success into

Table 3.1. Habitat type of 12 commercially important oyster species. From text of Carriker & Gaffney (1996). See Fig. 3.2 for global distribution. *May be an ecotype or subspecies of *Crassostrea virginica* (see text).

Species	Common name(s)	Habitat
Crassostrea virginica	Eastern oyster	Intertidal and subtidal regions of estuaries and coastal areas of reduced salinity
C. gigas	Pacific oyster; Japanese oyster	Shallow, sheltered, low-salinity (23–28 psu) waters
C. gasar	West African; mangrove oyster	Attached to roots and branches of mangroves in intertidal zone, and on muddy bottom of brackish regions of estuaries
C. columbiensis	Columbian oyster	Adheres to rocks or mangrove roots or other solid substrata in the mid-intertidal area
*C. rhizophorae**	Mangrove oyster; gureri	In the intertidal area attached to roots and branches of mangrove trees in high-salinity seawater
Ostrea edulis	European flat oyster	Shallow, sheltered estuarine waters
O. puelchana	Chilean flat oyster; mud oyster	Low tide to 15 m in estuarine conditions
Ostreola conchaphila	Native Pacific oyster; Olympia oyster	Usually solitary attached to living molluscs in estuaries and salt water lagoons
Saccostrea glomerata (formerly *S. commercialis*)	Sydney rock oyster	Attached to rock and shell in intertidal zone of estuaries
S. cucullata	Indian rock oyster; Bombay oyster; curly oyster, Red Sea oyster	Intertidal and shallow subtidal rocky or firm substrata
S. echinata	Black-bordered; black-edged; black-lipped oyster	Attached to intertidal and shallow subtidal rocks and other hard surfaces
S. palmula	Palmate oyster	Attached to mangrove roots or rocks in intertidal and shallow subtidal areas of estuaries and mangrove forests

western Europe. In contrast, the Manila clam *Ruditapes philippinarum*, indigenous to the western Pacific, was introduced first onto the west coast of North America and from there into western Europe. In Europe it has proved to be a hardy, fast-growing but generally non-reproducing species with substantial potential for commercial production (see Chapter 9). Giant clams (*Tridacna* sp.) are found throughout the tropical Indo–Pacific, generally inhabiting the shallow water of coral reefs. Unfortunately, stocks have become severely diminished, mainly due to overexploitation, and to a lesser extent, climate change taking place over many centuries. Efforts are now underway to conserve wild stocks and to set up breeding programmes for restocking purposes (Munro, 1993). Other clams that also have a broad geographic range in tropical waters are species of *Anadara*, *Arca*, *Mactra*, *Meretrix* and *Paphia* (Fig. 3.3).

Table 3.2. Habitat type of commercially important clam species. From text of Malouf & Bricelj (1989), Manzi & Castagna (1989) and Heslinga (1989). Habitat is given first, followed by preferred substrate type and burial depth. See Fig. 3.3. for global distribution.

Species	Common name(s)	Habitat
Mercenaria mercenaria	Hard clam	Estuarine intertidal to shallow subtidal; sand, sand-mud, sand-shell; 5–10 cm
Mya arenaria	Soft shell clam	Estuarine, upper intertidal to shallow subtidal to ~200 m depth; firm mud/sand; ~15 cm
Spisula solidissima	Surfclam	Open ocean to 30 m, sometimes much deeper; sand; 5–15 cm
Cerastoderma edule	Edible cockle	Estuarine to fully marine; mid-tide to low-water; sand, soft mud, gravel; <5 cm
Ruditapes philippinarum	Manila clam; Japanese carpet shell	Bays and protected coasts, intertidal to very shallow subtidal; sand, mud/gravel; 5–10 cm
Arctica islandica	Ocean quahog	Open ocean shelf to 150 m; sand; 5–10 cm
Saxidomus giganteus	Butter clam	Estuarine, low intertidal, subtidal to 20 m; sand/gravel; up to 30 cm
Protothaca staminea	Little neck clam	Estuarine, intertidal to shallow subtidal; mud/gravel; up to 20 cm
Panope abrupta	Geoduck	Protected bays, subtidal to 20 m but intertidal in north of range; sand, sand/mud; up to 100 cm
Siliqua patula	Pacific razor clam	Open coast, intertidal surf to 10 m; sand; 30–100 cm
Tridacna spp.	Giant clam	Shallow sunlit, fully saline waters of Indo-Pacific coral reefs

Of the four bivalve groups, clams occupy the broadest range of habitats (Table 3.2). They are found from open coast to sheltered, saline and estuarine locations. They extend from the upper intertidal to subtidal regions of shores, in some cases to depths of 200 m. They settle on a variety of substrates, e.g. mud, sand or gravel, or combinations of these, and bury themselves at depths ranging from 5 to 100 cm, depending on the species. Giant clams, *Tridacna* spp, do not burrow but byssally attach to coral reef early in life.

Scallops

Scallops are found in all waters of the Northern and Southern Hemispheres and show a more extensive global distribution than any of the groups mentioned so far. For example, one species, the Iceland scallop, *Chlamys islandica*, extends into arctic regions, as far as 75°N (Fig. 3.4). However, ranges of individual species of scallop are not as broad as those of some oyster and clam species (see Figs. 3.2 & 3.3).

Most scallop species are found at depths between 10 and 100 m in sheltered bays and open coast sites (Table 3.3). All scallops secrete a byssus when young

Table 3.3. Habitat type and typical densities of some commercially important species of scallop. From text of Ansell *et al.* (1991); Blake & Moyer (1991); Bourne (1991); Brand (1991); Bull (1991); Del Norte (1991); Luo (1991); Naidu (1991); Parsons *et al.* (1991); Rhodes (1991).

Species	Common name(s)	Habitat
Pecten maximus	Great or king scallop	Just below low water mark to ~180 m, most common at 20–45 m; in clean firm sand, fine or sandy gravel, sometimes with mud; recesses; densities 1–3 m^{-2}
P. novaezelandiae	New Zealand scallop	Semi-estuarine and coastal waters, low tide to ~100 m; wide variety of substrates; recesses; <1 per 5 m^2
Aequipecten opercularis	Queen scallop; queenie	At depths 20–100 m; does not recess
Chlamys islandica	Iceland scallop	At depths 10–250 m, but usually <100 m; attached to sand, gravel, shell and stones; does not recess, attached by byssus; densities 50–75 m^{-2}
C. farreri	Zhikong or Jicon scallop	From intertidal region to 60 m; attached by byssus to rocks and gravel; densities 2–4 m^{-2}
C. nobilis	Huagui scallop	Lower intertidal to subtidal depths of >350 m; on rocky or sandy substrate; does not recess; free or byssate; <1 m^{-2}
Amusium pleuronectes	Moon scallop	Found between 18 to 40 m; in soft bottoms, e.g. silt, mud and/or sandy mud; recesses
Argopecten purpuratus	Peruvian, Chilean or ostion scallop	Between 8–30 m in sheltered bays; sand is preferred substrate, but also found in mud and on rocks; recesses; densities ~0.2–5.0 m^{-2}
A. gibbus	Calico scallop	Shallow subtidal to 370 m, usually 20–50 m; hard sandy bottom; recesses; ~40 m^{-2}
A. irradians	Bay scallop	Protected bays and estuaries at depths <10 m; young attached by byssus to stones, algae, shell; adults on mud bottoms among eelgrass *Zostera*; sometimes recess in winter; 0.2–4.4 m^{-2}
A. circularis	Catarina scallop	Shallow waters from 6–35 m; very varied substrate, e.g. shell, eelgrass, algae, coral, gravel mixed with sand or mud; ~5 m^{-2}
Placopecten magellanicus	Sea or giant scallop	Found 10–100 m, sometimes in shallower water (2 m); mud, sand, gravel, pebbles, rocks, and even boulders; recesses, byssal attachment; 2–4 m^{-2}
Patinopecten yessoensis	Yesso scallop; primorye scallop; hotate gai	Found in shallow sheltered inlets and open deep-water sites; 0.5–25 m depths; in firm silt, sand and gravel; recesses; 1–8 m^{-2}
P. caurinus	Weathervane scallop	Found at depths 10–200 m; sand or mud; recesses; 1 per 65 m^{-2}

but most lose the byssus soon after metamorphosis and recess on sand or gravel bottoms. Species that retain the byssus throughout life need a firm substrate such as pebbles, rocks, shell or boulders.

Scallops are unique among bivalves in their ability to 'swim' (Chapter 2). They use this primarily to escape predation (see below) and for habitat selection. *Pecten maximus* that are normally recessed in sandy or sandy mud substrates swim more frequently and disperse widely when placed on hard substrates. In contrast, when placed in sand little or no swimming occurs (Baird, 1958). Similarly, when the bay scallop, *Argopecten irradians*, was released on sand at a distance of 25 cm from a bed of *Zostera* – its natural habitat – it swam towards the grass bed irrespective of the bed's direction relative to the scallop's facing direction (Hamilton & Koch, 1996). However, when placed at greater distances from the grassbed (>50 cm) scallop movement was random, which suggests that orientated behaviour in this species is based on visual information, and possibly chemical stimuli also. Swimming ability is so well developed in some species that it might be used for migration (see references in Wilkens, 1991). However, results from tagging experiments do not support this hypothesis. Tag returns on *P. magellanicus* showed that 52% of recaptures were within 5 km, 77% were within 10 km and 94% were within 25 km of the release point, and that movements were related to direction and velocity of water currents (Melvin *et al.*, 1985). Similar results have been observed for *P. maximus* and the yesso scallop, *Patinopecten yessoensis* (references in Brand, 1991).

Factors affecting distribution and abundance

Physical factors

As mentioned above, temperature and salinity not only set limits on the spatial distribution of bivalves but also affect every aspect of biology including feeding, reproduction, growth, respiration, osmoregulation and parasite-disease interactions (see Chapters 4, 5, 6, 7 & 11). When it comes to distribution on a large geographic scale it is generally recognised that temperature plays a more important role than salinity. However, in coastal and estuarine regions salinity is probably the most important limiting factor, particularly for oyster populations. The synergistic effect of temperature and salinity, acting in concert with other environmental variables such as water depth, substrate type, food availability, water turbidity and the occurrence of competitors, predators and disease, can have more profound consequences than either factor acting alone.

Temperature and salinity

Most marine bivalves live within a temperature range from −3°C to 44°C (Vernberg & Vernberg, 1972). Within this range the degree of temperature tolerance is species-specific, and within individual species early embryos and larvae have a narrower temperature tolerance than adults. In addition, the temperature required for spawning is invariably higher than the minimum temperature required for growth. All of these factors set limits on the natural distribution of individual species on both regional and local scales. A few pertinent examples will make this clear. The scallop *Placopecten magellanicus* is a cold–water species (Fig. 3.4) with a temperature optimum of about 10°C, and an upper lethal temperature between 20°C and 24°C, depending on acclimation temperature. At the northern end of its range (Newfoundland) the scallop inhabits shallow depths where the water is warmest. The distribution of the species in this part of its range is largely determined by low summer temperatures, which fail to reach the spawning threshold for the species, or which prolong larval development so that recruitment fails. At the southern end of its range (Cape Hatteras, North Carolina) the scallop occurs in much deeper water, usually >55 m, where the water remains cold, separated from the warmer upper layers by a thermocline (Brand, 1991). Cape Hatteras is also the southern limit for the mussel, *Mytilus edulis* (Fig. 3.1); the northward moving warm–water Gulf Stream meets the southward moving cold Labrador Current in the region of Cape Hatteras, and provides a temperature barrier for the distribution and survival of mussel larvae south of this point.

Environmental temperature is also an important determining factor in the geographic distribution of *Chlamys islandica*. This subarctic species (Fig. 3.4) occurs in temperatures from −1.3°C to 8°C but the cold Labrador Current flowing down along the east coast of Canada allows the species to penetrate as far south as 42°N. The disappearance of *C. islandica* from areas in Norway where it was once abundant is probably due to the steady rise in temperature in northern waters since the 1930s (Wiborg, 1963). In the western Pacific, *Patinopecten yessoensis* is a cold water species (Fig. 3.4) with a temperature range of 5–23°C. Attempts to cultivate this species in the south of Japan, where water temperatures in summer exceed 22°C, have resulted in very high mortality, demonstrating once again the important role of temperature in species distribution.

In the open oceans salinity varies between 32 psu and 38 psu, with an average of 35 psu (Kalle, 1971). In contrast, estuaries and bays are subject to pronounced salinity fluctuations because of evaporation, rainfall and inflow from rivers. Oysters are euryhaline (i.e. able to survive in a wide range of salinities) and so it is not surprising that they are successful colonisers of estuarine as well as fully saline waters (Table 3.1). *Crassostrea virginica* normally lives in waters of 5 to 40 psu with an optimum salinity in the range 14–28 psu (references in Shumway, 1996). Many mussels, in particular *Mytilus* spp., are euryhaline. *Mytilus edulis* has an extremely wide marine and estuarine distribution and a salinity tolerance that has been reported to range between 4 and 40 psu (Bayne, 1976). However, in the Baltic, where salinity is about 5 psu, the dwarfed mussels there are not in fact *M. edulis* but the rediscovered mussel, *M. trossulus* (see above). *Perna* also tolerates wide fluctuations in salinity. The normal salinity range for estuarine *P. viridis* is between 27 and 33 psu, but 50% survival was reported in this species after two weeks exposure to salinities of 24 and 80 psu (Sivalingham, 1977). Many clam species are also euryhaline. For example, the estuarine habitat in which *Mya arenaria* lives is constantly exposed to changes in salinity from about 10 to 25 psu, mainly as a result of freshwater run-off. Consequently, the adult salinity tolerance range is between 4 and 33 psu. Fig. 3.7 shows experimentally determined temperature and salinity tolerance ranges for adults and larvae of seven North American clam species. Not surprisingly, the data concur with regional and local distribution patterns for these species (Fig. 3.3 and Table 3.2). The narrower tolerance of larvae compared to adults is also highlighted in Fig. 3.7. In contrast, most scallop species live in fully saline waters and are unable to colonise low-salinity waters. For example, *Pecten maximus* and *Aequipecten opercularis* have similar geographic distributions (Fig. 3.4) in western Europe and both species extend as far as, but not into, the low salinity (5–6 psu) waters of the Baltic. Experiments have confirmed that *A. opercularis* cannot tolerate low salinity; 16–28 psu were lethal after a 24 h experimental exposure, depending on temperature and size of scallop (Paul, 1980; Fig. 3.8). In contrast, *Argopecten irradians*, living in shallow bays and estuaries (Table 3.3) in salinities ranging between 10 and 38 psu, is one of the few euryhaline scallop species (Brand, 1991 and references).

It is not surprising that temperature and salinity which play a key role in geographic distribution are also important in determining species distribution on a local scale. For subtidal species additional factors such as water depth, substrate type, currents, turbidity, as well as predation and competition play an important role. For intertidal species upper limits are generally set by tolerance to temperature extremes and desiccation, while lower limits are set primarily by biological factors such as predation and competition (see Rafffaelli & Hawkins, 1996 for discussion and references). For burrowing bivalves, additional physical factors such as substrate type and oxygen concentration come into play, while predation and competition once again are important biological factors. Needless to say, anthropogenic factors such as water borne contaminants, introduced species, and disease can also be significant in determining local distribution patterns.

Animals in the intertidal area of the shore have to cope with being out of water at regular intervals. For animals in the high intertidal emersion times are longest, and consequently these individuals are often subjected to temperature extremes and desiccation. Upper distribution limits for a species are set by its ability to tolerate such extremes by various physiological mechanisms. For example, in north-east Canada temperatures can drop to $-35°C$ in winter. Mussels (*Mytilus trossulus*) survive such low temperatures even when their tissue temperatures are as low as $-10°C$ (Williams, 1970). As much as 60% of their extracellular fluid is frozen at this temperature. The unfrozen extracellular fluid becomes more concentrated with solutes, and this process draws water by osmosis out of cells, thus lowering the intracellular freezing point. In addition, end-products of anaerobic respiration, e.g. strombine are utilised as cryoprotectants within cells (Loomis *et al.*, 1988). Both mechanisms act to prevent ice-crystal formation within cells. Ice formation is usually lethal, because as the crystals grow they rupture and destroy the cells; ice-crystal formation outside cells does little damage (Randall *et al.*, 1997). Unlike *M. trossulus*, *M. californianus* cannot tolerate freezing conditions and, consequently, it is usually restricted to intertidal pools, rock crevices and subtidal habitats (Seed & Suchanek, 1992).

High temperatures and associated desiccation also set upper limits for

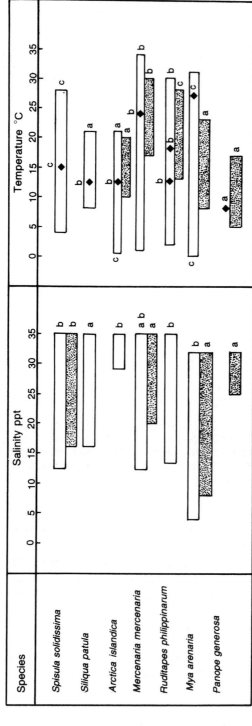

Fig. 3.7. Experimentally determined tolerance ranges and observed environmental limits of temperature and salinity for adults and larvae (stippled) of seven clam species. a: experimentally determined; b: approximate limits observed in nature; c: observed in nature or used in culture (not necessarily limits); approximate minimum spawning temperatures (♦) are also shown. See Fig. 3.3. for geographic distribution. *Siliqua patula* is found from Alaska to northern California, while *Panope generosa* has approximately the same geographic distribution as *Panope abrupta* in Fig 3.3. Modified from Malouf & Bricelj (1989 and references therein) with permission from Elsevier Science.

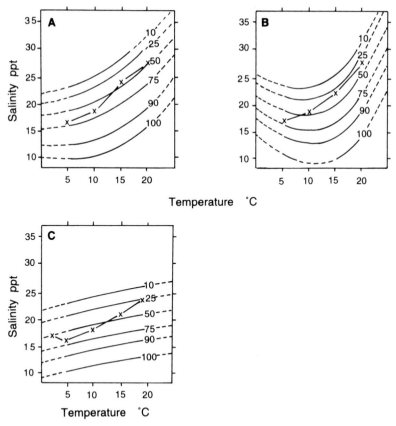

Fig. 3.8. Regression surfaces for percentage mortality in three size categories (A, >60 mm; B, 30–40 mm; C, 5–10 mm shell height) of *Aequipecten opercularis* at different salinity-temperature combinations. Isopleths are produced at intervals of mortality between 10 and 100%. The dashed lines represent extrapolations outside the test area; crosses indicate median lethal salinities. Modified from Paul (1980).

mussels. When the temperature rise is gradual mussels have sufficient time to modify rates of respiration, filtration and enzyme activity. Also, heat-stress proteins (hsp) play an important role in thermal acclimation. The primary function of these proteins is to prevent heat-damaged protein accumulation in cells during the process of thermal acclimation. Endogenous levels of stress-70 proteins in *M. edulis* vary seasonally and are positively correlated with environmental temperature (Chapple *et al.*, 1998). Also, when mussels were heat-shocked at 28.5°C – a temperature at which this species is unable to acclimate – there was a positive relationship between time of survival and levels of hsp. Seasonal variation in the levels of stress proteins have also been reported for *M. trossulus* and *M. galloprovincialis*, with higher summer levels in the more northerly species, *M. trossulus* (Hofmann & Somero, 1996). Secretion of hsp in response to heat shock has been reported in laboratory-based studies on *Crassostrea gigas* (Shamseldin *et al.*, 1997; Clegg *et al.*, 1998), but so far the only data from wild populations are from *Mytilus* (references in Chapple *et al.*,

1998). Incidentally, hsp are also induced by sublethal concentrations of a variety of environmental contaminants, and salinity fluctuations (Werner & Hinton, 1999).

Some species use a behavioural mechanism to increase tolerance to desiccation. For example, *Perna viridis*, which is exposed to year-round temperatures of about 28°C takes a bubble of air into the mantle cavity prior to valve closure during emersion (Davenport, 1983).

Bivalves on fully marine shores experience normal (*ca.* 35 psu) salinity most of the time. Tide pools high on the shore experience wide fluctuations in salinity through evaporative water loss or freshwater inflow. But, for the majority of the shore, these effects are minor. However, estuaries place significant limits on species distributions. Here mean salinity decreases and salinity variation increases with distance upstream, and both these factors have deleterious effects on bivalve distribution with the result that species diversity is significantly less in estuaries than on fully marine shores. However, oysters show a remarkable tolerance to low salinity and salinity variation. For example, the optimum salinity range for *Crassostrea virginica* is about 14–28 psu but this species can survive a salinity of 2 psu for a month, or even freshwater for several days when water temperatures are low. At very low salinity recruitment is poor and growth is slow. High salinities of 35–40 psu are also tolerated, but oysters usually show poor growth and do not reproduce (references in Shumway, 1996). Surprisingly, juveniles have the same salinity range as adults, and a similar physiological response to sub-optimal salinities.

Other factors

Most bivalves show a preference for a particular substrate. Oysters prefer hard rock, shell or sand bottoms and will not settle on muddy bottoms. The majority of scallops occur on hard substrates of gravel and coarse to fine sand. Such substrates are typical of areas with a strong current flow. Clams with their burrowing lifestyle have a preference for softer substrates of sand or mud, or sand/mud mixtures. In the hard clam, *Mercenaria mercenaria*, highest numbers occur in sand, moderate numbers in sand/mud, and fewest numbers in mud, indicating that substrate type is a significant factor in the distribution of this species (Wells, 1957). Muddy bottoms are characteristic of areas with poor current flow. Overlying water tends to be turbid due to suspended material of fine insoluble particles, either inorganic (clay, silt, sand) or organic, e.g. industrial or domestic waste. Oxidation of organic waste just above the sediment can reduce oxygen levels, especially in summer when water temperatures are high. Such conditions are believed to be a major source of mortality for scallop spat (*Patinopecten yessoensis*) when they move from byssal attachment sites to life on the seabed. Survival is increased if alternative sites, such as algae or bryozoans that are raised off the bottom, are available. It is clear that a lack of such sites is a major factor in determining local distribution of this species.

High turbidity is not detrimental to all bivalves. For example, evidence from both laboratory and field trials show that the surf clam, *Spisula subtruncata*, grows well in environments with a high sediment load. This is because the species is extremely efficient in selectively rejecting sediment particles so that only a small fraction of filtered algae is lost in pseudofaeces (Kiørboe & Møhlenberg, 1981). In contrast, in another clam, *Mercenaria mercenaria*, algal ingestion decreases significantly with increasing sediment load (Bricelj & Malouf, 1984), thus explaining the low abundance of this species on muddy bottoms (see above).

Exposure to wave action is another factor that influences bivalve distribution and abundance. Mussels are the only group that live on exposed shores and some species are better adapted than others. On the Pacific coast of North America where the ranges of *M. californianus* and *M. galloprovincialis* overlap (Fig. 3.1) the former dominates wave-beaten shores from mid-tide to below low water level, while *M. galloprovincialis* is the dominant mussel at sheltered sites. Witman & Suchanek (1984) have shown that byssal attachment strength was significantly greater in *M. californianus* than in *M. galloprovincialis*. However, when storms rip off patches of *M. californianus*, the bare patches are quickly colonised by *M. galloprovincialis* spat (regarded as *M. edulis* in publications prior to the early 1990s), which grow rapidly but are eventually out-competed by *M. californianus*. It is clear that both exposure to wave action and competition set limits on the distribution of *M. galloprovincialis* on the Pacific coast. Interestingly, further north in Alaska, where *M. californianus* is absent because of freezing temperatures, *M. trossulus*, which also has a much weaker byssal attachment strength than *M. californianus* (Bell & Gosline, 1997), dominates the region of the shore normally occupied by *M. californianus* further south (Seed & Suchanek, 1992).

The importance of large-scale ocean currents in the global distribution of bivalves has already been dealt with. Locally, areas with strong currents usually provide favourable feeding conditions for bivalves. However, very strong currents can have an inhibitory effect on feeding and consequently, growth (see Chapter 6). Also, strong currents may prevent larval settlement and byssal attachment of spat, ultimately resulting in local variability in recruitment. More data are needed to assess the importance of high water velocity as a factor in the local distribution of bivalves.

Finally, physical factors that effectively remove animal and plant communities from the substrate play a minor role in influencing abundance, i.e. patchiness at the local level. These disturbance factors include scouring by ice, storms that overturn boulders, wave-propelled logs and fishing gear. Most of the information on the effects of, and recovery from, disturbance comes from studies on *Mytilus* beds and their associated flora and fauna on exposed shore sites on the Pacific and north-east Atlantic coasts of North America (Seed & Suchanek, 1992 for review; Svane & Ompi, 1993; Wootton, 1993; Beukema & Cadee, 1996; Carroll & Highsmith, 1996; Hunt & Scheibling, 1998 & 2001).

Minchinton *et al.* (1997) monitored the recovery of intertidal algae and sessile macrofauna after a rare occurrence of scouring sea ice denuded the intertidal area of an exposed rocky shore in Nova Scotia, Canada. They found that barnacle cover was restored soon after the ice-scour, macroalgae cover in about two years, but that it took considerably longer (4–6 years) for the mussel beds to recover. Similar results have been reported by Brosnan & Crumrine (1994) who trampled 250 experimental plots in the intertidal zone for a period of one year and then monitored recovery. The algal-barnacle community recovered in the year following trampling but mussel beds had not recovered in the two years following cessation of trampling.

Wave or log damage occurs mostly during the winter and is typically responsible for removing 1–5% of *M. californianus* cover per month on exposed shores (Paine & Levin, 1981). The initial size of disturbance gaps can range from single mussel size to areas as large as $60\,m^2$. Subsequent enlargement of the gap (as much as 5000%) may occur, especially during winter months, primarily due to weaker byssal thread attachments (Witman & Suchanek, 1984).

Fishing methods can affect bivalve abundance directly by causing significant mortality, and, indirectly, by causing shell damage. For example, large numbers of razor clams (*Ensis*) were either killed or damaged by dredging operations in a sandy bay in Scotland (Eleftheriou & Robertson, 1992; Robinson & Richardson, 1998). Damage to shell valves slows down the clam's escape digging response, and thereby renders it more vulnerable to predatory attacks by crabs and fish.

Biological factors

Just as humans greatly appreciate the delicate flavour of bivalves so also do a whole range of other organisms from groups as diverse as fish, birds, mammals, crustaceans, echinoderms, flatworms, and even other molluscs. These predators are probably the single most important source of natural mortality in bivalve molluscs and have the potential to influence population size structure in addition to overall abundance and local distribution patterns (Seed & Suchanek, 1992).

In this section the main predators of bivalves will be dealt with, along with major pests, fouling organisms and competitors. For the sake of simplicity the groups, mussels, oysters, clams and scallops will be treated separately.

Predators, pests and competitors

Mussels

Gastropods are significant predators of mussels worldwide. The dogwhelk, *Nucella lapillus*, is widely distributed on exposed shores in northern Europe and on the east coast of North America, where it feeds extensively on barnacles and mussels. Predation is often seasonal with whelks remaining aggregated in pools and crevices over the wintertime. However, numbers on mussel beds on the low and mid shore start to increase in the spring, and densities as high as 300 whelks m^{-2} have been recorded over the summer months in north-east England (Fig. 3.9). Profitability (energy assimilated from a food item relative to handling time) for dogwhelks feeding on mussels increases with prey size (Fig. 3.10). Yet whelks prefer mussels smaller than the largest available. Hughes (1986) suggests that dogwhelks choose mussels with the maximum average profitability in the face of competition from other dog-

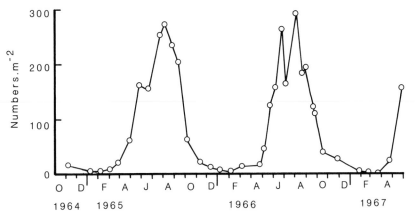

Fig. 3.9. Seasonal abundance of the dogwhelk *Nucella lapillus* on low shore *Mytilus edulis* beds in north-east England. From Seed (1969).

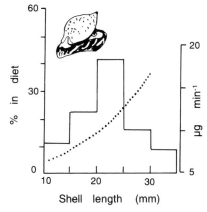

Fig. 3.10. The sizes of mussels (*Mytilus edulis*) eaten by the dogwhelk *Nucella lapillus* when given a choice of all sizes. The curve represents the profitability (energy assimilated from a food item relative to handling time) of mussels to *N. lapillus*. From Hughes & Dunkin (1984).

whelks that are attracted to the predator by olfactory stimuli from the damaged prey. Choosing whether to drill or reject a mussel takes time, often as long as one hour. The whelk uses the radula to drill a small hole through the thinnest part of the shell around the umbo or adductor muscle insertion regions (Seed, 1976), or though the shell area overlying the glycogen-rich digestive gland (Hughes & Dunkin, 1984). Prior to drilling, the whelk softens the area using a secretion from the foot. The proboscis is inserted through the hole and the flesh of the prey is rasped away by the radula, and then devoured. Alternatively, a more efficient mechanism, where the whelk inserts its proboscis through the valve gape and induces muscular paralysis by injecting toxins, may be used. Results from laboratory experiments have shown that an adult whelk can consume about two mussels (1–3 cm shell length) per week during the summer (Seed, 1969). Although this level of consumption may appear small, the high density of foraging whelks makes a serious impact on mussel coverage on exposed shores. Preference for mussels, as opposed to barnacles, appears to be fixed in early life, i.e. adult whelks transferred from sites with no mussel cover to those with a high coverage of mussels largely ignore mussels, preferring to feast on barnacles (Wieters & Navarette, 1998). Mussels respond by ensnaring and immobilising whelks in their byssus threads (Petraitis, 1987; Davenport *et al.*, 1998), or spat may use 'hairs' on the periostracum to reduce predation by young whelks (Dixon *et al.*, 1995).

On the west coast of the United States *Nucella canaliculata* and *N. emarginata* are major predators of mussels but seem to favour the thinner-shelled species, *M. galloprovincialis*, as opposed to the thicker shelled and less nutritious *M. californianus* (Suchanek, 1981). Other predatory gastropods such as *Ocenebra poulsoni*, *Acanthina sopirata*, *Ceratostoma nuttalli* and *Jaton festivus* also feed on *Mytilus* (Shaw *et al.*, 1988).

Starfish are also important predators influencing the distribution and abundance of mussels on the lower shore and in the sublittoral zone. In Washington state on the Pacific coast of the United States *M. trossulus*, which settles just below *M. californianus* in the intertidal zone, is soon rapidly eliminated by the sea star, *Pisaster ochraceus*. Physical removal of starfish extends the lower limits of not just *Mytilus* but also *Perna* and *Perumytilus* species. When the starfish return the lower limits revert to the previous state (Seed & Suchanek, 1992 and references). This starfish is a 'keystone' predator of mussels. Its removal exerts a very strong controlling influence on its prey, which, incidentally, can only be demonstrated through manipulative experiments.

In northern Europe *Asterias rubens* is a serious predator of *Mytilis edulis* (Fig. 3.11). This starfish aggregates seasonally on mussel beds in large numbers, sometimes as high as $450\,m^{-2}$, often completely destroying local mussel populations (Dare, 1982). In contrast to the results of laboratory-based experiments, *Asterias rubens* shows no size selectivity when feeding in the field. The solid structure of interconnected mussels forming the bed however, restricts predation to only those mussels situated at the bed surface, thus providing a refuge from predation for smaller mussels deeper down (Dolmer, 1998). Starfish prey upon mussels and other bivalves by either using force, or by secreting an anaesthetic from their stomach that numbs the bivalve and causes it to gape. The starfish then everts its stomach through its mouth into the shell opening and digests the prey. Not all mussels are equally susceptible to starfish predation. About 70% of *M. edulis* of North Sea origin were able to resist *A. rubens*, whereas all Baltic mussels were opened within one hour (Norberg & Tedengren, 1995). *M. edulis* cultured in close vicinity to *A. rubens* were significantly smaller in shell length, height and width but had significantly larger posterior adductor muscles, thicker shells and more meat/shell volume. These morphological changes have an adaptive value in that predator-exposed mussels have a significantly higher survival rate than unexposed mussels (Reimer & Tedengren, 1996). Behavioural changes were also evident; predator-exposed mussels in the laboratory formed larger aggregates, migrated less and sought structural refuges more often (Reimer & Tedengren, 1997).

Navarrete & Menge (1996) have studied the interactions between different predators that feed on mussels. On the Oregon coast of the United States, the

Fig. 3.11. Starfish preying on mussels. Photo courtesy of P. Dolmer, Ministry of Food, Agriculture and Fisheries, Charlottenlund, Denmark.

starfish, *Pisaster ochraceus* is a keystone predator of *M. trossulus*, which is also preyed upon by the whelks, *Nucella emarginata* and *N. canaliculata*. Predation intensity by the keystone predator was strong and was unaffected by the presence of whelks; the whelks only had important effects when *Pisaster* was absent.

Crabs (*Cancer*, *Carcinus* and *Pachygrapsus*) are also significant predators of mussels on the lower shore and sublittoral zone. Their effect on mussel abundance is seasonal, with reduced predation in winter when crabs migrate offshore. Results from laboratory and field experiments show that crabs employ size-selection of prey, with the upper size limit that can be opened being directly related to the size of the crab (references in Seed, 1976). Crabs will almost always choose small-sized prey when offered a range of sizes. It is handling time rather than the energetic costs of handling, estimated as a mere 2% of corresponding gains, that is the basis on which foraging crabs select their prey (Rovero *et al.*, 2000). During the handling period the crab is at risk from other predators, competitors and even claw damage. Small mussels are therefore particularly vulnerable to predation since they are easily crushed by most size classes of crabs. A mussel must attain a shell length of at least 45 mm before it is relatively safe from crab predation. Once again, mussels show several defence mechanisms. In laboratory experiments *M. edulis* increases byssus volume in response to waterborne cues from *Cancer pagurus* and *Carcinus maenas* (Cote, 1995; Leonard *et al.*, 1999). In addition, mussels subject to heavy predation develop thicker and more robust shells in response to not just crabs but also the broken shells of other mussels (Leonard *et al.*, 1999). Similar effects have also been reported in mussels subject to heavy whelk predation (Smith & Jennings, 2000). There is now growing evidence that inducible defences are

Fig. 3.12. The oystercatcher, *Haematopus ostralegus*, a significant predator of mussels. Photo courtesy of John Costello, Aquafact International Services Ltd., Galway, Ireland.

a widespread feature in other marine organisms such as algae, bryozoans, cnidarians and gastropods, (see Leonard *et al.*, 1999 for references).

Several bird species are predators of mussels. The most important group are the oystercatchers (*Haematopus* spp.). We now have a great deal of information on feeding habits, including aspects such as prey choice, size-selection of prey, competition (intra- and inter-specific), energetics and optimal foraging theory, and effects on fisheries. Predation on mussels is generally seasonal, with birds such as *Haematopus ostralegus* (Fig. 3.12) switching in the spring from mussels (and cockles) to deep-living prey such as *Scrobicularia* and *Macoma balthica*, and back to surface bivalves in autumn in order to maximise intake rate (Zwarts *et al.*, 1996b). The birds cannot survive if their diet is restricted to one or two prey species; they need to switch between three or four, and have to roam over feeding areas measuring at least some tens of square kilometres. Oystercatchers open mussels by stabbing into gaping mussels or prising open closed ones, or by hammering a hole in either the dorsal or ventral shell (Gosscustard *et al.*, 1993). Male birds have shorter but stronger bills than females and are thus better equipped to take large and thick-shelled cockles and mussels than females which concentrate on more deeply buried clams and polychaetes (Hulscher & Ens, 1992). All birds show size selection within the prey species; this is because flesh content increases more steeply with prey size than handling time (Zwarts *et al.*, 1996a). Removal of the largest mussels may reduce protected refuge for younger mussels, but may also allow younger mussels to grow at a faster rate (Gosscustard *et al.*, 1996).

Eider ducks (*Somateria* spp.) also prey extensively on mussels which often constitute as much as 60% of their diet (Nehls & Ruth, 1994). In the process of zoning in on their prey the ducks may remove whole mussel clumps, thus causing mussel mortality over and above that produced by direct predation (Raffaelli *et al.*, 1990).

Numerous studies have been undertaken to assess the impact of oyster-catchers and other bird predators on commercial mussel beds. In a sheltered bay in the Wadden Sea, Netherlands, annual production of mussels is about 500–600 g ash-free dry weight (AFDW) m^{-2}. Birds annually removed 30% of the standing stock (Nehls *et al.*, 1997). Eiders were by far the most important predators and consumed 346 g AFDW m^{-2}; these were followed by oyster-catchers with 28 g AFDW m^{-2} and herring gulls with 3.6 g AFDW m^{-2}. However, since other predators were absent, mussel production was sufficiently high to sustain such a high predation rate. In a separate study on tidal flats on the German coast, annual consumption of mussels in an area of 5.2 km^2 representing 311 metric tonnes (t) AFDW by these same predators was 165 t AFDW (Hilgerloh, 1997). This time, however, the highest proportion of total consumption was by oystercatchers (54%), while eiders consumed 39% and herring gulls 7%.

Other birds that feed on mussels include knots, *Calidris* sp. (Alerstam *et al.*, 1992), scoters, *Melanitta* sp. (Shaw *et al.*, 1988) and crows, *Corvus* sp. (Berrow *et al.*, 1992a). Indeed crows (*Corvus* spp.) are significant predators of mussels in the intertidal zone and show several interesting adaptations. The birds fre-quently cache mussels during low tide, and recover them during high tide some 2–3 days later. This behaviour is believed to be a response to short-term, daily fluctuations in food availability (Berrow *et al.*, 1992a). In order to break them open the crows drop mussels and other hard-shelled prey onto hard surfaces, e.g. road or rocky shore (Berrow *et al.*, 1992b). This behaviour peaks during October–February and usually involves only large-sized mussels, no doubt an adaptation to food shortages in winter.

On the west coast of North America the sea otter *Enhydra lutris* is an impor-tant predator of mussels. This species removes large clumps of mussels which it sorts and consumes on the sea surface by pounding the mussels on a flat stone on its chest, or against other mussels. So, although sea otters are selec-tive in terms of the size of prey they consume, they have profound effects on all size classes of mussels (Seed & Suchanek, 1992).

Other predators of mussels include sea urchins (*Strongylocentrotus droe-bachiensis*), lobsters (*Panulirus interruptus* and *Homarus americanus*), flatfish (*Platichthys flesus*, *Pleuronectes platessa* and *Limanda limanda*) and seals, walruses and turtles (see Seed & Suchanek, 1992 for references).

The most common pests of bottom-dwelling mussels are shell burrowing sponges (*Cliona* spp.), polychaetes (*Polydora* spp.) and pea crabs (*Pinnotheres* spp.). The detrimental effects of pea crabs and boring polychaetes will be described later in Chapter 11, while those of boring sponges will be covered below in the section on oysters.

Bivalves provide an excellent substrate for the settlement of many fouling organisms. In mussels, fouling appears to be a significant cause of mortality mainly due to dislodgement caused by the increased weight, especially by barnacles and seaweed. Fouling is a particular problem in suspended mussel culture and almost 100 invertebrate species, including gastropods, crustaceans, bivalves, polychaetes, ascidians, sponges and hydroids have been identified on mussel ropes (Hickman, 1992). These organisms cause reduced growth and productivity through competition for space, but are not a major cause of mortality in suspended culture.

Mussels are the most prominent competitors for space in mid to low-shore areas on gently sloping rocky shores (Lewis, 1977), but on steeper shores mussels tend to be replaced by barnacles or algae. Generally, where two species coexist there is competition but rarely elimination of one by the other. One exception is the case of *M. galloprovincialis*, accidentally introduced on the west coast of South Africa in the 1970s. It has more or less eliminated the slower growing mussel, *Aulacomya ater* (Griffiths *et al.*, 1992; C. McQuaid, personal communication 1999). *M. galloprovincialis* has now rapidly spread onto the south and east coasts, where it is a potential competitor of the indigenous mussel, *Perna perna*. The latter is heavily infected with the trematode *Proctoeces* but *M. galloprovincialis* is free of the parasite. This may give *M. galloprovincialis* a competitive advantage, which could result in the eventual displacement of *Perna perna* from the higher shore (Calvo–Ugarteburu & McQuaid, 1998). In general, however, intra-specific competition for space is a more serious problem than inter-specific competition in that heavy spatfall of mussels onto adult beds can cause the underlying mussels to suffocate, thus loosening the entire population from the rock surface (Seed, 1976).

Oysters

Gastropods are also the main predators of oysters. For *Crassostrea virginica* these are the drills, *Urosalpinx cinerea*, *Eupleura caudata* and *Thais haemastoma*. *U. cinerea* inhabits the intertidal and sublittoral zone to a depth of 30 m along the east coast of North America, and has been introduced with oysters onto the west coast. *E. caudata* has much the same distribution as *U. cinerea* but is not as abundant. *T. haemastoma* is found along the south-eastern and Gulf coasts of the United States. Like *Nucella* these species use their radula to drill, but in between drilling they soften the shell using a secretion from the foot (Carriker, 1961). Predation is seasonal with numbers of oysters consumed by *U. cinerea* and *E. caudata* reaching a maximum of 0.7 oysters week^{-1} snail^{-1} in late July (MacKenzie, 1981) when densities as high as 106 m^{-2} have been reported (Chestnut, 1956). *T. haemastoma*, while it is a significant predator of oysters, especially spat, if presented with a choice will select mussels in preference to oyster spat (see White & Wilson, 1996 for further details on the biology of oyster drills).

Another predator is the lightning whelk, *Busycon contrarium* which has been responsible for serious oyster mortality (as high as 80%) in Florida. The predator attacks oysters by chipping at the shell margin and wedging the valves apart with its foot so that it can insert its proboscis (Menzel & Nichy, 1958; Nichy & Menzel, 1960).

In Europe *Ocenebra erinacea* and *U. cinerea* (introduced from the United States) are the main predatory gastropods. The latter, and other species of *Urosalpinx*, feed mainly on oysters but also attack mussels, cockles and barnacles, and in the United States clam species can be added to this list (Hancock,

1960). *Ocenebra* is a much less serious predator of oysters than *Urosalpinx* as it appears to prefer mussels to oysters.

Several different methods are used to try and control gastropod predators on oyster beds. In the past a certain sum of money was paid to fishermen for every *Urosalpinx*, *Ocenebra* and *Crepidula* (a competitor of oysters; see below) that was collected on oysters beds in England and Wales (Hancock, 1960). This strategy was later replaced by physical methods such as trapping with baited traps or burying, neither of which are really effective as control methods (White & Wilson, 1996). Chemical methods have been tried, e.g. copper barriers around oyster beds, and pesticides. The former does not deter pelagic larvae of predator species from settling on oysters, and the latter not only kill the predator but also other benthic organisms, so their use is tightly regulated. Biological methods, where either parasites or predators of drills are used, have so far been ineffective.

Several crab species cause significant mortality of spat and juvenile oysters. On the Atlantic and Gulf coasts of the United States the main predators of *Crassostrea virginica* are the stone crab, *Menippe mercenaria*, the mud crab, *Panopeus herbstii*, and the blue crab, *Callinectes sapidus*. Stone crabs possess large and heavy claws (chelae) and thus are capable of crushing even market-size oysters. Typical annual consumption is about 200 oysters per crab (Menzel & Hopkins, 1956). *Panopeus herbstii* is a constant and serious threat for oyster farmers because of its wide salinity tolerance (10–34 psu), the large tooth on the major crushing chela that allows it to open oysters of a larger size than other crab species of similar carapace width, and the fact that it does not migrate offshore in winter (Bisker & Castagna, 1987). This species can consume as many as 20 oysters per day. Similar consumption rates have been reported for the blue crab, *Callinectes sapidus*, preying on oyster spat (Menzel & Hopkins, 1956). A detailed study of the foraging behaviour and predator-prey dynamics of this species has been published by Eggleston (1990).

Other predators of oysters include oystercatchers (Tuckwell & Nol, 1997a,b), starfish, fish (White & Wilson, 1996), and polyclad flatworms. Polyclad larvae (*Stylochus* spp.) settle in high densities onto oyster (*Crassostrea virginica*) beds, where they grow rapidly to adult size (25–50 mm length). The worms enter oyster spat through the gaping valves, eat the flesh, and are responsible for high mortality, especially under crowded culture conditions (Provenzano, 1961).

Similar to mussels the common pests of oysters are estuarine sponges (*Cliona* spp.), polychaetes (*Polydora* spp.) and pea crabs (*Pinnotheres* spp.); the last two will be covered in Chapter 11. Boring sponges excavate into the shell probably using a chemo-mechanical mechanism, chemical dissolution of the shell coupled with mechanical dislodgement of shell fragments (Hatch, 1980). The oyster shell comes to enclose the body of the sponge, except for papillae that extend from the shell to the outside environment. This pest greatly weakens the shell and probably makes the oyster more susceptible to predation. Also, the energy used in the continual effort by the oyster to repair its shell may have implications for somatic growth and gametogenesis, although Rosell *et al.* (1999) have found no evidence for this in infested populations of the European flat oyster, *Ostrea edulis*. Fortunately, boring sponge numbers are generally controlled because they are consumed by a range of predators such as molluscs, crustaceans, polychaetes and echinoderms (Guida, 1976).

Organisms that compete with oysters for space and food include algae, sponges, bryozoans, anemones, polychaetes, molluscs, arthropods and even oysters themselves. Some of these could also be regarded as fouling organisms. The main effects of competitors are: (1) consumption of oyster larvae before settlement by filter-feeders, (2) prevention of settlement by coverage of settlement area, (3) emission of a noxious chemical to repel settlement and (4) overgrowth or poisoning of settled spat, resulting in death (White & Wilson, 1996). Of these four, restricted space for settlement and overgrowth of the shell are the most common effects of competitors on oysters. But, even when oysters are not killed, competitors/foulers cause reduced survival and growth (Zajac *et al.*, 1989). In Europe the most important competitor is the exotic slipper limpet, *Crepidula* spp. introduced from the United States in the mid 1800s (see Minchin *et al.* (1995) for historical account). The limpets settle around the same time as oysters and grow more rapidly than oyster spat, causing high mortality (~60%) through overgrowth of the oyster shell (MacKenzie, 1970). In addition, the continuous production of faeces and pseudofaeces on the bottom leads to an accumulation of mud, making the substrate unsuitable for oyster settlement (Hancock, 1960).

Clams

Crabs are a serious predator of most species of clam. They gain access to the flesh by crushing small individuals, chipping the valve margins of large individuals, or forcing the valves apart (Gibbons, 1984). A wide variety of crab species prey on clams (Table 3.4) and a comprehensive picture of the foraging behaviour, patterns of predation and habitats use by blue crabs, in particular *Callinectes sapidus*, is emerging (Skilleter, 1994; Ebersole & Kennedy, 1995; Micheli, 1995, 1997a,b). Other crustaceans such as shrimp and lobster also prey on clams. In one locality in Sweden the shrimp, *Crangon crangon*, consumed 36% and 68% of the annual production of *Mercenaria mercenaria* and *Cerastoderma edule*, respectively (Møller & Rosenberg, 1983). Some species of snapping (*Alpheus* spp.) and mantis (*Squilla* spp.) shrimp can even crush *Mercenaria mercenaria* as big as 25 mm shell length (Beal, 1983). The American lobster, *Homarus americanus*, digs for hard clams and uses its crusher claw to crack the shell (Herrick, 1911).

Gastropod predators of clams tend to be the same species that prey on mussels and oysters, i.e. whelks (*Busycon* and *Murex* spp. on hard clams) and oyster drills (*Urosalpinx cinerea* and *Eupleura caudata* on hard clams, and *Ocenebra japonica* on the Manilla clam, *Ruditapes philippinarum*). Other predators include the horseshoe crab, *Limulus polyphemus*, starfish, turtles, various fish and bird species, and marine mammals such as sea otters and walrus.

The burrowing habit of clams gives them a certain amount of protection from some predators. For example, oystercatchers that feed on the clam, *Macoma balthica*, and the ragworm, *Nereis diversicolor*, both highly profitable prey, find it more difficult to locate buried clams than the less cryptic but mobile ragworms (Ens *et al.*, 1996). However, their burrowing habit also poses problems as portions of the exposed siphons are regularly eaten by fish, walrus and

Table 3.4. Species of crab that prey on clams in North America. Information from Carriker (1951); Ebersole & Kennedy (1995); Eggleston *et al.* (1992); Gibbons (1984); Gibbons & Blogosawski (1989); MacKenzie (1977).

Crab	Clam species	Geographic range of crab	Typical predation rates
Blue crab, *Callinectes sapidus*	*Mercenaria mercenaria, Macoma balthica Rangia cuneata, Mya arenaria*	Cape Cod, USA to Uruguay, S. America	*M. mercenaria*: 308 clams (≤40 mm shell length) crab^{-1} day^{-1}
Green crab, *Carcinus maenas*	*M. arenaria, M. mercenaria*	Nova Scotia, Canada to New Jersey, USA	
Rock crab, *Cancer irroratus*	*M. mercenaria*	Labrador, Canada to S. Carolina, USA	29 8–10 mm clams crab^{-1} (carapace width 55 mm) hour^{-1}; 30 1 mm clams crab^{-1} (carapace width 7 mm) hour^{-1}
Black-clawed crab, *Cancer productus*; Dungeness crab, *C. magister*; graceful crab, *C. gracilis*	*Ruditapes philippinarum*	Pacific coast of N. America	
Stone crab, *Menippe mercenaria*, Mud crabs, *Neopanope sayi*, *Panopeus herbstii*, *Europanopeus depressus*, *Rhithropanopeus harrissi*	*M. mercenaria* *M. mercenaria*	East coast USA East coast USA	*N. sayi*: 136 4 mm clams crab^{-1} day^{-1}

occasionally by birds (Gibbons & Blogoslawski, 1989 for references; Dame, 1996), and may have selected for a remarkable chemical defence mechanism in the butter clam, *Saxidomus giganteus* (Kvitek, 1991). This species sequesters a highly potent neurotoxin in its siphons to which the siphon-nipping fish predator, *Leptocotus armatus*, has developed an aversion. The fish shows no such aversion to toxic littleneck clams, *Protothaca staminea*, which, unlike *S. giganteus*, retain the toxin in their visceral mass. Vegetation such as seagrass cover also significantly reduces siphon-cropping and predation on clams (Irlandi, 1994). In the case of predators such as sea otters and walrus only deep-burrowing clams have spatial refuge (Kvitek *et al.*, 1988). Both of these predators eat large quantities of clams and other bivalves, and because of their abundance and large size, may also be responsible for large-scale disruption of soft bottom communities.

Attempts to control clam predators include many of the methods already described. Mechanical methods such as the use of starfish mops, suction dredges, nets traps and hand collection have all been tried with little success (see Gibbons & Blogoslawski, 1989 for references). Chemical methods, such as copper sulphate, quicklime, and insecticides, can be used as dips, incorporated in sand and heavy oils, spread over bivalve areas, or used as barriers to protect planted beds. But, since these chemicals are also harmful to endemic species as well as predators, their use is firmly regulated. Gibbons & Castagna (1985) have used the toadfish, *Opsanus tau*, along with crushed stone aggregate, to biologically control crab predation on juvenile *Mercenaria mercenaria*. The use of off-bottom culture, or on-bottom culture using gravel aggregate and nets, seems to be the most successful way to exclude and isolate predators from juvenile clams (Gibbons & Blogoslawski, 1989).

Pests of clams include the boring polychaete, *Polydora ciliata*, (see Chapter 11) and the boring sponge, *Cliona* spp. (see section on oysters above). The nemertean, *Malacobdella grossa*, has been found within the mantle cavity of several clam species on North American and European coasts. Numbers can be as high as 21 small worms per clam, and some species have infestation rates as high as 80% (Porter, 1964; Ropes & Merrill, 1967). The worms, while aesthetically displeasing to human consumers, cause no harm to their host. However, pea crabs (*Pinnotheres* spp.), which also inhabit the mantle cavity, are believed to cause gill damage in clams and other bivalves.

Fouling tends to be a significant cause of mortality in juvenile clams only. Various species of Vorticellidae, Entoprocta and blue green algae attach to shells, often suffocating newly settled clams. The amphipod, *Corophium cylindricum*, and the sea squirt, *Molgula manhattensis*, are also fouling pests of post-set hard clams (Gibbons & Blogoslawski, 1989).

There are few reports on the effect of competition for space or resources between species of clam. There is some evidence that dense assemblages of the gem clam, *Gemma gemma*, (Sanders *et al.*, 1962), or the deep-dwelling ghost shrimp, *Callianassa californiensis*, (Peterson, 1977), can limit recruitment of the hard clam, *Mercenaria mercenaria* and the butter clam, *Saxidomus nuttalli*, respectively. The presence of other large infaunal species, or dense plantings of the same species, can retard clam growth and, even in some cases, increase mortality (references in Gibbons & Blogoslawski, 1989).

Table 3.5. Predators of a selection of scallop species. Predation is usually highest on spat and juvenile stages. Common names of species in Table 3.3.

Scallop Species	Geographic range	Predator	Reference
Pecten maximus	Norway to Southern Spain	Fish, crabs, lobsters, and starfish	Orensanz *et al.*, 1991
Pecten fumatus Commercial scallop	South-east Australia	Starfish, angel sharks, octopus	Gwyther *et al.*, 1991
Chlamys islandica	Circumpolar in Northern Hemisphere	Starfish, eider ducks	Parsons *et al.*, 1991
Chlamys farreri	North coast China	Sea urchins and brittle stars	Luo, 1991
Chlamys tehuelcha Tehuelche scallop	East coast of Argentina	Starfish, gastropods, octopus, ratfish	Orensanz *et al.*, 1991
Argopecten irradians	Atlantic coast of North America	Starfish, gastropods, crabs, fish and birds	Rhodes, 1991
Amusium balloti Saucer scallop	East, north and west coasts of Australia	Turtles, slipper lobsters	Gwyther *et al.*, 1991
Argopecten circularis	California to Peru	Gastropods, crabs and stingrays	Felix-Pico, 1991
Patinopecten yessoensis	Pacific coasts of Russia and Japan	Starfish, crabs, octopus and benthic fish	Kalashnikov, 1991

Scallops

The most important predators of scallops are starfish, followed by crabs, lobsters, gastropods, sea anemones and octopus (Table 3.5). In continental shelf areas scallops are subject to heavy fish predation (Orensanz *et al.*, 1991 for references). Spat and juvenile scallops are most vulnerable to predation but this vulnerability varies between species, due to different behavioural and morphological adaptations. For example, where *Chlamys islandica* and *Pecten maximus* live together in the same habitat *C. islandica* is more heavily preyed on by plaice and flounder. This is due to its greater tendency to remain byssally attached, and also because of its weaker swimming escape response (Naidu & Meron, 1986). In contrast, *P. maximus* spat as small as 4 mm shell height can vacate byssal attachment sites to recess on the sea bed (Minchin, 1992). On the other hand, the thin and fragile shell of attached *P. maximus* spat makes this species more susceptible to a wider range and size of predators than same-size *Aequipecten opercularis* that have a stronger shell (Brand, 1991). Byssal attachment can in some cases provide a spatial refuge from benthic predators. The bay scallop *Argopecten irradians* in the wild attaches to shoots of the eelgrass, *Zostera marina*. In field experiments Pohle *et al.* (1991) found that 10–15 mm spat tethered to eelgrass 20–35 cm above the bottom experienced significantly reduced mortality compared to those placed on the sediment surface (Fig. 3.13). When this species detaches from eelgrass there is a critical window of high predatory risk between 15 and 40 mm, after which it attains a size refuge from predation.

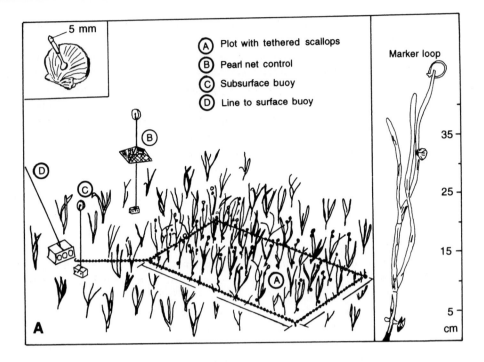

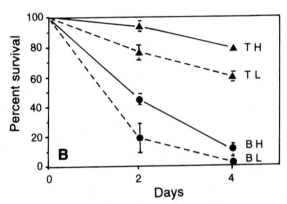

Fig. 3.13. (A) Drawing of a typical treatment plot used in scallop (*Argopecten irradians*) field tethering experiments. Ⓐ plot with tethered scallops; Ⓑ pearl net control; Ⓒ subsurface buoy; Ⓓ line to surface buoy. Also shown is the approximate vertical height of tethered scallops on an eelgrass shoot (inset at right), and a scallop with a plastic tether glued to its shell (inset in upper left corner). (B) Mean percentage survival over 4 days of juveniles *Argopecten irradians* tethered to eelgrass in Lake Montauk, Long Island, USA, at the base of the shoots (B, bottom) and in the upper canopy (T, top), <5 cm and 20–35 cm above bottom respectively, in plots of low (L) and high (H) density eelgrass. Error bars represent the standard error for 3 plots. From Pohle *et al.* (1991).

In the laboratory it seems that adults larger than a critical shell height of 70 mm are virtually immune from predation, particularly from crabs (Elner & Jamieston, 1979; Lake *et al.*, 1987). This size must be less in the field for species with a recessing habit and strong swimming escape response, as both of these are effective mechanisms for evading predation. For non-visual predators like starfish however, recessing affords little protection for scallops.

The escape response, either jumping or swimming, is elicited when the mantle tentacles of the scallop are touched by a predator. This suggests that the response is chemosensory, supported by the observation that non-predatory starfish species, for example, evoke little or no reaction (Thomas & Gruffydd, 1971). The ability to differentiate between potential predators and harmless species has an obvious adaptive value since it minimises feeding interruptions, and thus energy wastage (Brand, 1991). Escape responses of 3–4 m in the field, and more than 10 m under experimental conditions have been recorded (Wilkens, 1991 for references).

In an elegant series of field experiments Barbeau and colleagues have studied in detail the predatory effects of starfish, *Asterias* spp. and the rock crab, *Cancer irroratus* on juvenile *Placopecten magellanicus*, in Lundeburg Bay, Nova Scotia. Scallop survival was assessed using different size classes of scallops tethered at different sites and seasons, and in different densities of surrounding scallops and predators. Tethered scallops were much more susceptible to sea star than to crab predation (Barbeau *et al.*, 1994). This is because once encountered free scallops close their valves and are likely to be captured and consumed by crabs. In contrast, in sea star-scallop encounters free scallops readily escape by swimming or jumping, and therefore the probability of sea stars capturing free scallops is low (Figs. 3.14 & 3.15). Water temperature, site and scallop density affected crab predation. There was a significant interaction between temperature and site in that crab predation increased with temperature at one site, but was independent of temperature at another site. Predation also increased with scallop density, and to a lesser extent with crab density. Sea star predation increased with water temperature and decreased with scallop size, but was density independent (Barbeau *et al.*, 1998). These results indicate that crabs are more likely to have a greater impact than sea stars on seeded populations of scallops in bottom-culture operations. For example, *Asterias vulgaris* consumes on average less than one scallop (15–35 mm size range) per day as opposed to *Cancer irroratus* which consumes about 12 scallops per day (Nadeau & Cliché, 1998).

Pests of scallops are mostly the same as those mentioned for other bivalve groups: burrowing polychaetes (*Polydora* spp.), sponges (*Cliona* spp.) and pea crabs (Getchell, 1991).

Scallop shells often harbour a variety of fouling organisms, including algae, barnacles, tube worms, sponges, hydrozoans, bryozoans, and other molluscs. Heavy fouling can increase weight and drag on the scallop shell and thus hamper swimming ability. Winter & Hamilton (1985) have estimated that 6 g of epifauna could cause almost a 30% decrease in the distance travelled by *Argopecten irradians*. In some cases, however, fouling may have a beneficial effect. For example, epizoic sponges on the upper valve of various species of *Chlamys* greatly inhibit predation by starfish. The sponge reduces adhesion of the starfish tube-feet but tactile camouflage may also be involved (references in Brand, 1991). In addition, distasteful chemicals in the sponge may deter fish predators (Pitcher & Butler, 1987).

Competition occurs for scallops mainly at the settlement stage when they

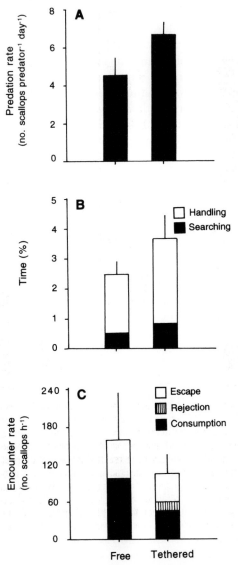

Fig. 3.14. Laboratory experiments examining the effect of tethering scallops (*Placopecten magellanicus*) on predation by crabs (*Cancer irroratus*). (A) Predation rate per crab, (B) percent foraging time (searching + handling time) of crabs, and (C) encounter rate between crabs and scallops (different outcomes of encounter have different shadings). The probability of capture upon encounter was calculated as (number of consumptions + rejections)/number of encounters, and the probability of consumption upon capture as number of consumptions/(number of consumptions + rejections). Crabs were offered either free or tethered scallops. Observation time in (B) and (C) = 420 min tank^{-1}. Mean shown for all variates, error bars = SE (for foraging time in (B) and for encounter time in (C)), and number of replicate tanks = 3. From Barbeau & Scheibling (1994).

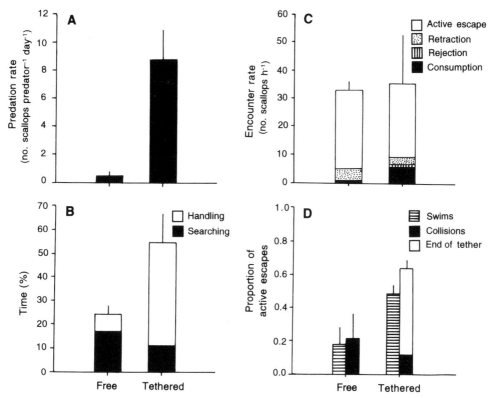

Fig. 3.15. Laboratory experiments examining the effect of tethering scallops (*Placopecten magellanicus*) on predation by sea stars (*Asterias vulgaris*). (A) Predation rate per sea star, (B) percent foraging time (searching + handling time) of sea stars, (C) encounter rate between sea stars and scallops (different outcomes of encounter have different shadings), and (D) proportion of active escapes in which scallops swam (active escapes = swims + jumps) and in which scallops collided with the tank walls or reached the end of their tether. The probability of capture upon encounter was calculated as (number of consumptions + rejections)/number of encounters, and the probability of consumption upon capture as number of consumptions/(number of consumptions + rejections). In (C) the number of free scallops rejected and consumed are 0.1 and 0.4 scallops h^{-1}, respectively. Sea stars were offered either free or tethered scallops. Observation time in (B), (C) and (D) = 510 min tank^{-1}. Mean shown for all variates, error bars = SE (for foraging time in (B) and for encounter rate in (C)), and number of replicate tanks = 3. From Barbeau & Scheibling (1994).

may compete with high densities of mussel spat, for example, for space and food. Once settled, either byssally attached or on the bottom, it would appear that scallops do not occur in sufficiently high densities for competition to play an important role.

Before concluding this section it should be pointed out that bivalve mortality during the planktonic larval stage is high, and Thorson (1950) has suggested that predation may be the single most important cause. Possible predators of larvae include sea squirts, ctenophores, sea anemones, barnacles, and larvae of various crustacean, echinoderm and fish species. The inadvertent consumption of larvae by filtering adults – of its own and other species – also contributes to mortality.

References

Alerstam, T., Gudmundsson, G.A. & Johannesson, K. (1992) Resources for long-distance migration intertidal exploitation of *Littorina* and *Mytilus* by knots *Calidris canutus* in Iceland. *Oikos*, **65**, 179–89.

Ansell, A.D., Dao, J.C. & Mason, J. (1991) Three European scallops: *Pecten maximus*, *Chlamys (Aequipecten) opercularis* and *C. (Chlamys) varia*. In: *Scallops: Biology, Ecology and Aquaculture* (ed. S.E. Shumway), pp. 715–51. Elsevier Science Publishers B.V., Amsterdam.

Baird, R.H. (1958) On the swimming behaviour of escallops (*Pecten maximus* L.). *Proc. Malacol. Soc. Lond.*, **33**, 67–71.

Barbeau, M.A. & Scheibling, R.E. (1994) Procedural effects of prey tethering experiments: predation of juvenile scallops by crabs and sea stars. *Mar. Ecol. Prog. Ser.*, **111**, 305–10.

Barbeau, M.A., Scheibling, R.E., Hatcher, B.G., Taylor, L.H. & Hennigar, A.W. (1994) Survival analysis of tethered juvenile sea scallops *Placopecten magellanicus* in field experiments: effects of predators, scallop size and density, site and season. *Mar. Ecol. Prog. Ser.*, **115**, 243–56.

Barbeau, M.A., Scheibling, R.E. & Hatcher, B.G. (1998) Behavioural responses of predatory crabs and seastars to varying density of juvenile Sea scallops. *Aquaculture*, **169**, 87–98.

Bayne, B.L. (ed.) (1976) *Marine Mussels: their Ecology and Physiology*. Cambridge University Press, Cambridge.

Beal, B.F. (1983) Predation of juveniles of the hard clam, *Mercenaria mercenaria* (Linné) by the snapping shrimp, *Alpheus heterochaelis* Say, and *Alpheus normanni* Kingsley. *J. Shellfish Res.*, **3**, 1–9.

Bell, E.C. & Gosline, J.M. (1997) Strategies for life in flow: tenacity, morphometry and probability of dislodgementof two *Mytilus* species. *Mar. Ecol. Prog. Ser.*, **159**, 197–208.

Berrow, S.D., Kelly, T.C. & Myers, A.A. (1992a) The mussel catching behaviour of hooded crows *Corvus corone cornix*. *Bird Study*, **39**, 115–19.

Berrow, S.D., Kelly, T.C. & Myers, A.A. (1992b) The diet of coastal breeding hooded crows (*Corvus corone cornix*). *Ecography*, **15**, 337–46.

Beukema, J.J. & Cadee, G. (1996) Consequences of the sudden removal of nearly all mussels and cockles from the Dutch Wadden Sea. *Mar. Ecol.*, **17**, 279–89.

Bisker, R. & Castagna, M. (1987) Predation on single spat oysters *Crassostrea virginica* (Gmelin) by blue crabs *Callinectes sapidus* Rathbun and mud crabs *Panopeus herbstii* Milne–Edwards. *J. Shellfish Res.*, **6**, 37–40.

Blake, N.J. & Moyer, M.A. (1991) The calico scallop, *Argopecten gibbus*, fishery of Cape Canaveral, Florida. In: *Scallops: Biology, Ecology and Aquaculture* (ed. S.E. Shumway), pp. 899–911. Elsevier Science Publishers B.V., Amsterdam.

Bourne, N. (1991) Fisheries and aquaculture: west coast of North America. In: *Scallops: Biology, Ecology and Aquaculture* (ed. S.E. Shumway), pp. 925–42. Elsevier Science Publishers B.V., Amsterdam.

Brand, A.R. (1991) Scallop ecology: distribution and behaviour. In: *Scallops: Biology, Ecology and Aquaculture* (ed. S.E. Shumway), pp. 517–84. Elsevier Science Publishers B.V., Amsterdam.

Bricelj, V.M. & Malouf, R.E. (1984) Influence of algal and suspended culture sediment concentrations on the feeding physiology of the hard clam, *Mercenaria mercenaria*. *Mar. Biol.*, **85**, 155–65.

Brosnan, D.M. & Crumrine, L. (1994) Effects of human trampling on marine rocky shore communities. *J. Exp. Mar. Biol. Ecol.*, **177**, 79–97.

Bull, M.F. (1991) Fisheries and aquaculture: New Zealand. In: *Scallops: Biology, Ecology and Aquaculture* (ed. S.E. Shumway), pp. 853–59. Elsevier Science Publishers B.V., Amsterdam.

Calvo–Ugarteburu, G. & McQuaid, C.D. (1998) Parasitism and introduced species: epidemiology of trematodes in the intertidal mussels *Perna perna* and *Mytilus galloprovincialis*. *J. Exp. Mar. Biol. Ecol.,* **220**, 47–65.

Carriker, M.R. (1951) Observations on the penetration of tightly closing bivalves by *Busycon* and other predators. *Ecology*, **32**, 73–83.

Carriker, M.R. (1961) Comparative functional morphology of boring mechanisms in gastropods. *Amer. Zool.*, **1**, 263–66.

Carriker, M.R. & Gaffney, P.M. (1996) A catalogue of selected species of living oysters (Ostracea) of the world. In: *The Eastern Oyster Crassostrea virginica* (eds V.S. Kennedy, R.I.E. Newell & A.F. Eble), pp. 1–18. Maryland Sea Grant, College Park, Maryland.

Carroll, M.L. & Highsmith, R. (1996) Role of catastrophic disturbance in mediating *Nucella-Mytilus* interactions in the Alaskan rocky intertidal. *Mar. Ecol. Prog. Ser.*, **138**, 125–33.

Chapple, P.J., Smerdon, G.R., Berry, R.J. & Hawkins, A.J.S. (1998) Seasonal changes in stress-70 protein levels reflect thermal tolerance in the marine bivalve *Mytilus edulis* L. *J. Exp. Mar. Biol. Ecol.,* **229**, 53–68.

Chestnut, A.F. (1956) The distribution of oyster drills in North Carolina. *Proc. Natl. Shellfish Assoc.*, **46**, 134–39.

Clegg, J.S., Uhlinger, K.R., Jackson, S.A., Cherr, G.N., Rifkin, E. & Friedman, C.S. (1998) Induced thermotolerance and the heat-shock-protein-70 family in the Pacific oyster *Crassostrea gigas*. *Mol. Mar. Biol. Biotechnol.*, **7**, 21–30.

Comesaña, A.S., Posada, D. & Sanjuan, A. (1998) *Mytilus galloprovincialis* Lmk. in northern Africa. *J. Exp. Mar. Biol. Ecol.,* **223**, 271–83.

Cote, I.M. (1995) Effects of predatory crab effluent on byssus production in mussels. *J. Exp. Mar. Biol. Ecol.,* **188**, 233–41.

Dame, R.F. (1996) *Ecology of Marine Bivalves: an Ecosystem Approach.* CRC Press, Boca Raton, Florida.

Dare, P.J. (1982) Notes on the swarming behaviour and population density of *Asterias rubens* L. (Echinodermata: Asteroidea) feeding on the mussel *Mytilus edulis*. *J. Cons. int. Explor. Mer.*, **40**, 112–18.

Davenport, J. (1983) A comparison of some aspects of the behaviour and physiology of the Indian mussel *Perna* (= *Mytilus*) *viridis* and the common mussel *Mytilus edulis* L. *J. Moll. Stud.*, **49**, 21–26.

Davenport, J., Moore, P.G., Magill, S.H. & Fraser, L.A. (1998) Enhanced condition in dogwhelks *Nucella lapillus* (L.) living under mussel hummocks. *J. Exp. Mar. Biol. Ecol.,* **230**, 225–34.

Del Norte, A.G.C. (1991) Fisheries and aquaculture: Philippines. In: *Scallops: Biology, Ecology and Aquaculture* (ed. S.E. Shumway), pp. 825–34. Elsevier Science Publishers B.V., Amsterdam.

Dixon, D.R., Sole–Cava, A.M., Pascoe, P.L. & Holland, P.W.H. (1995) Periostracal adventitious hairs on spat of the mussel *Mytilus edulis*. *J. mar. biol. Ass. U.K.*, **75**, 363–72.

Dolmer, P. (1998) The interactions between bed structure of *Mytilus edulis* L. and the predator *Asterias rubens* L. *J. Exp. Mar. Biol. Ecol.,* **228**, 137–50.

Ebersole, E.L. & Kennedy, V. (1995) Prey preferences of blue crabs *Callinectes sapidus* feeding on 3 bivalve species. *Mar. Ecol. Prog. Ser.*, **118**, 167–77.

Eggleston, D.B. (1990) Foraging behavior of the blue crab, *Callinectes sapidus*, on juvenile oysters, *Crassostrea virginica*: effects of prey density and size. *Bull. Mar. Sci.*, **46**, 62–82.

Eggleston, D.B., Lipcius, R.N. & Hines, A.H. (1992) Density-dependent predation by blue crabs upon infaunal clam species with contrasting distribution and abundance patterns. *Mar. Ecol. Prog. Ser.*, **85**, 55–68.

Eleftheriou, A. & Robertson, M.R. (1992) The effects of experimental scallop dredging on the fauna and physical environment of a shallow sandy community. *Neth. J. Sea Res.*, **30**, 289–99.

Elner, R.W. & Jamieson, G.S. (1979) Predation of sea scallops, *Placopecten magellanicus*, by the rock crab *Cancer irroratus* and the American lobster, *Homarus americanus*. *J. Fish. Res. Bd. Can.*, **36**, 537–43.

Ens, B.J., Bunskoeke, E.J., Hoekstra, R., Hulscher, J.B., Kersten, M. & Devlas, S.J. (1996) Prey choice and search speed: why simple optimality fails to explain the prey choice of oystercatchers *Haematopus ostralegus* feeding on *Nereis diversicolor* and *Macoma balthica*. *Ardea*, **84**A, 73–90.

FAO (2001) *Fishstat Plus* (v. 2.30). Food and Agricultural Organisation, United Nations, Rome.

Felix–Pico, E.F. (1991) Fisheries and aquaculture: Mexico. In: *Scallops: Biology, Ecology and Aquaculture* (ed. S.E. Shumway), pp. 943–80. Elsevier Science Publishers B.V., Amsterdam.

Getchell, R.G. (1991) Diseases and parasites of scallops. In: *Scallops: Biology, Ecology and Aquaculture* (ed. S.E. Shumway), pp. 471–94. Elsevier Science Publishers B.V., Amsterdam.

Gibbons, M.C. (1984) *Aspects of predation by crabs* Neopanope sayi, Ovalipes ocellatus *and* Pagurus longicarpus *on juvenile hard clams*, Mercenaria mercenaria. PhD. thesis, State University of New York at Stonybrook.

Gibbons, M.C. & Castagna, M. (1985) Biological control of predation by crabs in bottom culture of hard clams using a combination of crushed stone aggregate, toadfish and cages. *Aquaculture*, **47**, 101–04.

Gibbons, M.C. & Blogoslawski, W.J. (1989) Predators, pests, parasites and disease. In: *Clam Mariculture in North America* (eds J.J. Manzi & M. Castagna), pp. 167–200. Elsevier Science Publishing Company Inc., New York.

Gosling, E.M. (1992) Systematics and geographic distribution of *Mytilus*. In: *The Mussel Mytilus: Ecology, Physiology, Genetics and Culture* (ed. E.M. Gosling), pp. 1–20. Elsevier Science Publishers B.V., Amsterdam.

Gosscustard, J.D., West, A. & Durell, S.E.A. (1993) The availability and quality of the mussel prey (*Mytilus edulis*) of oystercatchers (*Haematopus ostralegus*). *Neth. J. Sea Res.*, **31**, 419–39.

Gosscustard, J.D., McGrorty, S. & Durell, S.E.A. (1996) The effect of oystercatchers *Haematopus ostralegus* on shellfish populations. *Ardea*, **84**A, 453–68.

Guo, X. & Luo, Y. (2003) Scallop culture in China, in press.

Griffiths, C.L., Hockey, P.A.R., van Erkom Schurink, C. & le Roux, P.J. (1992) Marine invasive aliens on South African shores: implications for community structure and trophic functioning. *S. Afr. J. Mar. Sci.*, **12**, 713–22.

Guida, V.G. (1976) Sponge predation in the oyster reef community as demonstrated with *Cliona celata* Grant. *J. Exp. Mar. Biol. Ecol.*, **25**, 109–22.

Gwyther, D., Cropp, D.A., Joll, L.M. & Dredge, M.C.L. (1991) Fisheries and aquaculture: Australia. In: *Scallops: Biology, Ecology and Aquaculture* (ed. S.E. Shumway), pp. 835–51. Elsevier Science Publishers B.V., Amsterdam.

Hamilton, P.V. & Koch, K. (1996) Orientation toward natural and artificial grassbeds by swimming bay scallops, *Argopecten irradians* (Lamarck, 1819). *J. Exp. Mar. Biol. Ecol.*, **199**, 79–88.

Hancock, D.A. (1960) The ecology of molluscan enemies of the edible mollusc. *Proc. Malacol. Soc. Lond.*, **34**, 123–43.

Hatch, W.I. (1980) The implications of carbonic anhydrase in the physiological mechanism of penetration of carbonate substrata by the marine burrowing sponge *Cliona celata* (Demospongiae). *Biol. Bull.*, **159**, 135–47.

Hedgecock, D. & Okazaki, N.B. (1984) Genetic diversity within and between populations of American oysters (*Crassostrea*). *Malacologia*, **25**, 535–49.

Herrick, F.H. (1911) Natural history of the American lobster. *Bull. US Bur. Fish.*, **29**, 149–408.

Heslinga, G.A. (1989) Biology and culture of the giant clam. In: *Clam Mariculture in North America* (eds J.J. Manzi & M. Castagna), pp. 293–322. Elsevier Science Publishing Company Inc., New York.

Hickman, R.W. (1992) Mussel cultivation. In: *The Mussel Mytilus: Ecology, Physiology, Genetics and Culture* (ed. E.M. Gosling), pp. 465–510. Elsevier Science Publishers B.V., Amsterdam.

Hilbish, T.J., Mullinax, A., Dolven, S.I., Meyer, A., Koehn, R.K. & Rawson, P.D. (2000) Origin of the antitropical distribution pattern in marine mussels (*Mytilus* spp.): routes and timing of transequatorial migration. *Mar. Biol.*, **136**, 69–77.

Hilgerloh, G. (1997) Predation by birds on blue mussel *Mytilus edulis* beds of the tidal flats of Spiekeroog (southern North Sea). *Mar. Ecol. Prog. Ser.*, **146**, 61–72.

Hofmann, G.E. & Somero, G. (1996) Interspecific variation in thermal-denaturation of proteins in the congeneric mussels *Mytilus trossulus* and *Mytilus galloprovincialis*: evidence from the heat-shock response and protein ubiquitination. *Mar. Biol.*, **126**, 65–75.

Hockey, P.A.R. & Schurink, C.V. (1992) The invasive biology of the mussel *Mytilus galloprovincialis* on the South African coast. *Trans. R. Soc. S. Africa*, **48**, 123–39.

Hughes, R.N. (1986) *A Functional Biology of Marine Gastropods*. Crook Helm, London.

Hughes, R.N. & Dunkin, S.B. de (1984) Behaviour components of prey selection by dogwhelks, *Nucella lapillus* (L.) feeding on mussels, *Mytilus edulis* L. in the laboratory. *J. Exp. Mar. Biol. Ecol.*, **77**, 45–68.

Hulscher, J.B. & Ens, B. (1992) Is the bill of the male oystercatcher a better tool for attacking mussels than the bill of the female? *Neth. J. Zool.*, **42**, 85–100.

Hunt, H.L. & Scheibling, R.E. (1998) Spatial and temporal variability of patterns of colonization by mussels (*Mytilus trossulus, M. edulis*) on a wave-exposed rocky shore. *Mar. Ecol. Prog. Ser.*, **167**, 155–69.

Hunt, H.L. & Scheibling, R.E. (2001) Predicting wave dislodgement of mussels: variation in attachment strength with body size, habitat and season. *Mar. Ecol. Prog. Ser.*, **213**, 157–64.

Ignacio, B.L., Absher, T.M., Lazoski, C. & Sole-Cava, A.M. (2000) Genetic evidence of the presence of two species of *Crassostrea* (Bivalvia: Ostreidae) on the coast of Brazil. *Mar. Biol.*, **136**, 987–91.

Inoue, K., Odo, S., Noda, T., Nakao, S., Takeyama, S., Yamaha, E., Yamazaki, F & Harayama, S. (1997) A possible hybrid zone in the *Mytilus edulis* complex in Japan revealed by PCR markers. *Mar. Biol.*, **128**, 91–95.

Irlandi, E.A (1994) Large-scale and small-scale effects of habitat structure on rates of predation: how percent coverage of seagrass affects rates of predation and siphon nipping on an infaunal bivalve. *Oecologia*, **98**, 176–83.

Jozefowicz, C.J. & Ó Foighil, D. (1998) Phylogenetic analysis of Southern Hemisphere flat oysters based on partial mitochondrial 16S rDNA gene sequences. *Mol. Phylogenet. Evol.*, **10**, 426–35.

Kalashnikov, V.Z. (1991) Fisheries and aquaculture: Soviet Union. In: *Scallops: Biology, Ecology and Aquaculture* (ed. S.E. Shumway), pp. 1057–82. Elsevier Science Publishers B.V., Amsterdam.

Kalle, K. (1971) Salinity: general introduction. In: *Marine Ecology* (ed. O. Kinne), pp. 683–88. Wiley–Interscience, New York.

Kiørboe, T. & Møhlenberg, F. (1981) Particle selection in suspension-feeding bivalves. *Mar. Ecol. Prog. Ser.*, **5**, 291–96.

Kvitek, R.G. (1991) Paralytic shellfish toxins sequestered by bivalves as a defense against siphon-nipping fish. *Mar. Biol.*, **111**, 369–74.

Kvitek, R.G., Fukayama, A.K., Anderson, B.S. & Grimm, B.K. (1988) Sea otter foraging on deep-burrowing bivalves in a California coastal lagoon. *Mar. Biol.*, **98**, 157–67.

Lake, N.C.H., Jones, M.B. & Paul, J.D. (1987) Crab predation on scallop (*Pecten maximus*) and its implication for scallop cultivation. *J. mar. biol. Ass. U.K.*, **67**, 55–64.

Leonard, G.H., Bertness, M.D. & Yund, P.O. (1999) Crab predation, waterborne cues, and inducible defenses in the blue mussel, *Mytilus edulis*. *Ecology*, **80**, 1–14.

Lewis, J.R. (1977) The role of physical and biological factors in the distribution and stability of rocky shore communities. In: *Biology of Benthic Organisms* (eds B. F. Keegan, P. O Céidigh & P.J.S. Boaden), pp. 417–24. Pergamon Press, London.

Loomis, S.H., Carpenter, J.F. & Crowe, J.H. (1988) Identification of strombine and taurine as cryoprotectants in the intertidal bivalve *Mytilus edulis*. *Biochim. Biophys. Acta*, **943**, 113–18.

Luo, Y. (1991) Fisheries and aquaculture: China. In: *Scallops: Biology, Ecology and Aquaculture* (ed. S.E. Shumway), pp. 809–24. Elsevier Science Publishers B.V., Amsterdam.

MacKenzie, C.L. Jr. (1970) Causes of oyster spat mortality, condition of oyster setting beds, and recommendations for oyster bed management. *Proc. Natl. Shellfish Assoc.*, **60**, 59–67.

MacKenzie, C.L. Jr. (1977) Predation on hard clam (*Mercenaria mercenaria*) populations. *Trans. Amer. Fish. Soc.*, **106**, 530–37.

MacKenzie, C.L. Jr. (1981) Biotic potential and environmental resistance in the American oyster (*Crassostrea virginica*) in Long Island Sound. *Aquaculture*, **22**, 229–68.

Malouf, R.E. & Bricelj, V.M. (1989) Comparative biology of clams: environmental tolerances, feeding, and growth. In: *Clam Mariculture in North America* (eds J.J. Manzi & M. Castagna), pp. 23–73. Elsevier Science Publishing Company Inc., New York.

Manzi, J.J. & Castagna, M. (1989) Introduction. In: *Clam Mariculture in North America* (eds J.J. Manzi & M. Castagna), pp. 1–21. Elsevier Science Publishing Company Inc., New York.

Mason, J. (1976) Cultivation. In: *Marine Mussels: their Ecology and Physiology* (ed. B.L. Bayne), pp. 385–410. Cambridge University Press, Cambridge.

McDonald, J.H., Seed, R. & Koehn, R.K. (1991) Allozymes and morphometric characters of three species of *Mytilus* in the Northern and Southern Hemispheres. *Mar. Biol.*, **111**, 323–33.

Melvin, G.D., Dadswell, M.J. & Chandler, R.A. (1985) *Movement of scallops* Placopecten magellanicus (*Gmelin, 1791*) (*Mollusca: Pectinidae*) *on Georges Bank*. CAFSAC Res. Doc. 85/30.

Menzel, R.W. & Hopkins, S.H. (1956) Crabs as predators of oysters in Louisiana. *Proc. Natl. Shellfish Assoc.*, **46**, 177–84.

Menzel, R.W. & Nichy, F.E. (1958) Studies of the distribution and feeding habits of some oyster predators in Alligator Harbor, Florida. *Bull. Mar. Sci. Gulf Carib.*, **8**, 125–45.

Micheli, F. (1995) Behavioral plasticity in prey-size selectivity of the blue crab *Callinectes sapidus* feeding on bivalve prey. *J. Anim. Ecol.*, **64**, 63–74.

Micheli, F. (1997a) Effects of predator foraging behaviour on patterns of prey mortality in marine soft bottoms. *Ecol. Monogr.*, **67**, 203–24.

Micheli, F. (1997b) Effects of experience on crab foraging in a mobile and a sedentary species. *Anim. Behav.*, **53**, 1149–59.

Minchin, D. (1992) Biological observations on young scallops, *Pecten maximus. J. mar. biol. Ass. U.K.*, **72**, 807–19.

Minchin, D., McGrath, D. & Duggan, C.B. (1995) The slipper limpet *Crepidula fornicata* (L.) in Irish waters, with a review of its occurrence in the north-eastern Atlantic. *J. Conch. Lond.*, **35**, 247–54.

Minchinton, T.E., Scheibling, R.E. & Hunt, H.L. (1997) Recovery of an intertidal assemblage following a rare occurrence of scouring by sea-ice in Nova Scotia, Canada. *Bot. Mar.*, **40**, 139–48.

Møller, P. & Rosenberg, R. (1983) Recruitment, abundance and production of *Mya arenaria* and *Cardium edule* in marine shallow waters, western Sweden. *Ophelia*, **22**, 33–55.

Munro, P. (ed.) (1993) Genetic Aspects of Conservation and Cultivation of Giant Clams. *ICLARM Conf. Proc.*, **39**, 47pp.

Nadeau, M. & Cliché, G. (1998) Predation of juvenile sea scallops (*Placopecten magellanicus*) by crabs (*Cancer irroratus* and *Hyas* sp.) and starfish (*Asterias vulgaris, Leptasterias polaris* and *Crossaster papposus*). *J. Shellfish Res.*, **17**, 905–10.

Naidu, K. (1991) Sea scallop, *Placopecten magellanicus*. In: *Scallops: Biology, Ecology and Aquaculture* (ed. S.E. Shumway), pp. 861–97. Elsevier Science Publishers B.V., Amsterdam.

Naidu, K. & Meron, S. (1986) *Predation of scallops by American plaice and yellowtail flounder*. CAFSAC Res. Doc. 86/62.

Navarrete, S.A. & Menge, B. (1996) Keystone predation and interaction strength: interactive effects of predators on their main prey. *Ecol. Monogr.*, **66**, 409–29.

Nehls, G. & Ruth, M. (1994) Elders, mussels, and fisheries in the Wadden Sea: continuous conflicts or relaxed relations. *Ophelia*, **56**, 263–78.

Nehls, G., Hertzle, I. & Scheiffarth, G. (1997) Stable mussel *Mytilus edulis* beds in the Wadden Sea: they are just for the birds. *Helgo. Wiss. Meersunters.*, **51**, 361–72.

Nichy, F.E. & Menzel, R.W. (1960) Mortality of intertidal and subtidal oysters in Alligator Harbor, Florida. *Proc. Natl. Shellfish Assoc.*, **51**, 33–41.

Norberg, J. & Tedengren, M. (1995) Attack behavior and predatory success of *Asterias rubens* L. related to differences in size and morphology of the prey mussel *Mytilus edulis* L. *J. Exp. Mar. Biol. Ecol.*, **186**, 207–20.

Ó Foighil, D., Marshall, B.A., Hilbish, T.J. & Pino, M.A. (1999) Trans-Pacific range extension by rafting is inferred for the flat oyster *Ostrea chilensis. Biol Bull.*, **196**, 122–26.

Orensanz, J.M., Parma, A.M. & Iribarne, O.O. (1991) Population dynamics and management of natural stocks. In: *Scallops: Biology, Ecology and Aquaculture* (ed. S.E. Shumway), pp. 625–713. Elsevier Science Publishers B.V., Amsterdam.

Page, H.M. & Hubbard, D.M. (1987) Temporal and spatial patterns of growth in mussels *Mytilus edulis* on an offshore platform: relationships to water temperature and food availability. *J. Exp. Mar. Biol. Ecol.*, **111**, 159–79.

Paine, R.T. & Levin, S.A. (1981) Intertidal landscapes: disturbance and the dynamics of pattern. *Ecol. Monogr.*, **51**, 145–78.

Parsons, G.J., Dadswel, M.J. & Rodtrom, E.M. (1991) Fisheries and aquaculture: Scandinavia. In: *Scallops: Biology, Ecology and Aquaculture* (ed. S.E. Shumway), pp. 763–75. Elsevier Science Publishers B.V., Amsterdam.

Paul, J.D. (1980) Salinity-temperature relationships in the queen scallop *Chlamys opercularis. Mar. Biol.*, **56**, 295–300.

Peterson, C.H. (1977) Competitive organization of the soft-bottom macrobenthic community of southern California lagoons. *Mar. Biol.*, **43**, 343–59.

Petraitis, P.S. (1987) Immobilisation of the predatory gastropod, *Nucella lapillus*, by its prey *Mytilus edulis*. *Biol. Bull.*, **172**, 307–14.

Pitcher, C.R. & Butler, A.J. (1987) Predation by asteroids, escape response, and morphometrics of scallops with epizoic sponges. *J. Exp. Mar. Biol. Ecol.*, **112**, 233–49.

Pohle, D.G., Bricelj, V.M. & García–Esquivel, Z. (1991) The eelgrass canopy: an above-bottom refuge from benthic predators for juvenile Bay scallops *Argopecten irradians*. *Mar. Ecol. Prog. Ser.*, **74**, 47–59.

Porter, H.J. (1964) Incidence of *Malacobdella* in *Mercenaria campechensis* off Beaufort Inlet, North Carolina. *Proc. Natl. Shellfish Assoc.*, **53**, 133–45.

Provenzano, A.J.J. (1961) Effects of the flatworm *Stylochus ellipticus* (Girard) on oyster spat in two salt water ponds in Massachusetts. *Proc. Natl. Shellfish Assoc.*, **50**, 83–88.

Raffaelli, D. & Hawkins, S. (1996) *Intertidal Ecology*. Chapman & Hall, London.

Raffaelli, D., Falcy, V. & Galbraith, C. (1990) Eider predation and the dynamics of mussel bed communities. In: *Trophic Relationships in the Marine Environment* (eds M. Barnes & R.N. Gibson), pp. 89–103. Aberdeen University Press, Aberdeen.

Randall, D., Burggren, W. & French, K. (1997) *Eckert Animal Physiology: Mechanisms and Adaptations*. W.H. Freeman, New York.

Rawson, P.D., Hayhurst, S. & Vanscoyoc, B. (2001) Species composition of blue mussel populations in the north-eastern Gulf of Maine. *J. Shellfish Res.*, **20**, 31–38.

Reimer, O. & Tedengren, M. (1996) Phenotypical improvement of morphological defenses in the mussel *Mytilus edulis* induced by exposure to the predator *Asterias rubens*. *Oikos*, **75**, 383–90.

Reimer, O. & Tedengren, M. (1997) Predator-induced changes in byssal attachment, aggregation and migration in the blue mussel, *Mytilus edulis*. *Mar. Freshwat. Behav. Physiol.*, **30**, 251–66.

Rhodes, E.W. (1991) Fisheries and aquaculture of the bay scallop *Argopecten irradians*, in the eastern United States. In: *Scallops: Biology, Ecology and Aquaculture* (ed. S.E. Shumway), pp. 913–24. Elsevier Science Publishers B.V., Amsterdam.

Robinson, R.F. & Richardson, C.A. (1998) The direct and indirect effects of suction dredging on a razor clam (*Ensis arcuatus*) population. *ICES J. Mar. Sci.*, **55**, 970–77.

Ropes, J.W. & Merrill, A.S. (1967) *Malacobdella grossa* in *Pitar morrhauna* and *Mercenaria campechiensis*. *Nautilus*, **81**, 37–40.

Rosell, D., Uriz, M.-J. & Martin, D. (1999) Infestation by excavating sponges on the oyster (*Ostrea edulis*) populations of the Blanes littoral zone (north-western Mediterranean Sea). *J. mar. biol. Ass. U.K.*, **79**, 409–13.

Rovero, F., Hughes, R.N. & Chelazzi, G. (2000) When time is of the essence: choosing a currency for prey-handling costs. *J. Anim. Ecol.*, **69**, 683–89.

Sanders, H.L., Goudsmit, E.M., Millis, E.L. & Hampson, G.E. (1962) A study of the intertidal fauna of Barnstable Harbor, Massachusetts. *Limnol. Oceanogr.*, **7**, 63–77.

Sanjuan, A., Zapata, C. & Alvarez, G. (1994) *Mytilus galloprovincialis* and *M. edulis* on the coasts of the Iberian Peninsula. *Mar. Ecol. Prog. Ser.*, **113**, 131–46.

Seed, R. (1969) The ecology of *Mytilus edulis* L. (Lamellibranchiata) on exposed shores. 2. Growth and mortality. *Oecologia*, **3**, 317–50.

Seed, R. (1976) Ecology. In: *Marine Mussels: their Ecology and Physiology* (ed. B.L. Bayne), pp. 13–65. Cambridge University Press, Cambridge.

Seed, R. & Suchanek, T.H. (1992) Population and community ecology of *Mytilus*. In: *The Mussel Mytilus: Ecology, Physiology, Genetics and Culture* (ed. E.M. Gosling), pp. 87–169. Elsevier Science Publishers B.V., Amsterdam.

Shamseldin, A.A., Clegg, J.S., Friedman, C.S., Cherr, G.N. & Pillai, M.C. (1997) Induced thermotolerance in the Pacific oyster, *Crassostrea gigas*. *J. Shellfish Res.*, **16**, 487–91.

Shaw, W.N., Hassler, T.J. & Moran, D.P. (1988) *Species profiles: life histories and environmental requirements of coastal fishes and invertebrates (Pacific southwest): California sea mussel and bay mussel.* U.S. Fisheries and Wildlife Services Biology Report 82(11.84)TR-EL-82–4.

Shumway, S.E. (1996) Natural environmental factors. In: *The Eastern Oyster* Crassostrea virginica (eds V.S. Kennedy, R.I.E. Newell & A.F. Eble), pp. 467–513. Maryland Sea Grant, College Park, Maryland.

Siddall, S.E. (1980) A classification of the genus *Perna* (Mytilidae). *Bull. Mar. Sci.*, **30**, 858–70.

Sivalingham, P.M. (1977) Aquaculture of the green mussel *Mytilus viridis* Linnaeus, in Malaysia. *Aquaculture*, **11**, 297–312.

Skilleter, G.A. (1994) Refuges from predation and the persistence of estuarine clam populations. *Mar. Ecol. Prog. Ser.*, **109**, 29–42.

Smith, L.D. & Jennings, J.A. (2000) Induced defensive responses by the bivalve *Mytilus edulis* to predators with different attack modes. *Mar. Biol.*, **136**, 461–69.

Suchanek, T.H. (1981) The role of disturbance in the evolution of life history strategies in the intertidal mussels *Mytilus edulis* and *Mytilus californianus*. *Oecologia*, **50**, 143–52.

Suchanek, T.H., Geller, J.B., Kreiser, B.R. & Mitton, J.B. (1997) Zoogeographic distributions of the sibling species *Mytilus galloprovincialis* and *M. trossulus* (Bivalvia: Mytilidae) and their hybrids in the North Pacific. *Biol. Bull.*, **193**, 187–94.

Svane, I. & Ompi, M. (1993) Patch dynamics in beds of the blue mussel *Mytilus edulis* L: effects of site, patch size, and position within patch. *Ophelia*, **37**, 187–202.

Thomas, G.E. & Gruffydd, L.D. (1971) The type of escape reactions elicited in the scallop *Pecten maximus* by selected sea-star species. *Mar. Biol.*, **10**, 87–93.

Thorson, G. (1950) Reproduction and larval ecology of marine bottom invertebrates. *Biol. Rev.*, **25**, 1–45.

Toro, J.E. (1998) PCR-based nuclear and mtDNA markers and shell morphology as an approach to study the taxonomic status of the Chilean blue mussel, *Mytilus chilensis* (Bivalvia). *Aquat. Living Resour.*, **11**, 347–53.

Tuckwell, J. & Nol, E. (1997a) Foraging behavior of American oystercatchers in response to declining prey densities. *Can. J. Zool.*, **75**, 170–81.

Tuckwell, J. & Nol, E. (1997b) Intra-specific and inter-specific interactions of foraging American oystercatchers on an oyster bed. *Can. J. Zool.*, **75**, 182–87.

Vakily, J.M. (1989) The biology and culture of mussels of the genus *Perna*. *ICLARM Stud. Rev.*, **17**, 1–63.

Vernberg, F.J. & Vernberg, W.B. (1972) *Environmental Physiology of Marine Animals.* Springer-Verlag, New York.

Waller, T.R. (1991) Evolutionary relationships among commercial scallops (Mollusca: Bivalvia: Pectinidae). In: *Scallops: Biology, Ecology and Aquaculture* (ed. S.E. Shumway), pp. 1–73. Elsevier Science Publishers B.V., Amsterdam.

Wells, H.W. (1957) Abundance of the hard clam *Mercenaria mercenaria* in relation to environmental factors. *Ecology*, **38**, 123–30.

Werner, I. & Hinton, D.E. (1999) Field validation of hsp-70 stress proteins as biomarkers in Asian clam (*Potamocorbula amurensis*): is down regulation an indicator of stress? *Biomarkers*, **4**, 473–84.

White, M.E. & Wilson, E.A. (1996) Predators, pests and competitors. In: *The Eastern Oyster* Crassostrea virginica (eds V.S. Kennedy, R.I.E. Newell & A.F. Eble), pp. 559–79. Maryland Sea Grant, College Park, Maryland.

Wiborg, K.F. (1963) Some observations on the Iceland scallop *Chlamys islandica* (Müller) in Norwegian waters. *Fisk. Skr. Havunders.*, **13**, 38–53.

Wieters, E.A. & Navarrete, S.A. (1998) Spatial variability in prey preferences of the intertidal whelks *Nucella canaliculata* and *Nucella emarginata*. *J. Exp. Mar. Biol. Ecol.*, **222**, 133–48.

Wilkens, L.A. (1991) Neurobiology and behavior of the scallop. In: *Scallops: Biology, Ecology and Aquaculture* (ed. S.E. Shumway), pp. 428–69. Elsevier Science Publishers B.V., Amsterdam.

Williams, R.J. (1970) Freezing tolerance in *Mytilus edulis*. *Comp. Biochem. Physiol.*, **35**, 145–61.

Winter, M.A. & Hamilton, P.V. (1985) Factors influencing swimming in Bay scallops *Argopecten irradians* (Lamarck, 1819). *J. Exp. Mar. Biol. Ecol.*, **88**, 227–42.

Witman, J.D. & Suchanek, T.H. (1984) Mussels in flow: drag and dislodgement by epizoans. *Mar. Ecol. Prog. Ser.*, **16**, 259–68.

Wootton, J.T. (1993) Size-dependent competition: effects on the dynamics versus the end-point of mussel bed succession. *Ecology*, **74**, 195–206.

Zajac, R.N., Whitlatch, R.B. & Osman, R.W. (1989) Effects of inter-specific density and food supply on survivorship and growth of newly settled benthos. *Mar. Ecol. Prog. Ser.*, **56**, 127–32.

Zwarts, L., Ens, B.J., Gosscustard, J.D., Hulscher, J.B. & Durell, S.E.A. (1996a) Causes of variation in prey profitability and its consequences for the intake rate of the oystercatcher *Haematopus ostralegus*. *Ardea*, **84A**, 229–68.

Zwarts, L., Wanink, J.H. & Ens, B.J. (1996b) Predicting seasonal and annual fluctuations in the local exploitation of different prey by oystercatchers *Haematopus ostralegus*: a 10-year study in the Wadden Sea. *Ardea*, **84A**, 401–40.

4 How Bivalves Feed

Introduction

Primitive protobranch bivalves such as *Nucula* feed primarily by means of ciliated fleshy extensions of the labial palps. These extend into the substrate and collect particles that are carried on to the labial palps for sorting, prior to ingestion. Thus, the gills are primarily respiratory organs. In contrast some protobranchs, such as *Solemya velum*, ingest suspended algae and are regarded as crude suspension-feeders. Such species rely on chemo-autotrophic bacteria in their gills for their main nutrient supply, with phytoplankton from suspension-feeding as a supplementary source (see below).

The vast majority of bivalves use the gills for feeding and these have become greatly enlarged to deal with their secondarily derived role. This method of feeding is called suspension or filter feeding because the gills with their different ciliary tracts remove suspended particles from the water pumped through the mantle cavity. The gills divide the mantle cavity into inhalant and exhalant chambers (Fig. 2.10A, Chapter 2). The water that enters through the inhalant opening or siphon, is driven from the inhalant to the exhalant chambers by cilia on the gills and mantle surface, and exits by the exhalant opening or siphon. Both openings possess a muscular velum, the inner fold of the mantle, which regulates water flow through the mantle cavity (Chapter 2).

This chapter describes feeding, digestion and absorption in bivalves, and the various mechanisms that they employ to control the quantity and quality of their diet. Such mechanisms include:

- varying the rate at which water is passed over the gills, and the rate and efficiency with which particles are removed from the feeding current;
- sorting of edible seston from material of low nutritional value;
- regulating the volume of material processed and ingested;
- modulating the digestion and absorption process, through control of the rate and passage of material through the gut, sorting in the stomach, and changes in digestive enzymes.

Filtration rate

Filtration or pumping rate is defined as the volume of water flowing through the gill in a unit of time. The clearance rate is that volume of water completely cleared of particles per unit of time. When all particles presented to the gill are cleared from suspension, then the clearance rate is the same as the filtration rate.

The two processes, clearance and filtration, are controlled independently of each other in bivalves. This makes absolute sense in view of the dual function of the gill in feeding and respiration. Bivalves are capable of altering the size of the gill ostia and the gill itself, and changing the beat frequency and pattern of the latero-frontal cirri to accommodate differing particle concentrations in the incoming water. However, the ventilation current created by the lateral cilia is unaffected by particle concentration, and thus gas exchange is unimpaired (Bayne *et al.*, 1976a). The general mantle surface is believed to play a significant role in gas exchange.

Estimation of filtration and clearance rates

Filtration rate in bivalves can be determined directly or indirectly. The direct method is most successful when applied to those species where the inhalant and exhalant currents can be easily separated. The set-up for the direct method is illustrated in Fig. 4.1. A bivalve is placed in a tank that is divided into two chambers C_1 and C_2 by a membrane with a slit to accommodate the bivalve and to separate its inhalant and exhalant siphons. A shunt connects the two chambers and this is closed at the start of the experiment. As the animal pumps water from C_1 to C_2 the water level in C_2 is monitored with a laser beam striking a mirror that is fixed to a floating ping-pong ball. The mirror reflects the laser beam onto a scale at about 8 m distance from the mirror. A deflection of 1 cm on the scale represents a 0.1 mm change in the water level in C_2.

Indirect methods are used to determine clearance rates. The bivalve is placed in a known volume of water containing a suspension of particles for a set period of time. The particles are usually algae cells, organic and inorganic powders, or even bacteria. Filtration results in a decrease in particle concentration over time, and clearance rate is determined from the exponential decrease (e.g. verified as a straight line in a semi-log plot, Fig. 4.2) in algal concentration as a function of time using the equation:

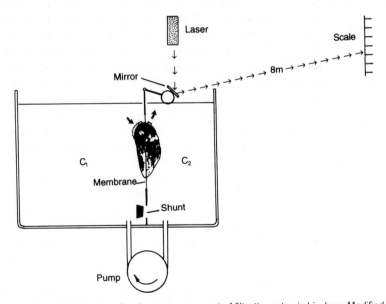

Fig. 4.1. Experimental set-up for direct measurement of filtration rates in bivalves. Modified from Famme *et al.* (1986).

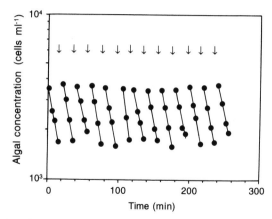

Fig. 4.2. Reduction of algal cell concentration during an experiment to measure clearance and filtration rates in mussels (*Mytilus edulis*). Arrows indicate additions of algal suspension. The slope of lines gives the mussels' filtration rate. From Clausen & Riisgård (1996).

$$Cl = (V/nt)\ln(C_0/C_t)$$

where C_0 and C_t are the initial and final particle concentrations, V is the volume of water cleared, t is the time and n is the number of animals. The slope of the lines (Fig. 4.2) express filtration rate.

There are several disadvantages associated with closed systems, such as accumulation of excretory products, reduction in oxygen concentration, declining particle concentration, all of which can, to varying degrees, alter normal filtration behaviour. These problems have been overcome in flow-through systems where the animal is kept in a chamber through which the particle suspension flows at a constant rate. Water is pumped at high speed through the chamber to prevent recirculation, and particle concentration in the inflow water is kept constant (Fig. 4.3). See Riisgård (2001) for a critique of the methods currently in use for measurements of filtration and clearance rates.

Clearance rate measurements cannot be easily used in field experiments without disturbing the bivalves. However, clearance rate can be measured from direct collection of bio-deposits according to the ratio: Cl = (inorganic matter egested as faeces and pseudofaeces)/(total inorganic matter in seawater). The method is easy to set up and has been used successfully for several field studies (see Pouvreau *et al.*, 2000 & Riisgård, 2001)

Up to the late 1960s particle concentration was estimated either photometrically or by the use of radioactively labelled particles. The introduction of the electronic particle Coulter Counter (Sheldon & Parsons, 1967) made it possible to obtain rapid and accurate determinations of particle density in flow-through systems. More recently, fluorometric sensing by flow cytometry has been introduced to record particle density (Cucci *et al.*, 1985; Shumway *et al.*, 1985). This technique sorts cells on the basis of shape and on the fluorescence of naturally occurring plant pigments (or absence of fluorescence in inert or detritus particles). It is now possible to examine differential use by bivalves of food resources even when those resources are composed of groups

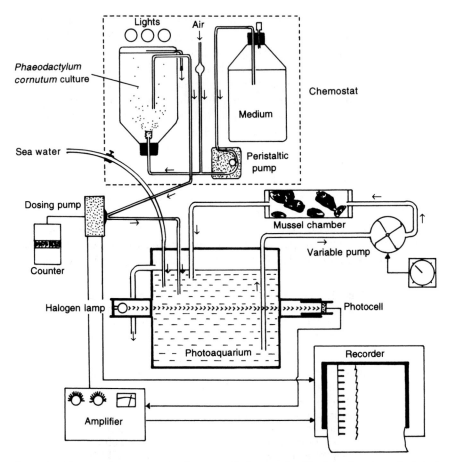

Fig. 4.3. Diagram of an open system for automatically recording filtration rates at constant algal concentration. From Riisgård & Møhlenberg (1979) and Jørgensen (1990).

having similar sized cells. In addition, flow cytometry overcomes the size limitation of the Coulter Counter, and very small particles, such as bacteria, can now be used in filtration experiments, provided that they have first been stained with fluorescent dyes (Shumway *et al.*, 1990).

Filtration rate and body size

Water pumping rate and thus feeding capacity scales closely with gill area and size in bivalves. The allometric relationship between filtration rate (*Fr*; lh⁻¹) and tissue dry weight (*W*; g) is described by the equation

$$Fr = aW^b$$

The constants *a* and *b* are fitted parameters, *a* represents the filtration rate of an individual of unit body weight (or length) and *b* is the power to which increase in size raises that rate. Generally speaking, weight-specific filtration rates decline with increase in body size and this decline occurs at a faster rate than metabolic rate. This causes a decline in growth efficiency, which limits the ultimate body size of bivalve species (see Chapter 6).

There is considerable variation in filtration rates both within and between species (Table 4.1). While some of this variation represents genuine differences, some can be attributed to the differing experimental conditions employed in the different studies. Bricelj & Shumway (1991) have suggested that since filtration rates are extremely sensitive to changes in food quality and quantity, inter-specific comparisons are best carried out from studies that employ identical experimental protocols on a wide variety of bivalve species. More recently, Riisgård & Larsen (1995) and Riisgård (2001) have pointed out the importance of performing filtration experiments under optimal conditions; for example, using algae concentrations similar to those to which the bivalve is adapted in nature. Sub-optimal conditions, e.g. unnaturally high algae concentrations, can result in partial valve closure leading to a reduction in filtration rate (Jørgensen *et al.*, 1988). In addition, Jones *et al.* (1992) have suggested

Table 4.1. Filtration rates measured by different methods under optimal laboratory conditions. Filtration rates (F, $1\,h^{-1}$) as a function of size (W, g body dry wt) are presented as allometric equations $F = aW^{b}$. For clarity temperature, size range etc. have been omitted. Data and details on methods in Riisgård (2001).

Method	Filtration Rate
Suction method	
Møhlenberg & Riisgård (1979)	
Cardium echinatum	$F = 4.22W^{0.62}$
Cerastoderma edule	$F = 11.60W^{0.70}$
Mytilus edulis	$F = 7.45W^{0.66}$
Modiolus modiolus	$F = 6.00W^{0.75}$
Arctica islandica	$F = 5.55W^{0.62}$
Photoaquarium method	
Riisgård & Møhlenberg (1979)	
Mytilus edulis	$F = 7.37W^{0.72}$
Clearance method	
Griffiths (1980)	
Choromytilus meridionalis	$F = 5.37W^{0.60}$
Berry & Schleyer (1983)	
Perna perna	$F = 8.85W^{0.66}$
Riisgård (1988)	
Crassostrea virginica	$F = 6.79W^{0.73}$
Guekensia demissa	$F = 6.15W^{0.83}$
Replacement method	
Coughlan & Ansell (1964)	
Mercenaria mercenaria	$F = 2.5W^{0.78}$
Thermistor method	
Meyhöfer (1985)	
Clinocardium nuttalii	$F = 3.1W^{0.80}$
Mytilus californianus	$F = 7.9W^{0.72}$
Chlamys hasata	$F = 8.7W^{0.94}$

that for population comparisons, where a size range is used to determine pumping rate, it should be determined for individuals, or collectively for animals with a very small size range. This is because water-pumping rates vary considerably with time for any single individual (Fig. 4.4).

Factors affecting filtration rate

Particle concentration

Pumping rates are largely dependent on the concentration of particulate material in the medium. In laboratory experiments Wilson & Seed (1974) have shown that mussel pumping rate rapidly decreased or stopped altogether when particle-free seawater was used. They concluded that particulate material present in natural seawater, in the form of plankton and fine detritus, stimulates and maintains pumping. Mussels do not filter in very dilute suspensions, and this undoubtedly conserves energy during periods when particulate food is scarce, as for example in winter months. In order for filtration to start particle concentration must reach a critical threshold level. In *Mytilus edulis* this is about 50 cells μl^{-1} for the algae *Phaeodactylum tricornutum*, and filtration rate remains high and constant (Fig. 4.5A) until particle concentration reaches about 70 cells μl^{-1} (Foster-Smith, 1975). However, results from a more recent study (Riisgård, 1991) have shown that mussels filter maximally at much lower particle concentrations than those given above. For example, mussels fed on a concentration of 3–10 cells μl^{-1} (*Rhodomonas baltica*) filter maximally, but when the concentration is elevated above 15 cells μl^{-1} the mussels reduce valve gape, which ultimately leads to a decline in filtration rate. Riisgård & Randløv (1981) have reported similar results using *Phaeodactylum tricornutum*. These

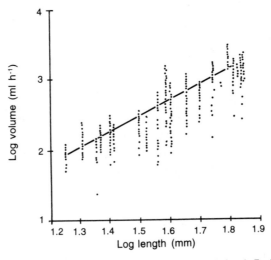

Fig. 4.4. Graph of log pumped volume (ml h^{-1}) against log length (mm). Each vertical array of points represents many measurements from a single animal. The fitted regression line through the points is also shown. Modified from Jones *et al.* (1992) and reprinted with permission from Elsevier Science.

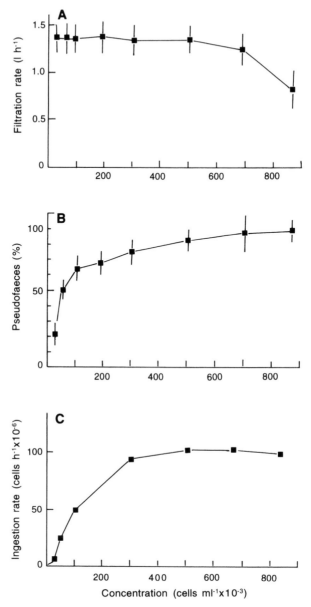

Fig. 4.5. Filtration rate (A), percentage of material rejected before ingestion (B) and rates of ingestion (C) in *Mytilus edulis* with increasing concentrations of the algae, *Phaeodactylum tricornutum*. From Foster-Smith (1975).

authors conclude that mussels may be adapted to exploit their filtration capacities at low algae concentrations only. Thus, the generally low growth rates obtained in laboratory studies may be due to the of unnaturally high concentrations that lead to valve closure, reduced metabolism and reduced biosynthesis/growth.

In juvenile scallops, *Argopecten irradians*, filtration declined by 85% when the concentration of *Thalassiosira weissflogii* was increased from 1.2 to 12 cells μl^{-1} (Kuenster, 1988), while a lower reduction (56%) was reported when a different alga species, *Isochrysis galbana*, was used (Cahalan *et al.*, 1989). Scallops are, therefore, not well adapted to high food levels and are not normally found in very turbid environments, where many mussel and oyster species thrive.

As particle concentration increases the rate of pseudofaeces (material cleared from suspension but rejected before ingestion) production rises. In the mussel, *Mytilus edulis*, pseudofaeces production rises rapidly when algae concentration is increased from 50 to 100 cells μl^{-1}, and increases more gradually at higher concentrations (Fig. 4.5B). Subtracting the rate of pseudofaeces production from filtration rate gives the true ingestion rate (Fig. 4.5C). For example, in *M. edulis* Foster-Smith (1975) found that ingestion rate gradually increased up to a concentration of 300 cells μl^{-1} and then remained constant up to 800 cells μl^{-1}. Thus, this species maintains a constant filtration rate over a wide range of cell concentrations, but rapidly increases pseudofaecal production to control ingestion ration. On the other hand, the clam, *Venerupis pullastra*, achieves the same result by decreasing its filtration rate at elevated particle concentrations, so producing less pseudofaeces, but all the while maintaining a fairly consistent ingestion rate (Foster-Smith, 1975).

Temperature

Temperature has a marked influence on filtration rates in filter-feeding bivalves. In the scallop *Argopecten irradians* filtration rates are fairly constant over a temperature range of 10 to 26°C but are markedly reduced at 5°C (Table 4.2). Since metabolic expenditure (measured by O_2 consumption) increases with rising temperatures (Fig. 4.6) *A. irradians* needs a concomitant increase in food levels in order to avoid rapid weight loss at high tempera-

Table 4.2. Weight-standardised filtration rates (Fr_s) in litres $h^{-1} g$ dry tissue weight^{-1} of the scallops *Aequipecten opercularis* and *Argopecten irradians* at different temperatures. $Fr_s = Fr_e (W_s/W_e)^b$, where Fr_e and W_e are the filtration rate and tissue dry weight, respectively, of the experimental animal, $W_s = 1 g$ and b, taken as 0.7, is the exponent of the allometric relationship between filtration rate and W (see text). Table modified from Bricelj & Shumway (1991).

Species	Temp °C	Suspension	Fr_s	Reference
A. opercularis	5	*Dunaliella euchlora*	1.64	McLusky, 1973
	10	8–10 cells μl^{-1}	3.23	
	20		5.90	
A. irradians	5	*Nitzschia*	1.75	Kirby-Smith, 1970
	10–26	100–500 cells μl^{-1}	5.82	
	22	*Thalassiosira weissflogii*		Kuenster, 1988
		1.2 cells μl^{-1}	10.3	
		4.8 cells μl^{-1}	4.71	
		12.0 cells μl^{-1}	1.39	
	22–26	*Nitzschia*	4.74	Chipman &
		0.85–8.0 cells μl^{-1}		Hopkins, 1954
		Chlamydomonas		
		28 cells μl^{-1}		

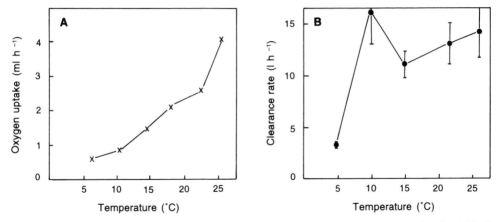

Fig. 4.6. Mean oxygen uptake (A) and filtration rate (B) of a standard scallop (*Argopecten irradians*) 20 g in wet tissue weight fed on *Nitzchia* sp. Modified from Bricelj & Shumway (1991).

tures. In *Aequipecten opercularis* feeding rates have also been shown to be relatively independent of temperature over the range 10 to 20°C but, unlike *A. irradians*, oxygen consumption remains constant over this temperature range, allowing conservation of energy at higher temperatures (McLusky, 1973). A similar strategy has been reported for *Mytilus edulis* by Bayne *et al.* (1976a). Jørgensen (1990) has proposed that the decreased viscosity of seawater with increasing temperature may account for most of the increase in pumping rates in *M. edulis*.

Temperature acclimation in filtration rates has been demonstrated for a variety of bivalve species. For example, Rao (1953) has shown that weight-specific filtration rates in *Mytilus californianus* from various localities on the west coast of the United States were identical at the minimum temperatures recorded for the respective latitudes, thus maintaining a potential for growth that is relatively independent of latitude. However, at temperatures above 20°C, approaching the lethal limits for the species, there is an apparent failure of the adaptive response, and this is reflected in a reduced scope for growth (see Chapter 6).

Salinity

Mussels are euryhaline, being able to survive in salinities ranging from 4–5 psu to fully marine conditions. When filtration rates were measured in Baltic (15 psu) and North Sea (30 psu) mussels Theede (1963) observed that despite differing salinity, filtration rates were similar for the two populations. However, when mussels were transferred to either higher or lower salinity filtration rate was markedly reduced. He concluded that the optimum salinity for filtration was that of the natural habitat, although some degree of acclimation was possible. When mussels were acclimated for 7 days to salinity ranging from 15–50 psu pumping rate and clearance efficiency were maximal at 34 psu (Wilson & Seed, 1974). No pumping occurred at the extreme salinities of 15 psu and 50 psu, although natural populations are found in salinities as low as 5 psu. Bayne *et al.* (1976a) have pointed out, however, that a period of up to several weeks is necessary for successful salinity adaptation. For example, when mussels were transferred from 34 psu to 18 psu their filtration rate, which was initially depressed, took more than seven weeks to return to the control value (Bøhle, 1972). During this period of acclimation, the length of which depends on the extent of salinity change, complete metabolic compensation takes place, culminating in a recovery of respiration and filtration rates to control values. Similar results have been reported for oysters. When *Crassostrea virginica* was exposed to a sudden reduction in salinity from 27 psu to 20, 15, 10 and 5 psu there was a decrease in filtration rate of 24%, 89%, 91% and almost 100%, respectively, for about six hours after transfer. Normal pumping activity resumed after this time and there were no long-term effects on filtration rate (Loosanoff, 1953).

Other factors

There is increasing evidence that marine bivalves orientate to maximise their filtration capacities and exploit their food supply (Dame, 1996; Sakurai & Seto, 2000). In hatchery-grown oysters, *Crassostrea gigas*, Walne (1974) has shown that filtration rate increased with increase in flow rate. As flow rate through the trays was increased from 50 to 100 ml min^{-1} filtration increased by 50% and by another 50% when the flow rate was increased from 100 to 200 ml min^{-1}. Increased water flow brings not just more food but also stimulates the animal to feed more rapidly. In the clam, *Ruditapes decussatus*, Sobral & Widdows (2000) observed a maximum clearance rate of 2.5 l h^{-1} individual^{-1} (0.3 g dry tissue mass) at current velocities up to 8 cm s^{-1}. Clearance rate declined with increasing current speed, especially above 17 cm s^{-1}. At high flow speeds there is a build-up of a pressure differential between inhalant and exhalant apertures that interferes with filtration, and ultimately results in decreased growth rates. In scallops growth inhibition in the laboratory occurs at current velocities exceeding 10 cm s^{-1}, but in the wild scallops are often exposed to current speeds far in excess of this (Wildish & Kristmanson, 1988).

Chemical stimuli also affect filtration rates in shellfish. The scallop *Placopecten magellanicus* increases its filtration rate in response to metabolites from the diatom *Chaetoceros muelleri*. The stimulus saturates at a low concentration of diatom extract equivalent to 5 cells μl^{-1} (Ward *et al.*, 1992). These authors have suggested that chemical cues from phytoplankton are important factors that allow scallops to adjust their feeding rates in the wild.

Dodgson (1928) observed that under similar temperature conditions mussels cleared water more rapidly in September–October than in February–March. While this may be related to gonad development, different food levels in the water at different times of the year could also explain the observation (Wilson & Seed, 1974).

Reduction in valve gape is accompanied by a retraction of the mantle edges and exhalant siphon in *Mytilus edulis*. The muscles of mantle and siphon are continuous with muscles running along the base of the gills at their attachment to the body wall. Retraction causes the gill axes to shorten leading to a reduction in the width of the inter-filament canals. Consequently, Jørgensen *et al.* (1988) have found that filtration rate is proportionately reduced with reduced valve gape. This is not seen as a mechanism for controlling filtration rate and thus food ingestion, but as a secondary effect of the mussel's response to sub-optimal conditions (Riisgård & Larsen, 1995).

Regulation of filtration rate

There are two opposing schools of thought on whether filtration rate is physiologically controlled. The most widely held belief is that the process *is* under physiological control, and that bivalves regulate pumping rate and retention efficiency as a function of particle concentration and gut satiation (Bayne 1998; Hawkins *et al.*, 1998a). The alternative less popular view, first mooted by Jørgensen in 1966 and elaborated on since in various publications (see Jørgensen, 1990, 1996 for reviews), is that accumulation of food particles on the gills is not regulated according to the nutritional needs of the individual, but is an automatic process determined by the capacity of the pump, the concentration of food particles in the water, and by the efficiency with which these particles are retained on the gills. The debate continues.

Energy costs

Suspension feeders must filter large volumes of water to meet their food requirements. Because of 'life in a nutritionally dilute environment' (Conover, 1968) they possess low-energy pumps that continuously pump the surrounding water through gills that are efficient in retaining small food particles (Riisgård & Larsen, 1995; Clausen & Riisgård, 1996). However, published estimates of the metabolic cost of pumping are somewhat variable. Several studies have inferred high energy costs in a wide selection of species. For example, in *Mytilus edulis* and *M. californianus*, Bayne & Newell (1983) estimated that 24% of an ingested ration of algae cells represented the cost of feeding, of which digestion and assimilation accounted for only 4–6%. A later study estimated the energetic costs of digestion to be about 17% of energy expenditure in *M. edulis* (Widdows & Hawkins, 1989). In contrast, low values (1–3% of total metabolic expenditure) have also been reported (Bernard & Noakes, 1990; Jørgensen, 1990; Hawkins & Bayne, 1992; Riisgård & Larsen, 1995). Jørgensen (1990) contends that high feeding costs are based on the mistaken assumption that increased rates of oxygen consumption with increased pumping rates reflect energetic costs of feeding, particularly water processing, despite the evidence that ciliary work is relatively inexpensive. These conflicting findings remain to be resolved.

Particle processing on the gills

During the 1990s considerable progress was made in understanding particle processing mechanisms in suspension-feeding bivalves. Several complementary techniques have made this possible. Video-endoscopy (Fig. 4.7) permits direct *in vivo* observations of whole intact structures, e.g. gills and labial palps, in relatively undisturbed animals (Ward *et al.*, 1991; Beninger *et al.*, 1992).

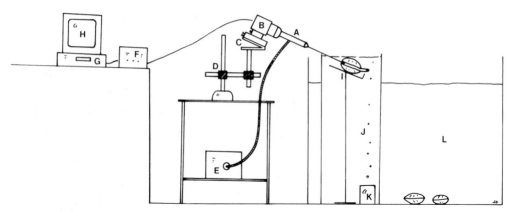

Fig. 4.7. Equipment used for endoscopic observations: (A) endoscope, (B) video camera, (C) micromanipulator, (D) adjustable stand, (E) light source, (F) camera control box, (G) Hi-8 video-cassette recorder, (H) monitor, (I) inclinable specimen stand, (J) bucket, (K) water-circulating pump, (L) aquarium). From Beninger *et al.* (1997b).

Combined with fluid sampling and subsequent chemical and histochemical analyses, histology and mucocyte (mucus–secreting cell) mapping, endoscopy has been responsible for a major leap forward in our understanding of the filter feeding process in bivalves. Confocal laser scanning microscopy (CLSM) has been used to good effect to examine the spatial relationships, structure and movement of cilia and cirri on living gill filaments (Silverman *et al.*, 1999).

The simplest gill structure is the homorhabdic filibranch type found only in mussels (Fig. 4.8A). Scallops also have this gill type but the gill filaments are differentiated into principal and ordinary filaments, so the gill type is termed heterorhabdic filibranch (Fig. 4.8B). The majority of bivalves possess the slightly more complex eulamellibranch structure (Fig. 4.8C). The pseudo-lamellibranch type is only found in oysters (Fig. 4.8D). In some bivalves, e.g. scallops and oysters, the surface area of the gill filaments is greatly increased by folds or plicae (Figs. 4.8B & D). Particle processing has been studied in detail in these four different gill types using representative species (see references below): the mussel, *Mytilus edulis* (homorhabdic filibranch), the scallop, *Placopecten magellanicus* (plicate heterorhabdic filibranch), the clam species *Spisula solidissima* and *Mya arenaria* (eulamellibranch), and the oyster *Crassostrea virginica* (plicate heterorhabdic pseudolamellibranch).

Bivalves employ a hydromechanical and mucociliary mechanism of particle transport. Lateral cilia on the gill filaments maintain a flow of water through the mantle cavity and the gills. The water is filtered at the entrance to the inter-filament spaces (leading to the ostia in the eulamellibranch gill) by the latero-frontal cilia, which strain particles from the water and throw them onto the frontal surface of the filaments. Cilia on the abfrontal surface, facing the excurrent flow, do not participate in pumping water, primarily because of their small size and sparse density, and also because of associated high mucocyte densities (Dufour & Beninger, 2001). There are two main types of frontal ciliary tracts, lateral tracts of fine cilia and median tracts of coarse cilia (Figs. 2.10B & C). The filaments, particularly the frontal surfaces, are well

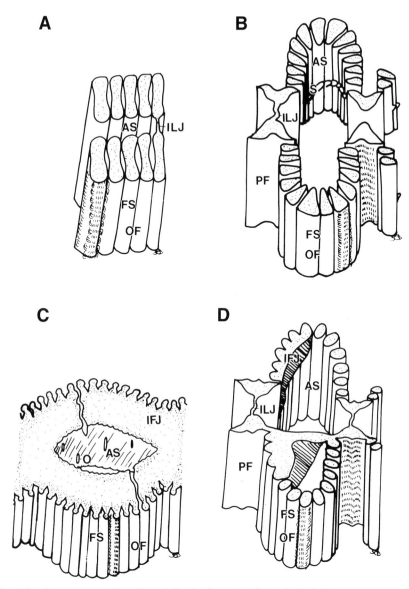

Fig. 4.8. Transverse sections through the demibranchs of the principal gill types in suspension-feeding bivalves. A. Homorhabdic filibranch: ordinary filaments (OF) connected by interlamellar junctions (ILJ). B. Heterorhabdic filibranch: principal filaments (PF) and ordinary filaments, joined by interlamellar junctions and ciliated spurs (S). C. Homorhabdic eulamellibranch: ordinary filaments, joined by interfilament junctions (IFJ); O: ostia. D. Heterorhabdic pseodolamellibranch: principal filaments and ordinary filaments, joined by interlamellar junctions and interfilament junctions; AS abfrontl surface, FS frontal surface. From Dufour & Beninger (2001).

supplied with mucocytes. The particles are trapped in a fine mucus layer (raft) overlying the frontal cilia and transported towards the ventral ciliated particle grooves (Beninger *et al.*, 1997a). There the material is incorporated into mucus strings that are transported along the grooves towards the labial palps (Fig. 4.9). In the ventral gill particle grooves material is never in suspension, even

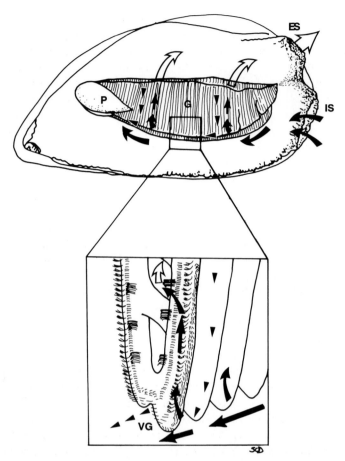

Fig. 4.9. Principal pathway of current flow and particle transport on the homorhabdic filibranch gill of *Mytilus edulis*. Water enters the animal through the inhalant siphon (IS). The frontal surface of the gill (G) is exposed to an antero-posterior flow at the ventral margin, and to a ventro-dorsal flow on the rest of the frontal surface (solid arrows). Water exits the gill from the abfrontal region, out through the exhalant siphon (ES: open arrows). Arrowheads represent particle transport to the ventral particle groove (VG); (P) labial palps. From Beninger & St-Jean (1997a).

at low particle concentration (Ward *et al.*, 1993). This mechanism of particle processing is seen in homorhabdic filibranchs (mussels) and eulamellibranchs (clams), i.e. the vast majority of bivalve species. However, heterorhabdic fili-branchs (scallops) employ a different strategy. Their gills do not have a ventral food groove so particles drawn into the pallial cavity move towards the dorsal region of the gill, and are deflected from the ordinary filaments into the plical troughs. A principal filament occupies each plical trough (Fig. 4.8B). Mater-ial exits the plical troughs and is incorporated into a mucus slurry in the dorsal ciliated tracts and is transported to the labial palps. The transport mechanism is hydrodynamic but the medium is a mucus slurry. At high particle concen-trations particles embedded in viscous mucus are ejected from the principal filaments onto the ordinary filaments plicae, and then directed to ventral cili-ated tracts. The ventral tract appears to be a rejection route that is activated when the ingestive or handling capacity of the scallop is saturated. The scallop gill therefore exhibits a division of labour, with the principal filaments involved in feeding and the ordinary filaments concerned with cleaning (see Beninger *et al.*, 1992 & 1993 for details). Pseudolamellibranchs (oysters) employ yet another strategy. Most particles captured on the frontal surfaces of ordinary filaments are directed to the ventral grooves and travel embedded in a mucus string towards the palps. Some particles however move from the ordinary filaments into the plical troughs. From there they are transported into dorsal ciliated grooves, and move anteriorly suspended in a slurry (Ward *et al.*, 1994).

The gills are not mere sieves but are able to control the rate at which suspended material is removed from the surrounding water. They do this either by altering the angle of ciliary beat, or by muscular expansion of the ostia. Most bivalves retain particles 3–4 µm diameter with an efficiency of 100% and retain particles of 1 µm diameter with a reduced efficiency of as much as 50% (Shumway *et al.*, 1985). Mussels can retain motile flagellates as small as 1–2 µm diameter, but can also remove bacteria (0.3–1.0 µm) from suspension, albeit with low efficiency. The mussel, *Guekensia demissa*, for example, retains particles as small as 0.4–0.6 µm with 86% of its efficiency for larger particle capture. This ability is related to the narrow space (~0.5 µm) between the latero-frontal cilia (Wright *et al.*, 1982). The mussel, *Perna perna*, also shows high retention efficiency for particles of this size and therefore both of these species seem specifically adapted to exploit the bacterial-size fraction. In contrast, the lower limit for effective retention of particles in scallops is about 5–7 µm and therefore bacterioplankton is not available as a food source (see below). But scallops are capable of ingesting relatively large particles; 10–350 µm diameter particles were described from the gut contents of *Placocopecten magellanicus* (Shumway *et al.*, 1987) and particles up to 950 µm from *Patinopecten yessoensis* (Mikulich & Tsikhon-Lukanina, 1981). There is now increasing evidence to indicate that particle retention efficiency depends not just on particle size but also on shape, motility, density, and chemical cues such as algae ectocrines (Hawkins & Bayne, 1992). Also, Defossez & Hawkins (1997) have suggested that preferential rejection of larger particles as pseudofaeces could have an adaptive value if larger particles are on average less nutritious than smaller particles.

It is generally believed that the latero-frontal cilia or cirri play a key role in bivalve feeding (Silverman *et al.*, 1999). Each latero-frontal cirrus consists of a double row of ~20–25 pairs of smaller cilia that beat in such a way as to form a meshwork between the cirri and between adjacent filaments. In *Mytilus edulis*, for example, the arrangement forms a filter with a mesh size of 2.7 × 0.6 µm, which would clearly explain the high retention of 1–2 µm diameter particles recorded for this species (Møhlenberg & Riisgård, 1979). Indeed, species with well-developed latero-frontal cirri show 90% retention efficiency of particles in the size range 2.0–3.5 µm, whereas species with either short (*Ostrea*) or undeveloped latero-frontal cirri (some scallop species) show only 50% efficiencies in this size range (Jørgensen, 1990). In species lacking latero-frontal cirri it seems likely that oscillatory currents produced by the lateral cilia are responsible for the transfer of suspended material onto the frontal surface of the filaments (references in Jørgensen, 1990). Ward *et al.* (1998)

contend, however, that it is the gill filaments themselves that are the main capture units, and not the latero-frontal cirri. They argue that the latter still play an important role in particle capture by producing vortices that redirect particles and flow away from the inter-filamentary spaces towards the frontal surfaces of filaments. Rather than acting as mechanical sieves (see above) the latero-frontal cirri may act as solid paddles and function in a manner very similar to the fine setules of small aquatic organisms.

The role of mucus in filter feeding has been the subject of considerable controversy for decades, the debate being whether its role is in feeding or just in the cleaning of feeding surfaces, i.e. production of pseudofaeces. A variety of different techniques have clearly demonstrated that mucus plays a key role in all aspects of particle processing, and that different types of mucus are used for different functions on gills, labial palps, mouth and mantle (Beninger & St-Jean, 1997a). Generally, high-viscosity mucus is used for particle transport on an exposed surface, or on an enclosed surface leading to an exposed surface; also on rejection pathways as well as initial transport on the gill for a sub-sequent 'decision' (pertinent only to mussels and clams with a single ventral particle groove). Transport in these cases is counter to the current flow and so highly viscous mucus greatly facilitates particle transport. ('Counter' is defined as an angle approximately 0–90° with the current flow.) Low-viscos-ity mucus is for particles destined for ingestion and moving with the current flow. Although mucus is an important element in suspension feeding there are no data on the energetic cost of its production. However, the costs are unlikely to be high since mucus produced in feeding is either reabsorbed or ingested, and the animal is generally in adequate energy balance to support the costs of mucus produced for rejection (Beninger *et al.*, 1993).

The labial palps and pseudofaeces production and transport

The labial palps are paired, fleshy, leaf-shaped structures on either side of the mouth. The inner surface of each palp faces the gill, and is folded into numer-ous ridges that carry a complicated series of ciliary tracts and various types of mucocytes. The outer surfaces of the palps are smooth, and between the inner and outer surfaces there is muscular-connective tissue. The palps may reach posteriorly for some one third to half of the length of the gills.

Bivalve gill type dictates the nature of particle processing on the labial palps. On the homorhabdic filibranch gill all captured particles move ventrally into the ventral particle groove and proceed anteriorly embedded in a viscous mucus cord towards the palps. The outer and inner palps enclose the anterior gill region (Fig. 4.10); the outer demibranch is applied to the ridged surface of the outer palp and the inner demibranch to the ridged inner palp surface. The gill rests against a specialised region of the palp called the dorsal fold. This is a smooth, non-ridged flap of densely ciliated tissue that covers approx-imately half of the ridged surface of each palp (Fig. 4.11). Its ventral margin is unattached to the underlying ridges, forming an antero-posterior ciliated tract called the palp particle groove. This part of the dorsal fold is contrac-tile allowing it to cover or expose areas of the ridged palp surface. If the ingestive handling capacity of the gill is not overloaded, a mucus–particle cord

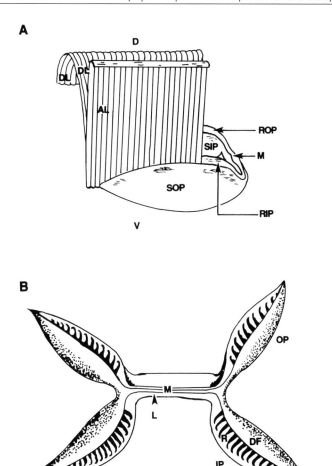

Fig. 4.10. Arrangement of the gill, labial palps and mouth in *Mytilus edulis*: A. AL ascending lamella of gill; D dorsal; DL descending lamella; M mouth; RIP ridged surface of inner palp; ROP ridged surface of outer palp; SIP smooth surface of inner palp; SOP smooth surface of outer palp; V ventral. B. DF dorsal fold; IP inner palp; L lip; M mouth; OP outer palp; R ridges; From Beninger *et al.* (1995).

arriving in the gill particle groove is pulled out by cilia on the dorsal fold and directed ventrally to the palp particle groove. The mucus-particle cord is subjected to mechanical stress through kneeding of palp ridges in the palp particle groove and to biochemical stress through mucus secretion (Beninger & St-Jean, 1997b). Consequently, the cord disintegrates and the material becomes less viscous and more flocculent. Some of this material travels anteriorly, presumably for ingestion, while some enters the palp troughs, exiting at the palp ventral margin for eventual rejection. When the ingestive handling capacity is overloaded the dorsal fold retracts, thus preventing the mucus cord from entering the palp particle groove. Instead the cord is moved to the ventral margins of the palps for rejection. Thus, the dorsal fold, which is unique to homorhabdic filibranchs, determines whether material from the gill is destined for ingestion or rejection.

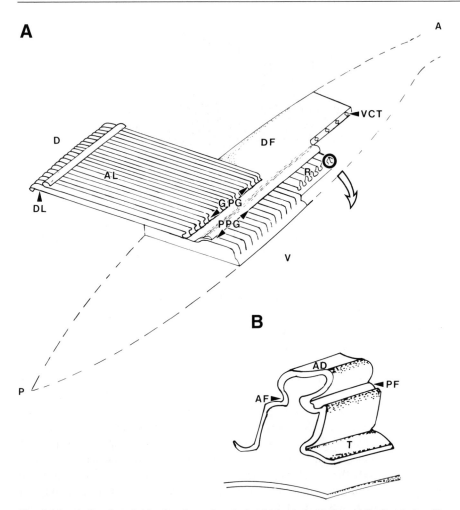

Fig. 4.11. A. Drawing of ridged surface of a single labial palp in *Mytilus edulis*: A anterior; AL ascending lamella of gill; D dorsal; DF dorsal fold; DL descending lamella of gill; GPG gill particle groove; P posterior; PPG palp particle groove; R ridges; V ventral; VCT vestigial ciliated tracts (derived from the fusion of the epithelium of the palp crests with that of the dorsal fold). B. Drawing of a single palp ridge; AD apical depression; AF anterior fold; PF posterior fold; T trough. From Beninger *et al.* (1995).

In eulamellibranchs, however, the palps do not have a dorsal fold and the inner and outer palps of each pair are fused along their dorsal margins. The inner and outer palps are thus a functional unit with the ridged surface of the inner and outer palp accepting material from the inner and outer demibranchs, respectively. The eulamellibranch gill is similar to the homorhabdic filibranch gill in that all captured particles move into the ventral particle groove and proceed embedded in a viscous mucus cord towards the palps (Fig. 4.12). There the cord detaches from the gill onto the ridged surface of the palps. Material for ingestion passes over the palp crests at right angles to the ridges (Fig. 4.13A). *En route* it is subjected to mechanical forces applied by the apposition and grinding motion of both palps. Consequently, the integrity

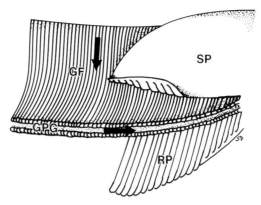

Fig. 4.12. Position of the eulamellibranch gill and labial palps in *Mya arenaria*: arrows indicate direction of mucus-particle masses along gill filaments (GF) and in gill particle groove (GPG); RP ridged palp surface; SP smooth palp surface. From Beninger *et al.* (1997b)

of the cord is destroyed and all material in the oral region is in the form of a mucus-particle slurry (Ward *et al.*, 1994; Beninger *et al.*, 1995). After a prolonged period of feeding mucus particle cords from the gill are instead transported obliquely over the palp crests to the palp ventral margin for rejection as pseudofaeces (Fig. 4.13B).

To date particle selection on either filibranch or eulamellibranch labial-palps has not been described in any detail. However, Ward *et al.* (1994) have given a very comprehensive account of the process in the pseudolamellibranch labial palps of the oyster, *Crassostrea virginica*. Labial-palp structure in pseudolamellibranchs is similar to that in eulamellibranchs. As described above, particles on the gills of pseudolamellibranchs are transported in the dorsal and ventral grooves towards the labial palps. Particles in the dorsal tracts are transported in a slurry towards the gill–palp junction. They leave this junction by one (or more) of four routes (Fig. 4.14):

1 into the oral groove on the palp, and onwards to the mouth;
2 onto the ridged sorting surface of the palp;
3 onto the smooth ciliated tract on the palp edge; and
4 onto the anterior margin of each demibranch for posterior transport.

Cilia on the palp ridges are organised into different functional tracts: rejection tracts that transport particles to the palp margins, oral acceptance tracts that transport particles towards the mouth, and resorting tracts that either reject or accept particles. The route that particles take is highly dependent on ambient particle concentration. Under most conditions particles are transported along the oral groove, or onto the ridged sorting surfaces of the palps. However, at high particle concentrations the rate of particle capture by the gills exceeds the animal's ingestive capacity, or the sorting capacity of the palps, and consequently the gill–palp junction becomes overloaded. Then, particles are rapidly removed via the smooth ciliated tract from where they are either carried onto the ridged palp surface, or rejected off the edge of the palp as pseudofaeces.

A

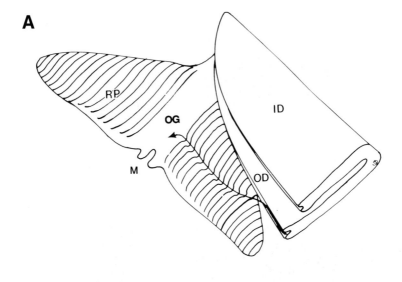

B

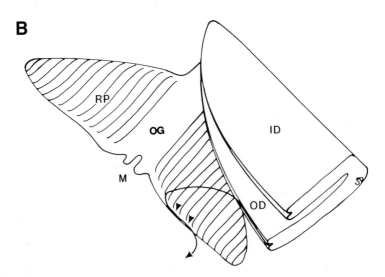

Fig. 4.13. Drawings of labial palps and gill of *Mya arenaria* showing mucus-particle trajectories during ingestion and rejection. Labial palps are spread to reveal ridged surface. A. Direction of mucus-particle masses destined for ingestion along ridged palp surface (RP). B. Direction of rejected mucus-particle masses along ridged palp surface. Arrowheads represent small mucus-particle masses exiting troughs and joining mucus cord on the palp ventral margin. ID inner demibranch; M mouth; OD outer demibranch; OG oral groove. From Beninger *et al.* (1997b).

When particle concentrations increase even more, particles are removed from the gill–palp junction onto the ventral ciliated indentation (continuation of ventral groove) on the anterior margin of each demibranch.

Particles in the ventral grooves are carried anteriorly in mucus strings until they reach a point where anteriorly directed ciliary action stops. At these

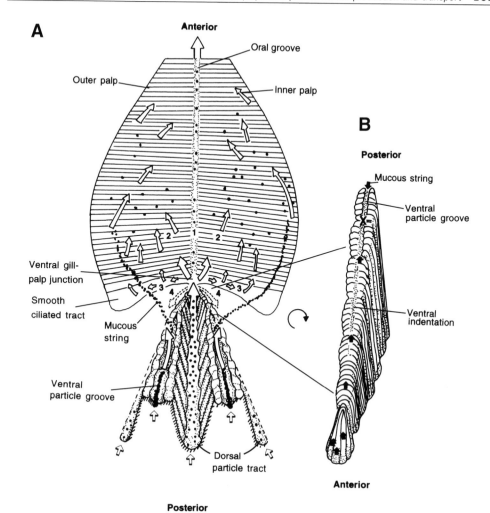

Fig. 4.14. Schematic diagram of the anterior portion of two demibranchs of *Crassostrea virginica* and their junction with one pair of labial palps. (A) The labial palps are folded open (not real-life position) to reveal the general direction of particle movement on the smooth ciliated tract and the ridged sorting surface. The movement of particles bound in two mucus strings from the ventral food grooves onto the inner surface of the palps is illustrated, and open arrows indicate the subsequent direction of movement of these particles (solid circles) on the ridged palp surface. The viscosity of the mucus strings is reduced by the mechanical action of the ridged palp surface so that the entrapped particles are dispersed and subsequently sorted. Solid circles and short open arrows indicate the movement of particles from the basal food tracts towards the gill-palp junction. These particles may either move directly into the oral food groove (marked 1), onto the palps (marked 2), or initially along the smooth ciliated tract and then onto the palps (marked 3). Transport of material away from the gill-palp junction via the ventral indentation on each demibranch is shown by broken arrows (marked 4). (B) An enlargement of the anterior termination of one demibranch illustrating how the ventral food groove becomes narrower and shallower until it forms the ventral indentation tract. Note that the orientation of (B) is opposite to (A) to more clearly illustrate the anterior end of the demibranch. Small black arrows indicate the transport of particles posteriorly, away from the gill-palp junction. X marks the point where these posteriorly beating cilia meet the anteriorly beating cilia in the main ventral groove. From Ward *et al.* (1994).

points (marked X in Fig. 4.14B) the mucus-bound particle strings coalesce forming balls with any particles being moved posteriorly along the ventral indentation. If large enough, these balls are picked up by the highly mobile palp tips and transported onto the ridged palp surface, resulting in a continuous particle string from the gills to the palps. Alternatively, the balls are sometimes swept off the gill by water currents, and rejected as pseudofaeces. The mucus particle strings are fluidised in a manner similar to that already described above. The particles released from the mucus strings, together with those from the dorsal grooves, are sorted on the palp ridges, and then either enter the oral groove, or are rejected off the palp. At high particle concentrations mucus strings from the ventral food grooves may not enter the palps at all but may be rejected at the gill-palp junction and combine with other rejected material to form pseudofaeces.

In heterorhabdic filibranchs, e.g. the scallop *Placopecten magellanicus*, particles destined for ingestion travel along dorsal tracts on the gill, whereas the ventral tract is concerned solely with material destined for rejection. The ventral tract is not a groove, so pseudofaeces can be dislodged at any point along its course (Beninger *et al.*, 1999). The mantle, therefore, does not possess ciliated rejection tracts, and pseudofaeces is expelled entirely by periodic valve clapping. Material arriving at the palps from the dorsal tracts is in the form of a low-viscosity slurry. When the ingestive capacity of the scallop is not overloaded this material moves to the mouth. When overloading occurs this material is rejected off the palps, and is transferred to the mantle until it is ejected at the next valve-clapping episode. Particle processing on the gill and labial palps in the four major gill types described above is summarised in Table 4.3.

With the exception of heterorhabdic filibranchs (see above) pseudofaeces is transported along well-defined rejection tracts on the ciliated mantle surface to the inhalant opening or siphon and periodic, sudden, and forceful closure of the shell valves ensures that it is carried away from the animal in the exhalant current. In bivalves with a gill ventral particle groove, i.e. homorhab-

Table 4.3. Summary of particle processing on the gills and labial palps of the four main gill types.

Gill type	Species	Function of		
		Gill	Labial palps	Mantle
Homorhabdic filibranch	*Mytilus edulis*	Indiscriminate transport; ventral groove only	Sorting and rejection	Rejection
Heterorhabdic filibranch	*Placopecten magellanicus*	Transport for ingestion (dorsal tracts) and pseudofaeces rejection (ventral tracts)	Sorting and rejection	No rejection function
Eulamellibranch	*Spisula solidissima, Mya arenaria*	Indiscriminate transport; ventral groove only	Sorting and rejection	Rejection
Pseudolamellibranch	*Crassostrea virginica*	Indiscriminate transport; ventral and dorsal grooves	Sorting and rejection	Rejection

dic filibranchs and eulamellibranchs, the tracts are characterised by composite cilia that are extraordinarily long compared to those on the general mantle surface (Beninger & Venoit, 1999; Beninger *et al.*, 1999). These effectively elevate and isolate the pseudofaeces from the rest of the mantle surface. In addition, high-viscosity mucus secreted in the tracts firmly binds the rejected material and anchors it to the cilia, thus ensuring counter-current transport to the inhalant siphon (see Figs. 4.15 & 4.16). In contrast, transport of pseudofaeces in pseudolamellibranchs is effected by short simple cilia on the top of specialised radial ridges on the mantle (Fig. 4.15[4A]). It is the ridges, rather than specific cilia types, that elevate the pseudofaeces from the general mantle surface. In all species with mantle rejection tracts, the tracts are positioned away from the gill ventral particle groove (Beninger *et al.*, 1999). This is to ensure that the pseudofaeces travelling in an antero-posterior direction does not entangle with the mucus strings moving in the opposite direction along the ventral gill particle groove. In eulamellibranchs, which possess small gills and a ventral particle groove that only extends one-third to one-half the distance between the hinge and the ventral margin of the shell, the tracts are situated in the most ventral part of the mantle (Fig. 4.15[1A & 2A]). In *Mytilus edulis*, however, where the gill extends almost to the mantle edge the tracts are situated along the middle of the mantle (Fig. 4.15[3A]). In pseudolamellibranchs where under high particle load mucus strings may be ejected from the gill ventral particle groove at any point, the radial arrangement of tracts over the entire inhalant portion of the mantle facilitates the rejection process (Fig. 4.15[4A]).

Ingestion volume regulation

Because the gills have both a respiratory and feeding role it is necessary that bivalves possess mechanisms that can uncouple these two functions. Beninger *et al.* (1992) refer to this as 'ingestion volume control'. In homorhabdic filibranchs, e.g. *Mytilus edulis*, all particles captured on the gill are carried forward to the labial palps along the ventral gill particle groove, even when particle concentrations are high (Beninger *et al.*, 1995). The only structures that are capable of producing pseudofaeces, and thus assisting in ingestion volume control, are the labial palps. In *Mytilus* there is evidence of pre-ingestive selection on the gill (see later). Eulamellibranchs also carry all captured particles on the gill forward to the labial palps, but these bivalves are able to close the ventral gill particle groove after prolonged exposure to a range of particle concentration, thus effectively shutting down the feeding function of the gill (Beninger *et al.*, 1997b). A similar mechanism operates in the pseudolamellibranch, *Crassostrea virginica* (Ward *et al.*, 1994). However in both eulamellibranchs and pseudolamellibranchs the labial palps are the main site of ingestion volume control. In contrast, the gills rather than the labial palps are the major site of ingestion volume control in heterorhabdic filibranchs (see above). In eulamellibranchs the lips are an additional site for ingestion volume control. Excess or rejected material from the palps accumulates in the region of the mouth and is eventually passed back onto the palps for rejection (Beninger *et al.*, 1997b).

Pre-ingestive particle selection

Bivalve food consists of a variety of suspended particles such as bacteria, phytoplankton, microzooplankton, detritus, but also dissolved organic material (DOM) such as amino acids and sugars (see later). However, this nutritious material is in suspension along with silt that is usually present in much higher

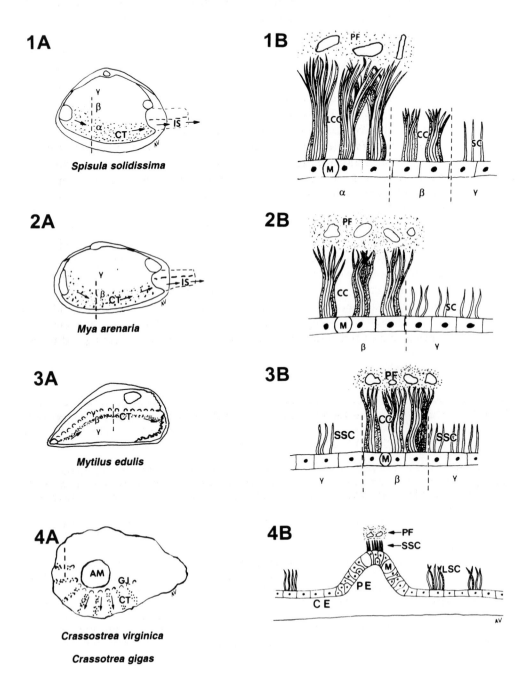

1A *Spisula solidissima*

2A *Mya arenaria*

3A *Mytilus edulis*

4A *Crassostrea virginica*

Crassotrea gigas

concentrations than the food particles. To compensate for the dilution of organic material in suspension bivalves are able to preferentially select nutritive particles, such as algae, and reject particles of poor nutritive value in their pseudofaeces (Hawkins *et al.*, 1996, 1998a). Kiørboe & Møhlenberg (1981) have determined the efficiency with which ten species of filter-feeding bivalves from temperate latitudes separated algae from silt when the algae and silt concentrations were 10–30 cells μl^{-1} and 10–20 mg l^{-1} respectively. Selection efficiency was expressed as the ratio of chlorophyll *a* in the suspension offered and pseudofaeces rejected by undisturbed filtering bivalves. This ratio varied considerably and was significantly correlated with the size of the labial palps (Fig. 4.17). In addition, intra-specific differences in selection efficiency and palp size were observed when populations of mussels (*Mytilus edulis*), exposed to differing silt concentrations, were compared.

Bivalves are also capable of discriminating between similar-sized algae cells in their diet (Shumway *et al.*, 1985). When mixed cell suspensions of the dinoflagellate *Prorocentrum minimum*, the diatom *Phaeodactylum tricornutum* and the crypyomonad flagellate *Chroomonas salina* were fed to bivalves, the oyster *Ostrea edulis* preferentially cleared *Prorocentrum minimum*, while the clam *Ensis ensis* and scallop species *Placopecten magellanicus* and *Arctica islandica* consistently rejected this species. Shumway *et al.* (1997) have reported similar results for other species of scallop. The site of selection differed between the different species. In *O. edulis* preferential selection took place on the gills, while in *E. ensis, P. magellanicus* (but see above) and *A. islandica* the labial palps were the site of pre-ingestive selection. Preferential selection on the gills has also been shown for other oyster species, but in the mussel *Mytilus trossulus* it is the labial palps that play the major role in particle sorting (see above and Ward *et al.*, 1998). The ability of bivalves to distinguish between different food types in mixed suspension could be a means whereby co-existing and poten-

---

Fig. 4.15. Schematic drawings of the distribution and types of cilia involved in pseudofaeces transport on the mantle of *Spisula solidissima*, *Mya arenaria*, *Mytilus edulis*, *Crassostrea virginica* and *C. gigas*. The first three species possess a ventral gill particle groove and siphons, while the last two possess both ventral and dorsal particle tracts but no siphons.

Spisula solidissima 1A: three ciliary bands are arranged in a ventro-dorsal sequence: α, long composite cilia (LCC); β, composite cilia (CC); γ, simple cilia (SC). Arrows show direction of pseudofaeces transport; dashed line shows plane of section for 1B in which transport of pseudofaeces (PF) atop the LCC cilia of the mantle (M) rejection tract is illustrated.

Mya arenaria 2A: two ciliary bands are found: β (CC) and γ (SC). Arrows show direction of pseudofaeces transport; dashed line shows plane of section for 2B in which transport of pseudofaeces atop the CC cilia of the rejection tract is illustrated.

Mytilus edulis 3A: three parallel ciliary bands extend dorso-ventrally: short simple cilia (γ), compound cilia (β) and short simple cilia (γ). Arrows show direction of pseudofaeces transport; dashed line shows plane of section for 3B in which transport of pseudofaeces atop the CC cilia of the rejection tract is illustrated.

Crassostrea virginica, C.gigas 4A: Ciliated tracts of mantle pseudofaeces rejection ridges, showing direction of pseudofaeces transport (arrows). AM, adductor muscle; GJ, gill junction with mantle. Dashed line shows plane of section for 4B in which short simple cilia (SSC) on rejection ridge, and isolated tufts of long simple cilia (LSC) in inter-ridge region are illustrated. From Beninger & Venoit (1999).

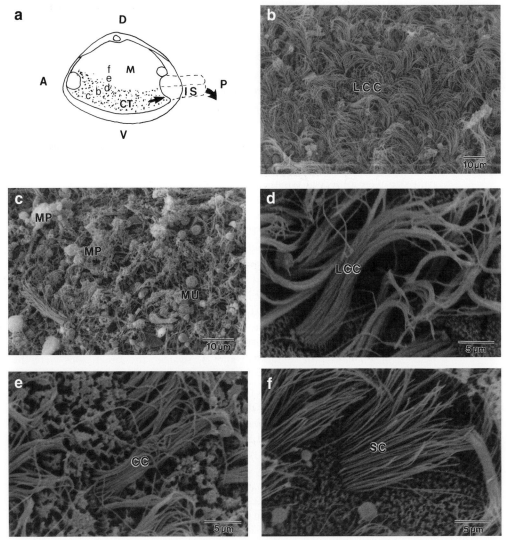

Fig. 4.16. Mantle ciliation in *Spisula solidissima*. (a) Location of ciliated mantle rejection tract (CT) in relation to general mantle surface (M) and inhalant siphon (IS). A, D, P, V: anterior, dorsal, posterior and ventral orientations, respectively. Specimen oriented as in subsequent micrographs. Lower case letters (b to f) designate locations of corresponding SEM; large arrows show direction of pseudofaeces transport. (b) Dense long composite cilia (LCC) cover of mantle rejection tract. (c) Mucus-particle masses (MP) and mucus balls (MU) representing dehydrated residue of mucus-particle raft characteristic of mucociliary transport in ciliated mantle rejection tract. (d) Detail of cilia within ciliated mantle rejection tract, showing them to be of the long composite cilia type (LCC). (e) Detail of cilia on the intermediate band of the general mantle surface, showing them to be of the composite cilia type (CC). (f) Detail of cilia in the dorsal region of the general mantle surface, showing them to be of the simple cilia (SC) type. From Beninger *et al.* (1999).

tially competing species partition the available food resource (Shumway *et al.*, 1990).

At low particle concentration all particles filtered from suspension may be ingested. However, above a certain threshold concentration, which corresponds

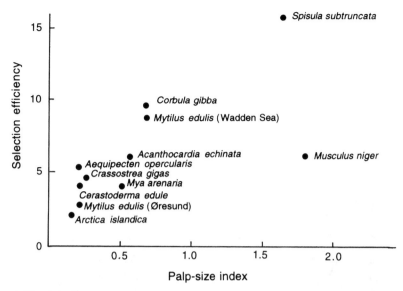

Fig. 4.17. Plot of selection efficiency (relative chlorophyll *a* concentration in inhalant water/relative chlorophyll *a* concentration in pseudofaeces) against palp-size index (palp area relative to clearance) in bivalves filtering a mixture of algae and silt. Wadden Sea *Mytilus edulis*, which are exposed to high concentrations particulate matter, have larger labial palps than mussels from Øresund, where the concentration of particulate matter is considerably lower. From Jørgensen (1990) from data in Kiørboe & Møhlenberg (1981).

with saturation of the animal's digestive capacity, ingested ration remains approximately constant and the surplus filtered material is rejected as pseudofaeces. The threshold concentration depends on the species, body size of the individual and on the particle type being ingested. In general, threshold concentrations lie between 1–6 mg seston (dry weight) per litre (Bayne & Newell, 1983). There is also a maximum value to the proportion of particles rejected; Bayne and Newell (1983) estimate this as 80–95%. The proportion rejected (R) as pseudofaeces can be estimated from:

$$R = R_{\max} \left[1 - e^{K\,(C-T)} \right]$$

where C is the concentration of dry particulate matter ($mg\,l^{-1}$), T is the pseudofaeces threshold concentration, $R_{\max}$ is the maximum proportion rejected, and K is a fitted parameter. According to Bayne *et al.* (1976b) T and K are related to body weight as follows:

$$K = 0.25\,W^{-0.40}$$
$$T = W^{0.40}$$

The dry matter ingested can then be calculated by subtracting the dry matter rejected from the dry matter filtered. Knowing the organic content of the ingested ration the organic matter intake per day can be calculated. However, not all of the organic material ingested is assimilated into the body (see later).

The alimentary canal and digestive process

The bivalve digestive system consists of a mouth, oesophagus, stomach, digestive gland, intestine and anus. For details on the anatomical structure of these organs see Chapter 2.

Mouth and oesophagus

Particles in suspension enter the mouth from an oral groove at the base of the labial palps. The mouth, and the oesophagus which leads into the stomach, have a ciliated epithelial lining which is well supplied with mucocytes that secrete both acid and neutral mucopolysaccharides, even when the animal is not feeding (Beninger *et al.*, 1991). The oesophagus does not have a digestive function, merely serving to propel material along ciliated tracts towards the stomach.

The stomach and extracellular digestion

The stomach is a flattened sac into which the oesophagus opens at the anterior end and from which the midgut leaves at the posterior end (Fig. 2.13). The crystalline style projects from the posterior end of the stomach across the floor of the stomach to rest against the gastric shield. The digestive gland opens into the stomach by several ducts. Extending from the duct openings are ciliary tracts that traverse the floor of the stomach to the intestinal groove. Purchon (1957) presents very detailed information on stomach structure in a wide variety of bivalves.

The co-ordinated ciliary beat of the stomach and intestinal epithelia cause the crystalline style to rotate against the gastric shield. As it does so it is abraded and dissolved, releasing digestive enzymes in the process. Ingested particles are mixed with the liberated digestive enzymes from the crystalline style. The rotation of the style helps this mixing process to take place. During the mixing and extracellular digestive processes the stomach contents come under the influence of ciliary tracts that cover large areas of the stomach. These tracts have fine ridges and grooves and act as sorting areas. Finer particles and digested matter are kept in suspension by cilia at the crests of the ridges, and this material is continually swept towards the digestive gland duct openings. Larger particles, and also small dense particles such as sand grains, are segregated out and channelled into the intestine along a deep rejection groove on the floor of the stomach.

The crystalline style is secreted by the stomach and is largely composed of protein, at least some of which is bound to sugars such as glucose, mannose, and galactose. But how these substances form the solid gel of the style is not yet known (Beninger & Le Pennec, 1991). The style sac secretes a variety of enzymes and these are incorporated into the crystalline style as it is being secreted. The stomach wall and gastric shield are also important in digestive enzyme secretion, and it is also likely that some enzymes from the digestive gland enter the lumen of the stomach and act extracellularly.

The following enzymes have been reported from bivalve stomachs (Reid, 1968; Mathers, 1973a,b; additional references in Bayne *et al.*, 1976a): esterases that are important in lipid digestion; acid and alkaline phosphatases that are believed to play a role in absorption and phagocytosis of material from the stomach; and endopeptidases, e.g. trypsin, that break down proteins. High to moderate activity of carbohydrate splitting enzymes such as α-amylase, α- and β-glucosidase, β-galactosidase, maltase, chitinase and cellulases have also been reported. Chitin is a structural component in diatoms and cellulose is a common component of algal cell walls. Not, surprisingly, the activity of cellulases is particularly high in bivalve stomachs.

When bivalves are actively feeding the pH of the style is about 6.0, but when food is not being ingested the pH is about 6.5. The pH of the stomach contents may fall as low as pH 5.5, due to the dissolution of the style and to acid secretions from the digestive gland (Owen, 1974).

The crystalline style is not a permanent structure in the stomach. The size of the style has been reported to vary systematically with the tidal cycle. The maximum size corresponds to the time when there is most food in the stomach, i.e. at high tide when the animal is feeding. Morton (1971) found that in the oyster, *Ostrea edulis*, the volume of the crystalline style decreased from 25 mm³ to 3–4 mm³ during each tidal cycle, although Langton (1972) reported a much smaller decrease in volume for the same species. Alternatively, Morton (1970 & 1971) has suggested that feeding in bivalves is rhythmic and correlated with this could be a rhythmic digestion pattern. He proposed that feeding occurs over the high-tide period and during the period of ebb and low tide extracellular digestion in the stomach takes place. Ingested material is not passed to the digestive gland until the next high-tide period. However, the evidence for rhythmic digestion has not been forthcoming since other investigators have since shown that material once ingested is rapidly passed to the digestive gland for intracellular digestion (see Owen, 1974 for references).

The digestive gland and intracellular digestion

The principal (primary) ducts of the digestive gland branch into smaller secondary ducts that ultimately lead into blind-ending digestive tubules (see Fig. 2.14). These tubules are of two cell types, digestive cells and basophil secretory cells. Digestive cells are the most abundant type. They are columnar and vacuolated and are responsible for intracellular digestion of food. The free surface of the digestive cell is extended into microvilli and the cytoplasm is characterised by the presence of numerous cytoplasmic vesicles types (Fig. 4.18A). Particulate matter is taken up at the base of the microvilli by pinocytosis and is initially stored within large vesicles called phagosomes. Digestion takes place within large vesicles called lysosomes that contain hydrolytic enzymes. The end products of digestion are released directly into the haemolymph system and waste products are contained in residual bodies within the digestive cells. The cells eventually rupture and the waste material is swept along the ciliated secondary and primary ducts of the digestive gland towards the stomach, and ultimately to the intestine. It is not known which cells, if any, act as the stem cells of digestive cells.

Secretory cells are pyramidal in shape (Fig. 4.18B) and like digestive cells they also have microvilli on the free surface. Much of the cytoplasm is filled

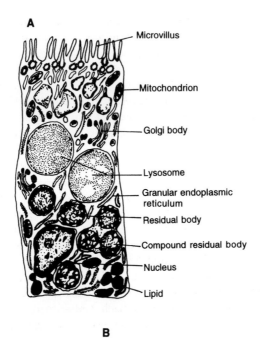

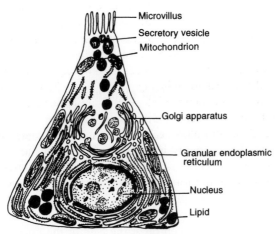

Fig. 4.18. Diagram of a bivalve (A) digestive and (B) basophil secretory cell. From Owen (1972).

with rough endoplasmic reticulum and Golgi bodies, which indicates that these cells play an extensive role in protein synthesis. The exact role of secretory cells is unclear; they may play a role in enzyme secretion for extracellular digestion (Weinstein, 1995). A third cell type, the flagellated basophil cell, has been identified and described in the oyster *Crassostrea virginica* (Weinstein, 1995). This cell is columnar and possesses a single long flagellum, but lacks the conspicuous organelles found in digestive and secretory cells. Although flagellated cells are as numerous as secretory cells their role in digestion has not so far been elucidated. Lack of mitotic activity in both types of basophil cell makes it unlikely that they serve as stem cells for digestive cells (Weinstein, 1995).

It is clear from the above that the digestive gland cells have a finite life span after which they disintegrate and are replaced. Various workers have suggested that there is a well-defined rhythm to the cycle of changes taking place in the digestive tubules, and some have linked this with tidal periodicity (Owen, 1972; Mathers, 1976; additional references in Beninger & Le Pennec, 1991). Owen (1972) has proposed that at low tide, when food is not available, the tubules are in 'holding phase'. When the animals are covered by the tide and feeding starts, food material rapidly reaches the tubules and is taken up by endocytosis for intracellular digestion. Breakdown and regeneration of the digestive cells then follows and the cells are then in the 'holding phase' for the next high tide. However, in many species no such synchronised structural changes have been observed (Langdon & Newell, 1996).

Mathers (1972) has shown that when ^{14}C-labeled algae cells were fed to oysters, within 10 min of commencement of feeding the cells had been broken down by extracellular digestion in the stomach, passed to the digestive gland, and the breakdown products had been phagocytosed by digestive cells. After 90 min waste material appeared in the ciliated gutter of the primary ducts, thus lending support for a two-way movement of material in the primary ducts of the digestive gland.

A wide range of enzymes has been recorded from the digestive tubules of bivalves (Table 4.4). Much of the work on both stomach and digestive gland enzymes has been either qualitative or semi-quantitative providing little indication of levels of enzyme activity. Recently, however, some carbohydrate-splitting enzymes and proteases have been purified and characterised from the digestive gland of *Perna viridis* and *Pecten maximus* (see Table 4.4 for references). The substrates for carbohydrate-splitting enzymes have been identified from a range of algae, fungi and bacteria (Kristensen, 1972).

The digestive gland also plays an important role as a storage site for carbohydrate and lipid. These energy reserves, mainly stored during the summer, are used to fuel gametogenesis during the autumn and winter (see Chapter 5), but can also be utilised during periods of physiological stress (Bayne *et al.*, 1976a).

The intestine

Rejected particles from the stomach as well as waste material from the digestive gland pass into the intestine. The rejected particles are not necessarily lost since wandering haemocytes that migrate freely into the gut from the haemolymph stream may phagocytose them, digest and transport them round the body. These cells, which also phagocytose diatoms and other food particles in the stomach, are, therefore, important in intracellular digestion and also play a unique role in the transport of food. That the intestine might play a role in enzymatic digestion was first mooted in the 1950s (Le Pennec *et al.*, 1991 for references). Mathers (1973a,b) reported the following enzymes from

Table 4.4. Carbohydrate, fat and protein-splitting enzymes in the digestive gland of bivalves.

Enzyme	Substrate	Reference
α-amylase	Starch, glycogen	Teo & Sabapathy, 1990*; Sabapathy & Teo, 1992*
Cellulases	Cellulose	Crosby & Reid, 1971; Teo & Sabapathy, 1990
Laminarinase	Laminarin	Wojtowicz, 1972
Maltase Sucrase	Maltose Sucrose	Mathers, 1973a; Teo & Sabapathy, 1990; Teo *et al.*, 1990*
Trehalase	Trehalose	Teo & Lim, 1991*, 1993*
α-glucosidase	Sucrose, maltose, trehalose	Teo & Sabapathy, 1990
β-glucosidase	Cellobiose	Kristensen, 1972
β-N-acetyl- glucosaminidase	Chitobiose	Sumner, 1969
Lipase	Large fat molecules	Mathers, 1973b
Esterase	Small fat molecules	Mathers, 1973b
Chymotrypsin-like	Protein, peptides	Teo & Sabapathy, 1990; Le Chevalier *et al.*, 1995*
Trypsins Cathepsins	Peptides Peptides	Reid, 1968; Reid & Rauchert, 1970, 1972, 1976

* Studies in which biochemical characterisation has been carried out.

the mid–gut wall (first part of intestine) of *Ostrea edulis* and *Crassostrea angu-lata*: α–amylase, maltase, trehalase, cellobiase and various glucosidases. Esterases, acid and alkaline phosphatases, and endopeptidases have also been reported from the mid–gut of a range of bivalves (Reid, 1968).

It is interesting to note that in the evolution of bivalves there is an increas-ing tendency for digestion to become extracellular, and for the intestine to function in digestion and absorption. Extracellular digestion reduces the cell volume that must be devoted to intracellular digestion, and also reduces the time taken to obtain absorbable products from the diet. The mid–gut func-tioning as a region of digestion and absorption allows more efficient utilisa-tion of food, since under optimum feeding conditions much of the ingested material is passed undigested through the stomach to the intestine (Reid, 1968). The intestine terminates in an anus and faeces in the form of faecal pellets are swept away through the exhalant opening.

Food

In the wild bivalves feed on a variety of suspended particles (seston) such as bacteria, phytoplankton, micro–zooplankton, detritus, but also on dissolved organic material (DOM), such as amino acids and sugars. Various studies have shown that there is both spatial and seasonal variation in the quantity and quality of seston in coastal waters. Bayne & Hawkins (1990) have estimated that the quantity commonly varies from <3 to >100 total dry $mg^{-1}l^{-1}$, of which 5–80% by mass may be organic. Despite such variability bivalves maintain relatively constant rates of nutrient acquisition by the various compensatory adjustments discussed below.

Information on the diet of bivalves has been obtained from examination of gut contents, by measuring isotopic enrichment or depletion within tissues (Bayne & Hawkins, 1992), or by analysing digestive and absorptive efficiencies (see below). Shumway *et al.* (1987) examined the gut contents of *Placopecten magellanicus* throughout the year from a shallow water (~20 m depth) and a deep water (~180 m depth) location (Tables 4.5 & 4.6). A total of 27 species of algae, ranging in size from about 10–350 μm, were identified, as well as considerable amounts of detritus and bacteria. Benthic and pelagic food species were equally well represented in shallow water scallops, but benthic species, as might be expected, outnumbered pelagic ones in the deeper-water population. Seasonal variations of food items occurred and coincided

Table 4.5. Gut contents of inshore scallops, *Placopecten magellanicus*. Adapted from Shumway *et al.* (1987). Category B = Bacillariophyceae (diatoms); category D = Dinophyceae (dinoflagellates). Benthic (Bn) or pelagic (P) habitat of food species indicated.

Species	Category	Habitat	Size (μm)
Nitzschia spp.	B	Bn	25–150
Navicula spp.	B	Bn	8–240
Pleurosigma sp.	B	Bn	200
Thalassiothrix sp.	B	Bn	50 (chain)
Amphora sp.	B	Bn	10–30
Licmophora sp.	B	Bn	25–180
Acnanthes sp.	B	Bn	40–90
Pinnularia sp.	B	Bn	40–80
Surirella sp.	B	Bn	15–25
Cylindrotheca closterium	B	Bn	80–100
Protogonyaulax resting cyst	D	Bn	35–40
unidentified cyst		Bn	25–35
Melosira sp.	B	Bn	30–55 (chain)
Striatella sp.	B	Bn	40–50
Coscinodiscus spp.	B	Bn/P	40–180
Ditylum brightwellii	B	P	50–150
Protoperidinium sp.	D	P	60–70
Eucampia zoodiacus	B	P	40–75 (chain)
Peridinium sp.	D	P	20–30
Prorocentrum micans	D	P	45–55
Skeletonema costatum	B	P	30–50 (chain)
Dinophysis acuminata	D	P	50–60
Dinophysis spp.	D	P	32–60
Thalassiosira rotula	B	P	20 (chain)
Thalassiosira nordenskioldii	B	P	20 (chain)
Thalassiosira spp.	B	P	10–200

Miscellaneous:
Silicoflagellate strew; pollen grains (30–40 μm); green filamentous algae (>1000 μm); ciliates; zooplankton tests; bacteria; detritus; unidentified, non-pigmented still active forms: (3 μm), multiflagellate (10 μm), ciliated mass (40–200 μm).

Table 4.6. Gut contents of offshore scallops, *Placopecten magellanicus*. Adapted from Shumway et al. (1987). Abbreviations as in Table 4.5.

Species	Category	Habitat	Size (μm)
Melosira sp.	B	Bn	50 (chain)
Protogonyaulax resting cyst	D	Bn	35–40
Navicula spp.	B	Bn	60–350
Nitzschia spp.	B	Bn	110
Thalassiothrix sp.	B	Bn	50 (chain)
Acnanthes sp.	B	Bn	40–90
Amphora sp.	B	Bn	10–30
Pleurosigma sp.	B	Bn	280
Licmophora sp.	B	Bn	120–180
Pinnularia sp.	B	Bn	70–100
Surirella sp.	B	Bn	15–25
unidentified dinoflagellate cyst	D	Bn	25–35
Coscinodiscus spp.	B	P/Bn	50–160
Prorocentrum micans	D	P	45–55
Dinophysis spp.	D	P	32–60
Thalassiosira spp.	B	P	35–50
Eucampia zoodiacus	B	P	100 (chain)
Ditylum brightwellii	B	P	150
Ditylum brightwellii resting spore	B	P	40

Miscellaneous:
Pollen grains (30–40 μm); zooplankton tests (100–250 μm); bacteria; detritus; unidentified, non-pigmented still active forms: uniflagellate (3 μm), multiflagellate (10 μm), ciliated mass (70–300 μm).

with bloom periods for the individual algae species. For example, in inshore scallops the most prominent pelagic food items were *Prorocentrum* (October/November; January), *Thalassiosira* sp. (January; March), and *Dinophysis* (October). Variable quantities of detritus were also found in the gut contents of the scallops. When scallops recess into the substrate the inhalant current is lowered to at least the level of the sediment surface and benthic material can be more easily drawn into the mantle cavity (Brand, 1991). Vigorous clapping of the shell valves helps to re-suspend the surface sediments (Davis & Marshall, 1961). The importance of detritus in the diet of scallops has been highlighted by the results of a laboratory-based study on *Placopecten magellanicus* (Cranford & Grant, 1990). When diets of cultured phytoplankton (*Isochrysis aff. galbana* and *Chaetoceros gracilis* Schutt), kelp powder and re-suspended sediment were fed to this species the results demonstrated that, although phytoplankton is the primary food source, detritus particles do contribute to energy gains during periods when phytoplankton is less available. Langdon & Newell (1990) have demonstrated the importance of detritus in the diet of mussels living in a salt marsh creek in Delaware Bay, United States. Detritus from the marsh-grass *Spartina alterniflora* provided about 50% of the carbon and sulphur in tissues of mussels (*Guekensia demissa*) living in the inner parts of the marsh, but less than 40% of those elements for mussels living at the mouth of the creek. *Mytilus edulis*, living further out in more marine conditions, obtained about 30% of its nutrition from material derived from *S. alterniflora*, suggesting export of marsh-grass into Delaware Bay.

There is little information available on the specific food items utilised by other bivalve species in the wild, although it is generally assumed that they rely on phytoplankton from the water column as their main energy source. The extent to which detritus is exploited as a food item has probably more to do with sorting mechanisms on the gill than on the particular life habit of the species concerned. Scallops are capable of ingesting relatively large particles, up to 950 μm (Mikulich & Tsikhon-Lukanina, 1981; see also Tables 4.5 & 4.6), but have poor retention efficiency of particles less than 5 μm, due to under-development of latero-frontal cirri on the gill filaments (see above). In contrast, the mussel *Mytilus edulis* (and many clam species) has well-developed latero-frontal cirri and a retention efficiency of 50% and 90% for particles of 1 μm and 3 μm diameter, respectively (Jørgensen, 1990). Oyster species such as *Ostrea edulis* and *Crassostrea gigas* have short latero-frontal cirri and fall in between mussels and scallops in their retention efficiency of small particles (Jørgensen, 1990).

Exciting new evidence from field and laboratory studies, shows that *M. edulis* can capture and ingest zooplanktonic and benthic animals, e.g. crustacean and bivalve larvae in the 10–1000 μm size range (Davenport *et al.*, 2000). Feeding experiments demonstrated the ability of *M. edulis* to extract energy from the larvae, showing that this species is partially carnivorous. Gastric processing of the prey is rapid (<40 min at 15–20°C), which probably explains why the phenomenon had not been reported previously.

There is increasing interest in the role of bacteria in bivalve nutrition. Bacterial concentrations can be high (4×10^6 bacteria ml^{-1}) and can reach an average of 20% of planktonic primary production (Langdon & Newell, 1990; Prieur *et al.*, 1990). Since most bacteria are smaller than 1 μm the retention efficiency of bivalves may not exceed 20–30%, although some bivalves, such as the mussel *Guekensia demissa* can retain bacteria with more than 85% efficiency. Assuming a 20% retention efficiency for most bivalves and a bacterial production of 20% of phytoplankton production, bacteria probably only meet about 4% of a bivalve's nutritional needs (Prieur *et al.*, 1990). Langdon & Newell (1990) have estimated that bacteria contribute about 4% to the metabolic carbon requirements of the oyster *Crassostrea virginica* while for *Guekensia demissa* the figure is about 25%. Because the C:N ratio of bacteria is lower than that of phytoplankton (3.5 compared with 6.6) bacteria may better meet the nitrogen needs of bivalves. Langdon & Newell (1990) have shown that during the summer the estimated contributions of bacteria in meeting metabolic nitrogen requirements are 17% for *Crassostrea virginica* and 60% for *Guekensia demissa*. However, these authors point out that it is only in certain environments such as eutrophic estuaries, kelp beds and marshes that bacterial concentrations are sufficiently large to contribute significantly to the nutrition of suspension-feeding bivalves. In open coastal and oceanic waters bacteria are generally less abundant and therefore are unlikely to be nutritionally significant for bivalves.

The total mass of dissolved organic matter (DOM) in the oceans is vast, outweighing the total biomass on the earth. There is increasing evidence that bivalves can actively transport DOM across the gills (references in Stewart & Bamford, 1975) and utilise it as a nutritional supplement. Most of the evidence has been concerned with the uptake of free amino acids (FAA) and sugars. Manahan *et al.* (1982) demonstrated that *Mytilus edulis* rapidly removed up to 94% of naturally occurring amino acids from seawater, and they estimated that uptake of amino acids at ambient concentrations in seawater could meet 34% of the metabolic requirements in this species (Manahan *et al.*, 1983). Other studies (cited in Hawkins & Bayne, 1992) have downgraded this figure to as low as 10%. However, the figures indicate that DOM could serve as a potential energy source for bivalves when other sources of nutrition are least abundant.

Another source of nutrition for bivalves, but one that is only exploited by a few species, is the utilisation of the autotrophic pathways of symbiotic algae to provide an additional carbon source. Take for example the giant clam, *Tridacna gigas*, that contains symbiotic zooxanthellae within the mantle tissue, and is the largest and fastest-growing of all bivalves. In small clams (10 mg dry tissue weight) filter feeding provides about 65% of the total carbon needed for respiration and growth, whereas large clams (10 g) acquire only 34% of their carbon from this source (Klumpp *et al.*, 1992). In addition, zooxanthellae also conserve and recycle all nitrogenous end products within *T. gigas*, giving clams a nutritional advantage over non-symbiotic bivalves (Hawkins & Klumpp, 1995; Mingoalicuanan & Lucas, 1996).

Since 1980, the role of endosymbiotic bacteria in the tissues of a specialised group of bivalves that live in sulphide-rich environments has come to our attention. Examples of such environment are deep-sea hydrothermal vents and seeps, and sulphide-rich littoral sediments. Bivalves with endosymbionts rely on them, to varying extents, as a carbon source (Le Pennec *et al.*, 1995 and references). For example, the littoral species *Lucinella divaricata* obtains ~40% of its metabolic requirements from gill endosymbionts, while another littoral species, *Solemya reidi*, living in an extremely rich sulphide habitat, obtains 60–80% from this source and the rest from dissolved organic matter. With increasing reliance on endosymbionts as a carbon source there is a progessive reduction in the size of the labial palps, stomach and digestive gland so that *Solemya reidi* totally lacks a digestive system.

Larval feeding and diet is covered in the hatchery culture section of Chapter 9.

Absorption efficiency

The efficiency by which ingested ration is absorbed is called the absorption efficiency (AE). To estimate the AE of a bivalve for a particular species of alga, the alga is radioactively labelled and the label is counted in the alga, faeces and pseudofaeces. Using ^{14}C-labeling Peirson (1983) found that adult *Argopecten irradians concentricus* absorbed most algal species tested with absorption efficiencies ranging between 78 and 90%. Assimilation efficiencies of 72 and 74% were reported for *Crassostrea virginica* fed on the diatom *Thalassiora pseudonana* and the alga *Isochrysis galbana* respectively (Romberger & Epifanio, 1981). However, it is likely that values between 30 and 60% are more typical for natural seston (Bayne & Newell, 1983). Not all algae species show efficiencies in the above range. For example, in *Argopecten irradians concentricus* Peirson (1983) observed a low AE value of 17% for *Chlorella autotrophica*, and in *Crassostrea virginica*, *Tetraselmis suecica* was digested and absorbed with an efficiency of only 6% (Romberger & Epifanio, 1981). These low AE values have been attributed to the thick indigestible cell walls of these chlorophyte species.

High absorption efficiencies cannot be used as an indicator of food value. A high AE (83%) was reported for *Dunaliella tertiolecta*, an algae known to support poor growth of oysters due to its deficiency in essential fatty acids (Peirson, 1983). In contrast, kelp detritus was absorbed with an efficiency of 70–80% but did not support growth of adult *Placopecten magellanicus* (Grant & Cranford, 1989).

In laboratory experiments using pure cultures of algae cells AE decreases with increasing food concentrations. For example, a four-fold increase in the density of *Thallassiosira weissflogii* from 3 to 12 cells μl^{-1} caused a reduction from 90 to 65% in the AE of *Argopecten irradians* (Kuenster, 1988). In general, AE has been found to increase with the organic content of the ingested matter (Hawkins *et al.*, 1998b). Species such as the mussel *Perna viridis*, that shows pre-ingestive selection of organic material, grows faster than species that are not as well adapted, e.g. the pearl oyster, *Pinctada margaritifera* (Hawkins *et al.*, 1998a).

Absorption efficiency is functionally interrelated with gut capacity, the residence time for food in the gut and the ingestion rate. Altering the rate of ingestion and gut residence time provides a bivalve with a powerful means of adapting to a variable food source (Bayne & Newell, 1983). Active regulation of AE has been confirmed for a number of bivalves. When the mussel *Mytilus edulis*, was fed diets of different organic content those on a diet of low organic content increased their filtration rate, rejected a higher proportion of filtered material as pseudofaeces, and increased the efficiency with which filtered matter of higher organic content was selected (Bayne *et al.*, 1993). In an earlier study Bayne & Hawkins (1990) observed that elevated rates of ingestion were accompanied by unchanging gut passage time so that total gut content increased, helping to maintain organic gut content independent of changes either in food quality or quantity. Mussels that are periodically immersed compensate for reduced feeding time by exhibiting a higher AE than mussels that are permanently covered by water (Charles & Newell, 1997). Physiological changes that result in increased AE are believed to occur over a period of days, whereas changes in filtration rate and rejection of particles as pseudofaeces occur over much shorter time scales (Bayne *et al.*, 1993).

Another mechanism by which AE can be enhanced is through increased synthesis of digestive enzymes. This has the effect of both increasing the rates of enzymatic breakdown of ingested material and creating more space in the gut for ingested material (Bayne *et al.*, 1993; Wong & Cheung, 2001).

References

Bayne, B.L. (1998) The physiology of suspension feeding by bivalve mollusks: an introduction to the Plymouth 'TROPHEE' workship. *J. Exp. Mar. Biol. Ecol.*, **219**, 1–19.

Bayne, B.L. & Newell, R.C. (1983) Physiological energetics of marine molluscs. In: *The Mollusca*, Vol. 4, *Physiology*, Part 1 (eds A.S.M. Saleuddin & K.M. Wilbur), pp. 407–515. Academic Press, New York.

Bayne, B.L. & Hawkins, A.J.S. (1990) Filter-feeding in bivalve molluscs: controls on energy balance. In: *Animal Nutrition and Transport Processes*, Vol. 1. *Nutrition in Wild and Domestic Animals* (ed. J. Mellinger), pp. 70–83. Karger, Basel.

Bayne, B.L. & Hawkins, A.J.S. (1992) Ecological and physiological aspects of herbivory in benthic suspension-feeding molluscs. In: *Plant-animal Interactions in the Marine Benthos*, Systematics Association Special Volume No. 46 (eds D.M. John, S.J. Hawkins & J.H. Price), pp. 265–88. Clarendon Press, Oxford.

Bayne, B.L., Thompson, R.J. & Widdows, J. (1976a) Physiology: I. In: *Marine Mussels: their Ecology and Physiology* (ed. B.L. Bayne), pp. 121–206. Cambridge University Press, Cambridge.

Bayne, B.L., Widdows, J. & Thompson, R.J. (1976b) Physiological Integrations. In: *Marine Mussels: their Ecology and Physiology* (ed. B.L. Bayne), pp. 261–91. Cambridge University Press, Cambridge.

Bayne, B.L., Iglesias, J.I.P., Hawkins, A.J.S., Navarro, E., Héral, M. & Deslous-Paoli, J.M. (1993) Feeding behaviour of the mussel *Mytilus edulis*: responses to variations in quantity and organic content of the seston. *J. mar. biol. Ass. U.K.*, **73**, 813–29.

Beninger, P.G. & Le Pennec, M. (1991) Functional anatomy of scallops. In: *Scallops: Biology, Ecology and Aquaculture* (ed. S.E. Shumway), pp. 133–223. Elsevier Science Publishers B.V., Amsterdam.

Beninger, P.G. & St-Jean, S.D. (1997a) The role of mucus in particle processing by suspension-feeding marine bivalves: unifying principles. *Mar. Biol.*, **129**, 389–97.

Beninger, P.G. & St-Jean, S.D. (1997b) Particle processing on the labial palps of *Mytilus edulis* and *Placopecten magellanicus* (Mollusca: Bivalvia). *Mar. Ecol. Prog. Ser.*, **147**, 117–27.

Beninger, P.G. & Venoit, A. (1999) The oyster proves the rule: mechanisms of pseudofaeces transport and rejection on the mantle of *Crassostrea virginica* and *C. gigas*. *Mar. Ecol. Prog. Ser.*, **190**, 179–88.

Beninger, P.G., Le Pennec, M. & Donval, A. (1991) Mode of particle ingestion in five species of suspension-feeding bivalve mollusks. *Mar. Biol.*, **108**, 255–61.

Beninger, P.G., Ward, J.E., MacDonald, B.A. & Thompson, R.J. (1992) Gill function and particle transport in *Placopecten magellanicus* (Mollusca, Bivalvia) as revealed using video endoscopy. *Mar. Biol.*, **114**, 281–88.

Beninger, P.G., St-Jean, S.D., Poussart, Y. & Ward, J.E. (1993) Gill function and mucocyte distribution in *Placopecten magellanicus* and *Mytilus edulis* (Mollusca, Bivalvia): the role of mucus in particle transport. *Mar. Ecol. Prog. Ser.*, **98**, 275–82.

Beninger, P.G., St-Jean, S.D. & Poussart, Y. (1995) Labial palps of the blue mussel *Mytilus edulis* (Bivalvia: Mytilidae). *Mar. Biol.*, **123**, 293–303.

Beninger, P.G., Lynn, J.W., Dietz, T.H. & Silverman, H. (1997a) Mucociliary transport in living tissue: the two-layer model confirmed in the mussel *Mytilus edulis* L. *Biol. Bull.*, **193**, 4–7.

Beninger, P.G., Dufour, S.C. & Bourque, J. (1997b) Particle processing mechanisms of the eulamellibranch bivalves, *Spisula solidissima* and *Mya arenaria*. *Mar. Ecol. Prog. Ser.*, **150**, 157–69.

Beninger, P.G., Veniot, A. & Poussart, Y. (1999) Principles of pseudofaeces rejection on the bivalve mantle: integration in particle processing. *Mar. Ecol. Prog. Ser.*, **178**, 259–69.

Bernard, F.R. & Noakes, D.J. (1990) Pumping rates, water pressure and oxygen use in eight species of marine bivalve molluscs from British Columbia. *Can. J. Fish. Aquat. Sci.*, **47**, 1302–06.

Berry, P.F. & Schleyer, M.H. (1983) The brown mussel *Perna perna* on the Natal coast, South Africa: utilization of available food and energy budget. *Mar. Ecol. Prog. Ser.*, **13**, 201–10.

Bøhle, B. (1972) Effects of adaptation to reduced salinity on filtration activity and growth of mussels. *J. Exp. Mar. Biol. Ecol.*, **10**, 41–9.

Brand, A.R. (1991) Scallop ecology: distribution and behaviour. In: *Scallops: Biology, Ecology and Aquaculture* (ed. S.E. Shumway), pp. 517–84. Elsevier Science Publishers B.V., Amsterdam.

Bricelj, V.M. & Shumway, S.E. (1991) Physiology: energy acquisition and utilization. In: *Scallops: Biology, Ecology and Aquaculture* (ed. S.E. Shumway), pp. 305–46. Elsevier Science Publishers B.V., Amsterdam.

Cahalan, J.A., Siddall, S.E. & Luckenbach, M.W. (1989) Effects of flow velocity, food concentration and particle flux on the growth rate of juvenile bay scallops, *Argopecten irradians. J. Exp. Mar. Biol. Ecol.*, **129**, 45–60.

Charles, F. & Newell, R.I.E. (1997) Digestive physiology of the ribbed mussel *Guekensia demissa* (Dillwyn) held at different tidal heights. *J. Exp. Mar. Biol. Ecol.*, **209**, 201–13.

Chipman, W.A. & Hopkins, J.G. (1954) Water filtration by the bay scallop, *Pecten irradians* as observed with the use of radioactive plankton. *Biol. Bull.*, **107**, 80–91.

Clausen, I. & Riisgård, U. (1996) Growth, filtration and respiration in the mussel *Mytilus edulis*: no evidence for physiological regulation of the filter-pump to nutritional needs. *Mar. Ecol. Prog. Ser.*, **141**, 37–45.

Conover, R.J. (1968) Zooplankton – life in a nutritionally poor environment. *Amer. Zool.*, **8**, 107–18.

Coughlan, J. & Ansell, A.D. (1964) A direct method for determining the pumping rate of siphonate bivalves. *J. Cons. int. Explor. Mer*, **29**, 205–13.

Cranford, P.J. & Grant, J. (1990) Particle clearance and absorption of phytoplankton and detritus by the sea scallop *Placopecten magellanicus* (Gmelin). *J. Exp. Mar. Biol. Ecol.*, **137**, 105–21.

Crosby, N.D. & Reid, R.G.B. (1971) Relationships between food, phylogeny, and cellulose digestion in the Bivalvia. *Can. J. Zool.*, **49**, 617–22.

Cucci, T.L., Shumway, S.E., Newell, R.C., Selvin, R., Guillard, R.R.L. & Yentsch, C.M. (1985) Flow cytometry: a new method for characterisation of differential ingestion, digestion and egestion by suspension feeders. *Mar. Ecol. Prog. Ser.*, **24**, 201–4.

Dame, R.F. (1996) *Ecology of Marine Bivalves: an Ecosystem Approach.* CRC Press, Boca Raton, Florida.

Davenport, J., Smith, R.J.J.W. & Packer, M. (2000) Mussels *Mytilus edulis*: significant consumers and destroyers of mesozooplankton. *Mar. Ecol. Prog. Ser.*, **198**, 131–37.

Davis, R.L. & Marshall, N. (1961) The feeding of the bay scallop, *Aequipecten irradians. Proc. Natl. Shellfish Ass.*, **52**, 25–29.

Defossez, J.-M. & Hawkins, A.J.S. (1997) Selective feeding in shellfish: size-dependent rejection of large particles within pseudofaeces from *Mytilus edulis, Ruditapes philippinarum* and *Tapes decussatus. Mar. Biol.*, **129**, 139–47.

Dodgson, R.W. (1928) *Report on mussel purification.* Fish. Invest. Ser II. **10**, 1–498 HMSO, Lond.

Dufour, S.C. & Beninger, P.G. (2001) A functional interpretation of cilia and mucocyte distributions on the abfrontal surface of bivalve gills. *Mar. Biol.*, **138**, 295–309.

Famme, P., Riisgård, H.U. & Jørgensen, C.B. (1986) On direct measurements of pumping rates in the mussel *Mytilus edulis. Mar. Biol.*, **92**, 323–27.

Foster-Smith, R.L. (1975) The effect of concentration of suspension on filtration rates and pseudofaecal production of *Mytilus edulis* L., *Cerastoderma edule* (L.) and *Venerupis pullastra* (Montagu). *J. mar. biol. Ass. U.K.*, **17**, 1–22.

Foster-Smith, R.L. (1978) The function of the pallial organs of bivalves in controlling ingestion. *J. Moll. Stud.*, **44**, 83–99.

Grant, J. & Cranford, P.J. (1989) The effect of laboratory diet conditioning on tissue and gonad growth in the sea scallop *Placopecten magellanicus*. In: *Reproduction, Genetics and Distribution of Marine Organisms* (eds J.S. Ryland & P.A. Tyler), pp. 95–105. Olsen & Olsen, Fredensborg, Denmark.

Griffiths, R.J. (1980) Filtration, respiration and assimilation in the black mussel *Choromytilus meridionalis*. *Mar. Ecol. Prog. Ser.*, **3**, 63–70.

Hawkins, A.J.S. & Bayne, B.L. (1992) Physiological interrelations and the regulation of production. In: *The Mussel Mytilus: Ecology, Physiology, Genetics and Culture* (ed. E.M. Gosling), pp. 171–222. Elsevier Science Publishers B.V., Amsterdam.

Hawkins, A.J.S. & Klummp, D.W. (1995) Nutrition of the giant clam, *Tridacna gigas* (L.). II. Relative contributions of filter feeding and the ammonium nitrogen acquired and recycled by symbiotic alga towards total nitrogen requirements for tissue growth and metabolism. *J. Exp. Mar. Biol. Ecol.*, **190**, 263–90.

Hawkins, A.J.S., Smith, R.F.M., Bayne, B.L. & Héral, M. (1996) Novel observations underlying the fast growth of suspension-feeding shellfish in turbid environments: *Mytilus edulis*. *Mar. Ecol. Prog. Ser.*, **131**, 179–90.

Hawkins, A.J.S., Smith, R.F.M., Tan, S.H. & Yasin, Z.B. (1998a) Suspension-feeding behaviour in tropical bivalve molluscs, *Perna viridis, Crassostrea belcheri, Crassostrea iredalei, Saccostrea cucculata* and *Pinctada margarifera*. *Mar. Ecol. Prog. Ser.*, **166**, 173–85.

Hawkins, A.J.S., Bayne, B.L., Bougier, S., *et al.* (1998b) Some general relationships in comparing the feeding physiology of suspension-feeding bivalve mollusks. *J. Exp. Mar. Biol. Ecol.*, **219**, 87–103.

Jones, H.D., Richards, O.G. & Southern, T.A. (1992) Gill dimensions, water pumping rate and body size in the mussel *Mytilus edulis*. *J. Exp. Mar. Biol. Ecol.*, **155**, 213–37.

Jørgensen, C.B. (1966) *Biology of Suspension Feeding*. Pergamon Press, Oxford.

Jørgensen, C.B. (1990) *Bivalve Filter Feeding*. Olsen & Olsen, Fredensborg, Denmark.

Jørgensen, C.B. (1996) Bivalve filter feeding revisited. *Mar. Ecol. Prog. Ser.*, **142**, 287–302.

Jørgensen, C.B., Larsen, P.S., Møhlenberg, F. & Riisgård, H.U. (1988) The mussel pump: properties and modelling. *Mar. Ecol. Prog. Ser.*, **45**, 205–16.

Kiørboe, T.F. & Møhlenberg, F. (1981) Particle selection in suspension-feeding bivalves. *Mar. Ecol. Prog. Ser.*, **5**, 291–96.

Kirby-Smith, W.W. (1970) *Growth of the scallops*, Argopecten irradians concentricus *(Say) and* Argopecten gibbus *(Linné), as influenced by food and temperature*. Ph.D. Thesis, Duke University, Durham, North Carolina.

Klumpp, D.W., Bayne, B.L. & Hawkins, A.J.S. (1992) Nutrition of the giant clam, *Tridacna gigas* (L). 1. Contribution of filter feeding and photosynthesis to respiration and growth. *J. Exp. Mar. Biol. Ecol.*, **155**, 105–22.

Kristensen, J.H. (1972) Carbohydrases of some marine invertebrates with notes on their food and on the natural occurrence of the carbohydrates studied. *Mar. Biol.*, **14**, 130–42.

Kuenster, S.H. (1988) *The effects of the 'Brown Tide' alga on the feeding physiology of* Argopecten irradians *and* Mytilus edulis. M.Sc. Thesis, State University of New York at Stony Brook.

Langdon, C.J. & Newell, R.I.E. (1990) Utilization of detritus and bacteria as food sources by two bivalve suspension feeders, the oyster, *Crassostrea virginica* and the mussel *Guekensia demissa*. *Mar. Ecol. Prog. Ser.*, **58**, 299–310.

Langdon, C.J. & Newell, R.I.E. (1996) Digestion and nutrition in larvae and adults. In: *The eastern oyster* Crassostrea virginica (eds V.S. Kennedy, R.I.E. Newell & A.F. Eble), pp. 231–69. Maryland Sea Grant, College Park, Maryland.

Langton, R.W. (1972) *Some aspects of the digestive rhythm in the oyster Ostrea edulis (L.).* M. Sc. Thesis, University College of North Wales, Bangor, UK.

Le Chevalier, P., Sellos, D. & Van Wormhoudt, A. (1995) Purification and partial characterisation of chymotrypsin-like proteases from the digestive gland of the scallop *Pecten maximus*. *Comp. Biochem. Physiol.*, **110**B, 777–84.

Le Pennec, M., Beninger, P.G., Dorange, G. & Paulet, Y.-M. (1991) Trophic sources and pathways to the developing gametes of *Pecten maximus* (Bivalvia: Pectinidae). *J. mar. biol. Ass. U.K.*, **71**, 451–63.

Le Pennec, M., Beninger, P.G. & Herry, A. (1995) Feeding and digestive adaptations of bivalve molluscs to sulphide-rich habitats. *Comp. Biochem. Physiol.*, **111**A, 183–89.

Loosanoff, V.L. (1953) Behavior of oysters in water of low salinities. *Proc. Natl. Shellfish Assoc.*, **43**, 135–51.

Manahan, D.T., Wright, S.H., Stephens, G.C. & Rice, M.A. (1982) Transport of dissolved amino acids by the mussel *Mytilus edulis*: demonstration of net uptake from natural seawater. *Science*, **215**, 1253–55.

Manahan, D.T., Wright, S.H. & Stephens, G.C. (1983) Simultaneous determination of net uptake of sixteen amino acids by a marine bivalve. *Am. J. Physiol.*, **244**, 832–38.

Mathers, N.F. (1972) The tracing of a natural algal food labelled with a carbon 14 isotope through the digestive tract of *Ostrea edulis*. *Proc. Malacol. Soc. Lond.*, **40**, 115–24.

Mathers, N.F. (1973a) Carbohydrate digestion in *Ostrea edulis* L. *Proc. Malacol. Soc. Lond.*, **40**, 359–67.

Mathers, N.F. (1973b) A comparative histochemical survey of enzymes associated with the processes of digestion in *Ostrea edulis* and *Crassostrea angulata* (Mollusca: Bivalvia). *J. Zool. Lond.*, **169**, 169–79.

Mathers, N.F. (1976) The effects of tidal currents on the rhythm of feeding and digestion in *Pecten maximus* L. *J. Exp. Mar. Biol. Ecol.*, **24**, 271–83.

McLusky, D.S. (1973) The effect of temperature on the oxygen consumption and filtration rate of *Chlamys (Aequipecten) opercularis* (L.) (Bivalvia). *Ophelia*, **10**, 114–54.

Meyhöfer, E. (1985) Comparative pumping rates in suspension-feeding bivalves. *Mar. Biol.*, **85**, 137–42.

Mikulich, L.V. & Tsikhon-Lukanina, E.A. (1981) Food composition of the yesso scallop. *Oceanology*, **21**, 633–35.

Mingoalicuanan, S.S. & Lucas, J.S. (1996) Bivalves that feed out of water: phototrophic nutrition during emersion in the giant clam *Tridacna gigas* Linné. *J. Shellfish Res.*, **14**, 283–86.

Møhlenberg, F. & Riisgård, H.U. (1979) Filtration rate, using a new indirect technique, in thirteen species of suspension-feeding bivalves. *Mar. Biol.*, **54**, 143–48.

Morton, B. (1970) The tidal rhythm and rhythm of feeding and digestion in *Cardium edule*. *J. mar. biol. Ass. U.K.*, **50**, 499–512.

Morton, B. (1971) The diurnal rhythm and tidal rhythm of feeding and digestion in *Ostrea edulis*. *Biol. J. Linn. Soc.*, **3**, 329–42.

Owen, G. (1972) Lysosomes, peroxisomes and bivalves. *Sci. Prog. (Oxford)*, **60**, 299–318.

Owen, G. (1974) Feeding and digestion in Bivalvia. In: *Advances in Comparative Physiology and Biochemistry* (ed. O. Lowenstein), pp. 1–35. Academic Press, New York.

Peirson, W.M. (1983) Utilisation of eight algal species by the bay scallop, *Argopecten irradians concentricus* (Say). *J. Exp. Mar. Biol. Ecol.*, **68**, 1–11.

Pouvreau, S., Bodoy, A. & Buestel, D. (2000) *In situ* suspension feeding behaviour of the pearl oyster, *Pinctada margaritifera*: combined effects of body size and weather-related seston composition. *Aquaculture*, **181**, 91–113.

Prieur, D., Nicolas, J.L., Plusquellec, A. & Vigneulle, M. (1990) Interactions between bivalve mollusks and bacteria in the marine environment. *Oceanogr. Mar. Ecol.*, **28**, 277–352.

Purchon, R.D. (1957) The stomach in the Filibranchia and Pseudolamellibranchia. *Proc. Zool. Soc. Lond.*, **129**, 27–60.

Rao, K.P. (1953) Rate of water propulsion in *Mytilus californianus* as a function of latitude. *Biol. Bull.*, **104**, 171–81.

Reid, R.G.B. (1968) The distribution of digestive tract enzymes in lamellibranchiate bivalves. *Comp. Biochem. Physiol.*, **24**, 727–44.

Reid, R.G.B. & Rauchert, K. (1970) Proteolytic enzymes in the bivalve mollusc *Chlamys hericius* Gould. *Comp. Biochem. Physiol.*, **35**A, 689–95.

Reid, R.G.B. & Rauchert, K. (1972) Protein digestion in members of the genus *Macoma. Comp. Biochem. Physiol.*, **41**A, 887–95.

Reid, R.G.B. & Rauchert, K. (1976) Catheptic endopeptidases and protein digestion in the horse clam *Tresus capax* (Gould). *Comp. Biochem. Physiol.*, **54**B, 467–72.

Riisgård, H.U. (1988) Efficiency of particle retension and filtration rate in 6 species of northeast American bivalves. *Mar. Ecol. Prog. Ser.*, **45**, 215–23.

Riisgård, H.U. (1991) Filtration rate and growth in the blue mussel, *Mytilus edulis* Linneaus 1758: dependence on algal concentration. *J. Shellfish Res.*, **10**, 29–35.

Riisgård, H.U. (2001) On measurement of filtration rates in bivalves – the stony road to reliable data: review and interpretation. *Mar. Ecol. Prog. Ser.*, **211**, 275–91.

Riisgård, H.U. & Møhlenberg, F. (1979) An improved automatic recording apparatus for determining the filtration rate of *Mytilus edulis* as a function of size and algal concentration. *Mar. Biol.*, **52**, 61–67.

Riisgård, H.U. & Randløv, A. (1981) Energy budgets, growth and filtration rates in *Mytilus edulis* at different algal concentrations. *Mar. Biol.*, **61**, 227–34.

Riisgård, H.U. & Larsen, P.S. (1995) Filter-feeding in marine macroinvertebrates: pump characteristics, modelling and energy cost. *Biol. Rev.*, **70**, 67–106.

Romberger, H.P. & Epifanio, C.E. (1981) Comparative effects of diets consisting of one or two algal species upon the assimilation efficiencies and growth of juvenile oysters, *Crassostrea virginica* (Gmelin). *Aquaculture*, **25**, 89–94.

Sabapathy, U. & Teo, L.H. (1992) A kinetic study of the α-amylase from the digestive gland of *Perna viridis* L. *Comp. Biochem. Physiol.*, **101**B, 73–77.

Sakurai, I. & Seto, M. (2000) Movement and orientation of the Japanese scallop *Patinopecten yessoensis* (Jay) in response to water flow. *Aquaculture*, **181**, 269–79.

Sheldon, R.W. & Parsons, T.R. (1967) *A practical manual on the use of the Coulter Counter.* Coulter Electronic Sales Company, Toronto, Canada.

Shumway, S.E., Cucci, T.L., Newell, R.C. & Yentch, C.M. (1985) Particle selection, ingestion, and absorption in filter-feeding bivalves. *J. Exp. Mar. Biol. Ecol.*, **91**, 77–92.

Shumway, S.E., Selvin, R. & Schick, D.F. (1987) Food resources related to habitat in the scallop *Placopecten magellanicus* (Gmelin, 1791): a qualitative study. *J. Shellfish Res.*, **6**, 89–95.

Shumway, S.E., Newell, R.C., Crisp, D.J. & Cucci, T.L. (1990) Particle selection in filter-feeding bivalves molluscs: a new technique on an old theme. In: *The Bivalvia. Proceedings of a Memorial Symposium in Honour of Sir Charles Maurice Yonge, Edinburgh 1986* (ed. B. Morton), pp. 152–165. Hong Kong University Press, Hong Kong.

Shumway, S.E., Cucci, T.L., Lesser, M.P., Bourne, N. & Bunting, B. (1997) Particle clearance and selection in three species of juvenile scallops. *Aquaculture Int.*, **5**, 89–99.

Silverman, H., Lynn, J.W., Beninger, P.G. & Dietz, T.H. (1999) The role of latero-frontal cirri in particle capture by the gills of *Mytilus edulis*. *Biol. Bull.*, **197**, 368–76.

Sobral, P. & Widdows, J. (2000) Effects of increasing current velocity, turbidity and particle-size selection on the feeding activity and scope for growth of *Ruditapes decussatus* from Ria Formosa, southern Portugal. *J. Exp. Mar. Biol. Ecol.*, **245**, 111–125.

Stewart, M.G. & Bamford, D.R. (1975) Kinetics of alanine uptake by the gills of the soft shelled clam *Mya arenaria*. *Comp. Biochem. Physiol.*, **52**A, 67–74.

Sumner, A.T. (1969) The distribution of some hydrolytic enzymes in the cells of the digestive gland of certain lamellibranchs and gastropods. *J. Zool. Lond.*, **158**, 277–91.

Teo, L.H. & Lim, E.H. (1991) Some properties of the trehalase from the digestive gland of the green mussel, *Perna viridis* L. *Comp. Biochem. Physiol.*, **99**B, 489–94.

Teo, L.H. & Lim, E.H. (1993) Effects of storage temperature, tris buffer and divalent cations on the activity of trehalase of *Perna viridis* L. *J. Singapore Natn. Acad. Sci.*, **20** & **21**, 61–66.

Teo, L.H. & Sabapathy, U. (1990) Preliminary report on the digestive enzymes present in the digestive gland of *Perna viridis*. *Mar. Biol.*, **106**, 403–407.

Teo, L.H., Lateef, Z. & Ip, Y.K. (1990) Some properties of the sucrase from the digestive gland of the green mussel *Perna viridis*. *Comp. Biochem. Physiol.*, **96**B, 47–51.

Theede, H. (1963) Experimentelle Untersuchungen über die Filtrationsleistung der Miesmuschel *Mytilus edulis* L. *Kieler Meeresforschungen*, **21**, 153–166.

Walne, P.R. (1974) *Culture of Bivalve Molluscs: 50 Years' Experience at Conwy*. Fishing News Books Ltd., Oxford.

Ward, J.E., Beninger, P.G., MacDonald, B.A. & Thompson, R.J. (1991) Direct observations of feeding structures and mechanisms in bivalve mollusks using endoscopic examination and video image-analysis. *Mar. Biol.*, **111**, 287–91.

Ward, J.E., Cassell, H.K. & MacDonald, B.A. (1992) Chemoreception in the sea scallop *Placopecten magellanicus* (Gmelin). I. Stimulatory effects of phytoplankton metabolites on clearance and ingestion rates. *J. Exp. Mar. Biol. Ecol.*, **163**, 235–50.

Ward, J.E., MacDonald, B.A. & Thompson, R.J. (1993) Mechanisms of suspension feeding in bivalves: resolution of current controversies by means of endoscopy. *Limnol. Oceanogr.* **38**, 265–72.

Ward, J.E., Newell, R.I.E., Thompson, R.J. & MacDonald, B.A. (1994) *In vivo* studies of suspension-feeding processes in the eastern oyster, *Crassostrea virginica* (Gmelin). *Biol. Bull.*, **186**, 221–40.

Ward, J.E., Sanford, L.P., Newell, R.I.E. & MacDonald, B.A. (1998) A new explanation of particle capture in suspension-feeding bivalve molluscs. *Limnol. Oceanogr.*, **43**, 741–52.

Weinstein, J.E. (1995) Fine structure of the digestive tubule of the eastern oyster, *Crassostrea virginica* (Gmelin 1791). *J. Shellfish Res.*, **14**, 97–103.

Widdows, J. & Hawkins, A.J.S. (1989) Partitioning of rate of heat dissipation by *Mytilus edulis* into maintenance, feeding, and growth components. *Physiol. Zool.*, **62**, 764–84.

Wildish, D.J. & Kristmanson, D.D. (1988) Growth response of giant scallops to periodicity of flow. *Mar. Ecol. Prog. Ser.*, **42**, 163–69.

Wilson, J.H. & Seed, R. (1974) Laboratory experiments in pumping and filtration in *Mytilus edulis* L. using suspensions of colloidal graphite. *Ir. Fish. Invest. Series B*, **14**, 1–20.

Wojtowicz, M.B. (1972) Carbohydrases of the digestive gland and crystalline style of the Atlantic deep-sea scallop (*Placopecten magellanicus*, Gmelin). *Comp. Biochem. Physiol.*, **43**A, 131–41.

Wong, W.H. & Cheung, S.G. (2001) Feeding rhythms of the green-lipped mussel, *Perna viridis* (Linnaeus, 1758) (Bivalvia: Mytilidae) during spring and neap tidal cycles. *J. Exp. Mar. Biol. Ecol.*, **257**, 13–36.

Wright, R.T., Collins, R.B., Ersing, C.P. & Pearson, D. (1982) Field and laboratory measurements of bivalve filtration of natural marine bacterioplankton. *Limnol. Oceanogr.*, **27**, 91–98.

5 Reproduction, Settlement and Recruitment

Introduction

The reproductive system in bivalves is extremely simple. The paired gonads are made up of branching tubules, and gametes are budded off the epithelial lining of the tubules. The tubules unite to form ducts that lead into larger ducts that eventually terminate in a short gonoduct. In primitive bivalves the gonoduct opens into the kidneys, and eggs and sperm exit through the kidney opening (nephridiopore) into the mantle cavity. In most bivalves the gonoducts are no longer associated with the kidneys but open through independent pores into the mantle cavity close to the nephridiopore. Fertilisation is external and the gametes are shed through the exhalant opening of the mantle, except in the case of oysters where many species brood eggs within the mantle cavity.

Differentiation of sexes

The majority of bivalves are dioecious, i.e. the sexes are separate, and there are usually equal numbers of males and females, although the sexes cannot be differentiated on external characters. The standard method for distinguishing the sexes is to fix gonadal tissue and then use histological processing and light microscope examination (see below). Jabber & Davis (1987) have described a simple colorimetric test for sex determination in mussels that is based on the substantial difference in energy reserves between eggs and sperm. Also in mussels, Mikailov *et al.* (1995) have identified a 'male-associated polypeptide' that is virtually absent in females. Both these methods, while cheaper and less time-consuming than histological processing, involve sacrificing animals. Burton *et al.* (1996) have suggested that squash preparations from fine-needle aspirates of mantle tissue are a preferable technique because it can be done on live bivalves.

Scallops are predominantly hermaphrodites, with distinct male and female portions of the gonad (Fig. 2.4B). These bivalves exhibit synchronous hermaphroditism, i.e. the gonad simultaneously produces male and female gametes either within the same tubules or in two distinct zones in the gonad. To prevent self-fertilisation male gametes are generally released first, a process known as protandry. In contrast, asynchronous hermaphrodites, such as some oyster (*Ostrea*, *Crassostrea*) or clam (*Mercenaria*) species, are either male or female for one or several annual cycles, after which a sex change occurs. In *Ostrea edulis*, when the young oyster reaches sexual maturity, the gonad normally develops as a male; after spawning the gonad changes to female and regular alternation may continue throughout life (Walne, 1974). A similar scenario is observed for *Crassostrea*. Until recently, the prevailing view was that sex in oysters was determined by environmental factors such as food supply and water temperature (Coe, 1936), or that hormones might be the major controlling mechanism (Kennedy, 1983; Thompson *et al.*, 1996). However, Guo *et al.* (1998) have shown that sex in *Crassostrea gigas* is determined by a single gene locus with a dominant maleness (M) allele and an allele for protandric femaleness (F). MF genotypes are true males that do not change sex, while FF are protandric females that mature as males at the juvenile stage but can change sex in later years. Other genes and/or environmental effects may regulate the rate of sex change in protandric females.

Clams are mostly dioecious but two types of sexual pattern other than separate sexes have been recognised, asynchronous and simultaneous hermaphroditism. Quahog clams, *Mercenaria mercenaria*, become sexually mature at less than one year old, developing first as males but changing to an equal sex ratio in the second year (Menzel, 1989). On the other hand, giant clams, *Tridacna*, are both male and female at the same time; sperm are released first, followed by eggs some hours later. There is no evidence that giant clams are capable of producing viable offspring through self-fertilisation (Heslinga, 1989).

Mussels are dioecious and the gonad develops within the mantle tissue (Fig. 2.1C). In ripe mussels the mantle containing the gametes is typically orange in females and creamy-white in males. In clams the gonad is situated at the base of the foot (Fig. 2.6B). In oysters the gonad covers the outer surface of the digestive gland (Fig. 2.3C). When *Ostrea edulis* is in a ripe condition the gonad forms a layer 2–3 mm thick and its creamy colour obscures the brown colour of the digestive gland. In ripe *Crassostrea gigas* the gonad may be up to 6–8 mm thick and can make up a third of the total body weight exclusive of shell (Walne, 1974). In hermaphrodites such as the scallop *Pecten maximus*, the mature gonad is divided into a dorsal testis that is white in colour and a ventral ovary that is orange-red, and both curve around the large central adductor muscle (Fig. 2.4B). In dioecious scallop species such as *Chlamys varia* and *Placopecten magellanicus*, the female gonad is pink while the male gonad is white.

Gametogenesis

Bivalves undergo an annual reproductive cycle which involves a period of gametogenesis followed by a single, an extended or even several spawning events, which in turn is followed by a period of gonad reconstitution (see below). In spermatogenesis, primary spermatogonia undergo repeated mitotic divisions to produce secondary spermatogonia, that undergo meiosis to become spermatocytes. These produce spermatids that differentiate into flagellated spermatazoa that measure 25–60 μm in length. Oogenesis follows a similar pattern to spermatogenesis in that primary oogonia undergo repeated mitosis to give secondary oogonia that enter meiosis but are then arrested at the prophase stage of meiosis I. The remaining meiotic stages are completed at fertilisation. The oocytes then undergo a period of vitellogenesis, which involves the accumulation of lipid globules and small quantities of glycogen. The deposition of these reserves is accompanied by an increase in oocyte size, reaching 70 μm in the mussel *Mytilus edulis* (de Zwann & Mathieu, 1992) and 5–120 μm in scallops, depending on the species (Beninger & Le Pennec, 1991). Lysis of oocytes occurs throughout the gonadal cycle but is particularly marked at the start of spawning and at the end of the breeding season. It is believed that lysed oocytes provide metabolic substrates for energy production at a time when energy reserves in adductor muscle are at their lowest (Beninger & Le Pennec, 1991).

Reproductive cycles

Methods of assessment

The age at which bivalves become sexually mature varies depending on the species. For example, the mussel *Mytilus edulis*, can mature in its first year (Seed & Suchanek, 1992) whereas in scallops it takes 2–3 years in *Patinopecten yessoensis*, and 3–6 years in the Arctic species, *Chlamys islandica* (references in Orensanz *et al.*, 1991).

The most reliable methods for assessing the course of the reproductive cycle in bivalves are those based on either histological or squash preparations of gonad. Gross visual examination of the gonad, for example in scallops, while simple and quick, provides only a rough estimate of overall gonadal development and absolutely no information on the development of gametes (Barber & Blake, 1991). Observations of spawning either in natural or laboratory populations, together with indirect methods based on appearance of larvae in the plankton, or the recruitment of spat (settled larvae), serve as valuable checks on information obtained from gonad preparations.

From histological or squash preparations of the gonad (Fig. 5.1) the proportion of developing, ripe, spawning and spent individuals can be estimated at regular intervals throughout the year. By multiplying the number of individuals at each development stage by the numerical ranking of that stage, and dividing the result by the total number of individuals in the sample, a mean gonad index (GI) is arrived at for each sampling interval. The gonad index increases during gametogenesis and decreases during spawning. Gonad classification schemes, too numerous to mention, abound in the literature. One such scheme for the dioecious Iceland scallop, *Chlamys islandica* (Thorarinsdóttir, 1993) is presented below:

Stage I.   *Maturing (recovering)*. First recognisable stage of recovery after spawning. Gonad growing, flabby and containing much free water. Follicles becoming larger and denser, connective tissue between them. Lumina filling with growing oocytes 20–30 μm in diameter. Genital ducts losing circular configuration.

Stage II.   *Maturing (filling)*. Gonad still larger and thicker, colouring brighter. Still flabby but containing little free water. Follicles larger and closer together. Radially arranged spermatazoa in male follicles, lumen of females contains half-grown oocytes, many of them attached to the follicle wall. Little connective tissue.

Stage III.   *Maturing (half-full)*. Gonad again larger and thicker, containing

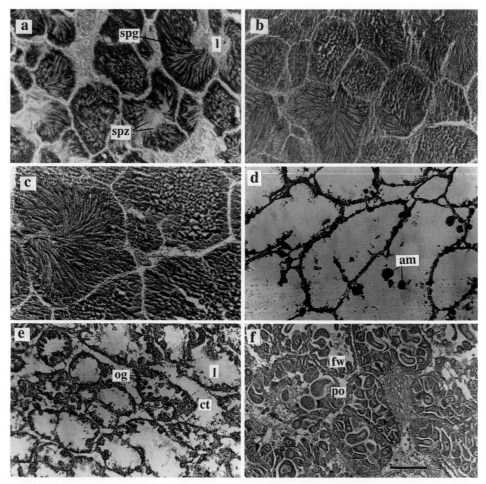

Fig. 5.1. Photomicrographs of transverse sections through male and female gonads of the scallop *Chlamys islandica* at various stages in gamete development. (a) Male, Stage II, maturing, (b) male, Stage III, maturing, (c) male, Stage IVA, mature, (d) male, Stage IVC, spent. (e) Female, Stage II, maturing, (f) female, Stage IVA, mature: spg, spermatogonia; spz, spermatozoa; ct, connective tissue; am, amoebocytes; og, oogonia; l, lumen of follicle; po, polygonal ova; fw, follicle wall. Bar is 100 μm. Photomicrographs courtesy of Gudrún Thorarinsdóttir, Hafrannsóknastofnunin, Reykjavík, Iceland and reprinted with permission from Elsevier Science.

very little free water. Follicles becoming packed together. Lumina becoming packed with spermatazoa or fully-grown oocytes 50–60 μm in diameter. Reduction in connective tissue.

Stage IVA. *Mature (full)*. Gonad has gained maximum size and contains no free water. Follicles highly coloured, and closely packed, testis cream-coloured, ovary usually orange to brick-red. Male follicles packed to periphery with spermatazoa. Sperm heads measuring about 3 μm. Follicle walls extremely thin. Female follicles crowded with polygonal or hexagonal oocytes. Connective tissue restricted to alimentary canal and gonoducts, which appear flattened.

Stage IVB. *Spawning*. Gonads retain differentiation. Tissue becoming dull, flabby and containing much free water depending on the number

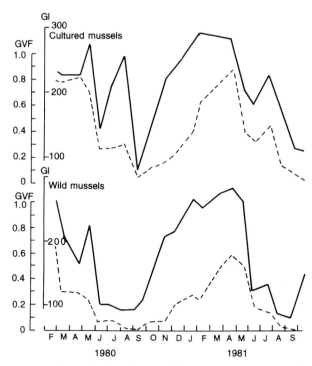

Fig. 5.2. Annual cycle of gonad volume fraction (GVF – broken line) and gonad index (GI – solid line) in cultured and wild mussels, *Mytilus edulis*. From Rodhouse *et al.* (1984).

	of follicles emptied. Zones containing un-spawned follicles exhibit ripe germ cells and are in Stage IV of development. Follicles of varying degrees of spawning are crowded by phagocytes. In some sections ova or sperm may be seen in ciliated ducts.
Stage IVC.	*Spent.* Gonad considerably shrunken in volume, generally dull, flaccid and fawn-coloured, containing much free water. Follicles empty, retaining few residual germ cells. Amoebocytes attack unspawned gametes, sometimes seen as cellular debris. In late stages some reorganisation of follicles and early gametogenic stages is apparent.

One of the main disadvantages of histological or squash methods is that classification of the different developmental stages tends to be subjective. Therefore, stereological techniques, that quantify changes in the volume fractions of different components within the gonad (gametes, nutritive storage cells, connective tissue) from point counts on test grids applied to random thin sections of gonad, are increasingly being used. Fig. 5.2 shows that there is good agreement between GI calculated from histological examination and gamete volume fraction (GVF) estimated by stereology.

Another method to assess reproductive stage is to measure mean gonad weight as a proportion of total body weight at regular intervals throughout the year. Both wet and dry tissue weights can be used, but the latter are preferable because water content varies seasonally and between different tissues. This method is particularly suitable for scallops where the gonad is anatomically distinct and therefore can be easily removed intact (Barber & Blake, 1991). Image analysis of histological sections to obtain a gonad area index (ratio of gonad area to entire visceral mass area ×100) is a technique reported as a sensitive indicator of gametogenic events in oysters (Barber *et al.*, 1991) while gonad weight as a function of shell length (Wilson, 1987a), changing DNA content of male gonad (Thompson, 1984) and changes in the density and size of oocytes in mantle sections (Seed & Suchanek, 1992; Lango-Reynoso *et al.*, 2000) have also been used.

All methods of assessing gametogenesis have their advantages and disadvantages and therefore the optimum strategy is to employ at least one qualitative (histology) and one quantitative (GI, GVF or gonad weight) method. Histology gives precise information on gamete development. Actual gamete weight can be calculated by multiplying GVF (obtained stereologically) by gonad weight, thus providing ecologically meaningful data between species and locations (Barber & Blake, 1991).

Annual cycles

The reproductive cycle in bivalves involves several stages: growth and ripening of gametes, spawning, and gonad redevelopment. Understandably, there is enormous variation in the timing and duration of each stage both within and between species. Therefore, in this section a few appropriate examples will be described to illustrate the breadth of variation.

Mussels in the genus *Mytilus* are widely distributed throughout the cooler waters of the northern and southern hemispheres (Fig. 3.1). *Mytilus edulis*, usually commences gonad redevelopment in October or November and the process proceeds over the winter months so that by February the gonads are ripe. Partial spawning occurs in the spring and a second period of gametogenesis takes place over the summer months, which culminates in spawning in early autumn. Under favourable feeding conditions, such as on the low shore, the number of gametes spawned in spring and autumn is comparable. This is not the case under sub-optimal conditions, e.g. the high shore, when fewer gametes are shed in the autumn. There are a number of variations on this pattern; for example, some populations exhibit a single short spawning period of a few weeks while others, e.g. at Strangford Lough, Ireland, have a spawning period from March to September that is sometimes extended into the winter months (Seed & Brown, 1975). On Irish mussel farms winter spawning (January–February), followed by 'normal' spawning later in the spring and summer, is a common phenomenon (Wilson, 1987b). This is a problem since winter spawning causes deterioration in flesh quality coincident with the time when mussels are harvested for the European market. It is ironic that the best growing sites are those that suffer most from winter spawning.

Differences in spawning patterns have been reported between coexisting species of mussel. On the West Coast of the United States *Mytilus galloprovincialis* spawns from October to February, whereas *M. californianus* dribbles gametes continuously throughout the year. *Perna perna* on the Venezuelan coast experiences very stable annual temperatures (28–32°C), and consequently exhibits year-round spawning with generally three prominent peaks (Carvajal, 1969; Table 5.1). For a detailed listing of spawning periods in *Mytilus* and other mussel species see Seed (1976). Mussels are therefore good examples of bivalves with a flexible reproductive strategy, adjusting their cycle according to prevailing environmental conditions.

The oyster, *Ostrea edulis*, inhabits waters from Norway to the Mediterranean and Adriatic seas (Fig. 3.2). The gonad remains dormant in the male or female phase over the winter months but development resumes once spring arrives (Walne, 1974). In Britain spawning occurs from the end of June to the beginning of July but in warmer waters several spawnings occur between May and October (Lubet, 1994). Another oyster, *Crassostrea gigas*, originates in the Pacific but was introduced into Europe in the 1970s. Unlike *O. edulis*, gametogenesis proceeds over the winter months, accelerates in spring, and several intensive spawnings occur between July and September, providing the temperature exceeds 20–22°C. Since there are areas in Europe (Ireland, parts of the UK) where these temperatures are seldom if ever reached in summer, *C. gigas* must be artificially conditioned and spawned in the hatchery (see Chapter 9). Otherwise, gametes are resorbed *in situ* in autumn, and the metabolites thus recovered are stored in the form of glycogen, which is then used to sustain growth over the winter months (Lubet, 1994).

On the east coast of the United States *C. virginica* from Long Island spawn about one month earlier and over a shorter duration than Delaware Bay oysters further south (Table 5.1). After six generations of inbreeding in Delaware Bay, Long Island oysters still maintain their characteristic pattern of gonadal development and spawning. This indicates that there is a strong genetic component to the reproductive cycle that allows oysters to complete reproduction and development over their entire range (Barber *et al.*, 1991). However, so far there is no evidence that different physiological races exist in this species (Gaffney, 1996).

In Europe the two clam species *Ruditapes decussatus* and *R. philippinarum* (the latter introduced from the Philippines), show inter–specific differences in breeding cycles (Laruelle *et al.*, 1994). When the two species were compared from the Bay of Brest, Brittany, France, *R. philippinarum* showed a more extended breeding period and a greater number of spawnings than *R. decussatus*. The latter spawned mainly in early July and in late August, while the former partially spawned in late May and in July, and completely in mid August to mid September (Table 5.1). Inter-site variations in breeding pattern were observed for each species, which appeared to be linked with environmental differences in temperature and food supply. In these two species sexual maturation begins at a small size; in *R. decussatus* it is between 10–21 mm, shell length (Lucas, 1968), while in *R. philippinarum* it starts at a shell length of 5 mm, when clams are still less than one year old (Holland & Chew, 1974). The hard-shell clam, *Mercenaria mercenaria*, usually produces gametes before they are one year old (Menzel, 1989), and in the surf clam, *Spisula solidissima*, sexual maturity occurs even earlier, within three months of settlement at a size of about 5 mm (Chintala & Grassle, 1995).

The king scallop, *Pecten maximus*, is distributed from northern Norway south

Table 5.1. Spawning periods in a selection of bivalve species.

Species	Locality	Spawning period	Reference
Mussels			
Mytilus edulis (wild)	Killary Harbour, Ireland	Partial, early spring; complete in summer	Rodhouse et al., 1984
M. edulis (cultured)	Killary Harbour	May–June; August–September	Rodhouse et al., 1984
M. galloprovincialis	Galicia, Spain	April	Villalba, 1995
		Spring to summer	Cáceres-Martínez & Figueras, 1998
M. trossulus	Alaska, USA	All summer	Blanchard & Feder, 1997
Mytilus edulis chilensis	Falklands Is. S. Atlantic	December–March	Gray et al., 1997
Perna perna	Venezuela, S. America	December–January; March and June–July	Carvajal, 1969
Cockles			
Cerastoderma edule	Aggersborg, Denmark	August	Brock, 1982
C. lamarkii	Aggersborg	July–August	Brock, 1982
Oysters			
Ostrea edulis	Murcia, Spain	Spring–early summer	Cano et al., 1997
	UK	June–July	Walne, 1974
Crassostrea gigas	Galicia, Spain	June–July; October	Ruiz et al., 1992
Crassostrea virginica	Delaware Bay	June and September	Barber et al., 1991
Clams			
Ruditapes decussatus	South coast of Ireland	September	Xie & Burnell, 1995
	Bay of Brest, France	July–October	Laruelle et al., 1994
	Galicia, Spain	April–August; complete August–September	Rodríguez-Moscoso & Arnaiz, 1998
	Atlantic coast of Morocco	May–October	Shafee & Daouadi, 1991
Ruditapes philippinarum	Bay of Brest	Partial late May and July; complete mid August–mid September	Laruelle et al., 1994
	South coast of Ireland	August	Xie & Burnell, 1995
	South coast of France	End June and mid-late August	Borsa & Millet, 1992

Table 5.1. *continued.*

Mercenaria mercenaria	Delaware, USA	August–October	Keck *et al.*, 1975
	Florida, USA	February–March and September–October, but partial year round	Dalton & Menzel, 1983
Mya arenaria	Washington, USA	May–August	Porter, 1974
	Maryland, USA	May–June and October–December	Pfitzenmeyer, 1965
Spisula solidissima	New Jersey, USA	October–November	Chintala & Grassle, 1995
Arctica islandica	Iceland	Year round; most intense June–August	Thorarinsdóttir, 2000
Scallops			
Pecten maximus	Isle of Man, Irish Sea	Partial April–May; complete July–August	Ansell *et al.*, 1991
Placopecten magellanicus	Iles de la Madeline, Canada	September	Giguere *et al.*, 1994
	Georges Bank, Canada	September–October (large and synchronised), May–June (partial, protracted; sometimes does not occur at all)	Dibacco *et al.*, 1995
Pecten fumatus	New Jersey, USA	November–December	MacDonald & Thompson, 1988
Pecten novaezelandiae	SE Australia	June–November	Gwyther *et al.*, 1991
Patinopecten yessoensis	New Zealand	August–April	Bull, 1991
Argopecten irradians	Japan	March–April	Ito, 1991
Argopecten gibbus	Massachusetts, USA	May–September	Rhodes, 1991
Chlamys islandica	Florida, USA	April–June, August–September (sometimes)	Blake & Moyer, 1991
Chlamys varia	West Iceland	End June–beginning July	Thorarinsdóttir, 1993
Aequipecten opercularis	West coast of Ireland	May–June and July–August	Ansell *et al.*, 1991
	Isle of Man, Irish Sea	Spring, and autumn	Ansell *et al.*, 1991
	Faroe Is., NE Atlantic	June	Parsons *et al.*, 1991
Chlamys glabra	Aegean Sea, Greece	March–April	Lykakis & Kalathakis, 1991
Chlamys farreri	Northern China	May–June and late September–October	Lou, 1991
Chlamys nobilis	Southern China seas	April–May, August–September	Lou, 1991

to the Iberian peninsula (Fig. 3.4), and inhabits a more stable environment than *Mytilus*, being found from the infra-littoral to 100 m depths (Ansell *et al.*, 1991). In the British Isles *P. maximus* shows two peaks of spawning, a partial spawning in spring (April or May), and a more complete autumn spawning (July or early August). Juvenile scallops only spawn in autumn. In Galway Bay on the west coast of Ireland, scallops show the same bimodal spawning pattern but, in addition, a low level of spawning continues throughout the summer months. At Loch Creran on the west coast of Scotland there is one main summer spawning, but this is not closely synchronous in all individuals, and there may also be considerable variation in the timing of the spawning period from year to year. In contrast, further south in the Bay of St. Brieuc on the northern coast of Brittany, scallops show a unique highly synchronous maturation of the population in the spring leading to a massive spawning in July, with secondary spawning until the end of August (Ansell *et al.*, 1991). However, scallops from the closely located Bay of Brest show little synchrony between individuals and there are repeated cycles of gamete maturation throughout the year, with much resorption of poorer quality gametes during the winter months. There are indications that the different reproductive strategies exhibited by these two populations are not just a consequence of differing temperature and food regimes but also reflect genetic adaptation on the part of two self-recruiting stocks (Ansell *et al.*, 1988).

The age of sexual maturity varies between scallop species; it is about 70 days in *Amusium gibbus* (20 mm shell length), one year in *Argopecten irradians* (55–60 mm shell length) and 3–5 years in *Chlamys islandica* (35–40 mm shell length) (see references in Barber & Blake, 1991).

Factors controlling reproduction

The control of reproduction involves the complex interplay of exogenous factors such as temperature, food, salinity and light, with endogenous factors such as neuro-endocrine cycles and genotype. The role of these factors in the initiation and duration of gametogenesis – a process that extends over several months – most likely differs from their role in the timing and synchronisation of spawning, a much briefer event (Seed & Suchanek, 1992). Therefore, in the following section gametogenesis and spawning are treated separately.

Exogenous regulation of gametogenesis

Temperature and food supply

Temperature is the single exogenous factor that is most often cited as influencing gametogenesis in bivalves. Bayne (1975) reported a linear relationship between the rate of gametogenesis in the mussel *Mytilus edulis* and the rate of temperature change measured as day degrees. Other authors (Lubet & Aloui, 1987) have suggested that a 'temperature window' exists for each species, the limits of which are set by the upper and lower lethal temperatures (LT_{50}) for the species. Inside this window gametogenesis occurs over a certain optimum temperature range. For example, the upper and lower lethal temperatures for

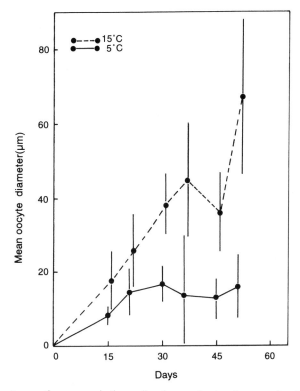

Fig. 5.3. Oocyte growth response in the scallop *Argopecten irradians* as a function of temperature. Scallops were injected with ^{14}C-leucine into the digestive gland and individuals held at 15°C incorporated more labelled leucine into the gonads than those held at 5°C. From Sastry & Blake (1971).

M. edulis are 23°C and 0°C respectively, and the optimum temperature for gametogenesis is between 2°C and 15°C. An increase in temperature within this range does not affect the speed of gametogenesis, except at temperatures close to the lower LT_{50}. However, in the scallop *Argopecten irradians*, Sastry & Blake (1971) have shown that temperature influences the initiation of oocyte growth by regulating the transfer of nutrient reserves to the gonads. The rate of nutrient transfer and hence oocyte growth, was observed to be markedly faster at 15°C than at 5°C (Fig. 5.3).

For many species, food rather than temperature is the major factor determining the timing of gametogenesis. This is not surprising since the process is energy demanding and therefore dependent on available food supply, stored energy reserves, or both. Sastry (1966, 1968 & 1970) and Sastry & Blake (1971) have examined these relationships in *A. irradians*. Scallops at different stages in their gametogenic cycle were held without food and exposed to various temperatures. In early gametogenic stages starvation at 10, 20 and 30°C resulted in a decrease in digestive gland tissue, gonad index and also resorption of oogonia. At a later reproductive stage scallops with minimal reserves released gametes at 25 and 30°C, but at 15 and 20°C a decrease in digestive gland tissue and gonadal index along with resorption of oocytes occurred. Scallops in the resting stage showed a decrease in digestive gland tissue and gonad index and failed to initiate gametogenesis at all experimental temperatures. Abundant food supply is thus necessary for gametogenesis, since pre-stored energy reserves are not sufficient by themselves. However, once minimal gonad reserves have accumulated gametes develop to maturity. But the deposition of minimal gonad reserves in this species is, as seen above, also temperature dependent; below 20°C *A. irradians* failed to accumulate gonad reserves and therefore did not initiate gametogenesis. These results emphasise the importance of food supply in the control of gametogenesis, but also highlight the complex interplay between food supply and temperature in this species.

There is little information on factors such as salinity, light, lunar phase or tides on bivalve gametogenesis. In a series of experiments Loosanoff (1953) noted that adult *Crassostrea virginica* held in salinities of ≤5 psu did not show normal gametogenesis but at ≥7.5 psu gamete development was normal in both male and female oysters. With regard to the effect of photoperiod Sastry (1979) has reported that in Massachusetts, United States, gametogenesis in *A. irradians* is initiated during the spring, a time of increasing day length, and maturity is attained in midsummer when day length is maximal. Further south in North Carolina gametogenesis commences in midsummer and maturation and spawning correlate with decreasing day length of late to early autumn. These differences could, however, be ascribed to the different temperature and food regimes experienced by these two populations.

Endogenous regulation of gametogenesis

Environmental factors such as temperature and food are believed to act through sensory receptors on nerve ganglia (see Chapter 2). Neurosecretory cells in the stimulated ganglia secrete neurohormones that exert their physiological effects on the gonad. The majority of neurosecretory cells are found in the cerebro-pleural ganglia. The activity of neurosecretory cells is low during the resting phase of gametogenesis, progressively increases in synchrony with the developing gonad, and reaches a maximum just before spawning (Fig. 5.4). Direct experimental evidence that neurosecretory cells control gametogenesis was furnished by Mathieu and colleagues (see de Zwann & Mathieu, 1992) when

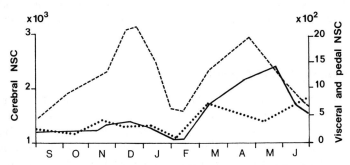

Fig. 5.4. Variation of active neurosecretory cell (NSC) numbers in ganglia of the mussel *Mytilus edulis* during the annual reproductive cycle. Cerebral ganglia, broken line; pedal ganglia, unbroken line; visceral ganglia, dotted line. From de Zwann & Mathieu (1992). Reprinted with permission from Elsevier Science.

they demonstrated that extracts of cerebropleural gangla stimulate *in vitro* mitosis in *Mytilus edulis* gonad. The active factor (oogonial-mitosis stimulating factor) has been identified as a peptide with a molecular weight of <5000 Da. Another (unnamed) factor, also of cerebral origin, plays an essential role in vitellogenesis (Lubet, 1994) and Sato *et al.* (1985) have reported, but not identified, a factor which controls oocyte maturation in the gonad of the clam *Spisula*.

Sastry (1975), Blake (1972) and Blake & Sastry (1979) have investigated the complex interplay between neuroendocrine and environmental factors in the reproductive cycle of the scallop *Argopecten irradians* (Fig. 5.5). Five neuro-secretory stages have been identified. In Stage I scallops secrete a neuro-endocrine that allows nutrient accumulation. Stages II and III act as 'on-off' mechanisms controlling the transfer of nutrients to the gonad. Stage IV coincides with growth of oocytes to maturity, and Stage V corresponds with the start of spawning. Initially, during the vegetative phase, this species is indifferent to environmental stimuli for the initiation of gonad growth. However, as oogonia and early oocytes develop scallops become responsive to the environmental stimuli. The change from Stage II to Stage III in the neurosecretory cycle seems to act as a switching mechanism, and growth of oocytes is initiated or delayed depending on ambient temperature and food conditions. With the initiation of oocyte growth reserves are transferred from the digestive gland to the gonad. After certain minimum reserves are accumulated within the gonad, oocyte development to maturation is independent of food supply. However, exposure to sub-threshold temperatures reverses the neurosecretory cycle to Stage II and oocyte growth is delayed. Provided temperatures are above the required minimum the neurosecretory cycle advances through Stage III and Stage IV when oocytes undergo cytoplasmic growth and vitellogenesis, respectively. Prolonged exposure to sub-threshold temperatures during vitellogenesis causes neuronal degeneration which in turn causes oocyte disintegration and resorption (Barber & Blake, 1991).

There is little information on the influence of genotype on gametogenesis. One report by Hilbish & Zimmerman (1988) shows that the reproductive cycle of *M. edulis* from Long Island, United States, is dependent upon geno-type at the leucineaminopeptidase (*Lap*) locus. *Lap* genotypes differ in the net rate of nitrogen accumulation – an indication of nutrient accumulation. Geno-types with lower rates of nutrient accumulation (*Lap*[95]) delay the initiation of gametogenesis by approximately six weeks relative to genotypes with higher rates of nutrient accumulation.

Exogenous regulation of spawning

Information on the influence of exogenous factors on spawning comes from field and laboratory-based studies. In the field, many species of scallop spawn when water temperatures are either increasing or decreasing (Barber & Blake, 1991). For example, the scallop *Argopecten irradians irradians* (the northern subspecies) spawned when water temperature was increasing, while *A. irradians concentricus* (the southern subspecies) spawned when water temperatures were decreasing. For *Mytilus edulis* the spring spawning coincides with rising water temperatures, while the autumn spawning occurs when

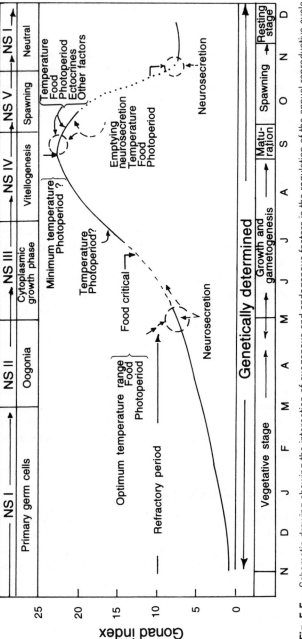

Fig. 5.5.   Schematic drawing showing the interaction of exogenous and endogenous factors in the regulation of the annual reproductive cycle of the scallop *Argopecten irradians* (see text for explanation). Redrawn from Sastry (1975).

temperatures are falling (Table 5.1). Raising scallops in lantern nets from deep to surface waters (a 1–2°C difference) caused spawning within 4 hours (Minchin, 1992a).

Physical stimulation by jarring or scraping the shell, pulling or cutting the byssus threads can stimulate spawning in *M. edulis*. Seed & Suchanek (1992) suggest that these are the types of environmental cues received during storms that could also signal the presence of storm-generated patches of bare rock onto which mussels can settle. Other natural stimuli include salinity change, lunar phase and tidal fluctuations. Once spawning is initiated the presence of gametes in the water provides a powerful chemical stimulus to other ripe individuals, and thereby enhances the chances of fertilisation.

Under laboratory conditions shellfish are often induced to spawn by temperature shock, either by raising the temperature by several degrees (e.g. from 26 to 30°C), or by using alternating cycles of high and low temperatures. For example, oysters in the hatchery are induced to spawn by raising the water temperature to 28–30°C. If no spawning occurs within 45 min then the water temperature is lowered to 24°C for 30 minutes. This temperature cycle is repeated until spawning is achieved. Additional, but less-used, methods of induction include cold shock, flowing seawater, mild electric shock, or injection of serotonin, ammonium chloride or potassium chloride into the mantle cavity. Protracted air exposure when ripe shellfish are being transferred between holding tanks can often stimulate unplanned spawning (Heasman *et al.*, 1995). If spawning does not occur gonad tissue may be dissected from a number of males and females. The tissue is blended with seawater for 5–10 seconds and the eggs are collected on a screen, washed, and a sample is then examined under the microscope to check fertilisation. Although this method produces fewer larvae than a natural spawn it is a common way of obtaining gametes from *Crassostrea gigas*, one of the major oyster species being cultured on a global scale.

Endogenous regulation of spawning

The neuro-endocrine control of spawning is well documented (see Barber & Blake, 1991 for references). Monoamines such as serotonin, dopamine and noradrenaline are neurosecretory products that are at high levels prior to spawning, but decrease on spawning (Martinez & Rivera, 1994). Various prostaglandins increase in the gonad before spawning, and are also believed to play a role in spawning (Lubet, 1994). This knowledge has been applied in the aquaculture industry where injection of dopamine combined with prostaglandins E_2 has been used to obtain viable gametes for fertilisation (Martinez *et al.*, 1996a,b).

Annual storage cycle

Gametogenesis is an energy demanding process. During the annual reproductive cycle nutrients are stored when food supplies are abundant, and gonad activity is minimal. Subsequently, these energy reserves are utilised to meet the energetic requirements of gametogenesis. The main energy reserve for gametogenesis is glygogen and the main storage tissue in most bivalves is the adductor muscle. In the scallop *Pecten maximus* the adductor muscle doubles in weight between March and November but declines as the gonad increases in weight over the winter (see references in Barber & Blake, 1991). Triploid (effectively sterile) scallops (*Argopecten irradians*) were reported to have adductor muscles that were 73% heavier and contained significantly more glycogen than normal diploid scallops (Tabarini, 1984). The explanation being that energy normally used to fuel gametogenesis was instead channelled into somatic growth.

Two types of cells are involved in energy storage: adipogranular (ADG) cells that store protein, fat and glygogen, and vesicular connective tissue (VCT) cells that store large quantities of glycogen. The uptake and release of metabolites by the two cell types is believed to be under hormonal control (de Zwann & Mathieu, 1992).

Changes in the nutrient status of shellfish can be estimated by calculating body condition index (CI) either on single individuals or whole populations, e.g:

$$CI = \frac{\text{Average dry weight of meat (g)}}{\text{Average volume between shell (ml)}} \times 1000$$

The volume is measured as the total displacement volume of a completely closed shell minus the displacement volume of the shell after removal of the body tissue (Gilek *et al.*, 1992). A sample size of 50–100 individuals is recommended, to allow for individual variation (Quayle & Newkirk, 1989). Using the above index Walne (1974) reports that an average sample of the European oyster, *Ostrea edulis*, will have an index of 90–100, while well-conditioned oysters have an index of 120 or more, and oysters in poor condition have an index of 70–80. Condition varies both seasonally and from site to site, being influenced by food levels (Smaal & van Stralen, 1990), local environmental conditions, parasitism (Barber *et al.*, 1988a; Gray *et al.*, 1999) and pollution (Pridmore *et al.*, 1990; Roper *et al.*, 1991). On a seasonal level, condition index increases during the period of energy storage and gametogenesis, and declines with the main spawning event.

Reproductive effort and fecundity

Young bivalves grow fast converting all available energy into somatic growth but with increasing size there is a gradual shift from somatic growth into reproduction, so that in the largest animals most production (>90%) is channelled into gamete synthesis (Fig. 5.6). Paine (1976) has suggested that this may be the result of selective pressure for early increase in size as a refuge from predation, coupled with selection for high fecundity (egg production) in larger individuals.

The term reproductive effort (RE) is used to refer to the level of energy allocated to reproduction and is commonly expressed as a proportion of total production:

$$RE = P_g/(P_g + P_s)$$

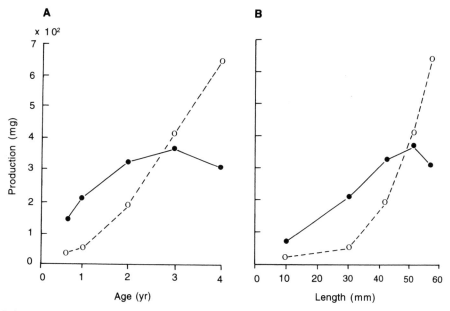

Fig. 5.6. Somatic (solid circles) and gamete (open circles) production as a function of age (A) and shell length (B) in low shore mussels (*Mytilus edulis*). From Rodhouse *et al.* (1986).

where P_g is gamete production and P_s is somatic production over the same time period, usually one year. There is substantial variation in RE between species (Fig. 5.7), most of which has been ascribed to environmental factors such as temperature and food supply. Some species, e.g. the scallops *Chlamys islandica* (Vahl, 1985) and *Argopecten irradians irradians* (Bricelj & Krause, 1992) actually show a decline in RE with age. There is also considerable variation in RE between conspecific populations. For example, RE in three-year-old mussels in different populations from south-west England ranged between 43% and 95% (Fig. 5.8). At Lynher mussels received a richer food supply and experienced little environmental stress compared to the population at Cattewater, which was located near a warm water outfall and was therefore subject to thermal stress during winter and spring (Bayne & Widdows, 1978; Bayne *et al.*, 1983).

Fecundity (reproductive output) is also age-related. In the mussel *Mytilus edulis*, individual females produce between 7 and 40 million eggs, depending on shell length (Thompson, 1979). In brooding oyster species, e.g. *Ostrea edulis* (see later), egg production varies from 1.0×10^5 to 1.5×10^6 eggs, depending on age (Walne, 1974). But environmental factors also play a role. For example, egg production in the scallop *Placopecten magellanicus* ranged from $3.1–6.6 \times 10^7$ eggs per female at a shallow site (13–20 m) to $1.4–2.4 \times 10^7$ at a deep site (170–180 m), almost a three-fold difference (Barber *et al.*, 1988b). Over a three-year period fecundity was significantly different in oysters in the James River, Virginia, with annual mean values per female of 5.6 million, 0.94 million and 0.12 million eggs, respectively (Mann *et al.*, 1994). The decline in fecundity was correlated with declining average salinity, the stress-

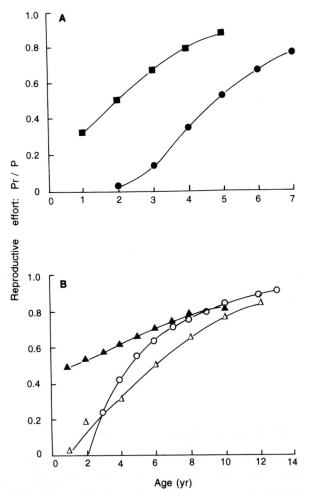

Fig. 5.7. Reproductive effort calculated as a proportion of total production in five bivalve species, related to age. A. ■, *Crassostrea virginica* (Dame, 1972, 1976) and ●, *Mytilus edulis* (Bayne & Widdows, 1978); B. ▲, *Choromytilus meridionalis* (Griffiths, 1980); ○, *Placopecten magellanicus* (R.J. Thompson, unpublished data); △, *Ostrea edulis* (Rodhouse, 1978). Modified from Bayne & Newell (1983).

ful effect of which may have reduced the energy available for reproduction in these oysters.

While environmental factors influence RE and fecundity, and are thus important in short-term adaptation, there is probably also a genetic component to RE and fecundity that allows adaptation to long-term changes in the environment. Evidence to support this comes from a study by Rodhouse *et al.* (1986), which reported a positive correlation between multiple-locus heterozygosity and fecundity in mussels, which had grown beyond the size at which gamete production exceeded somatic production. There was no relationship between heterozygosity and fecundity in younger individuals. Rodhouse *et al.* (1986) concluded that the higher scope for growth (see

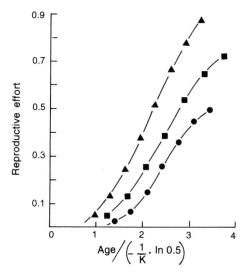

Fig. 5.8. Reproductive effort of the mussel, *Mytilus edulis* from three populations, related to age, corrected to exclude differences due to different rates of growth. K is the rate constant in the von Bertalanffy growth equation (see Chapter 6) for each site. ▲, Lynher; ■, Mothercombe; ●, Cattewater (all in south-west England). From Bayne *et al.* (1983).

Chapter 6) of more genetically variable individuals is translated into allocation of energy to somatic growth during early life, and into gamete production in later life.

Deleterious effects on the reproductive cycle

The presence of parasites has been correlated with the inhibition or retardation of gametogenesis. The most significant are those belonging to the Phylum Acetospora (protozoans) and Class Trematoda (flatworms) in the Phylum Platyhelminthes (see Chapter 11). The acetospore *Haplosporium nelsoni* (MSX) causes a lethal disease in the American oyster *Crassostrea virginica* on the mid-Atlantic coast of the United States. Oysters with systemic (general) MSX infections had a condition index that was 31% lower, and a relative fecundity that was 81% lower than uninfected oysters (Barber *et al.*, 1988a). When *C. virginica* and the closely related *Crassostrea gigas* were exposed to waters infected with MSX and another protozoan parasite, *Perkinsus marinus*, gametogenesis was unaffected in *C. gigas*, suggesting that the two species differ in susceptibility to the parasites (Barber, 1996). In *C. virginica* the observed reduction in fecundity is probably due to the effect of the parasite on carbohydrate storage. Selection for resistance to MSX improved gametogenesis, but only in oysters with few parasites (Ford *et al.*, 1990). Bucephalid trematodes parasitise many bivalve species. The larval stages (sporocysts) form a dense branching interwoven network that infiltrates most host organs, especially the gonad. Castration is often the outcome of heavy infestation and is caused by sporocyst consumption of storage tissue, especially glycogen (Houtteville & Lubet, 1975).

Gonadal neoplasia (tumours) is another condition that affects the reproductive cycle in bivalves. As the tumours grow they invade and destroy gonad follicles. An increased incidence has been reported in hybrids between *Mercenaria mercenaria* and *M. campechensis*, implicating a genetic component to the disease (Bert *et al.*, 1993; Eversole & Heffernan, 1995). The reduced fitness of the hybrids presents a good example of how a disease may act as a barrier to gene flow between species.

Ploidy level can also influence gametogenesis. In general, gametogenesis in triploid (3N) shellfish is abnormal and retarded, due to the inability of homologous chromosomes to pair during meiosis; this causes uneven separation of chromosome triplets in gametes. In triploid *Crassostrea gigas*, however, gametes are produced and are capable of fertilisation, but the relative fecundity of triploids is about 2% that of diploids, and survival of triploid larvae is extremely low (Guo & Allen, 1994). As a result triploids conserve nutrient reserves and theoretically should exhibit faster growth rate than diploids once sexual maturity is attained (see Chapter 10).

Various pollutants have also been shown to affect gametogenesis. Exposure to a number of metal and organic contaminants suppresses gamete development and/or enhances gamete breakdown in *Mytilus* mussel species. In addition, low-level hydrocarbon exposure reduces stored energy reserves, although the effect is reversible following depuration (see Livingstone & Pipe, 1992 for references).

Fertilisation

In most marine bivalves eggs and sperm are shed directly from the genital ducts into the water column, where fertilisation takes place. Sperm penetration of the egg is facilitated by the release of a substance capable of lysing the vitelline membrane around the egg. Meiosis is completed in the egg once fertilisation has taken place. In the mussel *Mytilus edulis*, eggs that are unfertilised for 4–6 hours at 18°C are not capable of developing further, and sperm lose their motility after 1–2 hours at this temperature (Bayne, 1976). The conditions for successful fertilisation vary depending on the species. In *M. edulis* fertilisation occurs successfully at temperatures from 5 to 22°C and salinities between 5 and 40psu (Bayne, 1965). In general, the limits for fertilisation are broader than the limits for successful larval development.

In oysters, e.g. *Ostrea edulis*, eggs and sperm are released through the genital ducts close to the exhalent opening in the mantle. When the oyster is functioning as a male the sperm are carried out into the sea, but in the female the eggs come to lie on the surface of the gills in the inhalant region of the mantle. The eggs are fertilised by sperm brought in with the feeding current, and the fertilised eggs develop into shelled larvae before being discharged into the sea (shell length 170–190µm) for further development (Walne, 1974). In *Crassostrea gigas* the eggs and sperm are released straight into the sea, where fertilisation takes place.

Although many scallop species are hermaphrodites, and frequently release both male and female gametes simultaneously, there is no direct evidence that self-fertilisation occurs. Under laboratory conditions *Pecten maximus* larvae from self-fertilised eggs had significantly reduced growth rates compared to larvae produced through cross-fertilisation, most likely because of reduced genetic variability in the inbred larvae (Beaumont & Budd, 1983). It is conceivable, however, that self-fertilisation may play a role in wild populations when scallop densities are low.

Pre- and post-reproductive isolating mechanisms exist which prevent cross-fertilisation between species in the wild. However, it is often possible to produce hybrids under laboratory conditions although the hybrids are invariably less fit than either of the parent species (see Chapter 10). Even within a species there are mechanisms which regulate the process of fertilisation. For example, gamete incompatibility systems are believed to prevent or regulate self-fertilisation in hermaphroditic species. Gaffney *et al.* (1993) also suggest that partial gamete incompatibility exists in the form of reduced fertilisation rates for between-population crosses, compared to within-population crosses of the oyster *Crassostrea virginica*.

Larval development

The fertilised egg rapidly divides to become a ball of cells that begins to swim once cilia appear some 4–5 hours after fertilisation. A ciliated trochophore stage is reached about 24 hours after fertilisation (Fig. 5.9A). A shell gland begins to secrete the first larval shell, the prodissoconch I. The shell is D-shaped in outline and the larva, now a veliger (100–120 µm shell length), immediately starts to secrete the second larval shell, or prodissoconch II. This shell is secreted by the mantle and exhibits growth lines. Veliger larvae possess a velum, a circular lobe of tissue bearing a ring of cilia, which serves as both a swimming and feeding organ (Fig. 5.9B). Small particles (1–2 µm diameter) caught by the cilia are swept towards the mouth and onwards into a simple gut. In *Mytilus edulis* this stage lasts several weeks and is characterised by rapid growth from 120 µm to 250 µm shell length (Bayne, 1976). As metamorphosis approaches pigmented eyespots and an extensible ciliated foot appear. The larva, now known as a pediveliger, is between 210 and 300 µm in length and is characterised by the features listed in Table 5.2.

Larvae vary in their response to light, gravity and pressure. During the veliger stage larvae are positively phototrophic and are sensitive to pressure. These responses tend to keep the shelled larvae in the surface waters. On the other hand, pediveligers are positively geotrophic and insensitive to pressure which encourages them to descend to the bottom in preparation for settlement.

Table 5.2. Morphological features of *Mytilus edulis* pediveliger larvae. After Bayne (1971) and Lutz & Kennish (1992).

Large velum for swimming and feeding
Ciliated palp that sorts food particles
A few pairs of gill filaments
Mouth, oesophagus, stomach with style sac and large digestive gland, simple intestine
Thin mantle that secretes shell
Foot used in crawling and byssus secretion
Cerebral, pedal and visceral ganglia, sensory pigment spots, pedal statocysts

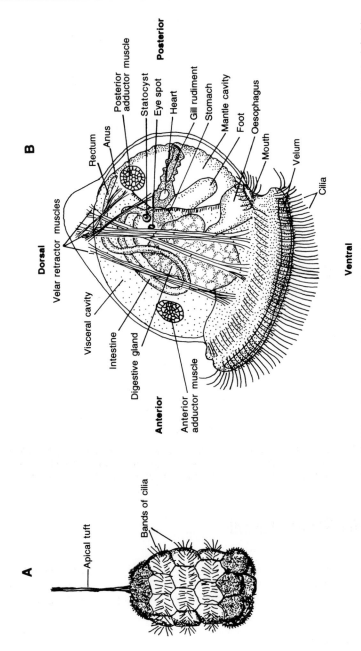

Fig. 5.9. A. Trochophore larva ×400. B. Veliger (prodissoconch II) larva of the oyster *Crassostrea virginica*, viewed from the left side ×380. Redrawn from Eble & Scro (1996) after Galtsoff (1964) and Elston (1980).

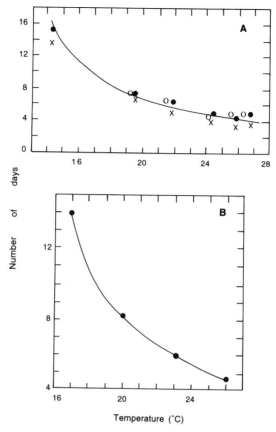

Fig. 5.10. The relationship between temperature and (A) the time taken for three broods of oyster larvae (*Ostrea edulis*) to grow from early cleavage to a mean shell length of 160 μm, and (B) the time taken for larvae to grow from 175 μm to 250 μm. Redrawn from Walne (1974).

Factors affecting larval growth

The factors important in larval development are, not surprisingly, those same factors influencing bivalve growth in general, namely temperature, salinity, and food ration (see Chapter 6 for details). Almost all information on the effects of these factors on growth comes from larval cultures produced in the laboratory.

Temperature and salinity

Temperature has a marked effect on larval growth. In the oyster *Ostrea edulis* development from early cleavage to 160 μm shell length took between 13–15 days at 14°C, but was reduced to 5 days at 24°C (Walne, 1974). In the free-swimming feeding larva of the same species the time taken to grow from 175 μm to 250 μm shell length was 14 days at 17°C but only 6 days at 23°C (Walne, 1974) (Fig. 5.10). This strong positive correlation between temperature and larval growth has also been reported for other bivalve species.

Bayne (1965) measured the effect of salinity on larval growth in *M. edulis* larvae from two populations where the parents experienced different salinity regimes. Larvae from North Wales, UK, did not grow at 19 psu and showed retarded growth at 24 psu, but at 30–32 psu growth was normal. In larvae from the Øresund, Denmark, close to the Baltic, growth occurred even at 14 psu, indicating that there is a genetic component to salinity (and probably temperature) tolerance.

It is now more usual to look at the combined effects of temperature and salinity on larval growth. Generally speaking, as the limits of salinity tolerance are approached for the larvae of a particular species the range of temperature tolerance is markedly narrowed. Also, shelled larva show a greater tolerance to salinity and temperature change than do the embryonic larval stages.

Food ration

About 70% of marine invertebrate species have planktotrophic larvae (that originate from small eggs with little yolk). These are produced in huge quantities, and once their feeding apparatus is functional are totally dependent on the plankton for food. Lecithotrophic larvae develop from large yolky eggs, feed exclusively on the energy reserves within the egg cell and generally have a short larval life.

The planktotrophic eggs of bivalves have reserves of lipid, protein and glycogen that fuel the early developmental stages. Feeding commences shortly after the development of the shell and velum. Although the diet of larvae in the wild is not known for certain, they probably depend on a ration of phytoplankton cells, dissolved organic material (DOM), detritus and bacteria for successful growth and development (Lutz & Kennish, 1992). There is now an impressive literature on the best hatchery foods for larval growth, but only a few pertinent examples will be mentioned at this point (see Chapter 10 for more detail). In considering the optimum diet for larval growth one must consider cell type and size, density, biochemical composition, and whether a mixture of cell types is desirable. The most widely used species for bivalve culture are the diatoms, *Phaeodactylum tricornutum*, *Chaetoceros calcitrans*, *Thalassiosira pseudonana*, *Skeletonema costatum* and the flagellates, *Isochrysis galbana* (Tahitian strain) and *Tetraselmis suecica*. A mixture of species gives best results; for example, with *O. edulis* a mixture of any two of the following: *Tetraselmis suecica*, *Isochrysis galbana* and the diatom, *Chaetoceros calcitrans*, will give larvae that are significantly larger and contain a greater proportion that are eyed after eight days than any one species on their own (Walne, 1974). Also, more spat are obtained on mixed diets, and these grow faster and have lower mortality than larvae fed on single foods, even when all receive the same diet after metamorphosis (Walne, 1974). Jesperson & Olsen (1982) recommend a concentration of 40–50 cells μl^{-1} of *Monochrysis* and *Isochrysis* for optimum growth in *M. edulis* larvae maintained at densities of 0.1–0.2 larvae ml^{-1}. In the wild phytoplankton concentrations are much lower than the concentrations recommended for optimal growth in the laboratory. However, it is likely that larvae in the wild may be feeding on a different, or wider, range of phytoplankton species, supplemented by detritus, bacteria and DOM.

Other determinants of larval growth are genetic factors, silt concentration,

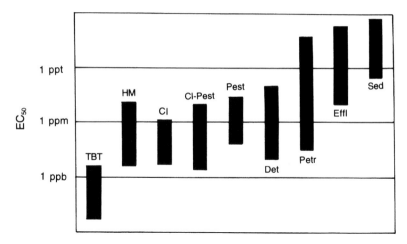

Materials

Fig. 5.11. Toxicity of various materials to bivalve embryos and larvae. EC_{50}: median effective concentration (concentration, that produces a response in 50% of individuals for a fixed exposure time). TBT, tributytin; HM, heavy metals; Cl, halogens and halogen-produced oxidants; Cl-Pest, organochlorinated pesticides; Pest, other pesticides; Det, surfactants and other components of detergents; Petr, petroleum products; Effl, industrial and urban effluents; Sed, sediments. From His *et al.* (1999).

and pollutants. Indeed, a larval bioassay, using oyster or mussel larvae, has been adopted as a method for testing toxicity of chemicals, and assessing water quality of effluents, dredge spoils etc. Larvae are incubated in the test water for a fixed period and at constant temperature, and the number of larvae developing and surviving to the D-larval stage are counted (Widdows & Donkin, 1992; His *et al.*, 1999; see also Chapter 6). Results indicate that the most toxic compound to bivalve embryos and larvae is TBT (tributyltin), the toxicity of which is one or two orders of magnitude greater than any other compound assayed to date. Next most toxic are heavy metals, especially mercury, silver and copper, followed by various pesticides, detergents, petroleum products, effluents and contaminated sediments (Fig. 5.11). Tolerance to silt concentration appears to be species specific. For example, in *Crassostrea virginica* larvae growth was retarded in water carrying $0.75 \, \text{g} \, \text{l}^{-1}$ of silt in suspension, but larvae of *Mercenaria mercenaria* grew normally in a concentration of $1.0 \, \text{g} \, \text{l}^{-1}$ (Loosanoff, 1962, cited in Bayne, 1976).

Larval dispersal

The main advantages of planktotrophy (as opposed to lecithotrophy) are:

1. better potential for wide dispersal and the invasion of new habitats;
2. more rapid recovery from damage to a population, as larvae colonise the area from outside;
3. rapid population expansion under favourable conditions;
4. planktonic feeding that allows a reduction in egg size and increased fecundity (Bayne, 1976).

The length of larval life (time to reach ~300 μm shell length) in marine bivalves varies between three and five weeks, depending on environmental factors such as temperature, salinity and food ration (see above). During the early embryonic phase of development dispersal is passive, but once feeding starts vertical dispersal is by active swimming, while lateral dispersal is by means of water currents. Recorded speeds for the vertical movement of bivalve larvae (0.15–10 mm sec^{-1}) indicate that they are able to control their vertical distribution to some extent (Bayne, 1976). Both field and laboratory studies also report a diurnal migration pattern in bivalve larvae and other invertebrate species. During darkness larvae move up into the surface waters, and during the day they move to less well illuminated, deeper waters, perhaps to avoid predation. This factor more than any other is responsible for massive larval mortality, and along with extremes of temperature, salinity, food shortage and chemicals (insecticides, herbicides, toxins), can result in mortalities as high as 99%. Therefore, the dispersal advantages accruing from a pelagic larval life of some 3–5 weeks are offset by high mortality, which is presumably positively correlated with duration of larval life.

Settlement and metamorphosis

Settlement

The act of settlement involves descent from the plankton to the sea bottom, followed by a sequence of swimming and crawling behaviour that culminates in attachment once a suitable substrate is chosen. Morphological change (metamorphosis) then follows and this heralds the end of the pelagic larval phase and the start of benthic life. Almost all bivalve larvae become competent to settle in the size range 250–300 μm shell length, although many are able to delay the process until they are considerably larger (see below).

Bivalve larvae are capable of discriminating between different substrates as evidenced by their behaviour prior to attachment. Pascual & Zampatti (1995) have described this behaviour in the oyster *Ostrea puelchana*: 'The pediveliger larva actively explores the substratum, crawling back and forth on its foot over each particle or surface, stopping on certain substrata for variable periods of time, occasionally resuming swimming, and crawling further again until final cementing'. Oysters are unique among bivalves in that settlement is an irrevocable process. Most bivalve larvae have an epifaunal post-larval stage, and for many there is a planktonic post-larval stage (see below). It is, therefore, not unexpected that oyster larvae exhibit clear responses to environmental cues during settlement. Oyster larvae favour oyster shells, preferably live ones. It has long been suggested that juvenile and adult oysters release pheromones, which serve as attractants for settling larvae. Not surprisingly, there is considerable interest in elucidating the chemical nature of these attractants. Bacterial films on the surface of oyster shells (Tamburri *et al.*, 1992; Satuito *et al.*, 1995), and various metabolites such as the neurotransmitters L-3, 4-dihydroxy-phenlyalanine (L-DOPA) and epinephrine (EPI), and ammonia (NH$_3$) have now been shown to act as attractants for settling *Crassostrea* larvae (Coon *et al.*, 1985, 1990a,b; Fitt & Coon, 1992). In contrast, various chemicals, e.g.

isothiocyanates and nicotinamide from plant extracts and neurotransmitter blockers such as phentolamine, act as repellents to settling larvae, and could be useful as antifouling agents (Yamashita *et al.*, 1989; Ina *et al.*, 1989; Yamamoto *et al.*, 1998). Physical cues may also be important. Baker (1997) has recently shown from field and laboratory studies that gravity is the primary settlement cue for *Crassostrea* larvae; larvae preferentially settled on the lower surfaces of shells. This strategy is most advantageous for larvae settling in estuaries with a high sediment load, where the lower surfaces of substrates are likely to be sediment-free.

Several studies have reported that mussel larvae settle initially on filamentous substrates such as algae and hydroids, and that post-larvae undergo a period of growth (up to 2 mm shell length) before finally recruiting onto adult mussel beds (Seed, 1976). This second pelagic phase is facilitated by the production of a long fine byssus-like thread (see below). Bayne (1964) was the first to describe this phenomenon of primary settlement and secondary recruitment in *M. edulis* in the UK. However, this pattern is by no means universal as direct settlement of *Mytilus* larvae onto adult mussel beds has been observed in Irish (McGrath *et al.*, 1988; McGrath & King, 1991), Norwegian (Bøhle, 1971), Spanish (Cáceres-Martínez *et al.*, 1993) and Baltic (Kautsky, 1982) waters. Direct settlement has also been reported in *Perna perna* from South Africa (Lasiak & Barnard, 1995) and *Perna canaliculus* in New Zealand (Buchanan & Babcock, 1997).

Scallops settle on a wide variety of algae species, hydroids, bryozoans, and empty shells. Minchin (1992b) lists some 47 species of algae to which *P. maximus* spat were attached. A chemical (jacaranone) that induces metamorphosis in *P. maximus* has been isolated from one of these species, *Delesseria sanguinea*, a rhodophyte alga (Yvin *et al.*, 1985). Most scallop species remain byssally attached for a variable period. Spat of *P. maximus* detach at a shell length of 4–13 mm and recess at 6–10 mm shell length (Minchin, 1992b). Spat of *Aequipecten (Chlamys) opercularis* remain byssally attached until they have reached a larger size than *P. maximus*, while *Chlamys varia* remain attached longer than either species and do not recess into the sediment (Ansell *et al.*, 1991).

Clams settle on a variety of substrates. For example, *Mercenaria mercenaria* settles on sand, sand-mud, stone, gravel and combinations of these, once the substrate is firm enough to prevent sinking. The small clam attaches to a pebble or a piece of shell by a single byssus thread and does not lose the ability to attach until it is 7–9 mm long (Menzel, 1989). The shell hinge of the settled clam is uppermost and parallel to the substrate with the siphons protruding upwards. The clams can migrate up and down in the substrate but there is little lateral movement. Giant clam spat, *Tridacna* spp., form byssal attachments to coral reef. The main function of the byssus is to maintain the clam in an upright position, thus ensuring a favourable orientation to sunlight for the symbiotic algae contained in the mantle edge (Heslinga, 1989). The larger species eventually lose the byssus and displacement is prevented by the weight of the shell valves. The smaller clam species remain strongly byssate throughout life and actively burrow into the reef by mechanical and chemical means (Yonge, 1980).

Various factors are known to affect the density of larval settlement; information on these is mainly derived from hatchery or laboratory-based studies and will be presented in more detail in Chapter 10. In the wild, Andre & Rosenberg (1991) observed a negative relationship between density of adult *Mya arenaria* and newly settled larvae of the same species. In contrast, the presence of dense assemblages of the gem clam, *Gemma gemma*, actually enhanced settlement in *Mercenaria mercenaria* (Ahn *et al.*, 1993).

Once a suitable substrate has been located the larva stops crawling and begins the process of attachment. In mussel, scallop and clam species attachment is by byssus. In oyster species however, a drop of 'cement' is squeezed from a gland in the foot, and the larva applies its left valve to the cement, which sets rapidly (Walne, 1974). The composition of the cement is remarkably similar to that of the oyster shell (Harper, 1992). In some species, e.g. *Crassostrea virginica*, the larvae attach to the substrate by means of a 'glue', a sticky polysaccharide, produced by the bacterium *Shewanella colwellii* (US Patent Number 5474933, University of Maryland). The polysaccharide also acts as an attractant for settling oysters.

Metamorphosis

Metamorphosis is a critical phase in the life history of a bivalve because the ability to move is lost, and there is massive re-organisation of body parts to suit a sedentary existence. When the pediveliger encounters a surface while swimming the velum is retracted and the larva explores the substrate by means of the foot. After a while the larva may swim off again and this cycle of activity may be repeated many times over a period of several days. Larvae of some species can delay metamorphosis for several weeks. For example, Bayne (1965) has reported that in the absence of suitable settlement surfaces mussel larvae can delay metamorphosis for up to five or six weeks in cool temperate waters. This is achieved by byssus drifting, using a long thread secreted from special glands in the foot. This thread, 1–3 µm in diameter and up to 10 cm in length, acts as a drogue enabling these post-larvae to be entrained into the water column (Lane *et al.*, 1985; Beaumont & Barnes, 1992; Buchanan & Babcock, 1997). They retain the ability to drift up to a size of 2.0–2.5 mm (Sigurdsson *et al.*, 1976), which probably represents an add-on period of one to two months for pelagic dispersal.

In all bivalves rapid morphological changes take place in the attached larva, now termed a plantigrade (Fig. 5.12). The extent of morphological re-organisation varies from species to species. In the oyster *Ostrea edulis* Walne (1974) describes the process as follows: 'within 48 hours of attachment, velum, foot, eyespot and anterior adductor muscle disappear, the mouth moves through an angle of 90°, and the posterior muscle moves centrally. The gills gradually acquire connections between the filaments, assuming the eulamellibranch gill type. The left attached shell valve starts to form the typical cup shape of the adult shell'. The entire process takes about three or four days, during which time the larva cannot feed, but stored nutrients are used as an energy source. After metamorphosis the adult dissoconch shell is secreted. This shell differs from the larval prodissoconch II shell in pigmentation, ornamen-

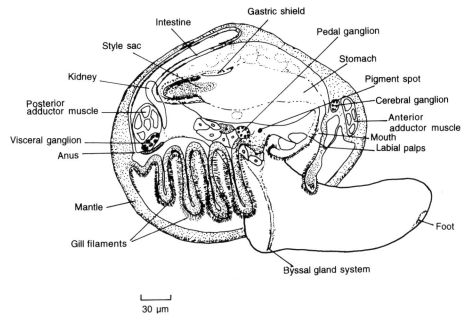

Fig. 5.12. Early plantigrade larva of the mussel *Mytilus edulis* immediately after metamorphosis and before secretion of the dissoconch shell has begun. Redrawn from Bayne (1971).

tation and mineral content. In addition, a remarkable change in shape takes place so that the dissoconch shell is more similar to the later adult form. The prodissoconch-dissoconch boundary provides a clear morphological trait that is useful in differentiating true juveniles from metamorphosing post-larvae (Lutz & Kennish, 1992). Shell shape and hinge and ligament structure of the larval shell are features that have proved useful in species identification. At the population level, measurements on the prodissoconch I and II, and pro-dissoconch-dissoconch boundary, provide data on ecology, biogeography and developmental history (see Lutz & Kennish, 1992 for further information).

Measurement of settlement

Settlement is generally difficult to measure in the field and so it is usually inferred from recruitment data measured some days, or even weeks, after settlement has actually occurred. Seed & Suchanek (1992) define recruitment as the process of successful colonisation after some specified period of time, during which some post-settlement mortality will generally have occurred.

From a commercial point of view it is useful to be able to predict when maximum recruitment can be expected. However, forecasting the timing can be difficult because peaks of reproductive activity in a population do not always correlate with subsequent recruitment in that area. Reasons for this include hydrography of the area, weather conditions prior to settlement, larval abnormalities, and pre- and post-settlement mortality through predation, inadequate settlement surfaces etc.

Table 5.3. Density (individuals m^{-2}) of spat *Ruditapes decussatus* in three locations in the lagoon of Thau, S. France from 1986 to 1989. A zero indicates no recruitment and a dash indicates no data (modified after Borsa & Millet, 1992).

Year	Location 1	Location 2	Location 3
1986	0	–	–
1987	177	390	107
1988	3	34	3
1989	1	–	–

Recruitment in the field, e.g. in clams, can be quantified by taking a delimited grab or fisherman's hand dredge sample from the surface sediment (upper 0.01 m) at a number of sites, and then sieving and counting the number of spat. Using this method Borsa & Millet (1992) reported dramatic spatial and temporal variations in recruitment of clam spat in the lagoon of Thau, southern France (Table 5.3). The poor recruitment of some years was attributed to weather conditions that swept larvae into intensive culture zones where the filtering activity of oysters and mussels may have caused massive larval mortality. This has indeed been demonstrated in a field-based study on the cockle *Cerastoderma edule*, where adult cockles reduced the settlement of larvae by 33% in an area immediately surrounding individual adults (Andre *et al.*, 1993). Over dense beds of cockles this figure rose to 75%.

Recruitment can also be quantified by counting numbers of spat on filamentous algae or on artificial substrates. To measure recruitment on filamentous algae sub-samples of algae species are examined at regular intervals for spat; the spat are removed, counted, and the algal material is dried to constant weight for estimation of settlement density (King *et al.*, 1989). Artificial substrata in use include rubberised hair, slate, asbestos, perspex, concrete, mortar, wood, fibreglass, polystyrene, rope and domestic pan scourers (see McGrath *et al.*, 1994 for references). Artificial substrates have the advantage over plant material in possessing a relatively constant surface area and textural composition. This makes it easier for recruitment to be quantified over fixed time intervals, unlike natural substrates where the numbers recorded represent cumulative results from recruitment over an unknown period prior to collection (McGrath *et al.*, 1994). Conditioning the substrata either in running or still water for 1–2 weeks prior to deployment often results in a greater recruitment density. The conditioning presumably removes soluble substances which might inhibit settlement, and gives time for colonisation by bacteria and small algae species. Parsons *et al.* (1993) found that substrates with a high biofilm coverage had significantly greater scallop recruitment than those without, but McGrath *et al.* (1994) found no difference in *M. edulis* recruitment between conditioned and unconditioned substrates.

We will later see how different artificial substrates are employed by the aquaculture industry to collect seed both for wild and hatchery populations (Chapter 10).

References

Ahn, I.Y., Malouf. R. & Lopez, G. (1993) Enhanced larval settlement of the hard clam *Mercenaria mercenaria* by the gem clam *Gemma gemma*. *Mar. Ecol. Prog. Ser.*, **99**, 51–59.

Andre, C. & Rosenberg, R. (1991) Adult-larval interactions in the suspension-feeding bivalves *Cerastoderma edule* and *Mya arenaria*. *Mar. Ecol. Prog. Ser.*, **71**, 227–34.

Andre, C., Jonesson, P. & Lindegarth, M. (1993) Predation on settling bivalve larvae by benthic suspension feeders – the role of hydrodynamics and larval behavior. *Mar. Ecol. Prog. Ser.*, **97**, 183–92.

Ansell, A.D., Dao, J.-C., Lucas, A., Mackie, L.A. & Morvan, C. (1988) *Reproductive and genetic adaptations in natural and transplant populations of the scallop*, Pecten maximus, *in European waters*. Report to the European Commission on research carried out under EEC Scientific Cooperation Contract No. ST2J-0058-1-UK (CD). 50pp.

Ansell, A.D., Dao, J.-C. & Mason, J. (1991) Three European scallops: *Pecten maximus. Chlamys (Aequipecten) opercularis* and *C. (Chlamys) varia*. In: *Scallops: Biology, Ecology and Aquaculture* (ed. S.E. Shumway), pp. 715–51. Elsevier Science Publishers B.V., Amsterdam.

Baker, P. (1997) Settlement site selection by oyster larvae, *Crassostrea virginica*: evidence for geotaxis. *J. Shellfish Res.*, **16**, 125–28.

Barber, B.J. (1996) Effects of gonadal neoplasms on oogenesis in softshell clams, *Mya arenaria*. *J. Invert. Pathol.*, **67**, 161–68.

Barber, B.J. & Blake, N.J. (1991) Reproductive physiology. In: *Scallops: Biology, Ecology and Aquaculture* (ed. S.E. Shumway), pp. 377–428. Elsevier Science Publishers B.V., Amsterdam.

Barber, B.J., Getchell, R., Shumway, S.E. & Shick, D. (1988a) Reduced fecundity in a deep-water population of the giant scallop, *Placopecten magellanicus*, in the Gulf of Maine, USA. *Mar. Ecol. Prog. Ser.*, **42**, 207–12.

Barber, B.J., Ford, S.E. & Haskin, H.H. (1988b) Effects of the parasite MSX (*Haplosporidium nelsoni*) on oyster (*Crassostrea virginia*) energy metabolism. II. Tissue biochemical composition. *Comp. Biochem. Physiol.*, **91**A, 603–08.

Barber, B.J., Ford, S.E. & Wargo, R.N. (1991) Genetic variation in the timing of gonadal maturation and spawning of the eastern oyster, *Crassostrea virginica*. *Biol. Bull.*, **181**, 216–21.

Bayne, B.L. (1964) Primary and secondary settlement in *Mytilus edulis* L. *J. Anim. Ecol.*, **33**, 513–23.

Bayne, B.L. (1965) Growth and delay of metamorphosis of the larvae of *Mytilus edulis* (L.). *Ophelia*, **2**, 1–47.

Bayne, B.L. (1971) Some morphological changes that occur at the metamorphosis of the larvae of *Mytilus edulis*. In: *Proceedings of the Fouth European Marine Biology Symposium* (ed. D.J. Crisp), pp. 259–80. Cambridge University Press, Cambridge.

Bayne, B.L. (1975) Reproduction in bivalve molluscs under environmental stress. In: *Physiological Ecology of Estuarine Organisms* (ed. F.J. Vernberg), pp. 259–77. University of South Carolina Press, Columbia, USA.

Bayne, B.L. (1976) The biology of mussel larvae. In: *Marine Mussels: their Ecology and Physiology* (ed. B.L. Bayne), pp. 81–120. Cambridge University Press, Cambridge.

Bayne, B.L. & Widdows, J. (1978) The physiological ecology of two populations of *Mytilus edulis* L. *Oecologia*, **37**, 137–62.

Bayne, B.L. & Newell, R.C. (1983) Physiological energetics of marine molluscs. In: *The Mollusca*, Vol. 4, *Physiology*, Part 1 (eds A.S.M. Saleuddin & K.M. Wilbur), pp. 407–515. Academic Press, New York.

Bayne, B.L., Salkeld, P.N. & Worrall, C.M. (1983) Reproductive effort and reproductive value in different populations of *Mytilus edulis* L. *Oecologia*, **59**, 18–26.

Beaumont, A.R. & Budd, M.D. (1983) Effects of self-fertilisation and other factors on the early development of the scallop, *Pecten maximus*. *Mar. Biol.*, **76**, 285–89.

Beaumont, A.R. & Barnes, D.A. (1992) Aspects of veliger larval growth and byssus drifting of the spat of *Pecten maximus* and *Aequipecten (Chlamys) opercularis*. *ICES J. Mar. Sci.*, **49**, 417–23.

Beninger, P.G. & Le Pennec, M. (1991) Functional anatomy of scallops. In: *Scallops: Biology, Ecology and Aquaculture* (ed. S.E. Shumway), pp. 133–223. Elsevier Science Publishers B.V., Amsterdam.

Bert, T.M., Hesselman, D., Arnold, W.S., Moore, W.S. Cruzlopez, H. & Marelli, D.C. (1993) High frequency of gonadal neoplasia in a hard clam (*Mercenaria* spp) hybrid zone. *Mar. Biol.*, **117**, 97–104.

Blake, N.J. (1972) *Environmental regulation of neurosecretion and reproductive activity in the bay scallop, Aequipecten irradians (Lamarck)*. Ph.D. Thesis, University of Rhode Island, Kingston, USA.

Blake, N.J. & Sastry, A.N. (1979) Neurosecretory regulation of oogenesis in the bay scallop, *Aequipecten irradians irradians* (Lamarck). In: *Cyclic Phenomena in Marine Plants and Animals* (eds E. Naylor & R.G. Hartnoll), pp. 181–90. Pergamon Press, New York.

Blake, N.J. & Moyer, M.A. (1991) The calico scallop, *Argopecten gibbus*, fishery of Cape Canaveral, Florida. In: *Scallops: Biology, Ecology and Aquaculture* (ed. S.E. Shumway), pp. 899–911. Elsevier Science Publishers B.V., Amsterdam.

Blanchard, A. & Feder, H.M. (1997) Reproductive timing and nutritional storage cycles of *Mytilus trossulus* Gould, 1850, in Port-Valdez, Alaska, site of a marine oil terminal. *The Veliger*, **40**, 121–30.

Bøhle, B. (1971) Settlement of mussel larvae *Mytilus edulis* on suspended collectors in Norwegian waters. In: *Proceedings of the Fouth European Marine Biology Symposium* (ed. D.J. Crisp), pp. 63–69. Cambridge University Press, Cambridge.

Borsa, P. & Millet, B. (1992) Recruitment of the clam *Ruditapes decussatus* in the lagoon of Thau, Mediterranean. *Est. Coast. Shelf Sci.*, **35**, 289–300.

Bricelj, V.M. & Krause, M.K. (1992) Resource allocation and population genetics of the bay scallop, *Argopecten irradians irradians*: effects of age and allozyme heterozygosity on reproductive output. *Mar. Biol.*, **113**, 253–61.

Brock, V. (1982) Does displacement of spawning time occur in the sibling species *Cerastoderma edule* and *C. lamarcki*? *Mar. Biol.*, **67**, 33–38.

Buchanan, S. & Babcock, R. (1997) Primary and secondary settlement by the green-shell mussel *Perna canaliculatus*. *J. Shellfish Res.*, **16**, 71–76.

Bull, M.F. (1991) Fisheries and aquaculture: New Zealand. In: *Scallops: Biology, Ecology and Aquaculture* (ed. S.E. Shumway), pp. 853–59. Elsevier Science Publishers B.V., Amsterdam.

Burton, S.A., Johnson, G.R. & Davidson, T.J. (1996) Cytologic sexing of marine mussels (*Mytilus edulis*). *J. Shellfish Res.*, **15**, 345–47.

Cáceres–Martínez, J., Robledo, J.A.F. & Figueras, A. (1993) Settlement of mussel *Mytilus galloprovincialis* on an exposed rocky shore in Ria de Vigo, NW Spain. *Mar. Ecol. Prog. Ser.*, **93**, 195–98.

Cáceres–Martínez, J. & Figueras, A. (1998) Long-term survey on wild and cultured mussels (*Mytilus galloprovincialis* Lmk) reproductive cycles in the Ria de Vigo (NW Spain). *Aquaculture*, **162**, 141–56.

Cano, J., Rosique, M.J. & Rocamora, J. (1997) Influence of environmental parameters on reproduction of the European flat oyster (*Ostrea edulis* L.) in a coastal lagoon (Mar-Menor, SE Spain). *J. Moll. Stud.*, **63**, 187–96.

Carvajal, R.J. (1969) Fluctuación mensual de las larvas y crecimiento del mejillón *Perna perna* (L.) y las condiciones ambientales de la ensenada de Guatapanare, Edo, Sucre, Venezuala. *Bol. Inst. Oceanogr. Universidad Oriente Venzuela*, **8**, 13–20.

Chintala, M.M. & Grassle, J. (1995) Early gametogenesis and spawning in juvenile Atlantic surfclams, *Spisula solidissima* (Dillwyn, 1819). *J. Shellfish Res.*, **14**, 301–06.

Coe, W.R. (1936) Environment and sex in the American oyster, *Ostrea virginica*. *Biol. Bull.*, **71**, 353–59.

Coon, S.L., Bonar, D.B. & Weiner, R.M. (1985) Induction of settlement and metamorphosis of the Pacific oyster, *Crassostrea gigas* (Thunberg), by L-DOPA and catecholamines. *J. Exp. Mar. Biol. Ecol.*, **94**, 211–21.

Coon, S.L., Walch, M., Fitt, W.K., Weiner, R.M. & Bonar, D.B. (1990a) Ammonia induces settlement behavior in oyster larvae. *Biol. Bull.*, **179**, 297–303.

Coon, S.L., Fitt, W.K. & Bonar, D.B. (1990b) Competence and delay of metamorphosis in the Pacific oyster *Crassostrea gigas*. *Mar. Biol.*, **106**, 379–87.

Dalton, R. & Menzel, W. (1983) Seasonal gonad development of young laboratory-spawned southern (*Mercenaria campechiensis*) and northern (*Mercenaria mercenaria*) quahogs and their reciprocal hybrids in northern Florida. *J. Shellfish Res.*, **2**, 11–17.

Dame, R.F. (1972) The ecological energetics of growth, respiration and assimilation in the intertidal American oyster, *Crassostrea virginica*. *Mar. Biol.*, **17**, 243–50.

Dame, R.F. (1976) Energy flow in an intertidal oyster population. *Estuarine Coastal Mar. Sci.*, **4**, 243–53.

Dibacco, C., Robert, G. & Grant, J. (1995) Reproductive cycle of the sea scallop, *Placopecten magellanicus* (Gmelin, 1791), on northeastern Georges Bank. *J. Shellfish Res.*, **14**, 59–69.

Eble, A.F. & Scro, R. (1996) General anatomy. In: *The Eastern Oyster* Crassostrea virginica (eds V.S. Kennedy, R.I.E. Newell & A.F. Eble), pp. 19–73. Maryland Sea Grant, College Park, Maryland.

Elston, R. (1980) Functional anatomy, histology and ultrastructure of the soft tissues of the larval American oyster, *Crassostrea virginica*. *Proc. Natl. Shellfish Assoc.*, **70**, 65–93.

Eversole, A.G. & Heffernan, P. (1995) Gonadal neoplasia in northern *Mercenaria mercenaria* (Linnaeus, 1758) and southern *M. campechiensis* (Gmelin, 1791) quahogs and their hybrids cultured in South Carolina. *J. Shellfish Res.*, **14**, 33–39.

Fitt, W.K. & Coon, S.L. (1992) Evidence for ammonia as a natural cue for recruitment of oyster larvae to oyster beds in a Georgia salt-marsh. *Biol. Bull.*, **182**, 401–08.

Ford, S.E., Figueras, A.J. & Haskin, H.H. (1990) Influence of selective breeding, geographic origin, and disease on gametogenesis and sex ratios of oysters, *Crassostrea virginica*, exposed to the parasite *Haplosporidium nelsoni* (MSX). *Aquaculture*, **88**, 285–301.

Gaffney, P.M. (1996) Biochemical and population genetics. In: *The Eastern Oyster* Crassostrea virginica (eds V.S. Kennedy, R.I.E. Newell & A.F. Eble), pp. 423–441. Maryland Sea Grant, College Park, Maryland.

Gaffney, P.M., Bernat, C.M. & Allen, S.K. Jr. (1993) Gametic incompatibility in wild and cultured populations of the eastern oyster, *Crassostrea virginica* (Gmelin). *Aquaculture*, **115**, 272–84.

Galstoff, P.S. (1964) The American oyster *Crassostrea virginica* Gmelin. *Fish. Bull.*, **64**, 1–480.

Giguere, M., Cliché, G. & Brulotte, S. (1994) Reproductive cycles of the sea scallop, *Placopecten magellanicus* (Gmelin), and the Iceland scallop, *Chlamys islandica* (Muller, O.F.), in Iles-de-La-Madeleine, Canada. *J. Shellfish Res.*, **13**, 31–36.

Gilek, M., Tedengren, M. & Kautsky, N. (1992) Physiological performance and general histology of the blue mussel, *Mytilus edulis* L., from the Baltic and North Seas. *Neth. J. Sea Res.*, **30**, 11–21.

Gray, A.P., Seed, R. & Richardson, C.A. (1997) Reproduction and growth of *Mytilus edulis chilensis* from the Falkland Islands. *Sci. Mar.*, **61**, 39–48.

Gray, A.P., Lucas, I.A.N., Seed, R. & Richardson, C.A. (1999) *Mytilus edulis chilensis* infested with *Coccomyxa parasitica* (Chlorococcales, Coccomyxacea). *J. Moll. Stud.*, **65**, 289–94.

Griffiths, R.J. (1980) Natural food availability and assimilation in the bivalve *Choromytilus meridionalis*. *Mar. Ecol. Prog. Ser.*, **3**, 151–56.

Guo, X. & Allen, S.K. Jr. (1994) Reproductive potential and genetics of triploid Pacific oysters, *Crassostrea gigas* (Thunberg). *Biol. Bull.*, **187**, 309–18.

Guo, X., Hedgecock, D., Hershberger, W.K., Cooper, K. & Allen, S.K. Jr. (1998) Genetic determinants of protandric sex in the Pacific oyster, *Crassostrea gigas* Thunberg. *Evolution*, **52**, 394–402.

Gwyther, D., Cropp, D.A., Joll, L.M. & Dredge, M.C.L. (1991) Fisheries and aquaculture: Australia. In: *Scallops: Biology, Ecology and Aquaculture* (ed. S.E. Shumway), pp. 835–51. Elsevier Science Publishers B.V., Amsterdam.

Harper, E.M. (1992) Postlarval cementation in the Ostreidae and its implications for other cementing Bivalvia. *J. Moll. Stud.*, **58**, 37–47.

Heasman, M.P., O'Connor, W.A. & Frazer, A.W.J. (1995) Induction of anaesthesia in the commercial scallop, *Pecten fumatus* Reeve. *Aquaculture*, **131**, 231–38.

Heslinga, G.A. (1989) Biology and culture of the giant clam. In: *Clam Culture in North America* (eds J.J. Manzi & M. Castagna), pp. 293–322. Elsevier Science Publishers B.V., Amsterdam.

Hilbish, T.J. & Zimmerman, K.M. (1988) Genetic and nutritional control of the gametogenic cycle in *Mytilus edulis*. *Mar. Biol.*, **98**, 223–28.

His, E., Beiras, R. & Seaman, M.N.L. (1999) The assessment of marine pollution bioassays with bivalve embryos and larvae. *Adv. Mar. Biol.*, **37**, 1–178.

Holland, D.A. & Chew, K.K. (1974) Reproductive cycle of the Manila clam (*Venerupis japonica*), from Hood Canal, Washington. *Proc. Natl. Shellfish Ass.*, **64**, 53–58.

Houtteville, P.H. & Lubet, P.E. (1975) The sexuality of pelecypod molluscs. In: *Intersexuality in the Animal Kingdom* (ed. R. Reinboth), pp. 179–87. Springer-Verlag, Berlin and New York.

Ina, K., Takasawa, R., Yagi, A., Etoh, H. & Sakata, K. (1989) Isothiocyanates as an attaching repellent against the blue mussel *Mytilus edulis*. *Agric. Biol. Chem.*, **53**, 3323–25.

Ito, H. (1991) Fisheries and aquaculture: Japan. In: *Scallops: Biology, Ecology and Aquaculture* (ed. S.E. Shumway), pp. 1017–55. Elsevier Science Publishers B.V., Amsterdam.

Jabbar, A. & Davies, J.I. (1987) A simple and convenient biochemical method for sex identification in the marine mussel, *Mytilus edulis* L. *J. Exp. Mar. Biol. Ecol.*, **107**, 39–44.

Jesperson, H. &. Olsen, K. (1982) Bioenergetics in veliger larvae of *Mytilus edulis* L. *Ophelia*, **21**, 101–13.

Kautsky, N. (1982) Quantitative studies on gonad cycle, fecundity, reproductive output and recruitment in a Baltic *Mytilus edulis* population. *Mar. Biol.*, **68**, 143–60.

Keck, R.T., Maurer, D. & Lind, H. (1975) A comparative study of the hard clam gonad development cycle. *Biol. Bull.*, **148**, 243–58.

Kennedy, V.S. (1983) Sex ratios in oysters, emphasizing *Crassostrea virginica* from Chesapeake Bay, Maryland. *The Veliger*, **25**, 329–38.

King, P.A., McGrath, D. & Gosling, E.M. (1989) Reproduction and settlement of *Mytilus edulis* on an exposed rocky shore in Galway Bay, west coast of Ireland. *J. mar. biol. Ass. U.K.*, **69**, 355–65.

Lane, D.J.W., Beaumont, A.R. & Hunter, J.R. (1985) Byssus drifting and the drifting threads of the young post-larval mussel *Mytilus edulis*. *Mar. Biol.*, **84**, 301–08.

Lango-Reynoso, F., Chávez-Villalba, J., Cochard, J.C. & Le Pennec, M. (2000) Oocyte size, a means to evaluate the gametogenic development of the Pacific oyster, *Crassostrea gigas* (Thunberg). *Aquaculture*, **190**, 183–99.

Laruelle, F., Guillou, J. & Paulet, Y.M. (1994) Reproductive pattern of the clams, *Ruditapes decussatus* and *R. philippinarum* on intertidal flats in Brittany. *J. mar. biol. Ass. U.K.*, **74**, 351–66.

Lasiak, T.A. & Barnand, T. (1995) Recruitment of the brown mussel *Perna perna* onto natural substrata – a refutation of the primary secondary settlement hypothesis. *Mar. Ecol. Prog. Ser.*, **120**, 147–53.

Livingstone, D.R. & Pipe, R.K., (1992) Mussels and environmental contaminants: molecular and cellular aspects. In: *The Mussel Mytilus: Ecology, Physiology, Genetics and Culture* (ed. E.M. Gosling), pp. 425–64. Elsevier Science Publishers B.V., Amsterdam.

Loosanoff, V.L. (1953) Behavior of oysters in water of low salinities. *Proc. Natl. Shellfish Assoc.*, **43**, 135–51.

Loosanoff, V.L. (1962) Effects of turbidity on some larval and adult bivalves. *Proc. Gulf Carib. Fish. Inst.*, **14**, 80–95.

Lou, Y. (1991) Fisheries and aquaculture: China. In: *Scallops: Biology, Ecology and Aquaculture* (ed. S.E. Shumway), pp. 809–24. Elsevier Science Publishers B.V., Amsterdam.

Lubet, P. (1994) Reproduction in molluscs. In: *Aquaculture: Biology and Ecology of Cultured Species* (ed. G. Barnabé), pp. 138–73. Ellis Horwood Limited, London.

Lubet, P. & Aloui, N. (1987) Limites letales thermmiques et action de la temperatur sur les gametogeneses et l'activité neurosecretrice chez la moule (*Mytilus edulis* et *M. galloprovincialis*, Mollusque Bivalve). *Haliotis*, **16**, 309–16.

Lucas, A. (1968) Mise en évidence de l'hermaphrodisme jevénile chez *Venerupis decussata* (L.) (Bivalia: Veneridae). *C.R. Acad. Sci. Paris*, Série D, **267**, 2332–33.

Lutz, R.A. & Kennish, M.J. (1992) Ecology and morphology of larval and early post-larval mussels. In: *The Mussel Mytilus: Ecology, Physiology, Genetics and Culture* (ed. E.M. Gosling), pp. 53–85. Elsevier Science Publishers B.V., Amsterdam.

Lykakis, J.J. & Kalathakis, M. (1991) Fisheries and aquaculture: Greece. In: *Scallops: Biology, Ecology and Aquaculture* (ed. S.E. Shumway), pp. 795–808. Elsevier Science Publishers B.V., Amsterdam.

MacDonald, B.A. & Thompson, R.J. (1988) Intraspecific variation in growth and reproduction in latitudinally differentiated populations of the giant scallop *Placopecten magellanicus* (Gmelin). *Biol. Bull.*, **175**, 361–71.

Mann, R., Rainer, J.S. & Morales-Alamo, R. (1994) Reproductive activity of oysters, *Crassostrea virginica* (Gmelin 1791) in the James River, Virginia, during 1987–1988. *J. Shellfish Res.*, **13**, 157–64.

Martinez, G. & Rivera, A. (1994) Role of monoamines in the reproductive process of *Argopecten purpuratus*. *Invertebr. Reprod. Dev.*, **25**, 167–74.

Martinez, G., Garrote, C., Mettifogo. L., Perez, H. & Uribe, E. (1996a) Monoamines and prostaglandin E(2) as inducers of the spawning of the scallop, *Argopecten purpuratus* Lamarck. *J. Shellfish Res.*, **15**, 245–49.

Martinez, G., Saleh, F., Mettifogo, L., Campos, E. & Inestrosa, N. (1996b) Monoamines and the release of gametes by the scallop, *Argopecten purpuratus*. *J. Exp. Zool.*, **274**, 365–72.

McGrath, D. & King, P.A. (1991) Settlement of mussels, *Mytilus edulis* L., on wave-exposed shores in Irish waters: a survey. *Proc. Roy. Ir. Acad.*, **91**B, 49–58.

McGrath, D., King, P.A. & Gosling, E.M. (1988) Evidence for the direct settlement of *Mytilus edulis* larvae on adult mussel beds. *Mar. Ecol. Prog. Ser.*, **47**, 103–06.

McGrath, D., King, P.A. & Reidy. M. (1994) Conditioning of artificial substrata and settlement of the marine mussel *Mytilus edulis* L.: a field experiment. *Proc. Roy. Ir. Acad.*, **94**B, 53–56.

Menzel, W. (1989) The biology, fishery and culture of quahog clams, *Mercenaria*. In: *Clam Culture in North America* (eds J.J. Manzi & M. Castagna), pp. 201–42. Elsevier Science Publishers B.V., Amsterdam.

Mikailov, A.T., Torrado, M. & Mendez, J. (1995) Sexual differentiation of reproductive tissue in bivalve molluscs: identification of male associated polypeptide in the mantle of *Mytilus galloprovincialis* Lmk. *Int. J. Dev. Biol.*, **39**, 545–48.

Minchin, D. (1992a) Induced spawning of the scallop, *Pecten maximus*, in the sea. *Aquaculture*, **101**, 187–90.

Minchin, D. (1992b) Biological observations on young scallops, *Pecten maximus. J. mar. biol. Ass. U.K.*, **72**, 807–19.

Orensanz, J.M., Parma, A.M. & Iribarne, O.O. (1991) Population dynamics and management of natural stocks. In: *Scallops: Biology, Ecology and Aquaculture* (ed. S.E. Shumway), pp. 625–713. Elsevier Science Publishers B.V., Amsterdam.

Paine, R.T. (1976) Size-limited predation: an observational and experimental approach with the *Mytilus-Pisaster* interaction. *Ecology*, **57**, 858–73.

Parsons, G.J., Dadswel, M.J. & Rodstrom, E.M. (1991) Fisheries and aquaculture: Scandinavia. In: *Scallops: Biology, Ecology and Aquaculture* (ed. S.E. Shumway), pp. 763–75. Elsevier Science Publishers B.V., Amsterdam.

Parsons, G.J., Dadswel, M.J. & Roff, J.C. (1993) Influence of biofilm on settlement of sea scallop, *Placopecten magellanicus* (Gmelin, 1791), in Passamaquoddy Bay, New Brunswick, Canada. *J. Shellfish Res.*, **12**, 279–83.

Pascual, M.S. & Zampatti, E.A. (1995) Evidence of a chemically mediated adult larval interaction triggering settlement in *Ostrea puelchana* – applications in hatchery production. *Aquaculture*, **133**, 33–44.

Pfitzenmeyer, H.T. (1965) Annual cycles of gametogenesis of the soft-shelled clam, *Mya arenaria*, at Solomons, Maryland. *Chesapeake Sci.*, **6**, 52–59.

Porter, R.G. (1974) Reproductive cycle of the soft-shell clam, *Mya arenaria*, at Skagit Bay, Washington. *Fish. Biol.*, **72**, 648–56.

Pridmore, R.D., Roper, D.S. & Hewitt, J.E. (1990) Variation in composition and condition of the Pacific oyster, *Crassostrea gigas*, along a pollution gradient in Manukau Harbor, New Zealand. *Mar. Environ. Res.*, **30**, 163–77.

Quayle, D.B. & Newkirk, G.F. (1989) *Farming Bivalve Molluscs: Methods for Study and Development*. The World Aquaculture Society and the International Development Research Centre, Baton Rouge, Louisiana, USA.

Rhodes, E.W. (1991) Fisheries and aquaculture of the bay scallop *Argopecten irradians*, in the eastern United States. In: *Scallops: Biology, Ecology and Aquaculture* (ed. S.E. Shumway), pp. 913–24. Elsevier Science Publishers B.V., Amsterdam.

Rodhouse, P.G. (1978) Energy transformations by the oyster *Ostrea edulis* L. in a temperate estuary. *J. Exp. Mar. Biol. Ecol.*, **34**, 1–22.

Rodhouse, P.G., Roden, C.M., Hensey, M.P., McMahon, T., Ottway, B. & Ryan, T.H. (1984) Food resource, gametogenesis and growth of *Mytilus edulis* on the shore and in suspended culture: Killary Harbour, Ireland. *J. mar. biol. Ass. U.K.*, **64**, 513–29.

Rodhouse, P.G., McDonald, J.H., Newell, R.I.E. & Koehn, R.K. (1986) Gamete production, somatic growth and multiple-locus enzyme heterozygosity in *Mytilus edulis. Mar. Biol.*, **90**, 209–14.

Rodríguez-Moscoso, E. & Arnaiz, R. (1998) Gametogenesis and energy storage in a population of the grooved carpet-shell clam, *Tapes decussatus* (Linné 1787), in northwest Spain. *Aquaculture*, **162**, 125–39.

Roper, D.S., Pridmore, R.D., Cummings, V.J. & Hewitt, J.E. (1991) Pollution related differences in the condition cycles of Pacific oysters *Crassostrea gigas* from Manukau Harbor, New Zealand. *Mar. Environ. Res.*, **31**, 197–214.

Ruiz, C., Abad, M., Sedano, F., Garciamartin, L.O. & Lopez, J.L.S. (1992) Influence of seasonal environmental changes on the gamete production and biochemical composition of *Crassostrea gigas* (Thunberg) in suspended culture in El-Grove, Galicia, Spain. *J. Exp. Mar. Biol. Ecol.*, **155**, 249–62.

Sastry, A.N. (1966) Temperature effects in reproduction of the bay scallop, *Aequipecten irradians* Lamarck. *Biol. Bull.*, 118–134.

Sastry, A.N. (1968) The relationships among food, temperature, and gonad development of the bay scallop, *Aequipecten irradians* Lamarck. *Physiol. Zool.*, **41**, 44–53.

Sastry, A.N. (1970) Reproductive physiological variation in latitudinally separated populations of the bay scallop, *Aequipecten irradians* Lamarck. *Biol. Bull.*, **138**, 56–65.

Sastry, A.N. (1975) Physiology and ecology of reproduction in marine invertebrates. In: *Physiological Ecology of Estuarine Organisms* (ed. F.J. Vernberg), pp. 279–99. University of South Carolina Press, Columbia, South Carolina, USA.

Sastry, A.N. (1979) Pelecopoda (excluding Ostreidae). In: *Reproduction of Marine Invertebrates* (eds A.C. Giese & J.S. Pearse), pp. 113–292. Academic Press, New York.

Sastry, A.N. & Blake, N.J. (1971) Regulation of gonad development in the bay scallop, *Aequipecten irradians* Lamarck. *Biol. Bull.*, **140**, 274–82.

Sato, E., Wood, D., Sahni, M. & Koide, S. (1985) Meiootic arrest in oocytes regulated by a *Spisula* factor. *Biol. Bull.*, **169**, 334–41.

Satuito, C.G., Natoyama, K., Yamazaki, M. & Fusetani, N. (1995) Induction of attachment and metamorphosis of laboratory cultured mussel *Mytilus edulis galloprovincialis* larvae by microbial film. *Fish. Sci.*, **61**, 223–27.

Seed, R. (1976) Ecology. In: *Marine Mussels: their Ecology and Physiology* (ed. B.L. Bayne), pp. 13–65. Cambridge University Press, Cambridge.

Seed, R. & Brown., R.A. (1975) The influence of reproductive cycle, growth and mortality on population structure in *Modiolus modiolus* (L.), *Cerastoderma edule* (L.), and *Mytilus edulis* (L.) (Mollusca: Bivalvia). In: *Proceedings of the Ninth European Marine Biology Symposium* (ed. H. Barnes), pp. 257–74. Aberdeen University Press, Aberdeen.

Seed, R. & Suchanek, T.H. (1992) Population and community ecology of *Mytilus*. In: *The Mussel Mytilus: Ecology, Physiology, Genetics and Culture* (ed. E.M. Gosling), pp. 87–169. Elsevier Science Publishers B.V., Amsterdam.

Shafee, M.S. & Daoudi, M. (1991) Gametogenesis and spawning in the carpet-shell clam, *Ruditapes decussatus* L. (Mollusca: Bivalvia), from the Atlantic coast of Morocco. *Aquac. Fish. Manag*, **22**, 203–16.

Sigurdsson, J.B., Titman, C.W. & Davies, P.A. (1976) The dispersal of young post-larval bivalve molluscs by byssus threads. *Nature*, **262**, 386–87.

Smaal, A.C. & van Stralen, M.R. (1990) Average annual growth and condition of mussels as a function of food source. *Hydrobiologia*, **195**, 179–88.

Tabarini, C.L. (1984) Induced triploidy in the bay scallop, *Argopecten irradians*, and its effect on growth and gametogenesis. *Aquaculture*, **42**, 151–60.

Tamburri, M.N., Zimmerfaust, R.K. & Tamplin, M.L. (1992) National sources and properties of chemical inducers mediating settlement of oyster larvae – a re-examination. *Biol. Bull.*, **183**, 327–38.

Thompson, R.J. (1979) Fecundity and reproductive effort of the blue mussel (*Mytilus edulis*), the sea urchin (*Strongylocentrotus droebachiensis*) and the snow crab (*Chionectes opilio*) from populations in Nova Scotia and Newfoundland. *J. Fish. Res. Board Can.*, **36**, 955–64.

Thompson, R.J. (1984) The reproductive cycle and physiological ecology of the mussel, *Mytilus edulis*, in a subarctic, non-estuarine environment. *Mar. Biol.*, **79**, 277–88.

Thompson, R.J., Newell, R.I.E., Kennedy, V.S. & Mann, R. (1996) Reproductive processes and early development. In: *The Eastern Oyster* Crassostrea virginica (eds V.S. Kennedy, R.I.E. Newell & A.F. Eble), pp. 335–70. Maryland Sea Grant, College Park, Maryland.

Thorarinsdóttir, G.G. (1993) The Iceland scallop, *Chlamys islandica*, (Müller) in Breidafjordur, West Iceland. 2. Gamete development and spawning. *Aquaculture*, **110**, 87–96.

Thorarinsdóttir, G.G. (2000) Annual gametogenic cycle in ocean quahog, *Arctica islandica* from north-western Iceland. *J. mar. biol. Ass. U.K.*, **80**, 661–66.

Vahl, O. (1985) Size-specific reproductive effort in *Chlamys islandica*: reproductive senility or stabilizing selection? In: *Proceedings of the Nineteenth European Marine Biology Symposium* (ed. P.E. Gibbs), 521–27. Cambridge University Press, New York.

Villalba, A. (1995) Gametogenic cycle of cultured mussel, *Mytilus galloprovincialis*, in the bays of Galicia (NW Spain). *Aquaculture*, **130**, 269–77.

Walne, P.R. (1974) *Culture of Bivalve Molluscs: 50 Years' Experience at Conwy.* Fishing News Books, Oxford.

Widdows, J. & Donkin, P. (1992) Mussels and environmental contaminants: bioaccummmulation and physiological aspects. In: *The Mussel Mytilus: Ecology, Physiology, Genetics and Culture* (ed. E.M. Gosling), pp. 383–424. Elsevier Science Publishers B.V., Amsterdam.

Wilson, J.H. (1987a). Spawning of *Pecten maximus* (Pectinidae) and the artificial collection of juveniles in two bays in the west of Ireland. *Aquaculture*, **61**, 99–111.

Wilson, J.H. (1987b) The problem of early spawning. *Aquac. Irl*, **30**, 20–21.

Xie, Q.S. & Burnell, G.M. (1995) The effect of activity on the physiological rates of two clam species, *Tapes philippinarum* (Adams & Reeve) and *Tapes decussatus* (Linnaeus). *Proc. Roy. Ir. Acad.*, **95**B, 217–23.

Yamamoto, H., Satuito, C.G., Yamazaki, M., Nato, K., Tachibana, A. & Fusetani, N. (1998) Neurotransmitter blockers as antifoulants against planktonic larvae of the barnacle *Balanus amphitrite* and the mussel *Mytilus galloprovincialis. Biofouling*, **13**, 69–82.

Yamashita, N., Sakata, K., Ina, H. & Ina, K. (1989) Isolation of nicotinamide from *Mallotus* leaves as an attaching repellent against the blue mussel, *Mytilus edulis. Agric. Biol. Chem.*, **53**, 3351–52.

Yonge, C.M. (1980) Functional morphology and evolution in the Tridacnidae (Mollusca: Bivalvia: Cardiacea). *Records of the Australian Museum*, **33**, 737–77.

Yvin, J.C., Chevolet, L., Chevolet–Magueur, A.M. & Cochard, J.C. (1985) First isolation of jacaranone from an alga, *Delesseria sanguinea*, a metamorphosis inducer of *Pecten* larvae. *J. Nat. Prod.*, **48**, 814–16.

Zwann, A. de & Mathieu, M. (1992) Cellular biochemistry and endocrinology. In: *The Mussel Mytilus: Ecology, Physiology, Genetics and Culture* (ed. E.M. Gosling), pp. 223–307. Elsevier Science Publishers B.V., Amsterdam.

6 Bivalve Growth

Introduction

Growth in bivalves is usually described in terms of an increase in some dimension of the shell valves. In mussels, for example, length – the maximum distance between the anterior and posterior margins of the shell – is the dimension of choice, whereas in scallops shell height – the maximum distance between the dorsal (hinge) and ventral shell margins – is used (Fig. 2.2). Growth rate can be measured in one of two ways: either the size of the whole organism is related to age (absolute growth), or the rate of growth of one size variable is related to that of another variable (allometric growth).

It should be pointed out at this stage that shell length as an indicator of size is chosen because it is more easily measured than the more appropriate variable, flesh weight, and also because growth history is recorded in the shell. In most bivalve species shell and flesh weight increase are uncoupled. For example, in the mussel *Mytilus edulis* shell growth is rapid during the spring and summer and slows down over the winter, while flesh weight is subject to seasonal fluctuations associated with the reproductive cycle (Fig. 6.1). In experiments designed to uncouple shell and soft tissue growth Lewis & Cerrato (1997) found that shell growth in the clam *Mercenaria mercenaria* was positively correlated with oxygen consumption, but that soft tissue growth was either not correlated, or negatively correlated, with oxygen consumption and shell growth. These authors suggest that growth-line patterns on the shell may be used to reconstruct metabolic rates from field-collected individuals.

Methods of measuring absolute growth

Size-frequency distributions

The biggest advantage of this method is that it measures growth of a population under undisturbed natural conditions. The accuracy of the analysis depends on how well the collected samples reflect the size structure of the actual population. Shell lengths are measured from a random sample of individuals and the data are divided into size frequency classes and plotted in the form of a histogram. Size-class intervals are normally small, from 1 to 5% of the size of the largest specimen in the population (Cerrato, 1980). Where recruitment is seasonal and the life span is short, and there is little variability in individual growth rates, individual year classes can often be identified as distinct modes. By following the position of these modes over time the mean growth rate of each year class can be estimated (Fig. 6.2). However, in long-lived species with extended recruitment and variable individual growth rates it is not possible to use size-frequency distributions to measure growth rate due to merging of year classes. This is particularly true for older populations where year classes overlap to such an extent that each loses identity because

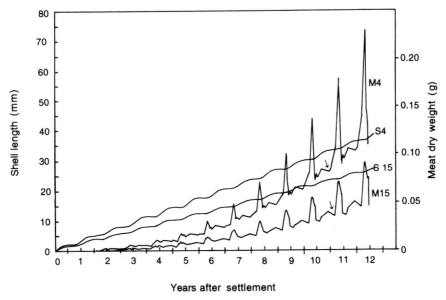

Fig. 6.1. Seasonal patterns of shell (S4, S15) and meat (M4, M15) growth over 12 years in the mussel *Mytilus edulis* growing at 4 and 15 m depths in the Baltic Sea. Arrows indicate examples of periods of negative meat growth before gonad build-up in the winter. From Kautsky (1982).

slow growers of one year class overlap with fast growers of the previous year class. Several methods are available that can separate a length-frequency distribution into groups, that are assumed to represent separate cohorts. For example, the Bhattacharya plot (Pauly & Caddy, 1985; Sparre & Venema, 1992) is a graphical method, derived from Bhattacharya (1967), which separates a length-frequency distribution into a series of normal distributions or pseudo-cohorts (Figs. 6.3 & 6.4).

In addition to information on age-specific growth, size-frequency distributions also provide information on recruitment patterns (Fig. 6.2) and size-selective mortality caused by predation or intra-specific competition for space.

Annual growth rings

Bivalve age can be determined from growth checks or rings on the external shell, or from growth lines in shell or ligament cross sections. Such rings or lines are usually produced annually during the winter period of suspended shell growth (Figs. 6.5 & 2.4A). Mark-recapture studies are used to confirm that growth lines are formed annually. Confirmation that the rings are annual is essential before they can be used to construct growth curves (see below). There are several difficulties associated with using rings on the shell surface to age bivalves. Firstly, while the method is successful for most scallop and clam species it is of little use in mussels and oysters because the rings are either absent or difficult to discern. Annual marks can often be difficult to distinguish from spawning or disturbance lines on the shell, caused as a result of storms, dredging, handling, transplanting or predation attempts (reviewed in

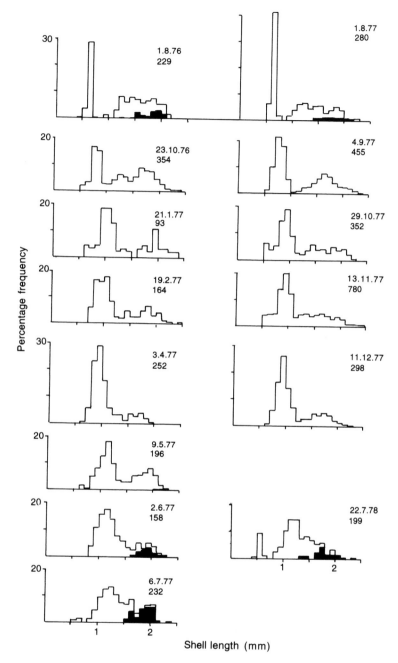

Fig. 6.2. Length-frequency distributions for the small, intertidal clam *Lasaea rubra* at Carnsore Point, SE Ireland. The species broods its young (6–22 embryos) for about two months. There is a single pulse of recruitment during August and this cohort remains distinct throughout the year, reaching a mean size of 1.2 mm shell length in twelve months. The older age classes (probably only 2 or 3) are more difficult to separate. From McGrath & Ó Foighil (1986).

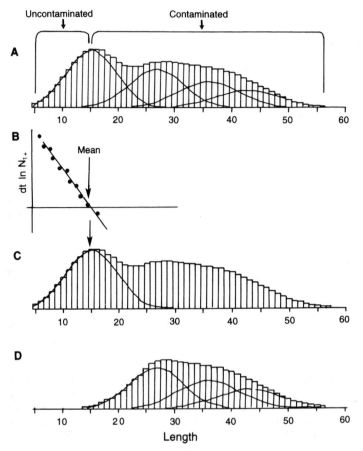

Fig. 6.3. The sequences of operation in the use of the Bhattacharya method of separating a length-frequency distribution into normal components. The method is based on approximating the assumed normal curve of a length-frequency distribution as a parabola, which is then converted to a straight line.

A) The left-hand side of the length-frequency distribution is assumed to consist only of individuals in the first pseudo-cohort, and is uncontaminated by individuals in other overlapping pseudo-cohorts to the right. B) A Bhattacharya plot of dt $(\ln N)$ against the upper limit of the preceding length class is used to estimate the mean of the first cohort from the point where the straight line crosses the length axis. dt $(\ln N)$ is the difference between the natural log of the number in one length class and the number in the preceding length class. C) By working backwards, the linear regression results are used to produce a normal curve that is assumed to contain the total number of individuals in the first pseudo-cohort. D) The number of individuals in the first cohort is subtracted from the total length frequency distribution. The remaining distribution has a new 'clean' left-hand side, from which a second cohort may be separated. From King (1995).

Richardson, 2001). In some scallop species winter rings are coincident with the spawning season but in other species winter and spawning rings are separate, although one type is usually more conspicuous than the other (Orensanz *et al.*, 1991). Age determination is often unreliable in species that do not mark a clear ring during the first year of life, and also in older bivalves where growth rings near the border of the shell are so close together that they are difficult or impossible to tell apart.

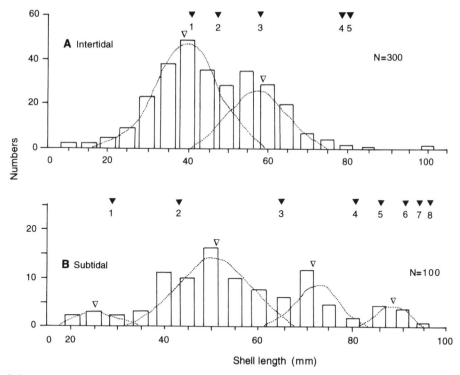

Fig. 6.4. Length-frequency distributions of (A) intertidal and (B) subtidal populations of the oyster *Tiostrea lutaria* at Tal-y-foel, N. Wales, UK. Size classes (dotted line) fitted using the method of Bhattacharya (1967); open symbols denote mean values of individual size classes, solid symbols the estimated age (years) determined from shell growth lines in population sub-samples. The intertidal and subtidal length-frequency distributions could be resolved into either two or four overlapping size classes, respectively. From Richardson *et al.* (1993b).

Fig. 6.5. The cockle *Cerastoderma edule* showing clear annual growth rings on the shell valve. © Craig Burton 2001.

The use of growth lines in shell sections is an alternative ageing method for species where external growth rings are absent or difficult to see (see method in Kennish *et al.*, 1980). The shell valves are sectioned whole, or embedded in epoxy resin and then sectioned along the axis of maximum growth (Fig. 6.6). Other parts of the shell, e.g. the umbone or ligament may also be used. The cut surfaces are ground, polished, and etched with a de-calcifying agent and then either viewed directly, or more commonly the etched shell surface is flooded with acetone and a thin sheet of acetate (~3 mm thick) is firmly applied to the surface. After about 30 min the acetate peel is removed, mounted on a glass slide and viewed under the microscope to determine the number of growth lines in the middle prismatic and inner nacreous shell layers. Nacreous lines are formed annually, while fine microgrowth bands within the prismatic layer have a tidal periodicity. The width and definition of these microgrowth bands is influenced by such factors as season, temperature at time of immersion, tidal amplitude and tidal cycle (Richardson, 2001). The annual lines are thin, translucent and darkly pigmented and first appear at the end of the winter. Between each winter line is a wider area that represents the spring-summer growth increment (Fig. 6.7). Counts should be made at or near the umbonal region of the shell, as this is the only region of the nacreous layer that contains a relatively complete record of growth (Lutz, 1976). Abrasion of the umbo or poor definition of the first winter ring can result in an under-

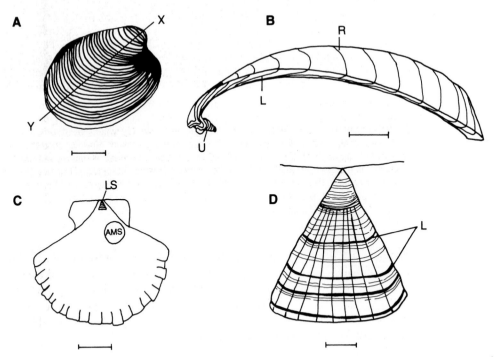

Fig. 6.6. Diagrammatic representation of A) the shell of the clam *Arctica islandica* with prominent external growth rings, X–Y line of cross-section. B) Radial shell section taken along the axis X–Y to reveal the internal growth lines (L) in the outer shell layers and umbone (U); R, surface growth rings. C) Inner surface of a scallop, *Pecten maximus* shell to show the position of the ligament scar (LS) and adductor muscle scar (AMS). D) Growth lines (L) on the inner surface of the ligament scar. Scale bar = 1 cm. From Richardson (2001).

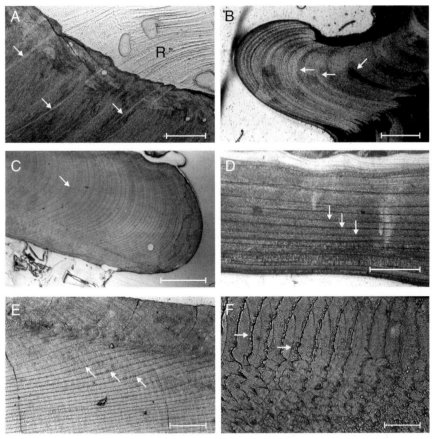

Fig. 6.7. Acetate peel replicas of shell sections of (A) the ligament scar in the oyster *Ostrea edulis*. R, resilium (enlargement of the ligament); arrows indicate positions of annual growth lines. (B), (C) Umbones of the mussel, *Modiolus modiolus*, and the cockle, *Glycymeris glycymeris*, respectively, showing annual growth lines (arrows). (D) Inner nacreous layer of *Modiolus modiolus* and annual growth lines (arrows). (E) Strongly defined growth bands (arrows) alternating with weakly defined bands in the crossed lamellar layer of the cockle, *Cerastoderma edule*. (F) Weakly defined endogenous bands in the prismatic layer of the clam, *Ruditapes philippinarum*. Scale bar = 100 μm, except D and E where scale bars = 0.5 mm. Photomicrographs courtesy of C. Richardson, University of Wales, Bangor, UK.

estimate of age, as for example in large, old oysters (Richardson *et al.*, 1993a). However, the method is generally reliable, but requires more time and expertise than the external shell ring method of aging. The method is particularly useful in habitats such as offshore platforms or water intake pipes where conventional methods involving repeated sampling of the population are difficult or impracticable (Richardson *et al.*, 1990).

Marking shells

The shells of animals *in situ* or in cages are marked and successive measurements are collected over a fixed period of time. Marking is done with indelible ink, by cemented tags, or if the shells are thick enough, by etching numbers into them, or tying tags through holes drilled into their margins. A filed notch (1 mm) on the edge of the shell can be used as a permanent marker in case these fail. However, filing the growing margin can cause a growth check, so chemicals such as tetracycline and calcein that are used as internal shell markers offer a better option (Dey & Bolton, 1978; Kaehler & McQuaid, 1999). A novel method that uses rust marking to tag shells involves placing bivalves on submerged rusted sieves for several weeks, after which time the shells become totally brown (Koshikawa *et al.*, 1997). The colour persists for about five months in the field. Shell growth after treatment can be clearly seen as a rust-free area on the shell margin. Shell marking gives valuable information on the effect of habitat and environmental conditions on growth but care must be taken with caged animals to avoid fouling as this can affect water movement and hence food supply to the bivalve.

Additional methods of measuring absolute growth include optical techniques such as laser diffraction, photographic techniques, and radionuclide uptake into shells. A sensitive laser diffraction technique was first developed by Strømgren (1975) to measure shell length increase in mussels over periods of days. Since then the method has been used to measure the effects of temperature, salinity, photoperiod, algal diets and heavy metals on shell growth of *Mytilus* (Seed & Suchanek, 1992 for references). Photographic techniques, which involve digitising negative images of shell outlines (see below), provide information about shape as well as size, information not registered by the laser technique (Davenport & Glasspool, 1987). More recently, Cochran and Landman (1993) have described a method of measuring growth in deep-sea slow-growing molluscs. It is based on the uptake and decay of radionuclides from seawater into shell. The nuclei of such chemicals are unstable and spontaneously change or decay into a more stable nucleus of another element. Once in the shell and isolated from its seawater source the radionuclide atoms undergo radioactive decay and the number of atoms remaining is a function of the time since the deposition of the shell. Radium (^{228}Ra) with a half-life of 5.7 years has been used not only to determine growth in slow-growing bivalves from the deep sea, but also in rapid growing bivalves from hydrothermal vents (Cochran & Landman, 1993).

Biochemical methods of growth estimation include measurement of protein synthesis using marked amino acids such as leucine, phenylalanine and tyrosine; RNA and DNA synthesis using radioactive uridine and thymidine respectively; and enzyme activity, e.g. ornithine decarboxylase, an enzyme important in cellular growth (Lubet, 1994).

Allometric growth

Bivalves exhibit progressive changes in the relative proportions of the shell with increasing body size. These changes are the result of differential growth vectors operating at different points around the mantle edge (Seed, 1980). The mantle plays a key role in shell secretion (Chapter 2). The relationship between any two size variables can be described by the allometric equation:

$$y = ax^b$$

when y is one growth variable, e.g. shell length, x is the other variable, e.g. shell height, and a and b are coefficients. The exponent b is the growth coefficient and represents the relative growth rate of the two variables, while a is the value of y when x is unity. In logarithmic form the equation becomes:

$$\log y = \log a + b \log x$$

The slope b and intercept a of the transformed data are estimated by regression analysis. If x and y have the same units of measurement a value of unity for b indicates that the rate of growth of the two variables is identical (isometric) and that geometrical similarity is maintained with increasing size. Values of b greater than unity indicate that y is increasing relatively faster than x (positive allometry), while values of b less than unity indicate the reverse (negative allometry). If the dimensions of the two variables x and y differ then different criteria for isometry and allometry apply (Seed, 1980).

Most bivalves exhibit some change in shape with increasing body size. Gradual allometric changes are usually associated with the maintenance of physiological favourable surface area to volume ratios rather than with changing environmental conditions. For example, in the mussel *Mytilus edulis* differential growth among shell variables (height, length, width, weight) is reflected in a gradual change in shell shape, and larger (older) mussels have relatively heavier and more elongated shells in which width frequently exceeds shell height (Seed, 1968). This may be particularly beneficial to mussels living in densely crowded conditions as a wedge-shaped profile effectively elevates the posterior current flow among conspecifics (Yonge & Campbell, 1968). Wave impact, trophic conditions and water depth are also believed to influence shell proportions in bivalves (Seed, 1980). The allometric equation is also employed in physiological investigations, and for obtaining estimates of seasonal variation in growth or productivity.

One disadvantage of the use of bivariate data in allometric studies is that only two variables are compared at any given time, so that this type of analysis deals only with changing proportions, but not changing shape of shell. Size-independent shape characterisation can now be rapidly and accurately assessed by a method known as Fourier analysis. Shell outlines are captured by video imagery and the Fourier series mathematically compartmentalises the shape outline into a series of standard sinusoidal components which when summated converge to the empirical shape of the outline. The method, which is capable of discerning rather subtle differences among very similar, nearly oval shapes has been used on mussel shells (Ferson *et al.*, 1985), and more recently in scallops. Comparisons of Fourier harmonics between four scallop beds separated by age and sampling year, and corrected for allometric effects, revealed significant differences in shape, which Kenchington & Full (1994) suggested might have a genetic basis.

Growth curves

Growth in bivalves is commonly represented by plotting changes in mean size or mean weight of individuals against age. Growth rate is usually rapid during the first years of life but slows down as size increases. Various equations are used to fit curves to the growth data but the one which best describes growth in bivalves and in fish is the von Bertalanffy (1938) growth equation:

$$l_t = L_\infty(1 - \exp[-k(t - t_0)])$$

where l_t is the length at age t, L_∞ is the theoretical maximum size attained under specific environmental conditions, and k is a growth constant reflecting the rate at which maximum size L_∞ is approached (Fig. 6.8). The assumption of equations such as this is that growth is determinate, i.e. some maximum attainable size exists for any given population. However, in many bivalves growth may be indeterminate, i.e. growth continues right through the entire life span (Seed, 1980), and because of this some workers use polynomial expressions to describe shell growth in preference to the von Bertalanffy or equivalent growth equations.

Both L_∞ and k can be estimated by using the Ford-Walford plot (Walford, 1946) where length at age t is plotted against length at $t + 1$ years. L_∞ is given where the line of best fit intercepts the 45° angle, or by:

$$y/1 - slope$$

where y is the intercept of best-fit line on the y axis. The growth constant is calculated as minus the natural log of the slope (Gulland, 1969).

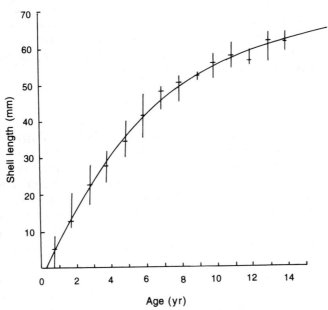

Fig. 6.8. A von Bertalanffy growth curve for a sample of mussels *Mytilus edulis* (N = 417) from Dungarvan Harbour, south coast of Ireland. The predicted maximum size (L_∞) for this population is 70.74 mm shell length, achieved after considerably more than 14 years of growth (k = 0.1598, t_0 = 0.423). This L_∞ value is fairly typical of Irish sheltered shore mussels. From Ottway & Ross (unpublished data).

While the von Bertalanffy equation has been widely used to describe growth in bivalves it does not take into account that many species, e.g. scallops, exhibit sigmoid (an inflection point on the growth curve) growth at a small size. The Gompertz equation, which is similar to the von Bertalanffy equation but uses the logarithms of length, has been used to describe growth in many scallops studies (Orensanz *et al.*, 1991), and also in slow growing populations of *Mytilus* (Seed & Suchanek, 1992). Neither of these equations takes into account seasonal variations in growth rate.

Scope for growth

An alternative to direct measurements of growth is a method based on the physiological energetics of the test individuals. Physiological energetics is concerned with the study of energy balance within individuals, not only in terms of the acquisition and expenditure of energy, but also with the efficiency with which it is converted from one form to another (Bayne & Newell, 1983).

The most common way to assess energy balance in an individual is to measure the various components of the balanced energy equation (Winberg, 1960):

$$C = P + R + F + U$$

Food consumption (*C*) is the primary source of energy input, while faecal losses (*F*), excretory products (*U*), respiratory heat loss (*R*), and energy invested in production of shell, soft tissue and gametes (*P*) represent energy expenditures. The absorbed ration (*A*) is the actual amount of material digested and is represented by $C - F$. The efficiency with which the ingested ration (*C*) is absorbed, A/C, is called the absorption efficiency (see Chapter 4).

By rearranging the energy balance equation we get:

$$P = A - (R + E)$$

where $E = F + U$. Production is, therefore, the difference between energy gains from the absorbed ration (*A*) and metabolic losses through respiration and excretion ($R + E$). An organism can only allocate energy to growth or reproduction if the energy gain from *A* exceeds total metabolic losses. This energy, surplus to metabolic demands, is referred to as 'scope for growth' or SFG (Fig. 6.9) (Warren & Davis, 1967), although a better term might be 'net energy balance' (Bayne, 1998). Note that in the majority of studies SFG does not distinguish between energy used for somatic growth and energy used for reproduction.

SFG is usually measured under laboratory conditions and over a short time-span. Thus, the effect of variables such as temperature, salinity, food type and ration on SFG can be quantified. Absorption efficiency (*AE*) is measured by comparing the proportion of the organic matter in the food and faeces of animals kept in individual experimental chambers (Conover, 1966). Oxygen consumption (*R*) is measured using an oxygen electrode to track the decline in oxygen partial pressure for individual animals maintained in closed glass

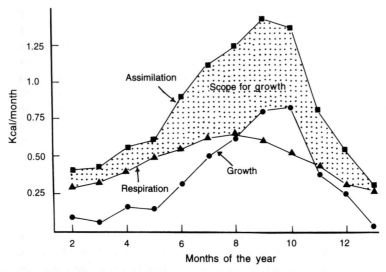

Fig. 6.9. Seasonal variations in energies of growth, respiration and assimilation in a 40 g inter-tidal oyster *Crassostrea virginica* over a 12 month period. The stippled area represents scope for growth. From Dame (1972) and Bayne & Newell (1983).

respirometers held in a temperature controlled water bath (Widdows & Johnson, 1988). Ammonia excretion (E) is measured by taking water samples at regular intervals from individual animals maintained in air-saturated filtered seawater (Widdows, 1985). Mucus should also be included under excretion. Bivalves may produce large quantities of mucus in pseudofaeces production yet loss of this potential energy source has not yet been documented (Hawkins and Bayne, 1992). All physiological rates are converted to mass-specific rates for animals of 1 g dry weight using appropriate weight exponents (b) in the allometric equation above. For mussels $b = 0.65$ for oxygen consumption and ammonia excretion (Widdows & Johnson, 1988). The physiological responses A, R and E are converted into energy equivalents, $J g^{-1} h^{-1}$ so SFG is expressed in these terms.

Generally, there is good agreement between growth rates estimated from energy budgets and growth rates measured by more direct methods (references in Bayne, 1998). However, some workers have suggested that in the scallop *Placopecten magellanicus* SFG overestimates actual growth under nitrogen-limited conditions (Grant & Cranford, 1991).

As seen above, SFG is equivalent to total production (P), and is the sum of the reproductive output (P_g) and growth of somatic tissue and shell (P_s). Both P_s and P_g are usually measured under controlled laboratory conditions or alter-natively, P can be measured indirectly as 'scope for growth'. Laboratory esti-mates of SFG, when made under appropriate conditions, have been shown to realistically predict production in the field (references in Griffiths & Griffiths, 1987).

Methods of estimating somatic growth have been described above. Repro-ductive output (P_g) can be measured directly by inducing animals to spawn and collecting the gametes, or from estimates of total gonad volume and counts of mature oocytes in serially sectioned gonads. Alternatively, P_g can be inferred indirectly from comparisons of gonad weight before and after spawning. Table 6.1 presents data on P_g as a proportion of total production (P) in a selection of bivalve populations. Mean P_g/P is 44%, but there is enormous inter-specific variation, with values ranging between 16 and 98%. Also, variation in P_g within a single species may be at least as great. For example, in the clam *Angulus* (*Tellina*) *tenuis* P_g/P varied between 8% and 62% over a three-year period (Trevallion, 1971). Similar variations have been reported in *Choromytilus meridionalis* populations, and to a lesser extent in *Perna perna* (Griffiths & Griffiths, 1987). Factors such as age (size) and changing environmental conditions contribute to this variation. Increase in body size for example, is generally accompanied by an increasing investment in gamete production (see Chapter 5 for more detail).

The methods that are used for estimating production in individual animals can equally well be applied to whole populations, or even to communities. Data on populations are normally derived by integrating field measurements on population density, age structure and production with laboratory estimates of C, A, R and E of different-sized individuals. The results are generally expressed in terms of energy flux per unit area of habitat $kJ\,m^{-2}\,yr^{-1}$. Production, expressed as a percentage of consumption, along with P_g and P_s (somatic growth) values for a variety of bivalve species are given in Table 6.2. Values for production vary from a low of 1–2% in the scallop *Chlamys islandica*, to over 20% for the two oyster species *Crassostrea gigas* and *C. virginica*, and the scallop *Patinopecten yessoensis*. Such variation reflects inter-specific growth differences (somatic and reproductive), and differences in recruitment, immigration, mortality and emigration. The variation in P_g and P_s estimates between species largely reflects differences in age structure. This can also be seen when individuals of the same species are analysed from a single location. For example, in Killary Harbour, a fjord on the Irish west coast, production of cultured mussels exceeded production of the wild population on the shore by an order of magnitude (Rodhouse *et al.*, 1985). However, the wild popula-

Table 6.1. Reproductive output (P_g) in relation to total production (P) in $kJ\,m^{-2}\,yr^{-1}$ in bivalves. Adapted from Griffiths & Griffiths (1987).

Species	$P_g/P \times 100$
Aulacomya ater	63
Mytilus edulis	35
Choromytilus meridionalis	75
Perna perna	16
Crassostrea gigas (mature)	98
Crassostrea virginica	16
Ostrea edulis	47
Mercenaria mercenaria	46
Chlamys islandica	28
Patinopecten yessoensis	16
Mean	44

Table 6.2. Energy budget parameters for a variety of bivalve populations: P_s (somatic growth), P_g (reproductive output) and production (P) are each expressed as a percentage of consumption (C; $kJ\,m^{-2}\,yr^{-1}$). Dashes indicate that no measurement was made. The enormous variation in C mainly reflects differences in population density (<10–10000 individuals m^{-2}). Adapted from Griffiths & Griffiths (1987).

Species	C	P_s	P_g	P
Aulacomya ater	27499	3.4	5.8	9.2
Chlamys islandica	55711	1.0	0.4	1.4
Choromytilus meridionalis	831890	1.6	4.8	6.4
Crassostrea gigas (mature)	10140	0.4	20.7	21.1
Crassostrea virginica	8811	17.7	3.4	21.1
Mercenaria mercenaria	5400	5.6	4.7	10.3
Mytilus edulis	–	8.9	4.8	13.7
Ostrea edulis	134–468	6.4	5.5	11.9
Patinopecten yessoensis	–	22.7	4.3	27.0
Mean		7.5	6.0	13.5

tion, which is dominated by large, older mussels (40–60 mm shell length) with a high reproductive output, makes a major contribution to total gamete production in the inlet.

Factors affecting growth

There are many factors that influence growth in bivalves. Of these, food supply is considered to be the most important since without this, sustained growth is impossible (Seed & Suchanek, 1992). But food supply to bivalves is influenced by factors such as temperature, aerial exposure, water depth and population density. Also some modulators of growth such as temperature and salinity interact, often synergistically. It is for these reasons that it is usually very difficult to quantify the precise influence of a single environmental factor on growth in natural populations of bivalves. In addition, endogenous influences that are inherent to the organism, e.g. genotype and physiological status, interact in a complex way with environmental factors. Although the bulk of published information deals with the effect of environmental factors on absolute growth, where possible their effect on scope for growth will also be dealt with.

Environmental modulators

Food

Food supply has consistently been shown to be the most important factor in determining bivalve growth in both hatchery (see Chapter 9) and wild populations. In the wild growth has been shown to be correlated with phytoplankton abundance (Utting, 1988; Smaal & van Stralen, 1990). For example, Page & Hubbard (1987) found that shell growth of *Mytilus edulis* on an offshore platform correlated with chlorophyll-*a* concentrations, while growth of

Crassostrea gigas on nets at ten different locations correlated with chlorophyll-*b* concentrations (Brown, 1988). However, there is growing evidence to show that growth may in fact be phytoplankton-limited owing to the enormous filtration capacity of bivalves. Consequently, in order to meet their energy requirements bivalve populations must exploit non-phytoplanktonic carbon in the form of re-suspended sediment, a complex mixture of benthic microflora, microalgae, fine organic detritus as well as quantities of inorganic material. Results from both laboratory and field experiments have shown that in appropriate concentrations sediment suspensions enhance growth. In the mussel *Mytilus edulis* an increase in growth has been reported at low or moderate seston concentrations of $1–25\,\text{mg}\,\text{l}^{-1}$ with a general decrease at high seston concentrations of $40–50\,\text{mg}\,\text{l}^{-1}$ (see Fréchette & Grant, 1991 for references). Similar results have been obtained from scope for growth experiments. When SFG in mussels (*Aulacomya ater*) fed on kelp detritus was compared with those fed on algae (Griffiths & King, 1979; Stuart, 1982), absorption of algal cells differed markedly from absorption of detritus. Also, mussels continued to maintain positive and increasing SFG at high ration levels of detrital material (Fig. 6.10). This is even more obvious when different size classes of mussel are compared (Fig. 6.11). On both diets small mussels exhibit faster relative growth rates largely because they are metabolically more efficient, consistently filtering larger volumes per unit of oxygen consumed than larger mussels (Griffiths & Griffiths, 1987). Response to variation in the concentration of suspended particles is also species-specific. In the scallop *Placopecten magellanicus* SFG at low particle concentrations equalled or exceeded SFG in the clam *Mya arenaria*, but the opposite was true at

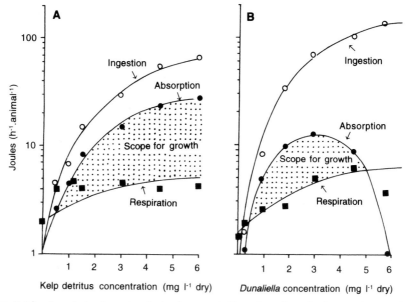

Fig. 6.10. Ingested ration, absorbed ration, respiration rate and resultant scope for growth in the mussel *Aulacomya ater* (50 mm shell length) as a function of increasing concentration of (A) *Dunaliella primolecta* and (B) kelp detritus. From Stuart (1982) with permission from Elsevier Science.

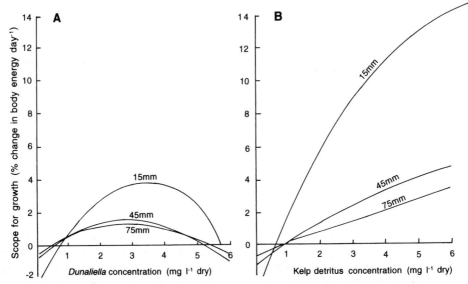

Fig. 6.11. Scope for growth in relation to food concentration in *Aulacomya ater* of differing shell lengths when fed on (A) *Dunaliella primolecta* and (B) kelp detritus. From Griffiths & Griffiths (1987) based on data in Griffiths & King (1979) and Stuart (1982).

high concentrations. The latter species appears to be better suited to cope with higher concentrations of seston, especially if it consists of relatively poor quality particles (MacDonald *et al.*, 1998).

The availability of re-suspended material as a source of food to bivalves is dependent on several factors. This source of food is more readily available to bivalves in suspended culture than to benthic-dwelling bivalves. The potential for re-suspension of organic matter is greater over mud than over sand but growth in bivalves is in fact poorer over mud than over sand due to excessive turbidity over muddy bottoms. The seston becomes diluted with inorganic particles, resulting in a decrease in seston quality and also to progressive clogging of the bivalve filtering mechanism. Other factors such as the hydrography of an area and weather patterns also affect re-suspension. In addition, the response to re-suspension is species-dependent. Growth enhancement by low additions (<5–$10\,\mathrm{mg\,l^{-1}}$) were reported for the mussel *Mytilus edulis* and the surf clam *Spisula subtruncata* but not for the hard clam *Mercenaria mercenaria* (references in Bricelj & Shumway, 1991). As re-suspended material plays an important role in bivalve diet Smaal & van Stralen (1990) have suggested that primary production data based on carbon turnover rather than chlorophyll values should be used in evaluating the feeding capacity of sites for bivalve culture (see Chapters 4 & 9 for additional information on feeding and food).

Tidal exposure

Bivalves only feed when they are submerged; the longer a species is out of water the less time it has to feed. It is not unexpected, therefore, that animals in the high intertidal zone exhibit markedly reduced growth rates compared to animals that are permanently covered by the tide. This is well illustrated for the oyster species *Tiostrea lutaria*, introduced into the UK from New Zealand in 1963. Growth of intertidal and subtidal oysters was similar over the first three years of life but after that growth of the intertidal population slowed appreciably (Richardson *et al.*, 1993b) (Fig. 6.12). Bivalves do, however, retain the potential for growth, even when they have been unable to exploit it due to unfavourable growing conditions. Seed (1968) found that when old slow-growing mussels were transferred down-shore they grew rapidly.

Species differ in their ability to tolerate aerial exposure. For example, Shurink & Griffiths (1993) found that of four mussel species *Mytilus gallo-provincialis* and *Perna perna* continue to grow at about 80% of the submerged rate even at 50% exposure level, compared to 66% for *Choromytilus meridion-alis* and only 54% for *Aulacomya ater*. But for most bivalve species a figure of around 50% exposure represents the point of zero growth, when the energy required for metabolism during aerial exposure exceeds that available during the feeding period (Seed & Suchanek, 1992). In reality this figure is an over-estimate since other factors, such as tolerance to extremes of temperature and desiccation may become limiting well before scope for growth declines to zero (Shurink & Griffiths, 1993).

Temperature

Water temperature varies with latitude and there is a general consensus in the literature that bivalves from low latitudes grow more rapidly at ambient tem-perature than do conspecifics from higher latitudes. However, growth rate is not a linear function of temperature. The reason for this is that several factors are superimposed and are often impossible to separate (Orensanz *et al.*, 1991). For example, growth rate and temperature are non-linearly related when a wide temperature range is considered. In mussels growth increases in a linear fashion between 3 and 20°C but declines sharply above 20°C and slows down

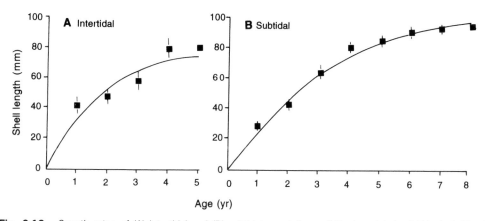

Fig. 6.12. Growth rates of (A) intertidal and (B) subtidal populations of *Tiostrea lutaria* at Tal-y-foel, Menai Strait, North Wales. Values are means (± SD). Curves were fitted using the von Bertalanffy equation: intertidal, $L_t = 79.89 (1 - \exp(-0.597t))$; subtidal, $L_t = 108.48(1 - \exp(-0.299t))$, where L_t is shell length at time t. From Richardson *et al.* (1993b).

at temperatures of 3–5°C (Almada-Villela *et al.*, 1982). Similar results have been observed for scallops (Mori, 1975). Also growth is affected by the reproductive cycle, which is temperature-dependent (Chapter 5).

Food supply, while predominantly light/nutrient limited, is also temperature-dependent. In temperate seas winter minimum temperatures and food availability are generally coincident, and growth in many species of bivalves seem to reflect changes in food availability rather than temperature. For example, growth rate in mussels on a submerged platform off the Californian coast was dependent on variations in phytoplankton concentrations, and temperature was virtually eliminated as an important growth regulator over the temperature range 10–18°C experienced by these mussels (Page & Hubbard, 1987). This is because bivalves (and other aquatic invertebrates) are able to adjust the components of energy balance (compensatory acclimation) in order to maintain positive scope for growth over a range of environmental temperatures. Widdows (1978) has shown that SFG in *Mytilus edulis* was virtually independent of temperature over the range 5–20°C (Fig. 6.13). However, at 25°C, which is close to the lethal temperature for this species, SFG declined sharply. This reflects a breakdown in the mechanism for metabolic compensation, resulting in an increase in metabolic costs but a decline in filtration rate (Griffiths & Griffiths, 1987). In contrast to the above results, Widman & Rhodes (1991) have shown that water temperature rather than food availability regulates growth in the

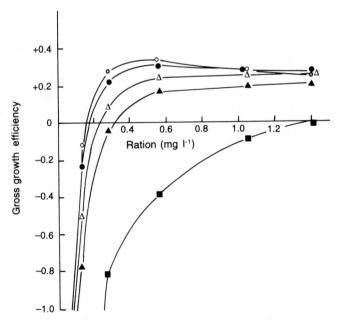

Fig. 6.13. The effects of acclimation temperature on the relationship between gross growth efficiency (growth per unit of ingested ration expressed as [A − R]/C) of the mussel *Mytilus edulis* (1 g dry flesh weight) and increasing concentration of the algae *Phaeodactylum tricornutum*. ○, 5°C; ●, 10°C; △, 15°C; ▲, 20°C; ■, 25°C. From Widdows (1978).

scallop *Argopecten irradians irradians*. The biochemical basis of temperature compensation is believed to involve the nervous system and both quantitative and qualitative changes in enzyme reaction rates (references in Hawkins & Bayne, 1992).

Salinity

Bivalves respond to changes in external salinity by closing the shell valves and by adjusting the intracellular concentrations of ions, amino acids and other small molecules to maintain a relatively constant cell volume. Initially rates of feeding and respiration are depressed but gradually recover as osmotic equilibration is reached. The period required for complete acclimation depends on the extent of the initial salinity change (Almada-Villela, 1984).

   Much of the information on the effect of salinity on growth pertains to mussel and oyster species. Mussels grow well in brackish estuaries but this is probably related to increased food levels in this habitat rather than to reduced salinity *per se* (Seed & Suchanek, 1992). Bayne (1965) has shown that mussel larvae from a North Wales population did not grow at 19 psu and had retarded growth at 24 psu, but at 30–32 psu growth was normal. However, in a population of larvae from the Øresund, where ambient salinity is lower than in Wales, growth occurred even at a salinity as low as 14 psu. This illustrates that larval growth seems to reflect the environmental conditions experienced by the adult and suggests that there may be a genetic component to salinity adaptation. In the Baltic Sea, however, where salinity is as low as 4–5 psu mussels survive but grow at a considerably reduced rate compared to North Sea mussels. It should be pointed out at this stage that the mussel in the Baltic is *Mytilus trossulus*, regarded by some as a distinct species but by others as an ecotype of *Mytilus edulis* (see Gosling, 1992 for review). Although there are differences in filtration rate and respiration between Baltic and North Sea mussels the slow growth of the former has been attributed to the increased excretion of ammonia and metabolism of free amino acids in this hypoosmotic environment. Tedengren & Kautsky (1986) have suggested that this probably represents a continuous energy loss for mussels in the Baltic.

   Loosanoff (1953) studied the effect of salinity on growth in laboratory maintained oysters (*Crassostrea virginica*) collected from Long Island Sound, New York. The results showed that growth was stunted at 7.5 psu and virtually non-existent at 5 psu, but at 12 psu growth was not significantly different from growth at 27 psu, the normal salinity experienced by these oysters. He suggested that 10 psu is the minimum salinity at which adult oysters exhibit normal growth rates. Similar results have been observed for cultured populations of this species (Bataller *et al.*, 1999).

   There is little published information on the effect of salinity on scope for growth, although Livingstone *et al.* (1979) have shown that in *M. edulis* SFG is maintained over a wide salinity range 20–30 psu, but is depressed at 15 psu. Once again, although the physiological mechanism by which SFG is linked to salinity has not been fully explained, excretory losses in low and fluctuating salinity must be considered as a potentially significant energy drain (Bayne & Newell, 1983).

Stock density

Most of the information on density as a modulator of growth has come from studies on scallops grown in suspended culture. Various workers have examined the effect of a range of stocking densities on scallop growth with a view to determining optimal stocking densities for different stages of the life cycle. Almost without exception there is an inverse relationship between density and growth. When *Placopecten magellanicus* were held at different stocking densities (15 to 90 individuals per net; initial shell height 10–15 mm) individuals held at lower density had faster growth rates, greater individual whole weights, higher individual meat weights and increased production compared to those held at higher stocking densities (Table 6.3). Similar results have been observed for other species of scallop (Parsons & Dadswell, 1992; Rheault & Rice, 1996; Román *et al.*, 1999; Maguire & Burnell, 2001), for rope-grown mussels (Rodhouse *et al.*, 1984), and for field and hatchery-cultured oysters (Holliday *et al.*, 1993; Rheault & Rice, 1996). One explanation is that a greater stocking density reduces food availability per individual. An alternative hypothesis is that growth is reduced at high densities because of the reduction in space; this leads to increased physical contact between individuals, with more frequent irritation and retraction of the mantle, or valve closure, resulting in less feeding (Côté *et al.*, 1993). To test these hypotheses Côté *et al.* (1994) grew *P. magellanicus* juveniles in nets at different densities in two series, one in which density was increased by adding living scallops, and a second in which density was increased by adding 'dummy' (non-living) scallops. The dummies occupied space but did not compete for food resources. For the first series, increasing the density of the scallops from 25 to 250 individuals per net caused a marked decrease in growth. In contrast, no significant decrease occurred in the dummy series. This clearly demonstrates that, for scallops at least, food depletion is the major factor causing decreased growth at high densities.

Wild populations also show a negative relationship between growth rate and density. Seed (1969) found that *Mytilus edulis* spat that recruited into a population of one-year-old mussels grew at less than half the rate than those recruiting onto an adjacent bare rock surface. Density also significantly reduced

Table 6.3. Average daily growth rates (mm day^{-1}) of scallops, *Placopecten magellanicus*, at different stocking densities for three different sampling intervals: I = August 1989–January 1990 (151 days); II = January–May 1990 (105 days); III = May–August 1990 (108 days), and for the whole year August 1989–August 1990. Adapted from Parsons & Dadswell (1992) and reprinted with permission from Elsevier Science.

Density	I	II	III	Year
15	0.158	0.054	0.100	0.111
30	0.154	0.044	0.084	0.101
45	0.148	0.050	0.060	0.094
60	0.141	0.051	0.047	0.087
75	0.140	0.042	0.051	0.085
90	0.115	0.055	0.070	0.084

growth rate in the clam *Spisula solidissima* and the effect was apparent in clams from 3–17 years of age (Weinberg, 1998). The effect of density on growth appears to be species specific. For example, in mussels, growth in *Choromytilus meridionalis* is more sensitive to density than is the case for *Aulacomya ater* or *Mytilus galloprovincialis*. This appears to correlate with natural packing densities in the wild: *C. meridionalis* usually occurs as a monolayer, while both *A. ater* and *M. galloprovincialis* form dense multi-layered beds (Schurink & Griffiths, 1993).

Water depth and flow

Once again, considerable information on the effect of water depth on growth derives from studies carried out on various scallop species in suspended culture. Scallops from inshore, shallower waters display higher growth rates and maximum sizes than those from deeper waters. It is the relatively higher temperatures and, more importantly, the higher food levels in the more productive shallower waters, rather than depth *per se*, that is responsible. However, the inverse relationship between growth and depth is not consistent for different tissues. For example, in the scallop *Placopecten magellanicus*, Emerson *et al.* (1994) have shown that although muscle and remaining soft tissue growth was significantly lower near the bottom, shell growth was almost uniform throughout the water column. These results suggest that high concentrations of seston near the bottom actually inhibit growth rather than providing an energetic benefit.

Another potentially important factor that varies with depth is fouling. Colonisation of scallop nets by mussels, barnacles and hydroids is greatest near the surface, and undoubtedly reduces the flow of water and suspended food particles to scallops. In addition, many of the fouling organisms are suspension feeders and most likely compete with the scallops for food resources. Therefore, fouling can negate the positive effects of increased temperature and abundance of suspended food at the surface (Côté *et al.*, 1993). The importance of fouling as a growth modulator was strikingly demonstrated by Claereboudt *et al.* (1994a) when they reported an increase of 68% in muscle mass of scallops in cleaned versus fouled nets that had been suspended at a depth of 9 m for 4 months.

Generally speaking, reduced growth rates are observed for suspension feeding bivalves in areas of low current speed and high population densities. This is attributed to a reduction in seston supply and consequent food limitation (Wildish & Kristmanson, 1985) and is one of the main reasons why bivalves held in suspended culture grow much more rapidly than those held under comparable conditions on the shore. Excessively high current speeds, however, reduce growth by inhibiting feeding activity. The underlying mechanism of growth inhibition is believed to be a build up of a pressure differential between inhalant and exhalant apertures that interferes with feeding (Wildish & Kristmanson, 1988). Yet some species do thrive in excessively turbulent wave-swept conditions. For example, on the south coast of South Africa the mussel *P. perna* grows twice as fast on wave-exposed as on sheltered shores (McQuaid & Lindsay, 2000).

To test the effect of current velocity on growth Claereboudt *et al.* (1994b) placed juvenile scallops (*Placopecten magellanicus*) glued to panels either inside or on the outside of pearl nets, and placed these at two sites where environmental conditions were similar except for a two-fold difference in the current velocity (mean velocity <0.9 m and >0.16 m sec^{-1}, respectively). For scallops on the outside of the pearl nets daily specific growth rates of the shell were similar at the two sites, but the mass of shell and flesh scaled to shell height was greater under weaker current conditions. At the strong current site tissue masses were greater for scallops inside than outside the net, probably because the net reduced the frequency of velocities that inhibit feeding. Pearl nets decreased water flow by 46–61%. In contrast, at the slow current site, tissue masses were less inside the pearl net, probably because flow was slowed to the extent that seston depletion was limiting growth. These results highlight the importance of adapting culture methods to hydrodynamic conditions at bivalve growing sites in order to optimise growth.

Under laboratory conditions currents far lower than those cited above have been shown to inhibit growth in scallops. Using a multiple-channel flume system Wildish *et al.* (1987) and Wildish & Kristmanson (1988) found that growth of adult *Placopecten magellanicus* was inhibited at current velocities exceeding 10 cm sec^{-1}. Filtration rates were reduced to 50% of the optimum at a flow of 40 cm sec^{-1} and ceased at velocities approaching 70 cm sec^{-1}. However, in the field scallops exhibit good growth in areas where surface current speeds are as high as 70 cm sec^{-1}, although current speeds on the bottom are much lower than this, and will be further avoided by the scallop's habit of recessing. Unlike adults, juvenile *Argopecten irradians* (Cahalan *et al.*, 1989) do not exhibit a sharp decline in growth above some threshold flow velocity, probably because in their natural habitat they live byssally attached to elevated substrates and experience flow speeds as high as 17 cm sec^{-1} (Eckman *et al.*, 1989).

Pollutants

Bivalves, in particular mussels, have been widely used to assess the effects of specific toxicants on such physiological responses as mortality, shell valve gape, scope for growth (see later), shell and tissue growth (Widdows & Donkin, 1992 for review). In addition, a variety of biosassays using bivalve embryos and larvae have been developed to monitor the toxic effect of marine pollutants on growth, and development both in the laboratory (His *et al.*, 1999) and in the field (Geffard *et al.*, 2002). The most common toxicity test is the 96 h LC$_{50}$, which determines the concentration of toxicant that results in a 50% lethal response over a period of 96 h exposure. Strømgren (1982, 1986) and Strømgren & Bonard (1987) have estimated water concentrations of various toxicants that produce a 50% reduction in shell growth rate in juvenile mussels (Table 6.4). In general, tissue growth is more sensitive to toxicants than shell growth. Also, the effect on growth depends on the type of pollutant being tested. For example, Widdows *et al.* (1997) have reported significant negative correlations between SFG and tissue concentrations of petroleum hydrocarbons, PCBs, DDT and HCH, but no association between

Table 6.4. Water concentration of various toxicants that induce a 50% reduction in shell growth rate of juvenile *Mytilus edulis*. Adapted from Widdows & Donkin (1992) and reprinted with permission from Elsevier Science.

Toxicant	Water concentration ($\mu g\,l^{-1}$)
Copper (Cu)	4
Cadmium (Cd)	100
Mercury (Hg)	<1
Zinc (Zn)	60
Lead (Pb)	>200
Nickel (Ni)	>200
Tributyltin (TBT)	<1
Oil	1500

growth and tissue levels of heavy metals (Cd, Co, Cu, Fe, Hg, Mn, Ni, Pb and Zn).

It might be expected that larval growth would be more sensitive to toxicants than juvenile or adult growth but this is not the case. Beaumont *et al.* (1987) found that EC_{50} (median effective concentration) values for shell growth in response to Cu were about 30 times lower for adult mussels (water concentration of $5\,\mu g\,l^{-1}$) than those reported for larvae ($150\,\mu g\,l^{-1}$ Strømgren, 1986), or juveniles (see Table 6.4). Generally, adults tend to be about 4–10 times more sensitive than larvae to a range of pollutants such as Cu, hydrocarbons, TBT and sewage sludge. The reduced sensitivity of larvae has been ascribed to a combination of their reliance on energy reserves, rather than direct feeding, and the absence of a developed nervous system, an important site of toxic action (Widdows & Donkin, 1992).

Toxicants often exhibit synergistic effects on growth. For example, when oyster spat (*Saccostrea commercialis*) were exposed to both Cu and TBT they suffered a greater reduction in growth than spat exposed to only one toxicant (Nell & Chvojka, 1992).

Other factors

Light appears to modulate growth, at least in mussels (see Seed & Suchanek, 1992 for references). Keeping mussels in continuous darkness, reduced levels of light, or photoperiods of 7 h or less, all significantly increased growth, probably due to increased feeding activity. Evidence of a circadian rhythm in shell gaping, with a greater duration of shell closure during the hours of expected daylight, lends support to this view (Ameyaw-Akumfi & Naylor, 1987).

Storms also can affect growth. Juvenile clams, *Mercenaria mercenaria*, maintained under simulated storm conditions showed up to a 40% reduction in shell growth compared to clams under gentle wave conditions (Turner & Miller, 1991). The decrease in shell growth was attributed to a reduction in filtration due to high concentrations of re-suspended sediment, coupled with energy costs of increased pseudofaeces production.

Other factors that have been shown to decrease bivalve growth are chlorinated water (Thompson *et al.*, 2000) low oxygen concentrations in water

(Baker & Mann, 1992), toxic algal blooms (Chauvaud *et al.*, 1998; see Chapter 12), parasites (see Chapter 11), manual handling of bivalves in land-based culture (Jakob & Wang, 1994), predation (Nakaoka, 2000) and inter-specific competition (Ahn *et al.*, 1993).

Endogenous growth modulators

Genotype

That variation for growth in bivalve has a genetic basis is without question. Evidence comes from reciprocal transfers between populations, breeding studies, allozyme analysis and genetic manipulation (see Chapter 10 for further information). One of these – allozyme analysis – has played a major role in helping to elucidate the mechanistic basis of genotype-dependent growth. Greater heterozygosity, calculated for individual animals as the mean number of loci that are polymorphic, is positively correlated with growth rate in many bivalves. The physiological basis to the relationship is that increased heterozygosity enables the individual to sustain its basal metabolism with lower expenditure of energy. The saving in energy can then be redirected to fuel other functions such as somatic or gonad growth. There is now considerable evidence that reduced energy expenditure in more heterozygous individuals is largely associated with decreased protein turnover – the continuous degradation and renewal or replacement of cellular proteins. Costs of protein turnover account for a major proportion (18–26%) of total metabolic expenditure (Hawkins, 1991). Therefore, less heterozygous individuals have higher energy costs of maintenance and less energy left over for other processes. It is important to note, however, that where a heterozygosity/growth correlation has been observed heterozygosity typically explains only a small percentage (2–4%) of the variation in size among individuals, although a higher figure of 27% has been reported by Bayne & Hawkins (1997). Also, the magnitude of the relationship appears to depend on factors such as age, reproductive state, environmental conditions, background genetic effects, and the number and type of loci analysed.

There are conflicting views regarding the genetic interpretation of the heterozygosity/growth correlation. One view is that the alleles under study are acting as indicators of heterozygosity at linked loci affecting metabolism. Another view is that the enzyme loci themselves are responsible for the correlation. However, to date it has not been possible to demonstrate any direct connection between an electrophoretic locus and growth (see Britten, 1996). The search for marker loci associated with nuclear loci that control economically important traits such as growth (quantitative trait loci, or QTLs) is underway for commercially important fish and shellfish species. When such markers are found they can be used in marker-assisted selection (Chapter 10). An alternative approach is to identify the physiological mechanisms underlying variability in growth. Bayne *et al.* (1999) compared various physiological parameters (clearance rate, absorption efficiency, rates of oxygen consumption and ammonia-nitrogen excretion) in a selected fast-growing line of oysters (*Saccostrea commercialis*) with a control (unselected) line. Faster-growing oysters fed more rapidly, showed higher net growth efficiency and allocated less energy per unit of tissue growth than slower-growing individuals. But the challenge ahead is to link these processes more directly to individual differences in specific genotypes (Bayne *et al.*, 1999).

Size

Bivalve growth is rapid in the first year(s) of life but progressively slows down with increasing age (see Fig. 6.8). For any animal the surface area available for oxygen diffusion limits general metabolism. Metabolism is proportional to a constant power of the body weight as described above by the allometric equation:

$$y = ax^b.$$

In this case y is the metabolic rate (measured as oxygen consumption), x is body size, b is the exponent and a denotes metabolic rate of an animal of unit weight. While the value of a varies according to factors such as temperature, b approximates 0.75 (Hawkins & Bayne, 1992). From this it is clear that as size increases the metabolic rate decreases. In addition, for reasons that are not clear at present, ingestion rate also declines with increasing size, but at a faster rate than metabolic rate. Both these factors contribute to a decline in growth efficiency (growth per unit of absorbed ration) with size. Also, there is evidence in mussels that metabolic efficiency *per se* decreases with increasing size (Fig. 6.14) but it is probably the increasing proportional allocation of available energy into gamete production that ultimately puts the brake on growth

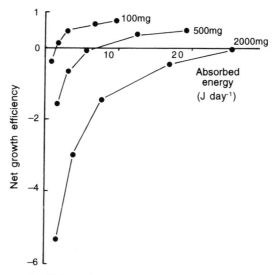

Fig. 6.14. Net growth efficiency (net energy balance/net energy absorbed) in relation to absorbed energy and body size (mg total dry tissue) in the mussel *Mytilus edulis*. After Hawkins & Bayne (1992) based on data in Thompson & Bayne (1974), with permission from Elsevier Science.

(Griffiths & Griffiths, 1987; Hawkins & Bayne, 1992). Because of various exogenous and endogenous influences, the proportional allocation of energy to somatic growth or reproduction varies widely between species, and even between populations of the same species (Figs. 5.7 & 5.8, Chapter 5).

Hormones and innate rhythms

Neurosecretory cells of the central nervous system play a role in the control of molluscan growth (Toullec *et al.*, 1992). In the mussel *Mytilus edulis* a low-molecular weight peptide has been purified from nervous ganglia that is believed to be a growth hormone. The molecule, protein-synthesis-stimulating factor (PSSF), stimulates DNA, RNA and protein synthesis (de Zwann & Mathieu, 1992). Another peptide, similar to human growth hormone, has also been identified in the haemolymph and digestive gland of *M. edulis* (Toullec *et al.*, 1992). To date there have been no reports of equivalent molecules in other bivalve species.

Bivalves display seasonal variation in digestion, absorption and excretion and until recently this was generally assumed to be the direct result of environmental changes in temperature and food availability. However, Hawkins & Bayne (1984) have shown that despite similar availability of the same unicellular algae, rates of ingestion by *Mytilus edulis*, acclimated in the laboratory during winter, were less than half those measured under identical conditions during summer. Similar results have been observed for other bivalve species. This suggests that net energy balance in bivalves is not maintained at a continuous steady maximum but that endogenous factors, at present unknown, regulate this inherent short-term rhythmicity (Hawkins & Bayne, 1992; Cranford & Hill, 1999).

References

Ahn, I.Y., Lopez, G. & Malouf, R. (1993) Effects of the gem clam *Gemma gemma* on early postsettlement emigration, growth and survival of the hard clam *Mercenaria mercenaria*. *Aquaculture*, **99**, 61–70.

Almada-Villela, P.C. (1984) The effects of reduced salinity on the shell growth of small *Mytilus edulis*. *J. mar. biol. Ass. U.K.*, **64**, 171–82.

Almada-Villela, P.C., Davenport, J. & Gruffydd, L.L.D. (1982) The effects of temperature on the shell growth of young *Mytilus edulis* L. *J. Exp. Mar. Biol. Ecol.*, **59**, 275–88.

Ameyaw-Akumfi, C. & Naylor, E. (1987) Temporal patterns of shell-gape in *Mytilus edulis*. *Mar. Biol.*, **95**, 237–42.

Baker, S.M. & Mann, R. (1992) Effects of hypoxia and anoxia on larval settlement, juvenile growth and juvenile survival of the oyster *Crassostrea virginica*. *Biol. Bull.*, **182**, 265–69.

Bataller, E.E., Boghen, A.D. & Burt, M.D.B. (1999) Comparative growth of the eastern oyster *Crassostrea virginica* (Gmelin) reared at low and high salinities in New Brunswick, Canada. *J. Shellfish Res.*, **18**, 107–14.

Bayne, B.L. (1965) Growth and delay of metamorphosis of the larvae of *Mytilus edulis* (L.). *Ophelia*, **2**, 1–47.

Bayne, B.L. (1998) The physiology of suspension feeding by bivalve molluscs: an introduction to the Plymouth 'TROPHEE' workshop. *J. Exp. Mar. Biol. Ecol.*, **219**, 1–19.

Bayne, B.L. & Newell, R.C. (1983) Physiological energetics of marine molluscs. In: *The Mollusca*, Vol. 4, *Physiology*, Part 1 (eds A.S.M. Saleuddin & K.M. Wilbur), pp. 407–515. Academic Press, New York.

Bayne, B.L. & Hawkins, A.J.S. (1997) Protein metabolism, the costs of growth, and genomic heterozygosity: experiments with the mussel *Mytilus galloprovincialis* Lmk. *Physiol. Zool.*, **70**, 391–402.

Bayne, B.L., Svensson, S. & Nell, J.A. (1999) The physiological basis for faster growth in the Sydney rock oyster *Saccostrea commercialis*. *Biol. Bull.*, **197**, 377–87.

Beaumont, A.R., Tserpes, G. & Budd, M.D. (1987) Some effects of copper on the veliger larvae of the mussel, *Mytilus edulis*, and the scallop, *Pecten maximus* (Mollusca, Bivalvia). *Mar. Environ. Res.*, **21**, 299–309.

Bertalanffy, L. von (1938) A quantitative theory of organic growth. *Hum. Biol.*, **10**, 181–213.

Bhattacharya, C.G. (1967) A simple method of resolution of a distribution into Gaussian components. *Biometrics*, **23**, 115–35.

Bricelj, V.M. & Shumway, S.E. (1991) Physiology: energy acquisition and utilization. In: *Scallops: Biology, Ecology and Aquaculture* (ed. S.E. Shumway), pp. 305–46. Elsevier Science Publishers B.V., Amsterdam.

Britten, H.B. (1996) Meta-analyses of the association between multilocus heterozygosity and fitness. *Evolution*, **50**, 2158–64.

Brown, J.R. (1988) Multivariate analysis of the role of environmental factors in seasonal and site-related growth variation in the Pacific oyster *Crassostrea gigas*. *Mar. Ecol. Prog. Ser.*, **45**, 225–36.

Cahalan, J.A., Siddall, S.E. & Luckenbach, M.W. (1989) Effects of flow velocity, food concentration and particle flux on the growth rate of juvenile bay scallops, *Argopecten irradians*. *J. Exp. Mar. Biol. Ecol.*, **129**, 45–60.

Cerrato, R.M. (1980) Demographic analysis of bivalve populations. In: *Skeletal Growth of Aquatic Organisms* (eds D.C. Rhoads & R.A. Lutz), pp. 417–68. Plenum Press, New York.

Chauvaud, L., Thouzeau, G. & Paulet, Y.M. (1998) Effects of environmental factors on the daily growth rate of *Pecten maximus* juveniles in the Bay of Brest (France). *J. Exp. Mar. Biol. Ecol.*, **227**, 83–111.

Claereboudt, M.R., Bureau, D., Côté, J. & Himmelman, J.H. (1994a) Fouling development and its effect on the growth of juvenile giant scallops (*Placopecten magellanicus*) in suspended culture. *Aquaculture*, **121**, 327–42.

Claereboudt, M.R., Himmelman, J.H. & Côté, J. (1994b) Field evaluation of the effect of current velocity and direction of the growth of the giant scallop, *Placopecten magellanicus*, in suspended culture. *J. Exp. Mar. Biol. Ecol.*, **183**, 27–39.

Cochran, J.K. & Landman, N. (1993) Using radioisotopes to determine growth rates of marine organisms. *J. Chem. Edu.*, **70**, 749–54.

Conover, R.J. (1966) Assimilation of organic matter by zooplankton. *Limnol. Oceanogr.*, **11**, 338–54.

Côté, J., Himmelman, J., Claereboudt, M. & Bonardelli, J.C. (1993) Influence of density and depth on the growth of juvenile sea scallops (*Placopecten magellanicus*) in suspended culture. *Can. J. Fish. Aquat. Sci.*, **50**, 1857–69.

Côté, J., Himmelman, J.H. & Claereboudt, M.R. (1994) Separating effects of limited food and space on growth of the giant scallop *Placopecten magellanicus* in suspended culture. *Mar. Ecol. Prog. Ser.*, **106**, 85–91.

Cranford, P.J. & Hill, P.S. (1999) Seasonal variation in food utilization by the suspension-feeding bivalve molluscs *Mytilus edulis* and *Placopecten magellanicus*. *Mar. Ecol. Prog. Ser.*, **190**, 223–39.

Dame, R.F. (1972) The ecological energetics of growth, respiration and assimilation in the intertidal American oyster, *Crassostrea virginica*. *Mar. Biol.*, **17**, 243–50.

Davenport, J. & Glasspool, A.F. (1987) A photographic technique for the measurement of short-term shell growth in bivalve molluscs. *J. Moll. Stud.*, **53**, 299–303.

Dey, D.D. & Bolton, E.T. (1978) Tetracycline as a bivalve shell marker. *Proc. Natl. Shellfish Assoc.* **66**, 77.

Eckman, J.E., Peterson, C.H. & Cahalan, J.A. (1989) Effects of flow speed, turbulence, and orientation on growth of juvenile bay scallops, *Argopecten irradians concentricus* (Say). *J. Exp. Mar. Biol. Ecol.*, **132**, 123–40.

Emerson, C.W., Grant J., Mallet, A. & Carver, C. (1994) Growth and survival of sea scallops *Placopecten magellanicus*: effects of culture depth. *Mar. Ecol. Prog. Ser.*, **108**, 119–32.

Ferson, S., Rohlf, F.J. & Koehn, R.K. (1985) Measuring shape variations of two-dimensional outlines. *Syst. Zool.*, **34**, 59–68.

Fréchette, M. & Grant, J. (1991) An *in situ* estimation of the effect of wind-driven resuspension on the growth of the mussel *Mytilus edulis* L. *J. Exp. Mar. Biol. Ecol.*, **148**, 201–13.

Geffard, O., His, E., Budzinski, H., Seaman, M. & Garrigues, P. (2001) Étude in situ de la qualité biologique de l'eau de mer à l'aode du test embryo-larvaire de bivalves marins, *Crassostrea gigas* (Thunberg) et *Mytilus galloprovincialis* (Lamarck). *C.R. Acad. Sci. Paris, Série III*, **324**, 1149–55.

Gosling, E.M. (1992) Systematics and geographic distribution of *Mytilus*. In: *The Mussel* Mytilus: *Ecology, Physiology, Genetics and Culture* (ed. E.M. Gosling), pp. 1–20. Elsevier Science Publishers B.V., Amsterdam.

Grant, J. & Cranford, P.J. (1991) Carbon and nitrogen scope for growth as a function of diet in the sea scallop *Placopecten magellanicus*. *J. mar. biol. Ass. U.K.*, **71**, 437–50.

Griffiths, C.L. & King, J.A. (1979) Some relationships between size, food availability and energy balance in the ribbed mussel *Aulacomya ater*. *Mar. Biol.*, **51**, 141–49.

Griffiths, C.L. & Griffiths, R.J. (1987) Bivalvia. In: *Animal Energetics*, Vol. 2, *Bivalvia through Reptilia* (eds T.J. Pandian & F.J. Vernberg), pp. 1–87. Academic Press, California.

Gulland, J.A. (1969) *Manual of Methods for Fish Stock Assessment: Part I. Fish Population Analysis*. FAO Manuals in Fisheries Science, 4. FAO, Rome.

Hawkins, A.J.S. (1991) Protein turnover: a functional approach. *Funct. Ecol.*, **5**, 222–33.

Hawkins, A.J.S. & Bayne, B.L. (1984) Seasonal variation in the balance between physiological mechanisms of feeding and digestion in *Mytilus edulis* (Bivalvia: Mollusca). *Mar. Biol.*, **82**, 233–40.

Hawkins, A.J.S. & Bayne, B.L. (1992) Physiological interrelations and the regulation of production. In: *The Mussel* Mytilus: *Ecology, Physiology, Genetics and Culture* (ed. E.M. Gosling), pp. 171–222. Elsevier Science Publishers B.V., Amsterdam.

His, E., Beiras, R. & Seaman, M.N.L. (1999) The assessment of marine pollution-bioassays with bivalve embryos and larvae. *Adv. Mar. Biol.*, **37**, 1–178.

Holliday, J.E., Allan, G.L. & Nell, J.A. (1993) Effects of stocking density on juvenile Sydney rock oysters, *Saccostrea commercialis* (Iredale and Roughley), in cylinders. *Aquaculture*, **109**, 13–26.

Jakob, G.S. & Wang, J. (1994) The effect of manual handling on oyster growth in land-based cultivation. *J. Shellfish Res.*, **13**, 183–86.

Kaehler, S. & McQuaid, C.D. (1999) Use of the fluorochrome calcein as an *in situ* growth marker in the brown mussel *Perna perna*. *Mar. Biol.*, **133**, 455–60.

Kautsky, N. (1982) Growth and size structure in a Baltic *Mytilus edulis* population. *Mar. Biol.*, **68**, 117–33.

Kenchington, E.L. & Full, W.E. (1994) Fourier analysis of sea scallop (*Placopecten magellanicus*) shells determining population structure. *Can. J. Fish. Aquat. Sci.*, **51**, 348–56.

Kennish, M.J., Lutz, R.A. & Rhoads (1980) Preparation of acetate peels and fractured sections for observation of growth patterns within the bivalve shell. In: *Skeletal Growth of Aquatic Organisms* (eds D.C. Rhoads & R.A. Lutz), pp. 597–601. Plenum Press, New York.

King, M. (1995) *Fisheries Biology, Assessment and Management.* Fishing News Books, Oxford.

Koshikawa, Y., Hagiwara, K., Lim, B.K. & Sakuri, N. (1997) A new marking method for short-necked clam *Ruditapes philippinarum* with rust. *Fish. Sci.*, **63**, 533–38.

Lewis, D.E. & Cerrato, R.M. (1997) Growth uncoupling and the relationship between shell growth and metabolism in the soft shell clam *Mya arenaria. Mar. Ecol. Prog. Ser.*, **158**, 177–89.

Livingstone, D.R., Widdows, J. & Fieth, P. (1979) Aspects of nitrogen metabolism of the common mussel, *Mytilus edulis*: adaptation to abrupt and fluctuating changes in salinity. *Mar. Biol.*, **53**, 41–55.

Loosanoff, V.L. (1953) Behavior of oysters in water of low salinities. *Proc. Natl. Shellfish Assoc.*, **43**, 135–51.

Lubet, P. (1994) Growth and reserves. In: *Aquaculture: Biology and Ecology of Cultured Species* (ed. G. Barnabé), pp. 119–37. Ellis Horwood, London.

Lutz, R.A. (1976) Annual growth patterns in the inner shell layer of *Mytilus edulis* L. *J. mar. biol. Ass. U.K.*, **56**, 723–31.

MacDonald, B.A., Bacon, G.S. & Ward, J.E. (1998) Physiological responses of infaunal (*Mya arenaria*) and epifaunal (*Placopecten magellanicus*) bivalves to variations in the concentration and quality of suspended particles. II. Absorption efficiency and scope for growth. *J. Exp. Mar. Biol. Ecol.*, **219**, 127–41.

Maguire, J.A. & Burnell, G.M. (2001) The effects of stocking density in suspended culture on growth and carbohydrate content of the adductor muscle in two populations of the scallop (*Pecten maximus* L.) in Bantry Bay, Ireland. *Aquaculture*, **198**, 95–108.

McGrath, D. & Ó Foighil, D. (1986) Population dynamics and reproduction of hermaphroditic *Lasaea rubra* (Montagu) (Bivalvia, Galeommatacea). *Ophelia*, **25**, 209–19.

McQuaid, C.D. & Lindsay, T.L. (2000) Effect of wave exposure on growth and mortality rates of the mussel *Perna perna*: bottom-up regulation of intertidal populations. *Mar. Ecol. Prog. Ser.*, **206**, 147–54.

Mori, K. (1975) Seasonal variation in physiological activity of scallops under culture in the coastal waters of Sanriku District, Japan, and a physiological approach of a possible cause of their mass mortality. *Bull. Mar. Biol. Sta. Asamushi*, **15**, 59–79.

Nakaoka, M. (2000) Nonlethal effects of predators on prey populations: predator-mediated change in bivalve growth. *Ecology*, **81**, 1031–45.

Nell, J.A. & Chvojka, R. (1992) The effect of bis-tributyltin oxide (TBTO) and copper on the growth of juvenile Sydney rock oysters *Saccostrea commercialis* (Iredale and Roughley) and Pacific oysters *Crassostrea gigas* Thunberg. *Sci. Total Environ.*, **125**, 193–201.

Orensanz, J.M., Parma, A.M. & Iribarne, O.O. (1991) Population dynamics and management of natural stocks. In: *Scallops: Biology, Ecology and Aquaculture* (ed. S.E. Shumway), pp. 625–713. Elsevier Science Publishers B.V., Amsterdam.

Page, H.M. & Hubbard, D.M. (1987) Temporal and spatial patterns of growth in mussels *Mytilus edulis* on an offshore platform: relationships to water temperature and food availability. *J. Exp. Mar. Biol. Ecol.*, **111**, 159–79.

Parsons, G.J. & Dadswell, M.J. (1992) Effect of stocking density on growth, production and survival of the giant scallop, *Placopecten magellanicus*, held in intermediate suspension culture in Passamaquoddy Bay, New Brunswick. *Aquaculture*, **103**, 291–309.

Pauly, D. & Caddy, J.F. (1985) *A modification of Bhattacharya's method for the analysis of mixtures of normal distributions*. FAO Fishery Circular, 781.

Rheault, R.B. & Rice, M. (1996) Food-limited growth and condition index in the eastern oyster, *Crassostrea virginica* (Gmelin 1791), and the bay scallop, *Argopecten irradians irradians* (Lamarck 1819). *J. Shellfish Res.*, **15**, 271–83.

Richardson, C.A. (1989) An analysis of the microgrowth bands in the shell of the common mussel *Mytilus edulis*. *J. mar. biol. Ass. U.K.*, **69**, 477–91.

Richardson, C.A. (2001) Molluscs as archives of environmental change. *Oceanogr. Mar. Biol. Ann. Rev.*, **39**, 103–64.

Richardson, C.A, Seed, R. & Naylor, E. (1990) Use of internal growth bands for measuring individual and population growth rates in *Mytilus edulis* from offshore production platforms. *Mar. Ecol. Prog. Ser.*, **66**, 259–65.

Richardson, C.A, Collins, S.A., Ekaratne, S.U.K., Dare, P. & Key, D. (1993a). The age determination and growth rate of the European flat oyster, *Ostrea edulis*, in British waters determined from acetate peels of umbo growth lines. *ICES J. Mar. Sci.*, **50**, 493–500.

Richardson, C.A., Seed, R., Al-Roumaihi, E.M.H. & McDonald, L. (1993b) Distribution, shell growth and predation of the New Zealand oyster, *Tiostrea* (= *Ostrea*) *lutaria* Hutton, in the Menai Strait, North Wales. *J. Shellfish Res.*, **12**, 207–14.

Rodhouse, P.G., Roden, C.M., Hensey, M.P., McMahon, T., Ottway, B. & Ryan, T.H. (1984) Food resource, gametogenesis and growth of *Mytilus edulis* on the shore and in suspended culture: Killary Harbour, Ireland. *J. mar. biol. Ass. U.K.*, **64**, 513–29.

Rodhouse, P.G., Roden, C.M., Hensey, M.P. & Ryan, T.H. (1985) Production of mussels, *Mytilis edulis*, in a suspended culture and estimates of carbon and nitrogen flow: Killary Harbour, Ireland. *J. mar. biol. Ass. U.K.*, **65**, 55–68.

Román, G., Campos, M.J., Acosta, C.P. & Cano, J. (1999) Growth of the queen scallop (*Aequipecten opercularis*) in suspended culture: influence of density and depth. *Aquaculture*, **178**, 43–62.

Schurink, C.v.E. & Griffiths, C.L. (1993) Factors affecting relative rates of growth in 4 South African mussel species. *Aquaculture*, **109**, 257–73.

Seed, R. (1968) Factors influencing shell shape in *Mytilus edulis*. *J. mar. biol. Ass. U.K.*, **48**, 561–84.

Seed, R. (1969) The ecology of *Mytilus edulis* L. (Lamellibranchiata) on exposed rocky shores. 2. Growth and mortality. *Oecologia*, **3**, 317–50.

Seed, R. (1980). Shell growth and form in the Bivalvia. In: *Skeletal Growth of Aquatic Organisms* (eds D.C. Rhoads & R.A. Lutz), pp. 23–67. Plenum Press, New York.

Seed, R. & Suchanek, T.H. (1992) Population and community ecology of *Mytilus*. In: *The Mussel Mytilus: Ecology, Physiology, Genetics and Culture* (ed. E.M. Gosling), pp. 87–169. Elsevier Science Publishers B.V., Amsterdam.

Smaal, A.C. & van Stralen, M.R. (1990) Average annual growth and condition of mussels as a function of food source. *Hydrobiologia*, **195**, 179–88.

Sparre, P. & Venema, S.C. (1992) *Introduction to tropical fish stock assessment*. FAO Fishery Technical Paper, **306**.

Strømgren, T. (1975) Linear measurements of growth of shells using laser diffraction. *Limnol. Oceanogr.*, **20**, 845–48.

Strømgren, T. (1982) Effects of heavy metals (Zn, Hg, Cu, Cd, Pb, Ni) on the length growth of *Mytilus edulis*. *Mar. Biol.*, **72**, 69–72.

Strømgren, T. (1986) The combined effect of copper and hydrocarbons on the length growth of *Mytilus edulis*. *Mar. Environ. Res.*, **19**, 251–58.

Strømgren, T. & Bonard, T. (1987) The effects of tributyltin oxide on growth of *Mytilus edulis*. *Mar. Pollut. Bull.*, **18**, 30–31.

Stuart, V. (1982) Absorbed ration, respiratory cost and resultant scope for growth in the mussel *Aulacomya ater* (Molina) fed on a diet of kelp detritus of different ages. *Mar. Biol. Lett.*, **3**, 289–306.

Tedengren, M. & Kautsky, N. (1986) Comparative studies of the physiology and its probable effect on size in blue mussels (*Mytilus edulis*) from the North Sea and northern Baltic proper. *Ophelia*, **25**, 147–55.

Thompson, I.S., Richardson, C.A., Seed, R. & Walker, G. (2000) Quantification of mussel (*Mytilus edulis*) growth from power station cooling water in response to chlorination procedures. *Biofouling*, **16**, 1–15.

Thompson, R.J. & Bayne, B.L. (1974) Some relationships between growth, metabolism and food in the mussel *Mytilus edulis*. *Mar. Biol.*, **27**, 317–26.

Toullec, J.Y., Robbins, I. & Mathieu, M. (1992) The occurrence and *in vitro* effects of molecules potentially active in the control of growth in the marine mussel *Mytilus edulis* L. *Gen. Comp. Endocrinol.*, **86**, 424–32.

Trevallion, A. (1971) Studies on *Tellina tenuis* (da Costa). III. Aspects of general biology and energy flow. *J. Exp. Mar. Biol. Ecol.*, **7**, 95–122.

Turner, E.J. & Miller, D.C. (1991) Behavior and growth of *Mercenaria mercenaria* during simulated storm events. *Mar. Biol.*, **111**, 55–64.

Utting, S.D. (1988) The growth and survival of hatchery-reared *Ostrea edulis* L. spat in relation to environmental conditions at the on-growing site. *Aquaculture*, **69**, 27–38.

Walford, L.A. (1946) A new graphic method of describing the growth of animals. *Biol. Bull.*, **90**, 141–7.

Warren, C.E. & Davis, G.E. (1967) Laboratory studies on the feeding, bioenergetics and growth of fish. In: *The Biological Basis of Freshwater Fish Production* (ed. S.D. Gerking), pp. 175–214. Blackwell, Oxford.

Weinberg, J.R. (1998) Density-dependent growth in the Atlantic surfclam, *Spisula solidissima*, off the coast of the Delmarva Peninsula, USA. *Mar. Biol.*, **130**, 621–30.

Widdows, J. (1978) Physiological indices of stress in *Mytilus edulis*. *J. mar. biol. Ass. U.K.*, **58**, 125–42.

Widdows, J. (1985) Physiological procedures. In: *The Effects of Stress and Pollution on Marine Animals* (eds B.L. Bayne, D.A. Brown, K. Burns *et al.*, pp. 161–78). Praeger Press, New York.

Widdows, J. & Johnson, D. (1988). Physiological energetics of *Mytilus edulis*: scope for growth. *Mar. Ecol. Prog. Ser.*, **46**, 113–21.

Widdows, J. & Donkin, P. (1992) Mussels and environmental contaminants: bioaccummmulation and physiological aspects. In: *The Mussel* Mytilus: *Ecology, Physiology, Genetics and Culture* (ed. E.M. Gosling), pp.383–424. Elsevier Science Publishers B.V., Amsterdam.

Widdows, J., Nasci, C. & Fossato, V.U. (1997) Effects of pollution on the scope for growth of mussels (*Mytilus galloprovincialis*) from the Venice Lagoon, Italy. *Mar. Environ. Res.*, **43**, 69–79.

Widman, J.C. & Rhodes, E.W. (1991) Nursery culture of the bay scallop, *Argopecten irradians*, in suspended mesh nets. *Aquaculture*, **99**, 257–67.

Wildish, D.J. & Kristmanson, D.D. (1985) Control of suspension feeding bivalve production by current speed. *Helgo. Wiss. Meeresunters.*, **39**, 237–43.

Wildish, D.J. & Kristmanson, D.D. (1988) Growth response of giant scallops to periodicity of flow. *Mar. Ecol. Prog. Ser.*, **42**, 163–69.

Wildish, D.J., Kristmanson, D.D., Hoar, R.L., DeCoste, A.M., McCormich, S.D. & White, A.W. (1987) Giant scallop feeding and growth responses to flow. *J. Exp. Mar. Biol. Ecol.*, **113**, 207–20.

Winberg, G.G. (1960) Rate of metabolism and food requirements of fish. *Trans. Ser. Fish. Res. Board Can.*, **194**, 1–202.

Yonge, C.M. & Campbell, J.I. (1968) On the heteromyarian conditions in the Bivalvia with special reference to *Dreissena polymorpha* and certain Mytilacea. *Trans. R. Soc. Edinb.*, **68**, 21–43.

de Zwann, A. & Mathieu, M. (1992) Cellular biochemistry and endocrinology. In: *The Mussel* Mytilus: *Ecology, Physiology, Genetics and Culture* (ed. E.M. Gosling), pp. 223–307. Elsevier Science Publishers B.V., Amsterdam.

7 Circulation, Respiration, Excretion and Osmoregulation

Circulation

A detailed description of the heart and associated haemolymph ('blood') vessels is presented in Chapter 2. Briefly, haemolymph flows from the gills into the heart, which contracts to drive the fluid into a single vessel, the anterior aorta, which divides into many arteries. The most important of these are the pallial arteries, which supply the mantle, and the visceral arteries (gastro-intestinal, hepatic and terminal) that supply the stomach, intestine, shell muscles and foot with haemolymph (Fig. 2.15). The arteries break up into a network of vessels in all tissues and these then join to form veins that empty into extensive spaces called sinuses. The haemolymph system is, therefore, an open circulatory system, the haemolymph in the sinuses bathing the tissues directly. As well as having a respiratory function these sinuses also serve as a fluid skeleton giving temporary rigidity to various parts of the body, e.g. labial palps, foot and mantle edges (Morton, 1967). From the sinuses the haemolymph flows to the kidneys for purification. From there a branch enters the gill circulation for oxygenation and returns to the heart via the kidney. In some bivalves oxygenated haemolymph from the gills returns directly to the heart. In all bivalves the mantle has an extensive system of haemolymph sinuses, and thus may function as the primary respiratory site.

The haemolymph plays an important role in gas exchange, osmoregulation, nutrient distribution, elimination of wastes and internal defence (see below and Chapter 11). The haemolymph contains cells called haemocytes that float in colourless plasma. There is no respiratory pigment, the haemolymph oxygen concentration being similar to seawater (Bayne *et al.*, 1976a).

Haemolymph plasma and haemocytes

The osmotic concentration of haemolymph is determined chiefly by inorganic ions and is equal to, or marginally greater than, the osmotic concentration of seawater (Table 7.1). The metabolic costs of maintaining a steep plasma-seawater osmotic gradient would be prohibitive because of the vast surface areas exposed to the external environment. However, as is evident from Table 7.1 the K^+ concentration of bivalve plasma is greater than seawater by a factor of 1–2 (Burton, 1983), which probably reflects the normal intracellular-extracellular K^+ gradient. Haemolymph plasma also contains numerous dissolved organic molecules (Table 7.2), and concentrations fluctuate depending on food ration, gametogenesis and short-term osmoregulation (see below).

Table 7.1. The ionic composition of the haemolymph of the mussel *Mytilus edulis* at two salinities. From Bayne *et al.* (1976a).

	Ionic concentration (mM per kg water)					
	Na$^+$	K$^+$	Ca^{++}	Mg^{++}	Cl$^-$	SO$_4^{2-}$
Seawater 37.5 psu	507	10.6	11.3	57.9	594	30.5
Haemolymph (pH 7.7)	502	12.7	12.6	55.8	586	30.7
Brackish water 16 psu	215	5.1	5.7	24.2	253	13.2
Haemolymph	213	7.5	5.8	24.5	253	13.2

Table 7.2. Organic constituents in the haemolymph of the mussel *Mytilus edulis* and the scallop *Placopecten magellanicus*. Values for *M. edulis* were over an annual cycle (Bayne, 1973a), while those for *P. magellanicus* (means and standard deviations) were collected after a single sampling event (Thompson, 1977).

	Concentration (mg 100 ml^{-1})	
	M. edulis	*P. magellanicus*
Protein	115–282	153 ± 28
Non-protein-nitrogen	3.5–23.4	no data
Ammonia	0.31–1.79	0.24 ± 0.12
Sugars	9.8–35.7	no data
Total Carbohydrate	no data	5.2 ± 1.2
Lipid	20.4–84.3	13.7 ± 2.4

Haemocytes are not confined to the haemolymph system but move freely out of the sinuses (haemolymph spaces) into surrounding connective tissue, mantle cavity and gut lumen. Several morphologically distinct categories of haemocytes have been described, but to date there is no universally accepted system of classification. Two groups of cells are recognised, granulocytes and agranulocytes (Hine, 1999). The majority of haemocytes belong to the former group and are characterised by granules of various sizes and shapes in the cytoplasm. Granulocytes phagocytose bacteria, algae, cellular debris and protozoan parasites. Agranular haemocytes have few, if any, cytoplasmic granules, and appear to be less phagocytic then granulocytes. There is evidence of at least three different types of agranulocytes in bivalves, and perhaps even two types of granulocytes. The general term 'haemocyte' will be used in the following sections to embrace both granulocytes and agranulocytes.

Haemocytes represent the most important internal defence mechanism in marine bivalves (see Chapter 11). The cells recognise and react to foreign substances by phagocytosis or encapsulation. In some species the process is non-specific. For example, Tripp (1992) has shown that haemocytes in the clam *Mercenaria mercenaria* avidly phagocytose not only biotic material like bacteria, yeast or red blood cells from mammalian species, but also abiotic material such as polystyrene spheres. However, the process in general appears to be specific and involves several stages: non-self recognition, cell proliferation, locomotion, binding and ingestion of particulate matter and intracellular degradation of particles.

The mechanism by which haemocytes recognise invading pathogens is unclear. Migration of haemocytes towards the invader is believed to be by chemotaxis, where the invading organism secretes chemicals – usually peptides or small proteins – that act as attractants for the haemocytes (Fawcett & Tripp, 1994). Infection is generally accompanied by intense haemocytosis (haemocyte proliferation). Indeed, Oubella *et al.* (1993) suggested that haemocyte density in bivalves could be used as a quantifiable determinant of the immuno-defence response to physiological or pathological stress.

After phagocytosis, foreign organisms are destroyed by haemocytes either through the release of lysosomal enzymes or other lysins, or by the release of highly reactive oxygen species (ROS) such as superoxide, hydrogen peroxide, and oxygen and hydrogen radicals (Pipe, 1990). To defend themselves against ROS, haemocytes contain antioxidant enzymes (Pipe *et al.*, 1993). Several anti-microbial peptides have been characterised from mussel haemocytes. In parasitic infections haemocytes migrate to the parasitised tissues and synthesise and secrete a specialised product for parasite encapsulation. Ultimately, these haemocytes are destroyed by the parasite, causing a collapse of haemolymph sinuses, which may induce a fall in respiratory capacity, and ultimately death for the host (Montes *et al.*, 1995a,b). In oysters infected by the parasite *Haplosporidium nelsoni* haemocytes do not phagocytose the parasite, even in strains selected for resistance to MSX, an often-fatal infection caused by the parasite. Kanaley & Ford (1990) have found similarities in cell surface receptors between host and parasite, which may help to explain why oyster haemocytes fail to phagocytose the parasite – they do not recognise it as foreign. More recent findings suggest, however, that *H. nelsoni* may produce a substance inhibitory to haemocytes but that disease resistance has developed even though host haemocytes fail to attack the parasite successfully (Ford *et al.*, 1993).

Various pollutants are known to exert adverse effects on immunity, and this can affect resistance to infection, thus influencing survival. Polycyclic aromatic hydrocarbons (PAHs), which enter the marine environment from a range of industrial sources, inhibit phagocytosis in *M. edulis* haemocytes, and membrane stability of the lysosomes, which play a central role in the degradation of phagocytosed material, is disrupted (Grundy *et al.*, 1996). Similar effects have been reported for other pollutants such as heavy metals, pesticides, fungicides, PCBs (polychlorinated biphenyls) and tributyltin (Fisher *et al.*, 1990; Alvarez & Friedl, 1992; Lowe *et al.*, 1995; Grundy *et al.*, 1996; Oubella, 1997; Ringwood *et al.*, 1998). Therefore, structural alterations of lysosomes can be used as a biomarker of exposure to pollutants and deleterious compounds (Moore *et al.*, 1996; Livingstone *et al.*, 2000). Other pollutants, such as crude oil, reduce the production of reactive oxygen species (ROS) by haemocytes – an important mechanism used by bivalves for killing microorganisms and parasites (Austin & Paynter, 1995; Dyrynda *et al.*, 1997). In contrast, Winston *et al.* (1996) reported enhanced production of ROS in response to accumulation of contaminants in haemocytes. The consequences of this might include release of hydrolytic enzymes into the haemocyte cytoplasm, resulting in enzyme degradation and organelle damage. Heightened defences, e.g. increase in the density, rate of locomotion, and superoxide generation of haemocytes have been reported in animals living in polluted environments (Fisher *et al.*, 2000). Chromosomal damage in haemocytes can be used to measure the carcinogenic effect of toxins, or the potential damaging effect of unknown compounds (Bihari *et al.*, 1990; Dopp *et al.*, 1996).

Haemocytes also play a role in wound repair by migrating in great numbers to the damaged area, and plugging the wound while the epithelium regenerates. Haemocytes are also involved in the early stages of shell repair, through transport of calcium and organic matrix material from the digestive gland to the repair site. In addition to their role in combating infection, certain disease organisms, e.g. the early life stages of the protozoan parasite *Bonamia ostreae* in *Ostrea edulis*, have been detected in the cytoplasm of haemocytes. Also, proliferative growth of abnormal haemocytes is responsible for a group of serious cell disorders, termed haemocytic neoplasia, in bivalves (see Chapter 11). The heart tissue itself may be infected with *Vibrio* or the trematode *Proctoeces maculatus* (Lauckner, 1983).

Haemocytes also have a digestive function, ingesting particles of nutritional value that are too large to enter the cells of the digestive gland. Various lipases, esterases, carbohydrases, proteases and peroxidases have been reported in the lysosomes, plasma membrane and granules of *Mytilus* haemocytes (Moore & Lowe, 1977; Carballal *et al.*, 1997). Glycogen and lipid are also present in haemocytes, but as quantities are small they are unlikely to function as energy reserves (Bayne *et al.*, 1976a). For more detail on haemocyte form and function see reviews by Cheng (1996) and Hine (1999).

Heart rate

Several methods have been used to study heart rate in molluscs. Investigators have drilled windows in the valves of bivalves to view the heart directly, a method that was later modified by covering the windows with glass or cellophane. Subsequently, an electrode of platinum, silver or stainless steel inserted through the shell into the pericardial cavity around the heart was used (references in Bayne *et al.*, 1976a). This method creates minimal disturbance to the animal and heart rate can be monitored from a distance. Depledge & Andersen (1990) developed a non-invasive method for continuously monitoring heart rate, which involved cementing transducers to arthropod and bivalve shells. More recently, Haefner *et al.* (1996) have used a non-invasive ultrasound technique to monitor heart rate in the mussel *Mytilus edulis*. This method involves positioning an ultrasound probe, parallel to the long axis of a submerged mussel. Dual video monitors provide a rectilinear image scan of the heart as well as its frequency of movement. These non-invasive methods are reliable, not just for physiological investigations, but can also enhance behavioural analysis (see below and Rovero *et al.*, 1999).

The frequency of heart beat increases linearly with ambient temperature (Fig. 7.1). The upper thermal limit varies for each species but is coincident with the temperature at which other physiological processes are disrupted. For *Mytilus edulis* this temperature is 25–27°C, whereas for *Perna perna* it is as high

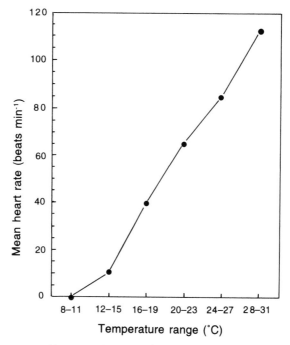

Fig. 7.1. Average rate of heart rate (beats min⁻¹) at various temperature ranges for the oyster *Crassostrea virginica*. From Menzel (1956) and Shumway (1996).

as 32°C (Bayne *et al.*, 1976a). Heart rate is controlled by thermoreceptor cells located in the mantle edge (Lowe, 1974). Salinity also affects heart activity by interfering with the osmotic balance of heart tissue, and also with the activity of the enzyme acetylcholinesterase (Dimock, 1967), an enzyme that carries out programmed destruction of the neurotransmitter acetylcholine. The heart rate of *Mytilus edulis* increases significantly when it is in the presence of effluent from the dogwhelk, *Nucella lapillus*, probably due to an olfactory-mediated perception of predation threat (Rovero *et al.*, 1999). When attacked through a drilled hole mussel heart rate increases significantly, but when attacked by penetration between the valves heart rate decreases throughout the attack, supporting the hypothesis that dogwhelks that penetrate between the valves induce muscular paralysis by injection of toxins.

Other factors such as body size, food, oxygen availability and aerial exposure affect heart rate. Since these same factors also affect oxygen consumption they will be discussed in more detail in the Respiration section below.

Respiration

The gills, as well as the mantle, play an important role in respiration. Their large surface area and rich supply of haemolymph make them both particularly well suited to this task. The function of gills in feeding is covered in Chapter 4.

The gill filaments are essentially hollow tubes within which the haemolymph circulates. Haemolymph flows from the kidney to the gill via the afferent gill vein, which gives off to each filament a vessel that descends one side and ascends the other. The ascending vessels join to form the efferent gill vein that returns to the kidney, or goes direct to the heart (Fig 2.15). As water passes through the gills, oxygen from the water diffuses into the haemolymph. However, the efficiency with which this occurs is low, usually in the range 1–13%, although this figure can be increased up to 30% or more by increasing heart rate (Bayne *et al.*, 1976b).

Oxygen consumption ($\dot{V}O_2$) can be measured by placing an individual animal in a closed chamber sealed with an oxygen probe that is connected to an oxygen meter. Oxygen concentration (PO_2) is recorded at regular intervals and is not allowed to drop below 80% saturation, unless the specific aim of the experiment is to examine the effects of declining PO_2 on $\dot{V}O_2$ (Shumway & Koehn, 1982). A control chamber without an animal is used to correct for bacterial respiration, electrode drift, etc. (Labarta *et al.*, 1997). To gain a full understanding of $\dot{V}O_2$ in a given species, measurements should be made at different times of the year, include animals of different sizes, and at different reproductive stages. Oxygen consumption can also be used as an indirect measure of metabolic rate, MR (see Chapter 6). Values for weight-specific oxygen consumption of a variety of bivalves are presented in Table 7.3. For some of these species values for aerial O_2 consumption (see later) are also included.

Factors affecting oxygen consumption

Size

The relationship between body size (usually expressed as weight) and the rate of oxygen consumption ($\dot{V}O_2$) is generally described by the allometric equation $y = ax^b$ where y is the rate of $\dot{V}O_2$, x is the body weight and a, the intercept, and b, the slope, are fitted parameters. The intercept a is a measure of $\dot{V}O_2$ of an individual of unit weight (or length), and its value changes depending on environmental conditions and on the species being tested. The slope b is a measure of the rate of increase in $\dot{V}O_2$ with size. Bayne & Newell (1983) reported that the value of b in eleven different bivalve species ranged between 0.44 and 1.09, with a mean value of 0.728 ± 0.130. This value is very similar to the mean value of 0.76 reported by Bricelj & Shumway (1991) for eight species of scallop. Table 7.4. presents data on the relationship between body size and oxygen consumption for some of these species. However, as $\dot{V}O_2$ has been measured under a range of environmental conditions it is difficult to examine the effect of size *per se* on $\dot{V}O_2$ from the data in this table (but see Table 7.3. for comparisons of $\dot{V}O_2$ in larvae versus adults of the oyster *Ostrea edulis*). For example, $\dot{V}O_2$ has been measured in three weight classes of *Argopecten irradians irradians* but each of these measurements have been taken at three different temperatures, making it impossible to separate the individual effects of size and temperature on $\dot{V}O_2$. Indirect evidence that $\dot{V}O_2$ is influenced by body size comes from MR studies where MR has been shown to decrease with increasing body size (see Chapter 6).

Table 7.3. Weight-specific oxygen consumption for a variety of bivalves. – indicates no data; *Mytilus edulis* can acclimate its routine oxygen consumption between 10 and 20°C; [†]experiments carried out in a flow-through system at near ambient summer temperature, Vancouver Island, British Columbia, Canada. Adapted from Bayne & Newell (1983), Bernard & Noakes (1990) and Bricelj & Shumway (1991).

| Species | Temp °C | O_2 uptake (O_2 ml h^{-1} g^{-1}) | | Reference |
		Water	Air	
Mytilus edulis	10	0.37	0.017	Vahl, 1973
	15	0.38*	–	Bayne, 1973b
	–1–4.8	0.12–0.28	–	Loo, 1992
M. galloprovincialis	25	0.33	0.044	Widdows *et al.*, 1979
M. californianus	13	0.23	0.17	Widdows *et al.*, 1979
Cerastoderma glaucum	15	0.15	0.012	Boyden, 1972
C. edule	15	0.20	0.13	Boyden, 1972
Crassostrea gigas	ambient[†]	0.51	–	Bernard & Noakes, 1990
C. virginica	10	0.171	–	Dame, 1972
	20	0.372	–	
	30	0.423	–	
Ostrea edulis	5	0.364	–	Rodhouse, 1978
	15	0.962	–	
	25	2.655	–	
O. edulis larvae	–	3.0–6.0	–	Holland & Spencer, 1973
Mya truncata	ambient[†]	0.63	–	Bernard & Noakes, 1990
Argopecten irradians	10.5	0.43	–	Bricelj *et al.*, 1987
Placopecten magellanicus	10	0.24	–	Shumway *et al.*, 1988
Patinopecten yessoensis	9	0.29	–	Fuji & Hashizume, 1974
Chlamys islandica	8	0.20	–	Vahl, 1978
C. hasata	ambient[†]	0.38	–	Bernard & Noakes, 1990
C. varia	10	0.34	–	Shafee, 1982

Food

In the absence of particulate food, $\dot{V}O_2$ declines to a steady state condition, called the 'standard (basal) rate', which is typical of an animal with shell valves open but showing minimal feeding activity. With a return to feeding there is a marked increase in $\dot{V}O_2$ to the 'active rate'. There can be as much as a five-fold difference between these two rates (Fig. 7.2). Between active and standard rates there are a variety of 'routine rates' that are dependent on variations in filtration or ventilation rates, that in turn are dependent on factors such as animal size, ration, season, and gametogenesis (Bayne *et al.*, 1976b). The relationship between $\dot{V}O_2$ and food supply is complex involving not only energy expenditure associated with the mechanical process of filtration, but also the physiological costs of digestion and excretion. In *Mytilus californianus* Bayne *et al.* (1976c) have estimated that the mechanical cost of feeding is 18% of ingested ration, while the physiological cost is about 6%. For *M. edulis* the equivalent values were 24% and 4%, respectively (Bayne & Scullard, 1977a). However, in *M. chilensis*, while the physiological costs of feeding were similar to those observed for *M. californianus* and *M. edulis*, the mechanical costs were

Table 7.4. Parameters *a* and *b* of the allometric equation $y = ax^b$ relating oxygen consumption ($\dot{V}O_2$ ml O_2 h^{-1}) and tissue weight (g) in a variety of bivalve species. In scallops weights are dry tissue weights; *value applies to all three size classes; – indicates no data. Adapted from Bayne & Newell (1983) and Bricelj & Shumway (1991 and references therein).

Species	Size range	Temp °C	*a*	*b*
Argopecten irradians irradians	0.47–2.99	17.4	0.931	0.725
	0.84–2.86	10.5	0.368	0.733
	0.87–4.37	1.5	0.065	0.986
Placopecten magellanicus				
10 m	1.8–42.0	10–12	0.339	0.78
31 m	0.5–25.0	5.5–7.2	0.234	0.79
Chlamys islandica				
Immature	0.02–0.9		0.145	0.486
Mature	0.5–5.0	5.7*	0.251	0.567
Mature	0.4–6.0		0.242	0.759
Patinopecten yessoensis	0.5–15.0	22.4	0.579	0.817
		14.8	0.398	0.777
		5.8	0.181	0.862
Crassostrea virginica	–	10	0.171	0.734
		20	0.372	0.710
		30	0.423	0.603
Ostrea edulis	–	5	0.364	0.899
		15	0.962	0.753
		25	2.655	1.090
Modiolus demissus	–	14	0.260	0.690
		22	0.629	0.798
Mytilus edulis	–	–		
Winter			0.549	0.744
Summer			0.339	0.702
Mytilus californianus	–	–		
Fed			0.540	0.650
Starved			0.230	0.650

about 5%, reflecting the tendency of this species to reduce its filtration rate in response to increased ration (Navarro & Winter, 1982).

Temperature

Temperature is one of the major factors influencing $\dot{V}O_2$ and MR in marine bivalves. Changes in temperature elicit two types of responses. If the animal is subject to a sudden change in temperature it responds by a change in $\dot{V}O_2$ which may initially result in overshoot, but which is followed by a period of stabilisation that takes place over a period of minutes or hours. If the temperature change persists over days or weeks there is a gradual adjustment in $\dot{V}O_2$ to a level comparable with that preceding the temperature change (Bayne & Newell, 1983). However, these responses are by no means universal among bivalves (see below).

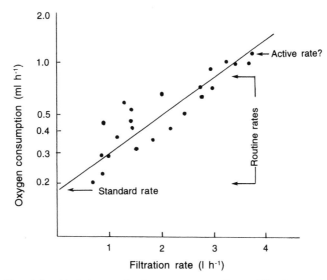

Fig. 7.2. The relationship between rate of oxygen consumption and filtration rate in the mussel *Mytilus californianus* (1 g dry flesh weight at 13°C). The metabolic cost of filtration increases logarithmically with filtration rate. The standard metabolic rate is estimated by extrapolation of the regression line to a point where filtration activity is zero (0.19 ml h⁻¹). However, it is not possible to identify active metabolic rate from plots such as this. From Griffiths & Griffiths (1987).

Acute temperature response

The effect of change in temperature on $\dot{V}O_2$ is often expressed in terms of temperature coefficient or Q_{10} values:

$$Q_{10} = \left[\frac{V_2}{V_1}\right]^{\frac{10}{t_2-t_1}}$$

where V_1 and V_2 are the rates of oxygen consumption at temperatures t_1 and t_2, respectively. The Q_{10} provides an index of the dependence of the physiological rate on temperature (Bayne *et al.*, 1976b). In bivalves Q_{10} values range from 1.0 (temperature independent) to more than 2.5, which represents over a doubling of rate with each 10°C increase in temperature. The values obtained are dependent on thermal history of the animal, body size, activity and reproductive condition.

Temperature effects on $\dot{V}O_2$ are usually depicted in the form of rate/temperature or R/T curves. One such curve is illustrated in Fig. 7.3. Here the acute response of standard, routine and active rates of $\dot{V}O_2$ in the mussel *Mytilus edulis* (1 g dry flesh weight, sexually mature, and acclimated to 15°C) are illustrated over the temperature range 5–25°C (Widdows, 1973). Between 10 and 20°C the Q_{10} values for standard, routine and active rates were 1.6, 2.4 and 1.9, respectively. Lower values (1.0–1.2) for the standard rate of oxygen consumption have been reported in young, sexually immature *M. edulis* (Newell & Pye, 1970), and in *M. californianus* after spawning (Bayne *et al.*, 1976d). The fact that standard, as opposed to routine $\dot{V}O_2$ is relatively

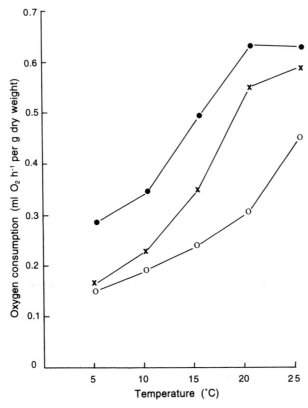

Fig. 7.3. The effects of temperature on the standard (O), routine (x) and active (●) rates of oxygen consumption of the mussel *Mytilus edulis* acclimated to 15°C. From Bayne *et al.* (1976b).

independent of temperature allows *M. edulis* resources to be conserved during stress periods, such as those of high temperature and low food availability experienced during aerial exposure in summer (Table 7.3). Routine rate is temperature dependent, but this is less crucial to the animal since increased ventilation is usually accompanied by increased filtration. However, in some species, e.g. the scallop *Chlamys varia*, both standard and routine rates are temperature dependent (Shafee, 1982), which means that the species experiences a metabolic deficit during periods of aerial exposure in summer (see below). Unlike *Mytilus* this species is limited to low-shore levels where environmental temperatures are more uniform.

The degree of separation (Fig. 7.3) of routine and standard rates at any given temperature is taken as an estimate of scope for routine activity, e.g. feeding (Newell, 1979). For *Mytilus* this tends to be greatest at the upper end of the normal thermal range, but declines at both the upper and lower limits of temperature tolerance.

Acclimation to temperature change
Some bivalves respond to temperature changes lasting days or weeks by adjusting $\dot{V}O_2$ to a level comparable with that that occurred before the tempera-

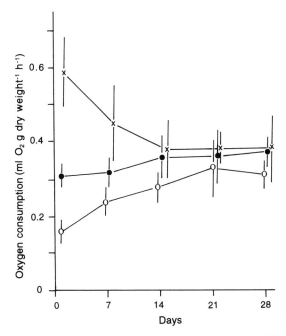

Fig. 7.4. Oxygen consumption in the mussel *Mytilus edulis* acclimated to 5°C (○) and 15°C (x) compared with mussels at 10°C (●; ambient temperature in November). From Widdows & Bayne (1971).

ture changed. This process of adjustment is called acclimation and has been demonstrated in many bivalve species to date. In *Mytilus edulis* collected from 10°C waters and acclimated to 5 and 15°C Widdows & Bayne (1971) recorded an initial 50% decrease in $\dot{V}O_2$ at 5°C and a doubling at 15°C. However, rates subsequently declined in the warm-adapted group and increased in the cold-adapted group so that after 14 days both groups had approximately the same $\dot{V}O_2$ as control mussels (Fig. 7.4). In *Mytilus* compensation is achieved by shifting the rate-temperature curve to the right following warm acclimation. This means that the animal tolerates higher temperatures but still maintains $\dot{V}O_2$ levels relatively constant. This acclimation process ensures that respiration, and also feeding and excretion, are maintained at approximately constant levels, but this is only possible over a temperature range that is species dependent. In oysters, e.g. *Ostrea edulis* (Newell *et al.*, 1977) and *Crassostrea virginica* (Shumway & Koehn, 1982) there is no evidence of metabolic adjustment following warm acclimation. When the oyster *Ostrea edulis* was acclimated to temperatures ranging between 5 and 25°C for 70 days $\dot{V}O_2$ was positively correlated with temperature (Fig. 7.5). The increase in energy expenditure with increasing temperature was offset by increased filtration rate, thus illustrating that this species uses a different compensatory strategy to *M. edulis*.

Thermal acclimation involves not just compensatory adjustments in $\dot{V}O_2$ but often involves the synthesis of different molecular forms of an enzyme, differential utilisation of metabolic pathways, alterations in protein turn-

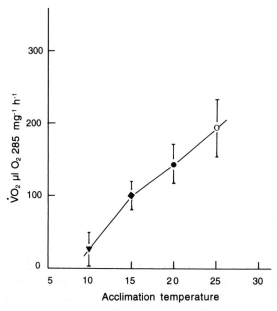

Fig. 7.5.   Routine rates of oxygen consumption in the oyster *Ostrea edulis* following acclimation to temperatures of 10°C (▼), 15°C (◆), 20°C (●) and 25°C (○) for up to 70 days. From Newell *et al.* (1977).

over, synthesis of stress proteins (see Chapter 3) and changes in membrane composition.

Oxygen availability

Bivalves frequently experience periods of reduced oxygen concentration. For example, low oxygen tension is a normal occurrence for species that live deep within soft sediments, or for intertidal species during shell closure on the low tide, or in crowded tide pools. Reduced oxygen tension can also be due to environmental conditions such as stagnation or pollution of the water column.

In some species, e.g. the mussel *Modiolus demissus*, the rate of oxygen consumption is directly dependent on the ambient partial pressure of oxygen (PO_2), which therefore declines as the PO_2 surrounding the gill surface is reduced. Such species are called 'conformers'. In the face of declining PO_2 other species are able to maintain a constant rate of oxygen consumption. Such species, termed 'regulators', do this by increasing ventilation rate, or by increasing the efficiency with which oxygen is removed from the water, or a combination of these two strategies (Bayne & Newell, 1983). For example, the clam *Arctica islandica* more than doubles the volume of water pumped per unit time as PO_2 falls from 160 to 60 mm Hg (Taylor & Brand, 1975); in *M. edulis* extraction efficiency can more than treble as PO_2 falls from 160 to 20 mm Hg (Bayne, 1971). Oysters (*Crassostrea* and *Ostrea* spp.) are also strong regulators (Shumway 1982, and references therein). The distinction between conformers and regulators is, however, more apparent than real. Species that regulate only maintain a constant rate of oxygen consumption down to some minimum 'critical' oxygen tension that is species dependent. Also, different individuals within a species may show different responses to PO_2, depending on body size (Wang & Widdows, 1991; Hole *et al.*, 1995), temperature, salinity or nutritional status (references in Griffiths & Griffiths, 1987). Despite intra-specific differences, one might still expect that species regularly exposed to low oxygen levels would regulate more efficiently than those from well-oxygenated habitats. Bayne (1973b) has found this to be the case, but in a study of 31 species of marine invertebrates Mangum & van Winkle (1973) found no relationship between ability to oxyregulate and environmental oxygen levels, although it must be pointed out that not all of these were molluscs, and many possessed blood pigments.

Aerial exposure

Many bivalves are capable of utilising oxygen in air during aerial exposure (Table 7.3). Their ability to do this depends on behavioural responses, such as intermittent air gaping. For example, when the mussel *Guekensia demissa* is emersed it opens its valves by a small amount (~2mm), sufficient to allow direct access by the air to a large surface area of water trapped in the mantle cavity (Lent, 1968). In contrast, during emersion the cockle *Cerastoderma edule* expels some water from the mantle cavity by rapid valve movement, and replaces it with a bubble of air, keeping the valves 50 to 100% agape. The cockle maintains a level of $\dot{V}O_2$ that is about 65% that of the aquatic rate (Boyden, 1972), and it does not accumulate anaerobic end products during exposure. However, in the closely related, but mainly sub-tidal species, *C. glaucum*, which keeps the shell valves tightly closed during exposure, this figure is only 8%, and anaerobic end products accumulate (see below). In all species, however, exposure to air probably results in metabolic stress through inability to feed, a limit to gas exchange and desiccation, all of which act to reduce scope for growth (Bayne *et al.*, 1976a; see also Chapter 6).

Anaerobiosis

Most bivalve species are facultative anaerobes i.e. they can live either aerobically, or anaerobically for short periods, but tend to use oxygen when it is available, since it allows more economical use of fuel molecules (de Zwann & Mathieu, 1992). Anaerobic respiration has a net yield of two energy-rich ATP (adenosine triphosphate) molecules per molecule of glucose, whereas the aerobic route may yield 36 or more ATPs.

Large fluctuations in temperature, salinity and oxygen availability can all induce anaerobic metabolism. During anaerobiosis there is almost a complete reduction in the rates of energetic processes such as digestion, absorption, muscular activity and growth. The outcome is energy conservation, with an ATP turnover in mussels of less than 10% of the resting aerobic rate (Widdows, 1987; de Zwann *et al.*, 1991). Anaerobiosis can also be involved in providing ATP when energy demand exceeds the capacity of aerobic energy production. For example, anaerobic pathways in scallops are predominantly utilised for energy production during sudden bursts of activity such as swimming or escape responses (Bricelj & Shumway, 1991). On a return to aerobic respiration there is a transient, above normal increase in $\dot{V}O_2$ for the oxidation of anaerobic end products, and possibly also re-oxygenation of the haemolymph.

A drastic temperature change that induces anaerobic metabolism has important implications in the survival of bivalves at freezing temperatures. The anaerobic end product strombine has been identified as a cryoprotectant in *M. edulis* (Loomis et al., 1988). Consequently, in the sub-Arctic where few species survive the winter in the upper intertidal zone, *M. edulis* is among the most freeze-tolerant of species. However, it has been suggested that other by-products of anaerobic metabolism, not yet identified, act as genotoxic agents, causing chromosomal damage in gill tissue cells (Brunetti et al., 1992).

Salinity

Many bivalves are euryhaline, i.e. they can tolerate an extremely wide range of salinity in their natural environment. The salinity range tolerated is, however, species dependent. For example, *M. edulis* is found from fully marine conditions to salinities as low as 4–5 psu, while the lowest salinities tolerated by *M. galloprovincialis* and *M. californianus* are 12 psu and 17 psu, respectively (references in Bayne et al., 1976b). Self-sustaining populations of the oyster *Crassostrea virginica* occur where salinities are as low as 0.2 to 3.5 psu for five consecutive months each year (Butler, 1952).

When the respiration rates of mussels from three different salinities (5–6 psu, 15 psu and 30 psu) were measured in the field they were found to be similar (Remane & Schlieper, 1971). However, if mussels are exposed to a sudden change in salinity the immediate response is to close the shell valves. The mussel *Guekensia demissa* may remain shut for up to 7–10 days, depending on the extent of the salinity change (Pierce, 1971). During this time the valves open periodically for short periods of time and eventually ventilation is resumed, or else death occurs. During this initial period of 'shock', respiration rate is depressed. However, should the new salinity regime persist there is gradual acclimation, which may take as long as 4–7 weeks, depending on the magnitude of the salinity change, the ambient temperature and the size of the animal (Bayne & Newell, 1983). The length of the acclimation period far exceeds the two weeks normally required for temperature acclimation, probably a reflection of the time needed for intracellular osmotic regulation (see osmoregulation below).

There is not much information on the combined effects of salinity and temperature on $\dot{V}O_2$. Shumway & Koehn (1982) measured the acclimated and acute rates of $\dot{V}O_2$ in *Crassostrea virginica* at three temperatures (10, 20 and 30°C) and three salinities (7, 14, and 28 psu) giving nine temperature-salinity regimes. They found that as acclimation salinity decreased, the effect of exposure temperature became more pronounced, and the effect of exposure salinity decreased. As acclimation temperature increased, the effect of exposure salinity decreased and the effect of exposure temperature increased (Table 7.5). They also demonstrated that oysters regulated $\dot{V}O_2$ when exposed to declining oxygen tension at all of the temperature-salinity combinations tested, but

Table 7.5. Oxygen consumption ($\dot{V}O_2$, ml O_2 $0.4\,g^{-1}h^{-1}$) in the oyster *Crassostrea virginica* acclimated to 9 salinity-temperature combinations, and exposed to the same 9 combinations; values in italics on the diagonal are acclimated rates. Each value is the mean of six determinations. Mean standard deviation for all determinations is 0.029. From Shumway & Koehn (1982).

	Experimental conditions								
	28 psu			14 psu			7 psu		
Acclimation conditions	10°C	20°C	30°C	10°C	20°C	30°C	10°C	20°C	30°C
28 psu									
10°C	*0.0531*	0.1211	0.2174	0.0951	0.1713	0.4122	0.1484	0.3603	0.5068
20°C	0.0335	*0.0962*	0.2196	0.1748	0.2100	0.5193	0.1590	0.2505	0.4248
30°C	0.1210	0.0594	*0.1783*	0.1122	0.3965	0.6742	0.0408	0.2344	0.5264
14 psu									
10°C	0.0809	0.2005	0.2916	*0.0933*	0.2298	0.4710	0.1072	0.2489	0.4349
20°C	0.0542	0.2166	0.4007	0.0892	*0.1845*	0.4538	0.0693	0.2740	0.4901
30°C	0.0652	0.2114	0.2351	0.0472	0.3171	*0.6216*	0.0351	0.3512	0.5532
7 psu									
10°C	0.0582	0.2849	0.4500	0.1040	0.2925	0.4849	*0.1371*	0.2752	0.5191
20°C	0.1224	0.2491	0.4907	0.1352	0.2542	0.4671	0.1114	*0.2464*	0.4624
30°C	0.0867	0.2296	0.5243	0.0402	0.2003	0.4351	0.0704	0.3001	*0.5031*

in general the degree of regulation decreased with increasing temperature or decreasing salinity. Oysters are thus able to utilise available oxygen over a wide range of temperature-salinity-oxygen combinations without resorting to anaerobiosis.

Other factors

Shumway *et al.* (1988) showed that seasonal changes in $\dot{V}O_2$ in the scallop, *Placopecten magellanicus*, were clearly related to changes in the gametogenic cycle, with highest rates of $\dot{V}O_2$ exhibited during the summer months when the gonads are ripening, and the lowest rates during the winter months. Similar results have been reported for other bivalves (references in Bayne *et al.*, 1976b; Sukhotin, 1992; Ekaratne & Davenport, 1993). While these changes in $\dot{V}O_2$ mirror environmental temperatures, Shumway *et al.* (1988) suggest that seasonal changes in food availability and reproductive stage have a greater effect on $\dot{V}O_2$ than temperature *per se*.

Other factors that affect $\dot{V}O_2$ in bivalves are suspended sediment (turbidity) or pollutants in the water column. Grant & Thorpe (1991) have found that in the clam *Mya arenaria* long-term exposures to sediment concentrations of $100-200\,mg\,l^{-1}$ for 35 days caused a significant decrease in $\dot{V}O_2$ but an increase in ammonia excretion, compared to controls. Reducing ventilation rate is an effective strategy for coping with intermittent turbidity in the wild, but this can lead to starvation if exposure is long-term. Pollutants such as copper, cadmium and zinc also significantly depress respiration rate in mussels (Akberali *et al.*, 1984, 1985; Cheung & Cheung, 1995) and oysters (Chen, 1994).

Table 7.6. Nitrogenous compounds in the urine of some bivalves; – signifies that the component was not measured. Adapted from Bayne *et al.* (1976a) and Griffiths & Griffiths (1987).

Species	Excreted components as % of total measured nitrogen (N)				Reference
	NH$_4$-N	Urea-N	Amino-N	Uric acid-N	
Crassostrea virginica	68	8	21	3	Hammen (1968)
Modiolus demissus	62–75	0	25–38	–	Lum & Hammen (1964)
Mercenaria mercenaria	66	0	30	4	Hammen (1968)
Mya arenaria	94	6	–	–	Allen & Garrett (1971)
Mytilus californianus	100	–	0	–	Bayne & Scullard (1977b)
Mytilus galloprovincialis					
Summer	37–58	–	42–63	–	Bayne & Scullard, 1977b
Autumn	100	–	0	–	Bayne & Scullard, 1977b
Mytilus edulis					
Summer	71	–	29	–	Bayne & Scullard, 1977b
Winter	97–100	–	0–3	–	Bayne & Scullard, 1977b

Mussels susceptible to summer mortality were found to have higher values of $\dot{V}O_2$ and lower levels of genetic variability than less susceptible mussels (Tremblay *et al.*, 1998). The higher metabolic demand associated with reduced heterozygosity (Chapter 6) may render these mussels more vulnerable to summer mortality.

Excretion and osmoregulation

The kidneys and pericardial glands are the major excretory organs (see Chapter 2 for details), although excretory products are possibly also lost across the general body surface and especially across the gills (Bayne *et al.*, 1976a). Waste accumulates in certain cells of the pericardial glands and this is periodically discharged into the pericardial cavity and from there it is eliminated via the kidneys. Other cells of the pericardial glands are probably involved in filtering the haemolymph, the first stage of urine formation. The filtrate then flows through the renopericardial canals to the glandular part of the kidneys, where the process of re-absorption occurs. The end result is urine that has a high concentration of ammonia, and smaller amounts of amino-nitrogen, urea and uric acid-nitrogen (Table 7.6). By analysing the water in which animals have been held for several hours these compounds can be identified. With this method one should be aware that some of the measured nitrogen may come from bacterial contamination, from faeces voided during the experiment, or from 'leakage' rather than true excretion (Bayne *et al.*, 1976a; Griffiths & Griffiths, 1987). Nevertheless, it is clear from Table 7.6. that the proportions of excreted ammonia to amino acid nitrogen differ considerably between species, and with season. The amounts of amino acids that are lost amount to as much as 63% of total measured excreted nitrogen in some species. This represents a loss in energy, which for *Mytilus edulis* is about 11% of routine metabolic rate, rising to much higher values during stress (Bayne 1973a). It is not

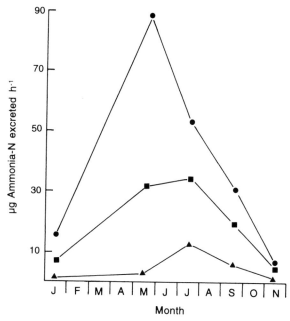

Fig. 7.6. Rates of excretion of ammonia (NH₄-N) by the mussel *Mytilus edulis* of 0.2 g (▲), 1.0 g (■) and 2.0 g (●) dry flesh weight during different months of the year. From Bayne & Scullard (1977b).

clear whether the amino acids are lost through leakage or through active excretion (Bayne *et al.*, 1976a).

Nitrogen excretion rates are extremely variable in bivalves; average values for a range of species varied from 9.6 to 94.7 µg ammonia–nitrogen $d^{-1}g^{-1}$ fresh weight (references in Bayne *et al.*, 1976a). Factors such as body size, food, temperature and salinity influence excretion rates. Bayne & Scullard (1977b) have investigated the relationship among these by means of the O:N ratio, the atomic ratio of oxygen consumed to ammonia excreted, which provides an index of the balance in an animal's tissues between the rates of catabolism of protein, lipid and carbohydrate substrates. A low value of ~10 indicates considerable protein catabolism, while higher values signify that greater proportions of lipid or carbohydrate are being metabolised (Bayne & Newell, 1983). It now appears that much of the reported variability in excretion rates can be attributed to feeding history and the gametogenic stage of the animals being analysed. For example, Bayne & Scullard (1977b) have shown that in *Mytilus edulis* nitrogen excretion peaks prior to spawning followed by a progressive decline to minimum rates in autumn and early winter (Fig. 7.6). Just before spawning glycogen reserves are at a minimum and mussels must catabolise protein to offset nutritional stress. The utilisation of protein reserves under stress conditions has been further substantiated by the results of Bayne & Thompson (1970) for starved *M. edulis* in the laboratory; and protein was utilised more rapidly at higher (16°C) than at lower temperatures (6°C). In the wild oxygen consumption tends to decline during starvation and this

Table 7.7. The concentration of amino acids in the adductor muscle of *Mytilus edulis* in 100% and 50% seawater. Adapted from Bayne *et al.*, 1976a; data from Bricteux-Grégoire *et al.* (1964).

Amino acid	Concentration (µmoles per g wet weight)	
	100% Seawater	50% Seawater
Alanine	18.40	13.0
Arginine	11.94	8.38
Aspartic acid	9.77	1.73
Glutamic acid	9.79	10.61
Glycine	61.87	18.40
Histidine	1.80	1.22
Isoleucine	0.31	0.39
Leucine	0.42	0.60
Lysine	2.81	1.71
Phenylalanine	0.51	0.28
Proline	6.29	2.80
Serine	5.99	3.05
Threonine	4.37	1.93
Tyrosine	0.77	0.48
Valine	0.94	0.50
Taurine	58.09	44.51
Betaine	62.99	40.66
Water content	69.80	80.20
Osmotic pressure (mOsm per kg H_2O) due to:		
Amino acids	216.54	90.19
Taurine	92.47	61.66
Betaine	100.27	56.23
Osmotic pressure of the medium (mOsm)	1180	573

results in a marked reduction in the O:N ratio. These changes are most pronounced in smaller individuals, probably because glycogen reserves are depleted more rapidly, and also because they have higher weight-specific metabolic rates than larger individuals. In the autumn, when glycogen reserves are high, but food is at a premium, ammonia excretion rates are low, which indicates that glycogen is preferentially being used to meet metabolic requirements (Fig. 7.6). Salinity also affects excretion rate, with increased amounts of ammonia and amino acids being excreted when animals are initially exposed to reduced salinities. However, excretion rates return to normal after a period of time, the duration of which depends on the extent of the salinity decrease (Allen & Garrett, 1971).

Bivalves exhibit wide variation in salinity tolerance. Euryhaline species, e.g. the mussel *Mytilus edulis*, tolerate a wide range of salinities (5–35 psu), whereas stenohaline species, e.g. most scallops, are confined to a much narrower range.

Despite such differences, all bivalves are osmoconformers possessing little if any capability for osmotic regulation of their extracellular fluid (haemolymph). Therefore, the cells bear the burden of volume regulation by adjusting the concentrations of intracellular free amino acids (FAAs) and other small organic molecules (Griffiths & Griffiths, 1987). Bivalves are thus able to maintain their cells iso-osmotic with their extra-cellular fluids while at the same time regulating their volume. The capacity for volume regulation is, however, species dependent. Euryhaline species such as mussels *M. edulis* and *Geukensia demissa* exhibit near-perfect volume regulation in comparison to the stenohaline species *Modiolus modiolus* (Gainey, 1994). Generally, a bivalve's first response to fluctuations in salinity is shell closure. As a result the tissues are isolated from osmotic changes in the external medium. This is a short-term response, however, and should salinity change persist the animal may be compelled to regulate cell volume by its demand for oxygen and food (Bayne *et al.*, 1976a). Alternatively, it may open, begin pumping, and then die due to its inability to regulate cell volume.

The amino acids alanine, arginine, aspartic and glutamic acids, glycine, taurine and betaine are particularly important in the process of cell volume regulation (Table 7.7 and Shumway *et al.*, 1977). Species, or even populations within a species (Pierce *et al.*, 1992), may differ in the size and content of their FAA pool. Hawkins & Hilbish (1992) have shown that the principle source of FAA is from breakdown of whole-body protein, and contributions from dietary sources, *de novo* synthesis, and direct uptake of dissolved amino acids are negligible. With an increase in salinity there is a rapid increase in the concentration of FAA and a corresponding decrease with reduced salinity. The decrease is effected by elevated excretion of FAA and ammonia. Bayne (1975) recorded rates of amino-nitrogen loss in *Mytilus edulis* of $0.42\,\mu g\,h^{-1}\,g^{-1}$ dry weight at 32.5 psu, which increased to $2.95\,\mu g\,h^{-1}\,g^{-1}$ dry weight within 3 h of transfer to water of 14.5 psu. This high excretion rate of amino-nitrogen gradually declined over the next 20 days. Such losses represent a major component cost of cell volume regulation, and Hawkins & Hilbish (1992) have suggested that this may help to explain stress and even mortality resulting from small but frequent fluctuations in salinity.

References

Akberali, H.B., Earnshaw, M.J. & Marriott, K.R.M. (1984) The action of heavy metals on the gametes of the marine mussel *Mytilus edulis* (L.). I. Copper-induced uncoupling of respiration in the unfertilized egg. *Comp. Biochem. Physiol.*, **77**C, 289–94.

Akberali, H.B., Earnshaw, M.J. & Marriott, K.R.M. (1985) The action of heavy metals on the gametes of the marine mussel *Mytilus edulis* (L.). II. Uptake of copper and zinc and their effects on respiration in the sperm and unfertilized egg. *Mar. Environ. Res.*, **16**, 37–59.

Allen, J.A. & Garrett, M.R. (1971) The excretion of ammonia and urea by *Mya arenaria* L. (Mollusca: Bivalvia). *Comp. Biochem. Physiol.*, **39**A, 633–42.

Alvarez, M.R. & Friedl, F.E. (1992) Effect of a fungicide on *in vitro* haemocyte viability, phagocytosis and attachment in the American oyster *Crassostrea virginica*. *Aquaculture*, **107**, 135–40.

Austin, K. & Paynter, K. (1995) Characterization of the chemiluminescence measured in hemocytes of the eastern oyster, *Crassostrea virginica*. *J. Exp. Zool.*, **273**, 461–71.

Bayne, B.L. (1971) Ventilation, the heart beat and oxygen uptake by *Mytilus edulis* L. in declining oxygen tension. *Comp. Biochem. Physiol.*, **40**A, 1065–85.

Bayne, B.L. (1973a) Physiological changes in *Mytilus edulis* L. induced by temperature and nutritive stress. *J. mar. biol. Ass. U.K.*, **53**, 39–58.

Bayne, B.L. (1973b) The responses of three species of bivalve mollusc to declining oxygen tension at reduced salinity. *Comp. Biochem. Physiol.*, **45**A, 793–806.

Bayne, B.L. (1975) Aspects of physiological condition in *Mytilus edulis* (L.) with special reference to the effects of oxygen tension and salinity. In: *Proceedings of the Ninth European Marine Biology Symposium* (ed. H. Barnes), pp. 213–38. Aberdeen University Press, Aberdeen.

Bayne, B.L. & Thompson, R.J. (1970) Some physiological consequences of keeping *Mytilus edulis* in the laboratory. *Helgo. Wiss. Meeresunters.*, **20**, 526–52.

Bayne, B.L. & Scullard, C. (1977a) An apparent specific dynamic action in *Mytilus edulis* L. *J. mar. biol. Ass. U.K.*, **57**, 371–78.

Bayne, B.L. & Scullard, C. (1977b) Rates of nitrogen excretion by species of *Mytilus* (Bivalvia; Mollusca). *J. mar. biol. Ass. U.K.*, **57**, 355–69.

Bayne, B.L. & Newell, R.C. (1983) Physiological energetics of marine molluscs. In: *The Mollusca*, Vol. 4, *Physiology*, Part 1 (eds A.S.M. Saleuddin & K.M. Wilbur), pp. 407–515. Academic Press, New York.

Bayne, B.L., Widdows, J. & Thompson, R.J. (1976a) Physiology II. In: *Marine Mussels: Their Ecology and Physiology* (ed. B.L. Bayne), pp. 207–60. Cambridge University Press, Cambridge.

Bayne, B.L., Thompson, R.J. & Widdows, J. (1976b) Physiology: I. In: *Marine Mussels: Their Ecology and Physiology* (ed. B.L. Bayne), pp. 121–206. Cambridge University Press, Cambridge.

Bayne, B.L., Bayne, C.J., Carefoot, T.C. & Thompson, R.J. (1976c) The physiological ecology of *Mytilus californianus* Conrad. 1. Metabolism and energy balance. *Oecologia*, **22**, 211–28.

Bayne, B.L., Bayne, C.J., Carefoot, T.C. & Thompson, R.J. (1976d) The physiological ecology of *Mytilus californianus* Conrad. 2. Adaptations to low oxygen tension and air exposure. *Oecologia*, **22**, 229–50.

Bernard, F.R. & Noakes, D.J. (1990) Pumping rates, water pressure and oxygen use in eight species of marine bivalve molluscs from British Columbia. *Can. J. Fish. Aquat. Sci.*, **47**, 1302–06.

Bihari, N., Batel, R. & Zahn, R.K. (1990) DNA damage determination by the alkaline elution technique in the haemolymph of mussel *Mytilus galloprovincialis* treated with benzo[a]pyrene and 4-nitroquinoline-N-oxide. *Aquat. Toxicol.*, **18**, 13–22.

Boyden, C.R. (1972) The behaviour, survival and respiration of the cockles *Cerastoderma edule* and *C. glaucum* in air. *J. mar. biol. Ass. U.K.*, **52**, 661–80.

Bricelj, V.M. & Shumway, S.E. (1991) Physiology: energy acquisition and utilization. In: *Scallops: Biology, Ecology and Aquaculture* (ed. S.E. Shumway), pp. 305–46. Elsevier Science Publishers B.V., Amsterdam.

Bricelj, V.M., Epp, J. & Malouf, R.E. (1987) Comparative physiology of young and old cohorts of bay scallops *Argopecten irradians* (Lamarck): mortality, growth and oxygen consumption. *J. Exp. Mar. Biol. Ecol.*, **112**, 73–91.

Bricteux-Grégoire, S., Duchâteau-Bosson, G., Jeuniaux, C. & Florkin, M. (1964) Constituants osmotiquement actifs des muscles adducteurs de *Mytilus edulis* adapté a l'eau de mer ou à l'eau saumâtre. *Arch. Int. Physiol. Biochim.*, **72**, 116–23.

Brunetti, R., Gabriele, M., Valerio, P. & Fumagalli, O. (1992) The micronucleus test:

temporal pattern of base-line frequency in *Mytilus galloprovincialis*. *Mar. Ecol. Prog. Ser.*, **79**, 89–98.

Burton, R.F. (1983) Ionic regulation and water balance. In: *The Mollusca*, Vol. 5. *Physiology*, Part 2 (eds A.S.M. Saleuddin & K.M. Wilbur), pp. 291–352. Academic Press, London.

Butler, P.A. (1952) Growth and mortality rate in sibling and unrelated oyster populations. *Proc. Gulf Carib. Fish. Inst.*, **4**, 71.

Carballal, M.J., Lopez, C., Azevedo, C. & Villaba, A. (1997) Enzymes involved in defense functions of hemocytes of mussel *Mytilus galloprovincialis*. *J. Invertebr. Pathol.*, **70**, 96–105.

Chen, I.M. (1994) The effects of copper on the respiration of oyster *Crassostrea gigas* (Thunberg). *Fish. Sci.*, **60**, 683–86.

Cheng, T. (1996) Haemocytes: form and function. In: *The Eastern Oyster* Crassostrea virginica (eds V.S. Kennedy, R.I.E. Newell & A.F. Eble), pp. 299–333. Maryland Sea Grant, College Park, Maryland.

Cheung, S.G. & Cheung, R. (1995) Effects of heavy metals on oxygen consumption and ammonia excretion in green-lipped mussels (*Perna viridis*). *Mar. Pollut. Bull.*, **31**, 381–86.

Dame, R.F. (1972) The ecological energetics of growth, respiration and assimilation in the intertidal American oyster, *Crassostrea virginica*. *Mar. Biol.*, **17**, 243–50.

Depledge, M.H. & Andersen, B.B. (1990) A computer-aided system for continuous, long-term recording of cardiac activity in selected invertebrates. *Comp. Biochem. Physiol.*, **96A**, 473–77.

Dimock, R.V. (1967) *An examination of physiological variation in the American oyster*, Crassostrea virginica. M.Sc. Thesis, Florida State University, Gainsville, Florida.

Dopp, E., Barker, C., Schiffmann, D. & Reinisch, C.L. (1996) Detection of micronuclei in hemocytes of *Mya arenaria*: association with leukemia and induction with an alkylating agent. *Aquat. Toxicol.*, **34**, 31–45.

Dyrynda, E.A., Law, R.J., Dyrynda, P.E.J. *et al.* (1997) Modulations in cell-mediated immunity of *Mytilus edulis* following the 'Sea Empress' oil spill. *J. mar. biol. Ass. U.K.*, **77**, 281–84.

Ekaratne, S.U.K. & Davenport, J. (1993) The relationships between the gametogenetic status of triploids or diploids of Manila clams, *Tapes philippinarum*, and their oxygen uptake and gill particle transport. *Aquaculture*, **117**, 335–49.

Fawcett, L.B. & Tripp, M. (1994) Chemotaxis of *Mercenaria mercenaria* hemocytes to bacteria *in vitro. J. Invertebr. Pathol.*, **63**, 275–84.

Fisher, W.S., Wishkovsky, A. & Chu, F. E. (1990) Effects of tributyltin on defence-related activities of oyster hemocytes. *Arch. Environ. Contam. Toxicol.*, **19**, 354–60.

Fisher, W.S., Oliver, L.M., Winstead, J.T. & Long, E.R. (2000) A survey of oysters *Crassostrea virginica* from Tampa Bay, Florida: associations of internal defense measurements with contaminant burdens. *Aquat. Toxicol.*, **51**, 115–38.

Ford, S.E., Ashton-Alcox, K.A. & Kanalay, S.A. (1993) *In vitro* interactions between bivalve hemocytes and the oyster pathogen *Haplosporidium nelsoni* (MSX). *J. Parasitol.*, **79**, 255–65.

Fuji, A. & Hashizume, M. (1974) Energy budget for a Japanese common scallop, *Patinopecten yessoensis* (Jay), in Mutsu Bay. *Bull. Fac. Fish. Hokkaido Univ.*, **25**, 7–19.

Gainey, L.F. (1994) Volume regulation in 3 species of marine mussels. *J. Exp. Mar. Biol. Ecol.*, **181**, 201–11.

Grant, J. & Thorpe, B. (1991) Effects of suspended sediment on growth, respiration and excretion of the soft-shell clam *Mya arenaria*. *Can. J. Fish. Aquat. Sci.*, **48**, 1285–92.

Griffiths, C.L. & Griffiths, R.J. (1987) Bivalvia. In: *Animal Energetics*, Vol. 2, *Bivalvia through Reptilia* (eds T.J. Pandian & F.J. Vernberg), pp. 1–87. Academic Press, California.

Grundy, M.M., Ratcliffe, N.A. & Moore, M.N. (1996) Immune inhibition in marine mussels by polycyclic aromatic hydrocarbons. *Mar. Environ. Res.*, **42**, 187–90.

Haefner, P.A., Sheppard, B., Barto, J., McNeil, E. & Cappellino, V. (1996) Application of ultrasound technology to molluscan physiology: noninvasive monitoring of cardiac rate in the blue mussel, *Mytilus edulis* Linnaeus, 1758. *J. Shellfish Res.*, **15**, 685–89.

Hammen, C.S. (1968) Aminotransferase activities and amino acid excretion of bivalve molluscs and brachiopods. *Comp. Biochem. Physiol.*, **26**, 697–705.

Hawkins, A.J.S. & Hilbish, T.J. (1992) The costs of cell volume regulation: protein metabolism during hyperosmotic adjustment. *J. mar. biol. Ass. U.K.*, **72**, 569–78.

Hine, P.M. (1999) The inter-relationships of bivalve haemocytes. *Fish & Shellfish Immunol.*, **9**, 367–85.

Hole, L.M., Moore, M.N. & Bellamy, D. (1995) Age-related cellular and physiological reactions to hypoxia and hyperthermia in marine mussels. *Mar. Ecol. Prog. Ser.*, **122**, 173–78.

Holland, D.I. & Spencer, B.E. (1973) Biochemical changes in fed and starved oysters, *Ostrea edulis* L., during larval development, metamorphosis and early spat growth. *J. mar. biol. Ass. U.K.*, **53**, 287–98.

Kanaley, S.A. & Ford, S.E. (1990) Lectin binding characteristics of hemocytes and parasites in the oyster, *Crassostrea virginica*, infected with *Haplosporidium nelsoni* (MSX). *Parasitol. Immunol.*, **12**, 633–46.

Labarta, U., Fernández-Reiríz, M.J. & Barbarro, J.M.F. (1997) Differences in physiological energetics between intertidal and raft cultivated mussels *Mytilus galloprovincialis*. *Mar. Ecol. Prog. Ser.*, **152**, 167–73.

Lauckner, G. (1983) Diseases of Mollusca: Bivalvia. In: *Diseases of Marine Animals*, Vol. 2 (ed. O. Kinne), pp. 477–961. Biologishe Anstalt Helgoland, Hamburg.

Lent, C.M. (1968) Air gaping by the ribbed mussel, *Modiolus demissus* (Dillwyn): effects and adaptive significance. *Biol. Bull.*, **134**, 60–73.

Livingstone, D.R., Chipman, J.K., Lowe, D.M. *et al.* (2000) Development of biomarkers to detect the effects of organic pollution on aquatic invertebrates: recent molecular, genotoxic, cellular and immunological studies on the common mussel (*Mytilus edulis* L.) and other mytilids. *Int. J. Environ. Pollut.*, **13**, 56–91.

Loo, L.O. (1992) Filtration, assimilation, respiration and growth of *Mytilus edulis* L. at low temperature. *Ophelia*, **35**, 123–31.

Loomis, S.H., Carpenter, J.F. & Crowe, J.H. (1988) Identification of strombine and taurine as cryoprotectants in the intertidal bivalve *Mytilus edulis*. *Biochim. Biophys. Acta*, **943**, 113–18.

Lowe, D.M. (1974) Effect of temperature change on the heart rate of *Crassostrea virginica* and *Mya arenaria* (Bivalvia). *Proc. Malacol. Soc. Lond.*, **41**, 29–36.

Lowe, D.M., Fossato, V.U. & Depledge, M.H. (1995) Contaminant-induced lysosomal membrane damage in blood cells of mussels *Mytilus galloprovincialis* from the Venice Lagoon: an *in vitro* study. *Mar. Ecol. Prog. Ser.*, **129**, 189–96.

Lum, S.C. & Hammen, C.S. (1964) Ammonia excretion of *Lingula*. *Comp. Biochem. Physiol.*, **12**, 185–90.

Mangum, C.P. & Winkle, W. van (1973) Responses of aquatic invertebrates to declining oxygen conditions. *Amer. Zool.*, **13**, 529–41.

Menzel, R.W. (1956) *The effect of temperature on the ciliary action and other activities of oysters*. Florida State Univ. Ocean. Inst., Contrib. **67**, 25–36.

Montes, J.F., Durfort, M. & García-Valero, J. (1995a) Cellular defence mechanism of the clam *Tapes semidecussatus* against infection by the protozoan *Perkinsus* sp. *Cell Tissue Res.*, **279**, 529–38.

Montes, J.F., Durfort, M. & García-Valero, J. (1995b) Characterization and localization of a Mr 225 kDa polypeptide specifically involved in the defence mechanism of the clam *Tapes semidecussatus*. *Cell Tissue Res.*, **280**, 27–37.

Moore, M. & Lowe, D. (1977) The cytology and cytochemistry of the hemocytes of *Mytilus edulis* and their responses to injected carbon particles. *J. Invertebr. Pathol.*, **29**, 18–30.

Moore, M.N., Wedderburn, R.J., Lowe, D.M. & Depledge, M.H. (1996) Lysosomal reaction to xenobiotics in mussel haemocytes using BODIPY-FL-Verapamil. *Mar. Environ. Res.*, **42**, 99–105.

Morton, J.E. (1967) *Molluscs*. Hutchinson University Library, London.

Navarro, J.M. & Winter, J.E. (1982) Ingestion rate, assimilation efficiency and energy balance in *Mytilus chilensis* in relation to body size and different algal concentrations. *Mar. Biol.*, **67**, 255–66.

Newell, R.C. (1979) *Biology of Intertidal Animals*. Marine Ecology Surveys Ltd., Faversham, England.

Newell, R.C. & Pye, V. (1970) Seasonal changes in the effect of temperature on the oxygen consumption of the winkle *Littorina littorea* (L.) and the mussel *Mytilus edulis* (L.). *Comp. Biochem. Physiol.*, **34**A, 367–83.

Newell, R.C., Johnson, L.G. & Kofoed, L.H. (1977) Adjustment of the components of energy balance in response to temperature change in *Ostrea edulis*. *Oecologia*, **30**, 97–110.

Oubella, R. (1997) Immunomodulation in populations of bivalve mollusks from the Bay of Brest. *Ann. Inst. Oceanogr.*, **73**, 77–87.

Oubella, R., Maes, P., Paillard, C. & Auffret, M. (1993) Experimentally-induced variation in haemocyte density for *Ruditapes philippinarum* and *Ruditapes decussatus* (Mollusca, Bivalvia). *Dis. Aquat. Org.*, **15**, 193–97.

Pierce, S.K. (1971) Volume regulation and valve movements by marine mussels. *Comp. Biochem. Physiol.*, **39**A, 103–17.

Pierce, S.K., Rowlandfaux, L.M. & O Brien, S.M. (1992) Different salinity tolerance mechanisms in Atlantic and Chesapeake Bay conspecific oysters: glycine, betaine and amino-acid pool variations. *Mar. Biol.*, **113**, 107–15.

Pipe, R.K. (1990) Hydrolytic enzymes associated with the granular haemocytes of the marine mussel *Mytilus edulis*. *Histochem. J.*, **22**, 595–603.

Pipe, R.K., Porte, C. & Livingstone, D.R. (1993) Antioxidant enzymes associated with the blood cells and haemolymph of the mussel *Mytilus edulis*. *Fish & Shellfish Immunol.*, **3**, 221–33.

Remane, A. & Schlieper, C. (1971) *Biology of Brackish Water*. Wiley-Interscience, New York.

Ringwood, A.H., Conners, D.E. & Hoguet, J. (1998) Effects of natural and anthropogenic stressors on lysosomal destabilization in oysters *Crassostrea virginica*. *Mar. Ecol. Prog. Ser.*, **166**, 163–71.

Rodhouse, P.G. (1978) Energy transformations by the oyster *Ostrea edulis* L. in a temperate estuary. *J. Exp. Mar. Biol. Ecol.*, **34**, 1–22.

Rovero, F., Hughes, R.N. & Chelazzi, G. (1999) Cardiac and behavioural responses to risk of predation by dogwhelks. *Anim. Behav.*, **58**, 707–14.

Shafee, M.S. (1982) Variations saisoniéres de la consommation d'oxygene chez le pétoncle noir *Chlamys varia* (L.) de Lanvéoc (rade de Brest). *Oceanol. Acta*, **5**, 189–97.

Shumway, S.E. (1982) Oxygen consumption in oysters: an overview. *Mar. Biol. Lett.*, **3**, 1–23.

Shumway, S.E. (1996) Natural environmental factors. In: *The Eastern Oyster* Crassostrea virginica (eds V.S. Kennedy, R.I.E. Newell & A.F. Eble), pp. 467–513. Maryland Sea Grant, College Park, Maryland.

Shumway, S.E. & Koehn, R.K. (1982) Oxygen consumption in the American oyster *Crassostrea virginica*. *Mar. Ecol. Prog. Ser.*, **9**, 59–68.

Shumway, S.E., Gabbott, P.A. & Youngson, A. (1977) The effect of fluctuating salinity on the concentrations of free amino acids and ninhydrin-positive substances in the adductor muscle of eight species of bivalve molluscs. *J. Exp. Mar. Biol. Ecol.*, **29**, 131–50.

Shumway, S.E., Barter, J. & Stahlnecker, J. (1988) Seasonal changes in oxygen consumption of the giant scallop, *Placopecten magellanicus* (Gmelin). *J. Shellfish Res.*, **7**, 77–82.

Sukhotin, A.A. (1992) Respiration and energetics in mussels (*Mytilus edulis* L.) cultured in the White Sea. *Aquaculture*, **101**, 41–57.

Taylor, A.C. & Brand, A.R. (1975) A comparative study of the respiratory responses of the bivalves *Arctica islandica* L. and *Mytilus edulis* L. to declining oxygen tension. *Proc. Roy. Soc. Lond., Series B*, **190**, 443–56.

Thompson, R.J. (1977) Blood chemistry, biochemical composition and the annual reproductive cycle in the giant scallop, *Placopecten magellanicus*, from Southeast Newfoundland. *J. Fish. Res. Board Can.*, **34**, 2104–16.

Tremblay, R., Myrand, B., Sevigny, J.M., Blier, P. & Guderley, H. (1998) Bioenergetic and genetic parameters in relation to susceptibility of blue mussels, *Mytilus edulis* (L.) to summer mortality. *J. Exp. Mar. Biol. Ecol.*, **221**, 27–58.

Tripp, M.R. (1992) Phagocytosis by hemocytes of the hard clam, *Mercenaria mercenaria*. *J. Invertebr. Pathol.*, **59**, 222–27.

Vahl, O. (1973) Pumping and oxygen consumption rates of *Mytilus edulis* L. of different sizes. *Ophelia*, **12**, 45–52.

Vahl, O. (1978) Seasonal changes in oxygen consumption of the Iceland scallop (*Chlamys islandica* (O.F. Müller)) from 70°N. *Ophelia*, **17**, 143–54.

Wang, W.X. & Widdows, J. (1991) Physiological responses of mussel larvae *Mytilus edulis* to environmental hypoxia and anoxia. *Mar. Ecol. Prog. Ser.*, **70**, 223–36.

Widdows, J. (1973) The effects of temperature on the metabolism and activity of *Mytilus edulis* L. *Neth. J. Sea Res.*, **7**, 387–98.

Widdows, J. (1987) Application of calorimetric methods in ecological studies. In: *Thermal and Energetic Studies of Cellular Biological Systems* (ed. A. M. James), pp. 182–215. Wright, Bristol.

Widdows, J. & Bayne, B.L. (1971) Temperature acclimation of *Mytilus edulis* with reference to its energy budget. *J. mar. biol. Ass. UK.*, **51**, 827–43.

Widdows, J., Bayne, B.L., Livingstone, D.R., Newell, R.C. & Donkin, P. (1979) Physiological and biochemical responses of bivalve molluscs to exposure to air. *Comp. Biochem. Physiol.*, **62**A, 301–308.

Winston, G.W., Moore, M.N., Kirchin, M.A. & Soverchia, C. (1996) Production of reactive oxygen species by haemocytes from the marine mussel, *Mytilus edulis*: lysosomal localization and effect of xenobiotics. *Comp. Biochem. Physiol.*, **113**C, 221–29.

Zwann, A. de, & Mathieu, M. (1992) Cellular biochemistry and endocrinology. In: *The Mussel Mytilus: Ecology, Physiology, Genetics and Culture* (ed. E.M. Gosling), pp. 223–307. Elsevier Science Publishers B.V., Amsterdam.

Zwann, A. de, Cortesi, P., Thillart, G. van den, Roos, J. & Storey, K.B. (1991) Differential sensitivities to hypoxia by two anoxia-tolerant marine molluscs: a biochemical analysis. *Mar. Biol.*, **111**, 343–51.

8 Fisheries and Management of Natural Populations

Introduction

Between 1950 and 1970 world marine and inland capture fisheries production increased on average by as much as 6% per year, trebling from 18 million tonnes in 1950 to 56 million tonnes in 1969 (SOFIA, 2000). By the 1970s and 1980s the rate of increase had fallen to 2% per year, and by the 1990s it had dropped to zero. It would appear that most of the world's fishing areas have reached their maximum potential, with the majority of stocks now being fully exploited. In contrast, since 1990 inland and marine aquaculture production is growing at the rate of 10% per year (see Chapter 9).

Although bivalves contribute a small percentage (~2%) to global capture fishery landings their generally high unit price compensates for the smaller landed weight when compared with the combined categories of fish, crustaceans and other molluscs (Tables 8.1 & 8.2). The economic importance of bivalves, therefore, calls for efficient approaches to the conservation and management of wild populations. Much of the methodology developed for finfish stock assessment and management has also been applied to bivalve molluscs, although much is inappropriate. Unlike fish, bivalve molluscs are sedentary, or almost so, and this has important implications for their population biology, conservation and management.

The first part of this chapter deals briefly with the main elements of bivalve population dynamics: abundance, mortality, growth, reproduction and recruitment. Strategies for the development and management of bivalve fisheries are also discussed. The focus thereafter is on commercial fisheries of selected species, described under: quantity and landed value from the main fishing regions, types of fishing methods utilised, practices for handling, processing and marketing. In each case management practices and constraints are also explored.

Population dynamics

What is a stock?

It is almost impossible to get universal agreement on what constitutes 'a stock' (see Carvahlo & Hauser (1995) and Grant et al. (1999) for discussion of the stock concept in fisheries). Some define it as a production or management unit, where differences within the group or exchanges with other groups are disregarded. Others regard a stock as a genetically discrete population, or a mixture of populations with limited genetic exchange (Cobb & Caddy, 1989; see also Chapter 10). One definition that does not demand reproductive

Table 8.1. Global landings from fisheries (live weight, tonnes × 1000) and monetary value (millions US$) of (A) marine bivalves and (B) combined categories of fish, crustaceans and molluscs, including bivalves for 1997–1999. Data from Food & Agriculture Organisation (FAO) *Fishery Statistics Yearbook* (1999). There are small discrepancies between these data and FAO (2001) fisheries data in Table 9.1.

A

Species	1997		1998		1999	
	t	$	t	$	t	$
Mussels	240	102	250	90	238	88
Oysters	185	157	160	122	158	116
Scallops	478	691	522	740	568	793
Clams	816	852	827	819	813	761
Total	1719	1802	1759	1771	1777	1758

B

	1997		1998		1999	
	t	$	t	$	t	$
All fish & shellfish combined	93766	80399	86933	76509	92867	76791

Table 8.2. The monetary value (US$ per tonne) of marine bivalves in comparison to various categories of fish for 1999. Data FAO *Fishery Statistics Yearbook* (1999). These data exclude landings from aquaculture production.

Species	$/t
Mussels	370
Oysters	738
Scallops	1398
Clams	936
Mean	861
Tilapia	1035
Salmon	912
Flatfish	2100
Cod	1017
Herring	256
Mackerel	400
Mean	953

isolation or genetic discontinuity, and which approaches the functional concept of a stock that a fisheries manager must deal with, is 'a group of individuals that sustains itself over time, and that responds in a similar way to environmental changes within a discrete geographic area' (Campbell & Mohn, 1983).

For sedentary organisms 'geographic area' can be as large as a sea, major off-shore bank or estuary, or as small as a single bivalve bed. Cobb & Caddy (1989) feel that Campbell & Mohn's definition is a reasonable working definition, and suggest that unless stocks are severely overexploited it may not be vitally important to know the extent of gene flow from adjacent populations. If characteristics such as growth rate, size at maturity etc. are sufficiently different to allow populations from different geographic areas to be clearly discriminated, then this allows the corresponding stocks to be managed separately, whether or not they are genetically distinct.

Distribution and abundance

A feature of most sedentary invertebrate populations is the non-uniform distribution of individuals and this must be taken into account in the development of properly designed sampling schemes. Methods for surveying the distribution and abundance of bivalve stocks involve the use of quadrats, benthic corers, dredging and trawling gear, or underwater photography and TV. The study site is first divided into blocks, the dimensions of which are dictated by the extent of the area involved. Blocks are numbered and the ones to be surveyed are selected using a table of random numbers or a string of computer-generated random numbers. Other methods such as mark-recapture techniques and fishing success methods, that are routinely used in fish stock-assessment, provide information on abundance but are essentially uninformative on spatial patterns (Orensanz *et al.*, 1991a). A brief account of some of these methods follows. Those wishing for more detailed information on sampling methodology may consult the many texts available e.g. Holme & McIntyre (1984), Caddy (1989a), King (1995) and Sutherland (1996).

Quadrats

To measure abundance on rocky shores a square metal or rigid plastic frame is laid on the substratum, and the animals within the frame are counted, weighed or estimated in terms of percentage cover of the surface. Good estimates of percentage cover can be made subjectively using sub-divisions of the quadrat as a guide. This may be done *in situ* or, alternatively, all organisms within the frame may be scraped off the substrate for subsequent analysis. Scraping should be avoided whenever possible as it invariably disrupts the shore community and subsequent recolonisation of the cleared area can take years (Hawkins & Jones, 1992). Quadrat size will vary depending on the abundance and size of animals, as well as the topology of the rock surface. For mussels it is best to use a quadrat size of $1.0\,m^2$ for estimating abundance on sheltered rocky shores, but on exposed shores where mussels are smaller, a $0.50\,m^2$ quadrat is best. Sampling stations should be spaced out from high to low water mark and several samples (4–5) should be taken at each tidal height, spanning the full range of microhabitats, e.g. rock pool, crevice, rocks exposed to or sheltered from sun or wind. At each station quadrat locations can be assigned by using random distances on either side of the transect line. For repeat sampling the positions of sample points can be marked with paint, chisel marks or drill holes; on soft substrates (see below) posts driven deeply into the sediment may be used.

An alternative to making many replicated counts in quadrats is to use abundance scales, semi-quantitative estimates of density or cover which fall into 5–7 broad categories: from 'extremely abundant' through to 'common' to 'not found' (Hawkins & Jones, 1992). Estimates are made in an area covering a few square metres around a station. Such estimates are particularly useful for the rapid assessment of abundance of a particular species along a stretch of coast, a headland or an island, where shore topography can make the use of quadrats a daunting, if not an impossible, task.

For small-scale underwater studies quadrat-based individual counting and cover estimates can sometimes be useful. However, patchy distribution, which is a characteristic of sublittoral scallop and clam species, reduces the likelihood of a randomly placed quadrat finding the species. This problem may be overcome by increasing the number of replicates per station, and by using larger quadrats that can be assembled *in situ*. Quadrat-based counts are impractical in large-scale studies where estimates on the size of whole populations, fishing grounds or stocks are needed.

Benthic corers, suction samplers and grabs

The corer is a quick, easy and effective sampling tool for estimating abundance and distribution of burrowing bivalves. For intertidal areas or in very shallow water, it can be a tube or a pipe of rigid material that is manually pushed into the substrate to a depth greater than the burrowing depth of the species being sampled. The dimensions of the corer are determined by the size of the animals being sampled, their expected abundance, and the maximum depth at which the animals are found. If the substrate is firm enough the core is simply lifted up, but in sandy areas a core retainer in the form of a thin piece of metal or wood is slid across the bottom to prevent material falling out as the corer is lifted. Sampling can be done at different levels by inserting a series of horizontal plates introduced through slots in the core tube.

This sampling method was employed by the Washington Department of Fish and Wildlife to determine the density and size of the Manila clam, *Tapes* (*Ruditapes*) *philippinarum*, before and after graveling, a process in which gravel and crushed oyster shell are mixed and used as habitats to improve recruitment, growth and survival in clams (Thompson, 1995). A randomised block design of nine plots (3 control, 3 gravel and 3 gravel + shell) was used and core samples were randomly selected in relation to a central line running the length of each plot. The sampling device was a piece of PVC pipe (10 cm diameter and 20 cm long) that was inserted into the substrate to a depth of ~15 cm. The core was removed and placed in a 1 mm mesh bag, and later sieved through a series of screens to a minimum diameter of 1 mm. Clams were measured counted and weighed for recruitment, survival and biomass data. Recruitment was higher on the control plots, while survival and biomass were enhanced by graveling. Recruitment improved on gravel, and gravel and shell, as the plots aged and accumulated a layer of fine sediment and organic debris.

Bivalves in deep water are usually sampled from a boat using a mechanical corer that is longer and larger than manually operated corers. The one illustrated in Fig. 8.1 is a rectangular box corer supported in a pipe frame, that samples an area 20 × 30 cm to a depth of 45 cm. The advantage of box corers is that they provide deep and relatively undisturbed samples from a variety of sediments, but a large vessel and calm weather are essential for the safe deployment of this big and heavy piece of equipment.

Suction samplers employ a coring tube which is forced into the substrate and held for a set period of time to draw up sediment and associated fauna into a mesh collection bag (Fig. 8.2). Most suction samplers are mechanically-operated, using either pumped water or compressed air to suck up samples. Grabs are lowered vertically from a stationary ship, and usually sample a surface area of 0.1–0.5 m², depending on the type of grab. The van Veen grab (Fig. 8.3) or the Day grab, a modified Smith–McIntyre grab, are the most common type used in sampling macrofauna. Holme & McIntyre (1984) give details on a whole range of corers, suction samplers and grabs, and evaluate their general performance and efficiency in capturing macrofauna.

Dredges

The dredge is a type of fishing gear that is towed behind a boat fishing for bivalves, e.g. oysters or scallops. It has a heavy metal frame to which is attached a steel-mesh bag and the leading edge of the dredge has a heavy chain, a sharp-edged blade or a toothed bar that scrapes or digs the animals from the substrate (Fig. 8.4A & B). The dredge is hauled in and put out by hand-operated or powered hydraulic winches. Alternatively, a suction pump is attached to the top of the dredge bag, through which the catch is continuously pumped aboard the fishing vessel, thus eliminating the need to haul the dredge on board. However, dredges are environmentally destructive because they disrupt the substrate and associated fauna, and can also damage the shells of the bivalves left behind (Bradshaw *et al.*, 2000; Jenkins *et al.*, 2001; Veale *et al.*, 2000, 2001)

Abundance surveys are performed by towing a dredge over a measured area for a set time and taking samples at a predetermined number of stations (tows) at set intervals. It is assumed that the number of bivalves caught per tow directly assesses local density. However, efficiencies of dredges are often below 20%, and are affected by the nature of the bottom, operation conditions, dredge design, gear leaping, gear saturation, skill of the operator, and size and behaviour of the species being sampled. A dredge can be calibrated by comparing dredged samples with those taken by scuba divers from marked quadrats. Using this method Rodhouse (1979) found the mean efficiency of a hand dredge, used to survey oyster density in the estuary of the Beaulieu River, Hampshire, UK, to be about 10%. An estimate of efficiency is needed to convert catch-per-tow figures into actual abundance (Orensanz *et al.*, 1991a). Catches are numbered and either the whole sample or a sub-sample is measured to get an estimate of the size distribution of the population at each station. Since 1965 the US National Marine Fisheries Service (NMFS) has been carrying out surveys to estimate the distribution, relative abundance and age structure

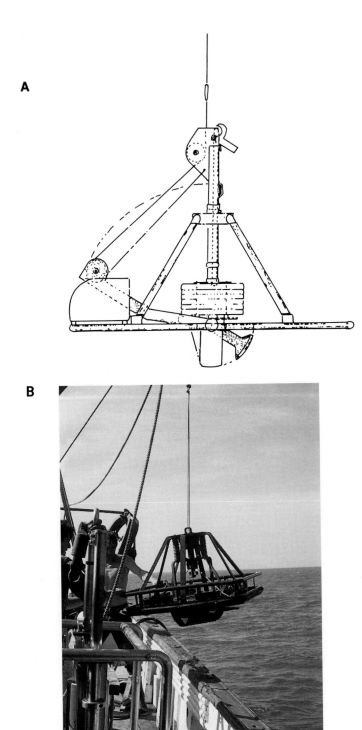

Fig. 8.1. Reineck box sampler. (A) The rectangular coring tube is closed by a knife edge actuated by pulling on the lever on the left. An attachment can be fitted to show the inclination and compass orientation of the core. Redrawn from Holme & McIntyre (1984). (B) The corer being lifted back on board. Photo courtesy of John Costello, Aquafact International Services Ltd., Galway, Ireland.

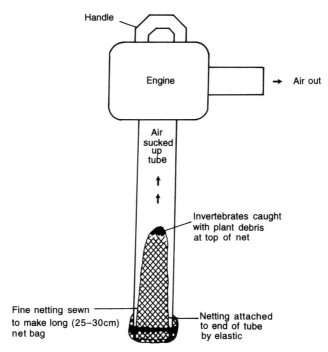

Fig. 8.2. A small lightweight suction machine for sampling invertebrates. Redrawn from Sutherland (1996).

of the surf clam, *Spisula solidissima*, off the north-east coast of the United States. The NMFS have divided the area into almost 100 strata based on bathymetry of the ocean floor, but have also broken the entire area up into regions (lines perpendicular to bathymetry, Fig. 8.5A). The strata remain fixed from survey to survey but the locations that are sampled within strata are randomised for each survey. A hydraulic dredge is towed for a set time and speed within each stratum, and the catch is counted, measured and a random sub-sample of each size class is retained for ageing (Figs. 8.5B & C). The number of allocated tows per stratum is roughly proportional to the area of the stratum and/or the density of clams, based on prior surveys. For more details see Weinberg (1999) and NEFSC (2000).

The most commonly used index of abundance in fisheries is catch per unit effort (CPUE). It may be recorded as the number or weight of bivalves caught per metre of dredge width per hour of towing. If CPUE is twice as large in area A than in area B the inference is that there are twice as many individuals in area A. Provided that the same type of fishing method is used, and that the distribution of bivalves in the two areas is similar, then CPUE can provide a good index of abundance in relative terms. The relationship of CPUE to stock size (N) is linear:

$$\text{CPUE} = qN$$

where q is the catchability coefficient (see below). However, in commercial bivalve operations CPUE has severe limitations as an abundance indicator;

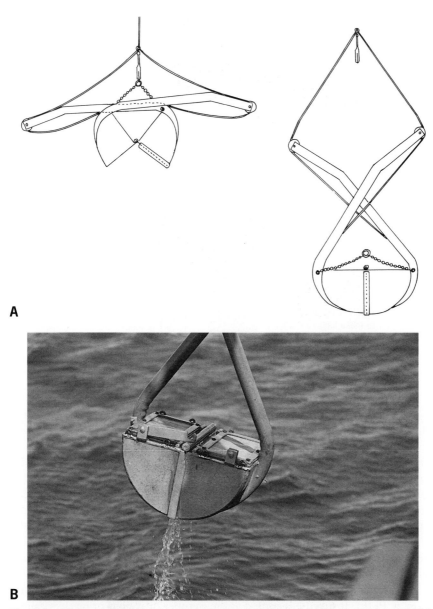

Fig. 8.3. (A) The van Veen grab, open and closed. Redrawn from Holme and McIntyre (1984). (B) The grab being lifted back on board. Photo courtesy of John Costello, Aquafact International Services Ltd., Galway, Ireland.

because bivalves are sedentary they do not mix after each fishing operation. The spatial structure of a bivalve stock is persistent and fishermen do not fish at random over the fishing ground, but tend to fish a bed until the density drops to some threshold level before they move on to another bed. Because of this sequential pattern of patch depletion CPUE is rarely a good index of abundance in bivalve populations. The method is adequate only when dense

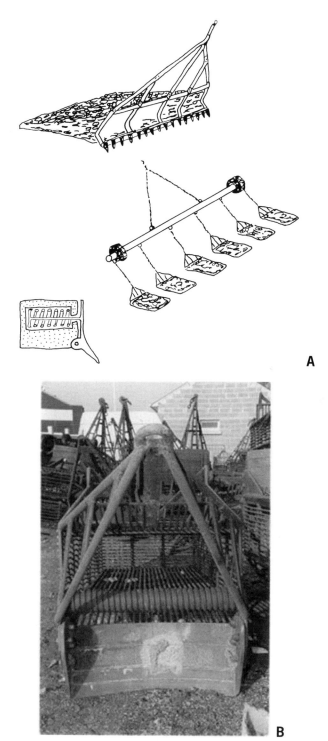

A

B

Fig. 8.4. (A) Scallop dredge with spring teeth and mounted six to a tow. From Hardy (1991). (B) Breton clam dredge (2 m long, 50 cm wide) used in NW France. This box-like dredge is made of long metal rods, and a sharp blade, shearing downwards, can be clearly seen at the lower part of the opening. Boats (10–15 m long) tow one dredge each side. Photo courtesy of M. Wilkins, Channel Marine Ltd., London.

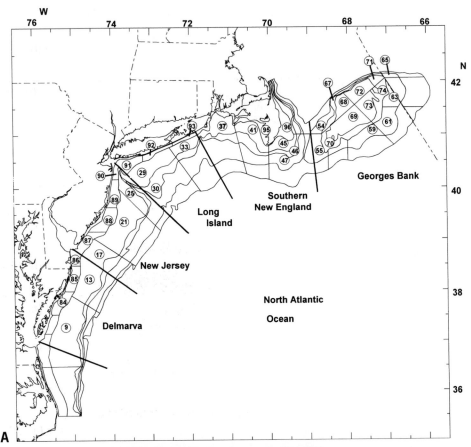

Fig. 8.5. (A) Regions, survey strata and stratum numbers of the National Marine Fisheries Service (NMFS) along the coast of the eastern United States, where the clam *Spisula solidissima* is fished. (B) *Spisula solidissima* size-frequency distributions from 1978–1997 for Stratum 88, off New Jersey. Survey date (mo/yr) and number of tows (n) are included. Vertical dotted lines facilitate comparisons among years. (C) Age-frequency distributions from 1978–1997 for Stratum 88. Solid diagonal lines connect values from the 1976 year class. A hypoxic event in the summer of 1976 caused mass mortality of *S. solidissima*, but in December 1978, when the dredge had smaller openings, samples from Stratum 88 were dominated by 2-year-old clams, which settled just after the hypoxic event in 1976. In 1997 there are about 18 year classes represented in the population. Average annual adult mortality, Z, (see text) was about 25%. From Weinberg & Hesler (1996) and Weinberg (1999).

patches of bivalves are small relative to the length of the tow, i.e. when they are invisible to the fishermen, and when the boundaries of the fishing ground are well defined (Orensanz *et al.*, 1991a).

Underwater visual methods

One advantage of working underwater is that the diver, with little effort, can get a preliminary qualitative overview of the area to be sampled, thus leading to precise quantification at a later stage. Underwater visual methods have been mainly used in surveys of scallop species. Recordings are made by divers or

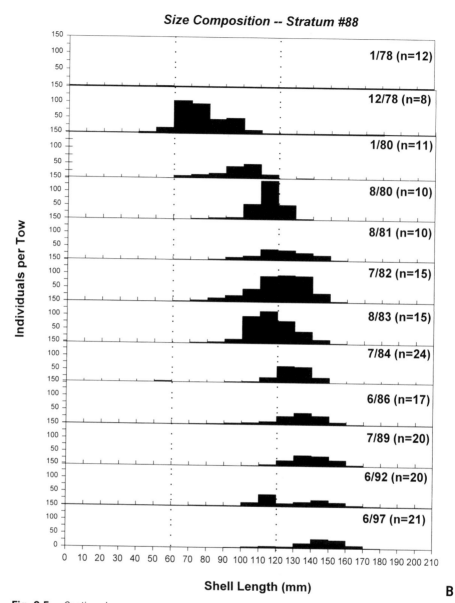

Fig. 8.5. *Continued*

by a towed underwater video camera. The diver swims along a fixed rope or chain, laid on the substrate, and which is marked at 1–5 m intervals. Numbers of individuals lying within a fixed distance (1–2 m in temperate waters; greater in clear tropical waters) of the transect line are recorded. The diver's swimming speed must be slow and constant to ensure optimum accuracy. This method is generally considered to be close to 100% efficient.

Alternatively, a diver may be carried on an underwater vehicle towed at a set speed behind a boat. If the diver needs to manually operate the vehicle

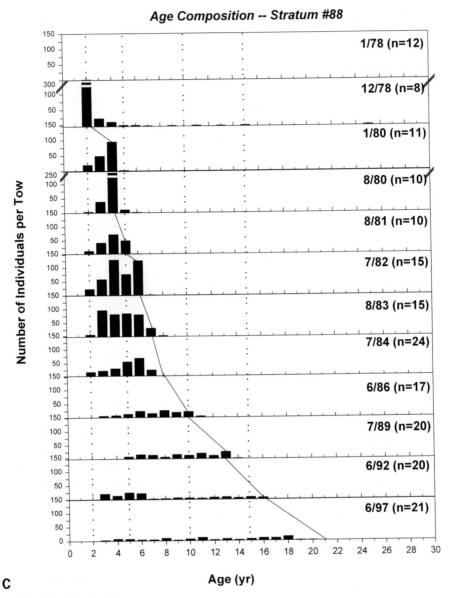

Fig. 8.5. *Continued*

then a tape recorder is used to record data. Bivalve densities can also be assessed by video camera, either hand-held by a diver on an underwater vehicle, or towed on a sledge or dredge. However, photographic methods have been found to have a much lower efficiency (36%) in estimating densities of the scallop (*Pecten maximus*), in comparison to methods that use divers to collect samples (Mason *et al.*, 1982). Thouzeau & Lehay (1988) have reported a higher efficiency (>80%) using the AQUAREVE device, a TV mounted on a dredge with an odometer, to estimate density of the same species in the Bay of Saint-Brieuc, France.

Age determination and growth

Direct measurements of growth rates can be made by marking individual shells and measuring them at fixed intervals. Alternatively, a time series of size-frequency histograms can be used to follow the position of individual modes over time; from this growth rates of individual cohorts can be estimated. Growth rate can also be determined from growth checks or rings on the external shell, or from growth lines in shell or ligament cross sections. The mean size of individuals for each year class is then calculated and a direct plot of size versus age gives a growth curve for the population. It should be pointed out that these methods are not equivalent in that shell marking provides data on individual growth rates, while growth rings and size-frequency histograms provide estimates of population growth rate (see Chapter 6).

The life expectancy of bivalves varies enormously depending on the species. This is well illustrated using scallops (Orensanz *et al.*, 1991a and references), which fall into two main groups:

- Long-lived species that live in temperate waters; large-sized species, i.e. >100 mm usually live more than 12 years, while medium-sized species (60–100 mm) live less than 10 years. Examples of the former are *Patinopecten caurinus* (>25 years), *Pecten maximus* (~22 years) and *Placopecten magellanicus* (>12 years). Examples of the latter are *Aequipecten* (*Chlamys*) *opercularis* (6–9 years) and *C. varia* (4 years).
- Short-lived species that live in tropical seas, grow to more than 100 mm but seldom live for more than three years, e.g. species of *Amusium* or warm-temperate species like *Argopecten irradians* and *A. gibbus*, that do not reach sizes above 80 mm, and rarely live beyond two years. Species in this group have a shorter larval life, a faster growth rate and higher natural mortality than long-lived species.

Estimates of mortality

Mortality in bivalves is caused by physical factors such as extremes of temperature and salinity, and biological factors such as predation, disease and fouling, plus interactions between these factors (see Chapters 3 and 11). For commercial species fishing is an additional source of mortality. The larval stages of bivalves are especially vulnerable to predation and consequently suffer extremely high mortality rates. With increasing size natural mortality rates decrease, but not in a constant manner because of seasonal influences and random catastrophic events such as oxygen depletion, algal blooms, disease etc. But as animals increase in size they become more and more susceptible to fishing mortality.

The loss of individuals in a population can be estimated in terms of the percentage of individuals that survive (survival rate) over a particular time period, or the percentage that die (mortality rate). The total mortality rate is

Table 8.3. Hypothetical age-composition data over three consecutive years under conditions of (A) constant recruitment, and (B) variable recruitment; a constant annual mortality rate of 60 per cent has been applied (see text). From King (1995).

	A) Constant recruitment			B) Variable recruitment		
	Age 1	Age 2	Age 3	Age 1	Age 2	Age 3
Year 1	1000	400	160	2000	2200	1100
Year 2	1000	400	160	1000	800	880
Year 3	1000	400	160	1500	400	320

referred to as Z, and is the sum of the instantaneous rate of fishing mortality (F) caused by the fishing operation, and the instantaneous rate of natural mortality (M), which includes deaths due to all other factors (King, 1995).

Total mortality (Z)

An indirect method to estimate mortality entails plotting the natural logarithms of the numbers of individuals surviving by age as a catch curve. In an exploited species, assuming a constant rate of mortality, numbers surviving tend to decline exponentially with time or age. This is expressed as follows:

$$\ln N_t = \ln N_0 - Z_t$$

N_t is the numbers surviving at time t, N_0 is the initial number of individuals at time zero, and Z_t is the total mortality rate at time t. When the natural logarithm of N_t is plotted over successive years the result is a straight line referred to as a catch curve. An estimate of Z is given by the slope of the line of best fit through these data (King, 1995). The utility of this type of analysis assumes that the age composition of the sample truly represents the age composition of the stock, and that recruitment and total mortality rates are constant for each age group. If recruitment and mortality rates are constant across year classes (Table 8.3A) then a large sample of several year classes may be used to construct a catch curve. Fig. 8.6 is a catch curve constructed using the data in Table 8.3A. Total mortality Z is 0.92 and using the equation (King, 1995):

$$\text{Mortality (\%)} = 100(1 - \exp[-Z])$$

this works out at 60%. If recruitment is variable from year to year, the more usual situation, then a single cohort is followed diagonally from upper left to lower right over three years to provide an estimate of Z (Table 8.3B), which gives 2000 individuals at age 1 in the first year, 800 at age 2 in the second year, and 320 at age 3 in the third year. Plotting the natural log of these numbers against age provides a catch curve (not illustrated) which has the same slope as in Fig. 8.6, and therefore the same estimated mortality rate of 60%.

In bivalves that show no annual growth rings on the shell, such as many oyster and mussel species, catch curves from length-frequency distributions can

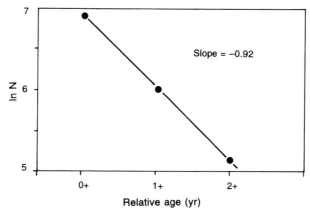

Fig. 8.6. A catch curve, natural logarithms of the number of individuals against age for the data in Table 8.3A. The absolute value of the slope, and therefore total mortality, is $0.92\,\mathrm{yr}^{-1}$ which is equivalent to 60% (see text). From King (1995).

be constructed to give estimates of Z. Once growth data for the species is available a length-frequency distribution can be converted to an age-frequency distribution. Details of how this is done are covered in King (1995).

Catch per unit effort (CPUE) data – commonly used as an index of abundance – can also be used to provide an estimate of Z. The natural logarithm of CPUE values for a particular cohort can be plotted over several years as a catch curve. This requires both age composition data as well as CPUE data. If the former are not available CPUE data may be used on their own, provided that recruitment occurs during well-defined periods. The decrease in total CPUE over the period between one recruitment and the next gives an estimate of the average total mortality for all age groups combined. The assumptions here are that all age groups, including the newly-recruited group, are equally vulnerable to the fishing gear, and that mortality is constant (King, 1995).

Mark-recapture data – normally used in growth studies – have been used to estimate survival and mortality in fish and crustacean species (Caddy, 1989b). However, mortality during the actual tagging operation, tag-induced death after tagging, loss of tags and possible emigration, in the case of scallops, from the survey area, are just some of the reasons why the technique has not been useful for bivalves, or indeed for fish or crustaceans also.

Natural mortality (M)

Natural mortality (M) may be estimated by following change in abundance of a cohort (or year class), or by analysis of dead shells. The former involves taking intertidal or subtidal quadrat samples (see above and Fig. 6.2), and following the decay in the numbers of a particular cohort or age class over a set period of time. This is the most direct method for estimating natural mortality but is really only suitable for small-scale studies.

Analysis of dead shells is only useful in species groups such as scallops and clams with well-defined annual growth lines on the shell. The use of dead shells in estimating M rests on the assumption that empty shells, held together by the ligament – sometimes referred to as 'cluckers' or 'clocks' – are a result of natural mortality alone, fished bivalves being either exported from the fishing grounds or their shells separated at shucking. The ratio of cluckers to living animals of a particular cohort can give an estimate of natural mortality for that cohort, provided that mean clucker life (the time elapsed between death of the bivalve and decomposition of the ligament) is known. This ranges from 28 days for *Chlamys tehuelcha* (Orensanz, 1986) to one year or more for *Placopecten magellanicus* (Dickie, 1955). It is important to have an accurate estimate of mean clucker life for individual species because an underestimate will inflate natural mortality rates (Orensanz *et al.*, 1991a). Also, bivalve damaged by fishing gear which are subsequently eaten by predators will also lead to inflated natural mortality rates, while the opposite bias will be seen if a proportion of cluckers are separated by fishing gear (Caddy, 1989b).

Another method for estimating M is to use the ratio of live shells to cluckers of a given cohort, where all have the same last growth ring marked on the shell (Caddy, 1989b). The number of cluckers counted is taken to be an estimate of the total number of natural deaths that have occurred since the formation of the last growth ring. However, the life of a clucker could often be shorter than the time elapsed since the formation of the last ring, in which case both cluckers and separated valves that have this growth ring should be counted. This will, however, give an overestimate of M since some of the separated valves could be due to fishing gear!

Fishing mortality (F)

Fishing mortality (F) may be estimated directly by obtaining density estimates before and after harvesting, or by swept-area methods. Both rely on having a reliable estimate of fishing gear performance.

In swept-area methods annual fishing mortality for a particular age group is calculated by using the formula:

$$F = fe(a/A)$$

where f is the annual total of effort units exerted, e is the efficiency of the fishing gear, a is the average area swept by a unit of fishing effort, and A is the total area of the fishing ground. The assumption of the formula is that units of fishing effort are distributed at random within the total fishing area A. However, because of the patchy nature of bivalve distribution, fishermen do not fish at random over the fishing ground, but tend to fish a bed until the density drops to some threshold level before moving to another bed. Thus fishing effort tends to match the non-homogeneous distribution of the stock, thereby resulting in a mosaic of fishing mortalities. Serious underestimates of F will result unless stock and effort contagion are taken into account (Caddy, 1975).

Indirect methods for estimating F can be obtained by subtracting the natural from the total mortality rate ($F = Z - M$), or if M is known, by cohort analysis; or calculated from total removals (C) over mean stock numbers (N) derived from surveys, i.e. $F = C/N$.

Reproduction and recruitment

Of most interest to fishery biologists/managers are the size/age at sexual maturity, and the timing of spawning and recruitment. The term 'recruitment' needs to be carefully defined as it has different meanings in different contexts. To a fisheries biologist the term refers to addition of new individuals to the exploited population.

From a commercial point of view it is useful to be able to predict when maximum recruitment can be expected to take place. However, forecasting the timing in a particular year can be difficult because peaks of reproductive activity in a population do not always correlate with subsequent recruitment in that area. This is because a variety of factors may affect the pre-recruit, such as hydrography of the area, weather conditions, larval abnormalities, and pre- and post-settlement mortality through predation and inadequate settlement surfaces. As well as the unpredictability of recruitment time in a particular area there is often an enormous variability in the strength of recruitment from one year to the next. This is not correlated with stock size as one might intuitively expect. Indeed, even if local stock numbers are very low recruitment can be high because of larval influx from other populations. Year-class strength variability most often reflects inter-annual variation in reproductive effort and success, the survival of pelagic larvae, or the survival of spat.

Year-class strength variability in scallops is well documented. Exceptional sets of *Patinopecten yessoensis* were observed in Mutsu Bay, Japan in 1948, 1952, 1954, 1957 and 1960. Spawning success in this species is dependent on a sudden rise in temperature to a minimum of 8–8.5°C and high numbers of pelagic larvae were observed in the years where this abrupt change in temperature was recorded (Yamamoto, 1964 cited in Orensanz *et al.*, 1991a). In *Argopecten purpuratus* in Peru a population explosion followed the 1983 El Niño, possibly due to increased spawning activity of adults or to increased larval survival associated with temperatures 6–8°C higher than normal (Wolff, 1987). Sometimes there is no obvious climatic cause for variability in year-class strength, in which case hydrographic conditions may be responsible. The importance of hydrography is illustrated by the Georges Bank fishery off the north-east United States, where the presence of a persistent tidal gyre permits retention of pelagic larvae in the vicinity of the adult stock, thus explaining why this area represents the world's largest single scallop resource (Caddy, 1989b; see also p. 251 this chapter).

While the size of whole populations, or subpopulations, is largely determined by environmental and hydrographic factors that influence spawning and the fate of pelagic larvae, once larvae are ready to settle it would seem that density-dependent processes are major influences on settlement, growth and mortality at the smaller scale of single grounds or beds (Orensanz *et al.*, 1991a). Details on the effects of density on these processes are in Chapters 5 (settlement), 6 (growth), and 3 plus 11 (mortality).

Fisheries assessment and management

The economic importance of bivalve fisheries calls for efficient methods of management to ensure sustainable production over time. Current practices are based on methods that have been developed for finfish stock assessment and management. Some useful references are Beverton & Holt (1957), Ricker (1975), Gulland (1983), Sinclair *et al.* (1985), Caddy (1989a), Hilborn & Walters (1992) and King (1995).

Stock assessment

Before management measures can be implemented an assessment of the stock(s) in question should be undertaken. Typically, total catch statistics along with measures of fishing effort are gathered; catch per unit effort (CPUE) data can be used to provide a measure of abundance over time and area (see above). These data are most often collected by commercial fishing operations. This is a cheap and comprehensive way of collecting data, provided that samples are reasonably large and representative of the population. But, as described above, fishermen often concentrate their effort on areas of high bivalve density so that CPUE can remain high over time, even when the total stock size is reduced substantially, e.g. as fewer areas of high density are left. Alternatively, CPUE can decline rapidly as a few local concentrations are depleted, but with little change in total stock size if only a small percentage of individuals are in the local concentrations.

Overfishing

In a stock that is fished at a low level losses due to mortality are balanced by gains through recruitment, and stock abundance will fluctuate around a mean level. But in an overexploited stock, the number of adults may be reduced to a level where reproduction is unable to replace the numbers lost (recruitment overfishing), or large numbers of individuals may be being caught at too small a size to maximise yield (growth overfishing). There is a lack of data to support (or refute) whether sedentary bivalve stocks are self-sustaining. Recruitment for a certain geographical area is assumed to be from surrounding contiguous areas. To verify this lengthy field studies on spawning, larval distributions and spat settlement are needed, but to date these are sadly lacking. However, the persistence of widely separated aggregations of scallops of characteristic absolute abundance in precise geographic locations strongly implies – at least for several scallop species – that many aggregations are self-sustaining (Sinclair *et al.*, 1985). To avoid recruitment overfishing a minimum spawning number should be maintained. The problem is that there is no information at present on what constitutes a minimum spawning number, so arbitrary measures have to be incorporated into management plans to avoid recruitment overfishing (see below and Sinclair *et al.*, 1985).

To understand the growth overfishing concept, and to devise management measures to prevent it happening, it is necessary at this stage to present a brief introduction to analytical yield models, the best known of which is probably the classical model of Beverton & Holt (1957). This model predicts the total yield to be obtained from a cohort over its entire life span, as a function of the year-class abundance, age at recruitment, growth and mortality. The model assumes that the total yield in any one year from all age classes is the same as that from a single cohort over its whole life span. The yield (catch in weight) of a single year class from age at first capture (t_c) to some maximum age (t_{max}) is an integral of the fishing mortality (F), the number of bivalves present (N) and their mean weight (W) which is:

$$Y = \int_{t_c}^{t_{max}} F_t N_t W_t dt$$

This model, usually referred to as the yield-per-recruit (Y/R) model, has been used to regulate fish and shellfish fisheries with the aim of maximising yield. Although the Y/R model only addresses the problem of growth overfishing, Y/R data can be tied to some index of mean reproductive potential per recruit, e.g. mean gonad weight per recruit (Orensanz et al., 1991a). Different harvesting regimes can be evaluated in terms of reproductive gains or losses. Mason et al. (1981) found that for Scottish scallop stocks an increase in fishing effort would only marginally increase Y/R, but would result in a considerable decrease in biomass per recruit.

Management measures

Ideally, a fishery should be managed from the beginning but in the majority of cases management measures are only applied when there is already evidence for overexploitation (overfishing) of the resource. The goal of fishery management is to maximise yield, either in weight or value, and to maintain a particular stock level to provide a buffer against poor recruitment years, or to maintain a minimum spawning stock (King, 1995).

Optimising yield can be achieved by regulating the size or age that bivalves can be harvested, and/or controlling fishing mortality. The former involves implementing regulations governing gear selectivity, minimum meat weights or meat counts, while the latter involves controlling effort or catch quotas. Size limits are usually regulated through selectivity of the fishing gear. In dredges, for example (Fig. 8.4), there are two potential routes for escaping individuals, the inter- and intra-ring spacing, the latter being larger and thus the most likely exit. Retention characteristics of dredges are regulated by stipulating minimum ring size and also through spacing of teeth in front of the dredge (see Table 8.4). However, this does not guarantee that small individuals escape entrapment, since exit from the dredge is often blocked by large animals and debris. It may be necessary to cull the catch on board, or land the catch unsorted.

In addition to regulating gear selectivity a minimum legal landing size is usually applied. If the aim is to protect breeding individuals it should be taken into account that reproduction is often age- rather than size-dependent. Where environmental conditions are less than optimal for a species individuals may grow more slowly and mature at a smaller size, and a lower minimum legal size would be justified in these areas. Meat counts (number of meats permit-

Table 8.4. Annual landings for 1997 (live weight, tonnes) together with information on fishing gear, fishing vessels and current regulations in the scallop *Pecten maximus* fishery in Europe. Maximum shell length is 150 mm. Adapted from text in Ansell et al. (1991) and Dao et al. (1993). The figure for England and Wales ([†]) is estimated from FAO (1997) figure for the UK, assuming that most UK landings (>90%) are from Scotland.

Fishery	Fishing gear	Fishing fleet	Current regulations
Scotland ~17000t	Spring-loaded toothed dredge (0.61–0.76 m wide) with belly rings 75–80 mm.	120 boats.	Minimum legal landing size 100 mm length (larger dimension than shell height); some sea lochs closed to mobile gear for various periods of the year.
Isle of Man 933t	Sets of toothed spring-arm dredges (0.61–0.73 m wide).	60 boats (11–25 m length).	Minimum legal landing size of 110 mm length; closed season (June–Oct inc.); boats >15 m length debarred from fishing within 3-mile Isle of Man territorial limit.
England and Wales ~2000t[†]	Spring-loaded toothed dredge (0.61–0.75 m wide) and beam trawl.		Minimum legal landing size 110 mm in Irish Sea and 100 mm length elsewhere; closed season in Irish Sea (June–Oct inc.); French dredges (0.61 m wide) in western English Channel restricted to vessels with a total beam length of <8 m and to offshore zone (>12 nautical miles).
Ireland 633t	Twin dredges 1.2 m wide with tooth spacing of 110 mm.	Boats <10 m length; number not available.	Minimum legal landing size 110 mm length; closed season in Irish Sea (June–Oct inc.).
France 12171t	Dredges (2 m wide) with tooth spacing of 100 mm and rings of 72 mm (national regulation dimensions).	800 boats (8–13 m length)	Minimum legal landing size 100 mm length; closed season 16 May–30 Sept, may sometimes be extended; limited entry, and duration of fishing limited in certain areas.

ted per unit weight) may be used when shell size is not a good predictor of muscle weight (see Naidu, 1991 and section on scallop fisheries below).

Yield is optimised by controlling fishing mortality by means of effort limitation and catch quotas. Effort is controlled by setting limits on the size, type and number of fishing gear per boat, the size of fishing boats and/or power of engines, entry to the fishery, number of fishing hours, fishing days, and by employing temporal and spatial closures. Sinclair *et al.* (1985) have proposed a *modus operandi* for controlling fishing mortality through catch quotas. This requires that the fishery be divided into management units, that population abundances are estimated for each unit at the beginning of each year, that the appropriate fishing mortality for each unit is calculated, and that this chosen value is applied to the selectivity-at-age (or size) vector. Catch quotas for a management unit may then be divided into quotas for individual fishers. (More information on management measures, and enhancement procedures such as habitat improvement, restocking, reseeding and predator control for fisheries can be found in the sections on specific fisheries later in this chapter.)

Regulations must be enforced for the management of a fishery to be effective. This should be done through public education (public meetings, radio and TV, press articles and poster displays), rather than by coercion. While prosecution should be regarded as a measure of last resort, regulations must be seen to be enforced. Penalties must be significant to the offender and appropriate to the offence. Enforcement staff must be trained in, for example, public relations, fishery management, evidence collecting and court procedures. Economic but effective ways of enforcing regulations should be explored. For example, to enforce a legal minimum size it may be more economic, and just as effective, to inspect the bivalves at point of sale rather than at point of capture (King, 1995). A regulation making it illegal to sell rather than catch undersized bivalve would be easier to enforce.

Scallop fisheries

Although there are about 350 scallop species most of the commercial harvest comes from just two of these, *Patinopecten yessoensis* and *Placopecten magellanicus* (see Fig. 3.4 for global distribution of the main commercial scallop species). Global landings from scallops fisheries have remained relatively stable over the last decade averaging about 520 000 t annually (Table 8.1A). Landings for *Placopecten magellanicus* have decreased by ~50% over this period, but this decrease has been compensated for by a corresponding increase in the landings of *Patinopecten yessoensis*, mainly from Japan. In 1999 about 63% of total global landings of scallops were from aquaculture operations and this is probably an irreversible trend as more and more countries move increasingly towards semi- or total-cultivation methods, not just for scallops but for the other bivalve groups as well (see Table 9.1).

This section reviews the main capture methods used in scallop fisheries, describes the major treatment and processing methods, and gives an account of the *Pecten maximus*, *Aequipecten opercularis* and *Chlamys islandica* fisheries in Europe, and the *Placopecten magellanicus* fishery on the east coast of North America. Although *Patinopectan yessoensis* accounted for about 55% of total global landings of scallops in 1999, the fishery is not a wild fishery in the generally accepted sense in that it depends on semi-cultivation methods. For this reason the species is not considered in this section but will be dealt with in Chapter 9, Aquaculture. As far as possible the same approach will be taken in the sections on oyster, mussel and clam fisheries.

Fishing methods

The main methods for catching scallops are by dredging and diving, the former being the one most commonly used.

Dredges

The type of dredge that is used varies depending on whether the seabed consists of mud, sand, pebble, rocks or boulders. In the offshore fishery in Canada and the United States, the New Bedford scallop dredge is used to capture *Placopecten magellanicus*. The dredge consists of a heavy metal frame, about 4 m wide, attached to a bag made from steel rings (10–12 cm diameter) with interconnecting chains as reinforcement (Bourne, 1964). On smooth terrain the bag may last several trips, but on rough ground it may not even last one trip (Naidu, 1991). In shallow inshore waters vessels may use as many as 13 small dredges (~1 m wide) individually shackled to a single tow bar. The efficiency of this system is only 5 or 12% for rocky and smooth bottoms, respectively (Dickie, 1955); the value for the New Bedford dredge is somewhat higher at 15–20% (Caddy, 1971). In the British Isles (English Channel and Irish Sea), the standard dredge for *Pecten maximus* is a 0.75 m wide spring-loaded Newhaven dredge. If the dredge becomes caught, the whole tooth bar folds backwards on a spring, thus allowing the dredge to pass over the obstacle, and regain its original position once free (Hardy, 1991; Fig. 8.4A). Most UK boats pull 4–12 Newhaven dredges per side i.e. 8–24 per boat; a few larger vessels pull 20 dredges per side (A. Brand, personal communication 2001). A depletion experiment where an area of the seabed was fished repeatedly, and run concurrently with a diver survey of the dredge tracks on the same fishing ground, found dredge efficiency to range between 24 and 30%, consistently lower than dredge-efficiency estimates from the diver surveys of 38–41% (Beukers-Stewart *et al.*, 2001).

Two components contribute to overall dredge efficiency (e): efficiency of capture (E) and gear (mesh) selectivity (s); $e = E \times s$, where E is the number of scallops entering the dredge divided by the number in dredge path, and s is the number of scallops caught divided by the number entering the dredge. One way of measuring s is by releasing marked scallops of known size from fragile plastic bags that burst as the dredge starts to move. On sea bottoms with abundant epifauna, dredge selectivity can be reduced to zero when the mesh becomes clogged with shells and debris. Efficiency of capture can be measured by comparing catches with densities photographed in front of the dredge (Caddy, 1989b). Capture efficiency tends to increase with increasing shell size; smaller scallops are able to swim away from the approaching dredge, or swim out of the dredge if captured. Many factors (some already mentioned) affect overall gear efficiency (e): nature of the sea bed, speed of tow, weather conditions, time of day (gear less visible to scallops at night), swimming ability and endurance of scallops.

Diving

Divers are limited to working in shallow waters (<30 m depth) because of safety and economic considerations. Usually teams of two to three divers operate from a small boat with an outboard engine. Each diver systematically covers lucrative spots within a given area, the choice invariably based on a prior working knowledge of the fishing ground. The scallops are picked off the bottom and placed in a net bag that is attached to a plastic drum filled with water, making it only partially buoyant. As the bag fills up the diver adds more air to compensate for the increasing weight of the bag. When the bag is full the diver fills the drum with air and the drum rises to the surface for collection by the boatman (Hardy, 1991).

The diver only picks scallops of market-size, leaving smaller individuals and all other epibenthic fauna undisturbed. This contrasts with the damaging effects of dredges already mentioned. Some regions have actually banned the use of dredges in favour of commercial diving. In the San José Gulf, Argentina, for example, dredges were banned in 1970 when the nearby San Matías Gulf fishery for *Chlamys tehuelcha* collapsed. Since 1975 the fishery has been opened every year but only to commercial divers (Orensanz *et al.*, 1991b).

In the past, before intensive fishing was introduced, scallops were fished using various types of hand-operated devices. One of these, the dip net, is used to this day in some parts of the world, e.g. the east coast of the United States. The net consists of a pole, about 8 m long attached to a small net bag. The fisherman works in shallow water (<8 m deep), lifting the scallops with the net off the bottom, often using a glass-bottomed box placed on the water surface to help spot the scallops. About 500 scallops per hour can be gathered from a well-stocked bed (MacKenzie, 1997a). Dip netting requires a strong and steady arm, a keen eye and calm and clear waters.

Treatment and processing

Sorting the catch usually takes place on board. Scallops are size-graded, packed into weight lots and transported either direct to the consumer or to a processing plant. For example, in the *Pecten jacobeus* and *Aequipecten opercularis* fishery in the Adriatic Sea the dredged catch is sorted by size and packed into 10–20 kg crates ready to be sold in the fish markets as soon as the boat reaches port (Renzoni, 1991). In the case of *Chlamys islandica* only large scallops (>60 mm) are kept and held in 300–500 kg lots in tanks on board. The scallops are transported to processing plants where mechanical shuckers remove the adductor muscle from the shell (Parsons *et al.*, 1991; Eiríksson, 1997). The meats are size and quality graded, then packed and quick-frozen. If both meat and gonad are required these are manually removed from the shell – a costly and labour-intensive process (Venvik & Vahl, 1985). The main market for this species is the United States. In New Zealand legal-sized 100 mm scallops (*Pecten novaezelandiae*) are sorted from the catch, packed in wooden boxes or sacks, and under-sized scallops and shell are returned to sea. Scallops are transported on a daily basis to processing factories where they are hand-shucked. Meats (muscle plus gonad) are either frozen in layer packs or free-flow form, or further processed by adding breadcrumbs or batter (Bull, 1991). About 50% supplies the home market and the rest is exported mainly to Australia, the United States and France. In the case of the *Argopecten gibbus* fishery on the coasts of Florida and Gulf of Mexico, the unsorted catch is taken to a processing plant where a shaker removes broken and dead shell, sand and unwanted species. The scallops are steamed to remove the meat and other tissues from the shell. A system of rollers separates out the adductor muscles, which are then rapidly cooled and packaged (Blake & Moyer, 1991). The adductor muscle, the only part consumed by Americans, is small compared

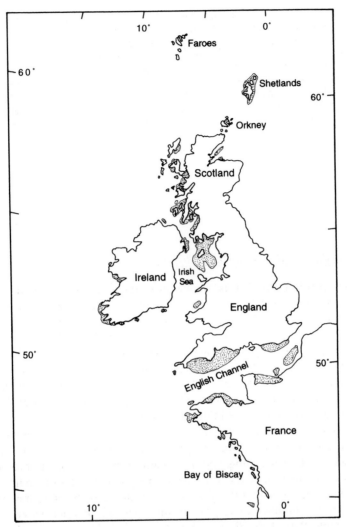

Fig. 8.7. Distribution of the major fishing areas (stippled) for the scallop *Pecten maximus* in western Europe. Redrawn from Mason (1983) and Brand (1991).

with other scallop species, and thus a high percentage (up to 70%) of the viscera is discarded. In all of the fisheries mentioned so far processing of the catch takes place on land, but in the case of *Placopecten magellanicus*, by far the most important commercial species on a global scale, shucking takes place at sea (Naidu, 1991).

European scallop fisheries

The major contributors to the European scallop fishery are *Pecten maximus*, *Aequipecten opercularis* and *Chlamys islandica*. Species of minor importance are *C. varia*, and *P. jacobeus* in the Mediterranean Sea.

Table 8.5. Annual landings in (live weight, tonnes) of the scallop, *Pecten maximus*, in Europe. UK landings include the Channel Islands and the Isle of Man. All landing data in chapter tables are from FAO (2001) and exclude landings from aquaculture production. Asterisked values are FAO estimates.

Country	1990	1991	1992	1993	1994	1995	1996	1997	1998	1999
Belgium	235	83	90	115	163	137	163	208	50	68
France	5077	8887	14059	13348	13503	12288	12100	12171	12866	12745
Ireland	1611	791	1029	543	918	423	560	633	693	1497
Spain	203	311	313	206	282	300*	300*	320*	299	86
UK	5257	4929	5721	6214	10024	10372	11053	19745	20939	20105
Netherlands							228	188	408	306
Norway				3	100	68	70	–	98	425
Area total	12383	15001	21212	20429	24990	23588	24474	33265	35353	35232

Pecten maximus and Aequipecten opercularis

Commercial fishing for *P. maximus* began in the 1930s and is based on stocks in the waters on the Atlantic coasts of western Europe (Fig. 8.7 & Fig. 3.4). Total landings (1990–1999) for *P. maximus* are shown in Table 8.5, together with the contributions made to the total catch by individual countries. Landing data for *P. maximus* and for all other species covered in this chapter are from the FAO (2001) and exclude landings from aquaculture production. Some aquaculture production of *P. maximus* is underway, mainly in Scotland and Ireland, but to date this represents a small percentage (<5%) of the combined landings of wild and cultivated scallops.

Table 8.4 gives information on current fishing methods and fishery regulations in the European *P. maximus* fishery. In Scotland and England there is no evidence of over-fishing and the major research effort is focussed on improving methods of stock assessment and predicting future trends in the fishery. The Isle of Man fishery is much smaller and therefore, not surprisingly, a greater number of regulatory measures are in place to control the fishery (Table 8.4). The effectiveness of one regulatory measure, i.e. closed areas, is illustrated by the following example. In an experimental closed area, set up by the Isle of Man Government off the south-west coast of the Isle of Man in 1989, there was a wide range of age classes up to 13 years old, while just outside the experimental area, where scallops were being fished, there were only 2, 3 and 4-year olds. In the closed area the density of scallops (>110 mm shell length) and the exploitable biomass (adductor muscle and gonad weight) was 15 and 23 times higher respectively, than on the adjacent fishing ground (A. Brand, personal communication 2001)

The decline in the Atlantic coast fishery in France in the 1970s, followed by a similar decline in English Channel stocks in the 1980s, both a result of over-exploitation, has led to a major national research programme on seeding techniques for direct capture at commercial size.

The main *Aequipecten opercularis* fishing areas and their individual contributions to the total catch (1988–1997) are presented in Table 8.6 (see Fig. 3.4 for geographic distribution). European landings peaked in 1990 at about

Table 8.6. Annual landings (live weight, tonnes) of the scallop *Aequipecten opercularis* in the European fishery. UK landings include the Isle of Man and Channel Islands.

Country	1990	1991	1992	1993	1994	1995	1996	1997	1998	1999
Faeroe Is.	8310	3300	3500	3320	3854	2781	3559	3581	4751	5993
France	948	772	1245	1399	2223	926	278	312	637	2088
Ireland	7	35	35	55	27	11	3	7	5	29
UK	7435	10154	11243	10485	4434	4322	3310	7260	9093	7150
Area total	16700	14261	16023	15259	10538	8040	7150	11160	14486	15260

Table 8.7. Annual landings in (live weight, tonnes) of the scallop *Chlamys islandica* in the major fishing countries, Iceland, Norway and Canada.

Country	1990	1991	1992	1993	1994	1995	1996	1997	1998	1999
Iceland	12117	10297	12429	11466	8401	8381	8976	10403	10098	8858
Norway	7387	7414	6805	10252	7916	8315	3	0	0	0
Canada			14	20	8	2	221	239	6637	3140
Area total	19504	17711	19248	21738	16325	16698	9202	10642	16735	11998

17000 t but declined to 43% of that figure by 1996, although landings since then show a good improvement. The species is fished by dredge, similar in design to that used for *Pecten maximus*, but with smaller teeth and belly rings because of the smaller maximum shell height (90 mm) of *A. opercularis*. In recent years many boats have changed to a different type of dredge called 'skids'. These resemble a small beam trawl with runners (hence the term 'skids') and have tickler chains across the mouth instead of the tooth bar (A. Brand, personal communication 2001). Otter trawls are also very effective on some grounds in the summer months when the scallop swims to escape gear. In general, regulations controlling the fishery are less stringent than those pertaining to *P. maximus* (Ansell *et al.*, 1991).

Chlamys islandica

The main fishing grounds are in waters around Iceland and Norway. The fishery started in Iceland in the late 1960s with an annual catch of 400 t (0.3% of world scallop catch), quickly increased to 17000 t in the mid 1980s, and is now about 10000 t per annum (Table 8.7). The main fishing grounds are in Breidafjördur on the west coast, where catches represent about 90% of total Icelandic landings. Fishing boats are converted trawlers or gill-netters, and scallops are caught by dredge (1.5–2.7 m wide with a system of chain links instead of a fixed bar) and processed as described above. All scallop meats are exported, with France as the largest market taking over 80%, and the United States importing the remainder (Eiríksson, 1997). The fishery is strictly regulated by management measures such as area quotas, seasonal closures, and annual dredge surveys as well as the use of CPUE data from skippers' catch reports (Parsons *et al.*, 1991; Eiríksson, 1997).

The Norwegian fishery which started in the 1980s is concentrated in waters from 70°–80°N, with large beds at Jan Mayen, Bear Island and Spitsbergen, and fjords along the coast of Norway from Lofoten to the Russian border. Boats are large (29–69 m length) and scallops were initially fished by a similar type dredge to that used in Iceland, but more recently a modification of the United States and Canadian dredge types is used (Strand & Vølstad, 1997). The boats process and freeze the catch on board and each vessel can handle up to 40 t of scallops per day. The main markets for the scallops are France and the United States. Regulations controlling the fishery include area closures, quotas, a closed season from 2 March–31 July, limits on the number of boat licences, and a minimum legal landing size of 65 mm shell height (Strand & Vølstad, 1997). Maximum shell height in this species is around 90 mm. There is now clear evidence that Norwegian stocks have been over-exploited with the annual catch in 1996 less than 0.05% of the 1990 catch (Table 8.7). The fishery has since collapsed and cultivation of *C. islandica* is being considered as a way of increasing production in Norwegian waters. In contrast, landings have increased in Canada from a low of <20 t in 1992 to about 3000 t in 1999.

North-west Atlantic fishery

This fishery is based on the sea scallop, *Placopecten magellanicus*, which is by far the most important species in wild fisheries, accounting for 23% of the total annual global scallop catch in 1999. The main fishing grounds for *P. magellanicus* are the Gulf of Maine, the Bay of Fundy, and waters off Nova Scotia and Newfoundland. In the Gulf of Maine most of the catch comes from inshore United States territorial waters, while the offshore fishery is concentrated on Georges Bank (Fig. 8.5A), situated in both United States and Canadian territorial waters, about 200 km south west of Cape Sable in Nova Scotia. This bank supports the world's largest, single natural scallop resource (Caddy, 1989b). Annual landings for the United States and Canada are in Table 8.8. Landings in Canada fell from 92 078 t in 1992 to 54 756 t in 1999, a decrease of 40%. A larger decrease of 68% occurred in the United States fishery, where landings fell from 133 859 t to 43 030 t between 1990 and 1998, although an improvement occurred in 1999. There are increasing efforts to culture *P. magellanicus* using the technology that has so successfully been employed by Japan and China (see Chapter 9).

Table 8.8. Annual landings (live weight, tonnes) of the scallop *Placopecten magellanicus* from Canada and the United States. The Georges Bank fishery contributes between 50–90% and 20–70% to Canadian and United States landings, respectively (figures calculated from Naidu, 1991).

Country	1990	1991	1992	1993	1994	1995	1996	1997	1998	1999
Canada	83 278	79 589	92 078	86 929	89 449	58 567	47 628	53 630	56 402	54 756
USA	133 859	131 569	107 638	56 702	61 188	62 456	61 552	48 120	43 030	77 206
Area total	217 137	211 158	199 716	143 631	150 637	121 023	109 180	101 750	99 432	131 962

The type of gear used to fish for *P. magellanicus* has been described above. For inshore fishing small wooden boats, about 20 m long, are used, while larger steel vessels, up to 46 m long, operate in offshore waters. The latter are technically well equipped and are capable of fishing on a 24-hour, year-round basis (see Naidu, 1991 for details).

The Georges Bank fishery is independently reviewed each year by American and Canadian scientists. In Canada there is limited entry to the fishery, mandatory logs for vessels of 19.8 m and over, and regulations governing trip catch limits and trip duration (not exceeding 12 consecutive 24-h periods). Both Canada and the United States have imposed a minimum shell height of 83 mm; shell height in this species is usually between 100 and 150 mm but can be as large as 200 mm (Naidu, 1991). Meat count regulations have also been imposed to protect young scallops; as the average scallop size increases the meat count declines. The meat count is currently at 33 meats per 500 g, but compliance with this regulation is achieved by blending large and smaller meats, a strategy that legitimises the harvesting of a considerable number of small scallops, thus leading to growth overfishing (Naidu, 1984). By using this meat count the fishery is selecting primarily 4, 5 and 6-year olds, an age span relatively low on the growth curve of this species (see p. 237). Inshore Canadian stocks are generally not subjected to a minimum legal size or meat count regulations. However, where regulations are in force, minimum shell height is 105 mm and meat count varies from 33 per 500 g to 72 per 500 g, depending on location and time of year. In some regions, e.g. the Bay of Fundy, there are management measures such as closed seasons, licence restrictions, and regulations governing boat size, gear type and gear dimensions (Sinclair *et al.*, 1985; Naidu, 1991). Several seeding programmes are underway in Nova Scotia and Newfoundland but most of these are at the R & D stage (Dao *et al.*, 1993).

In the United States there is open access to inshore stocks, a permit being the only requirement. In general, the only regulations are those that are self-imposed by the industry (crew size, trip duration). The State of Maine is the exception; a licence for commercial fishing is required and eligibility is limited to Maine residents. Only scallops of ≥76 mm shell height may be landed but no quotas are imposed. Massachusetts and New Hampshire also have this minimum size restriction. Regulations also specify areas where harvesting by dredge and trawl is authorised and a closed season operates between mid-April and the end of October.

Starting in 1994 three large areas (17 000 km²) on Georges Bank and southern New England (Fig. 8.5A) were closed year-round, primarily to protect groundfish stocks (see Murawski *et al.*, 2000 for details). Because of groundfish by-catch, the areas were closed to dredges used to fish for *P. magellanicus*. Enforcement is achieved through ship and airforce patrol, severe penalties for violators and a requirement that all scallop boats carry a vessel monitoring system. After four years of closure total and harvestable scallop biomasses were 9 and 14 times denser, respectively, in closed than in adjacent open areas. In 1999 portions of one closed area were reopened for scallop dredging, with severe restrictions on gear and areas fished to protect groundfish. Results were so encouraging that there are plans to implement a formal 'area rotation' scheme for scallops. Closed areas may prove to be a far more effective tool to increase yield per recruit than other methods such as minimum meat regulations or minimum dredge-ring sizes (Murawski *et al.*, 2000; also see above).

Summary information on a world-wide selection of scallop fisheries of commercial importance is presented in Table 8.9.

Oyster fisheries

The history of oyster production in both Europe and North America is well documented. In the nineteenth century flat oyster (*Ostrea edulis*) production in Europe was booming, but from the end of the century there was a gradual decline in production in the British Isles, most likely due to over-fishing and an increase in pollution. To reverse the trend non-native *Crassostrea virginica* were imported for relaying, carrying with them slipper limpets and predators such as oyster drills, which decimated much of the remaining native oyster beds. Much the same occurred in France but was exacerbated by disease that affected not just the indigenous flat oyster stocks but also the Portuguese oyster, *Crassostrea angulata*, introduced into France in the 1860s (Dore, 1991). Flat oyster production has fallen from 28 000 t in 1960 to 12 t in 1999 (Goulletquer & Héral, 1997; FAO, 2001). Concomitant with this decline has been the spectacular increase in the aquaculture production of *C. gigas* in France (134 800 t in 1999). A similar situation has occurred, or is occurring, in other European countries.

A corresponding decline in oyster production also occurred in the *Crassostrea virginica* fishery on the east coast of the United States. The fishery, which once extended from the Gulf of St. Lawrence to the Gulf of Mexico, is now concentrated in Long Island Sound, Chesapeake Bay and the Gulf area. Annual landings have fallen from 160 million pounds (73 000 t) of meat in the early 1900s to about 40 million pounds (18 000 t) in 1996, and most of this comes from the Gulf region. The initial decline, which started in the late 1880s, was probably due to over-fishing, but anthropogenic factors such as pollutants and disease (see Chapter 11) have accelerated the decline in recent years. For those interested in the history of the fishery the fascinating and well-illustrated account by MacKenzie (1996b) is highly recommended.

Out of 200 oyster species there are only about six that are fished from wild stocks in commercial quantities (Table 8.10). Total global landings from wild stocks for the years 1997 to 1999 are presented in Table 8.1A; more than 80% of annual landings are from *Crassostrea virginica* and about 8% from *C. gigas*. This table does not include oyster landings from aquaculture, which presently account for 96% of all oyster production (FAO, 2001).

In the following section the *Crassostrea virginica* fishery in North America will be described in some detail. While overfishing, disease, pollution and severe weather conditions have steadily reduced oyster stocks in many regions, this fishery still remains a wild fishery, although the situation will probably change with the development of hatchery-produced, disease-resistant strains for relaying (see Chapter 10).

Table 8.9. Summary information on a world-wide selection of scallop fisheries. Annual landing are less than 2000t except in the case of *Argopecten purpuratus* and *Patinopecten caurinus*. Maximum shell height is below species name. [a]The ecological preferences of the majority of the species are presented in Table 3.5. [b]minimum legal shell height; [†]areas where scallops are fished in commercial quantities.

Species	Main fishing areas[†]	Fishing methods	Ecology[a]	Management measures	Reference
Pecten jacobaeus 140mm	Adriatic Sea	Dredge-like gear (rapido or rampone)	Preference for 40–50m depth.	Shell size 80mm[b]; sale of shelled produce prohibited; some artificial seeding.	Renzoni, 1991
Pecten novaezelandiae 100mm	Around New Zealand	One or two dredges/boat (10–15m long)	Beds at 19–25m on soft muddy substrate or hard sand.	Shell size 100mm; restrictions on number and size of dredges; closed season (Feb–July); daily quotas and limit on licences and fishing hours; seeding program introduced in 1989.	Bull, 1991
Argopecten irradians 100mm	In waters off Massachusetts, New York and Rhode Island, USA	One or two dredges/boat		Regulations governing licensing, season of harvest, catch limits and minimum shell size; preliminary trials in aquaculture underway.	Rhodes, 1991
Argopecten gibbus 100mm	Cape Canaveral, Florida, USA	Two modified shrimp trawls/boat		Regulated by the industry; area not fished until 75% stock has reached 38mm shell height (size at first spawning).	Blake & Moyer, 1991
Argopecten circularis 100mm	Off the west and east coasts of the Baja Peninsula, California, USA	Diving; one or two divers/boat		Shell size 60mm; 330 meats/kg to limit capture of small scallops; few additional restrictions.	Felix-Pico, 1991

Table 8.9. *continued.*

Argopecten purpuratus 100 mm 30 141 t in 1999	Off the coast of N. Chile	Diving		Closed seasons of several years depending on resource. Expansion of the fishery expected when culture techniques to produce spat are developed.	Navarro Piquimil *et al.,* 1991
Chlamys patagonia 120 mm	Off the coasts of S. Chile and Argentina	Diving	2–40 m depth preferred habitat in protected bays and inlets	Off season from September to December; extended in certain areas depending on resource.	Navarro
Chlamys tehuelcha 100 mm	San José Gulf, Argentina	Diving	Shallow shelf bottoms, ≤15 m depth	Shell size 60 mm; closed season which may be extended in some areas to protect beds of juveniles.	Orensanz *et al.,* 1991b
Patinopecten caurinus 250 mm 2642 t in 1999	Gulf of Alaska, USA; much smaller landings in British Columbia and Washington waters	Dredge		Closed season (1 June–31 March); area closures; rings on dredge must be either 76.4 mm or 101.6 mm depending on locality; in British Columbia shell limit of 120 mm and gear width limit of 2 m	Bourne, 1991
Amusium pleuronectes 80 mm	Lingayan Gulf and Visayan Sea, Philippines	Trawl by-catch where main catch is finfish, shrimp and squid		Few regulations	Del Norte, 1991

Table 8.10. Global landings (live weight, tonnes) of the major commercial oyster species. *FAO estimate. †Probably includes *Crassostrea gigas*, *Saccostrea cucullata* and *S. echinata*. The geographic distribution of individual species is in Fig. 3.2. Species for which annual landings are less than 1000 t are omitted.

Species	1990	1991	1992	1993	1994	1995	1996	1997	1998	1999
Crassostrea gigas	21487	20438	21000	30802	23691	20123	25147	25696	12102	12271
C. virginica	117006	80436	99382	107511	128472	158430	148540	155166	135222	132207
C. rhizophorae	5877*	2136	2691	3229	4428	5224	4110	4241	4895	3906
Ostrea edulis	7469*	8658*	8711*	4065*	4770*	4368*	3270*	3329*	2765	2455
Others†	13637*	22515*	20258*	6679*	2484*	2755*	3019*	1562*	2950	3108
Total	165476	134183	152042	152286	163845	190900	184086	189994	157934	153947

The *Crassostrea virginica* fishery in the United States

The distribution of *C. virginica* extends from the St Lawrence River in Canada to the Gulf of Mexico, Caribbean, and may even extend onto the coasts of Brazil and Argentina (Fig. 3.2). There have been numerous introductions of the species onto the Pacific coast of N. America but none of these plantings have survived, with the exception of a small, but not commercially viable, population in British Columbia (Carlton & Mann, 1996).

The fishery (Fig. 8.8) is centred on Cape Cod (Connecticut), Chesapeake Bay (Maryland and Virginia), Florida, Louisiana and Texas, which altogether contribute 80–90% to annual landings of *C. virginica* in North America (Table 8.11). Mexico, on the Gulf coast, also lands substantial quantities, contributing between 20 and 30% to the annual global catch of this species. Canada, on the other hand, contributes only about 2%.

The main fishing methods for catching oysters are hand tongs, dredges and diving. In some areas such as the southern US states oyster reefs are harvested by hand. Hand tongs comprise two long flexible wooden poles, about 5–5.5 m long, joined like scissors towards one end. Metal basket rakes are attached at this end, and periodically the contents of the baskets are lifted and spilled out onto a culling board on deck for sorting. Market-size oysters are separated from under-sized oysters, shell and debris, which are returned to the sea to allow further growth in the case of the oysters, or to serve as cultch for settling larvae. The deeper the water the more difficult it is to use tongs with two rigid arms, so at depths greater than seven metres heavy metal tongs, operated by a winch system, are used. In dredge fishing, small dredges, called hand scrapes are used from small boats. Alternatively, larger heavier dredges similar to those used in scallop fisheries are towed, usually in twos, on alternate sides of a boat, and the contents are winched up and brought onto the deck for culling. In deep waters, where clumps of oysters may be widely separated from each other, diving is a very efficient method for collecting oysters, particularly large ones, that cannot be reached by tongers or dredgers (Kennedy, 1989).

The oysters are bagged on board and brought to the packing house on shore for grading. Unlike other oyster fisheries, most oysters in the United States are sold as shucked meats rather than as shellstock. When sold as shellstock the oysters are graded by count or by weight range in ounces (1 ounce = 28.4 g), and packed in sacks, bags, baskets or boxes. Fresh shucked meats are packed in containers (US half-pint to a US gallon – 0.236–3.785 l), or individually quick frozen (IQF) for coating with breadcrumbs or batter, block frozen for stews and soups, or canned (see Dore, 1991 for details).

As already mentioned, the fishery saw a downward turn at the end of the 1880s and this trend continued over the last century, especially in the Chesapeake area, once one of the world's largest oyster producers. Although mismanagement of the fishery, pollution from sewage, fouling control programmes, and economic factors have played a role, two disease organisms, *Haplosporidium nelsoni* (MSX) and *Perkinsus marinus* (see Chapter 11) have had a devastating effect on stocks from Chesapeake Bay southwards to Florida. In contrast, other areas such as Louisiana and Texas have not been so hard-hit and have managed to maintain production.

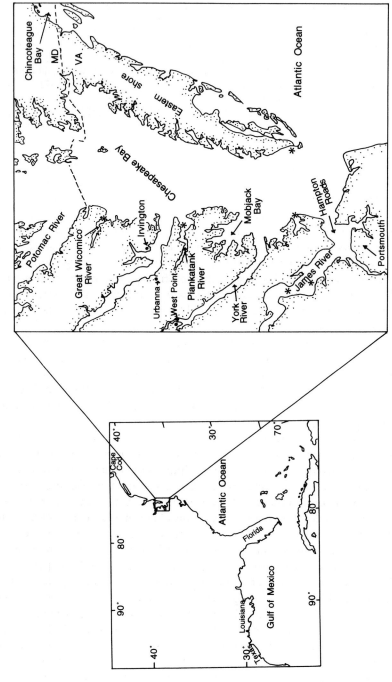

Fig. 8.8. Major fishing areas for the oyster *Crassostrea virginica* on the east and south coasts of the United States with enlargement showing the Virginia (VA) portion of the Chesapeake Bay. Indicated are major oystering areas (arrowed), and some of the localities where reef restoration programmes are underway (asterisked). Redrawn from MacKenzie (1997b) and based on additional information in Wesson *et al.* (1999), Mann (2000) and O'Beirn *et al.* (2000).

Table 8.11. Global landings (live weight, tonnes) of the oyster *Crassostrea virginica*.

Country	1990	1991	1992	1993	1994	1995	1996	1997	1998	1999
Canada	2644	1902	1739	621	2636	2208	2132	1717	3124	3225
USA	66017	43820	68752	84069	92189	128613	111682	114934	101383	89714
Mexico	48345	34714	28891	22821	33647	27609	34726	38515	30715	39268
Total	117006	80436	99382	107511	128472	158430	148540	155166	135222	132207

Each state has its own fishery regulations. Entry to the fishery is unrestricted once the applicant resides in the state and pays the requisite licence fee. For tonging and diving the fishing season extends from September 15 to 31 March; for dredging the season is usually shorter. Oysters are generally not harvested during the summer months, the months without an 'R', the period when oysters are spawning or recovering after spawning. Additional regulatory measures to control the fishery include: limits on dredge weights and dimensions, daily catch limits, and a ban on fishing at night and during weekends. Regulations can vary from one state to the next and even within a fishing area. For example, in 1994 all public beds in the Chesapeake Bay were closed to permit the excellent spatfall of 1992 to grow and reproduce. The St. James River fishery south of Chesapeake Bay is still open but in 1994 additional restrictions on tong size and daily time and season limits were also imposed (Wesson *et al.*, 1999). In all fisheries the minimum market size is 3 inches (76 mm shell length) and shell must be returned to the seabed to serve as cultch (see Kennedy, 1989 and MacKenzie *et al.*, 1997b for further details on fishing regulations).

These restrictions are not very effective in increasing oyster production unless additional measures to restore habitats and improve water quality are also put in place (Castagna *et al.*, 1996). Improvements involve enhancing sites for setting oyster larvae, controlling mortality, and transplanting oysters to better growing environments (MacKenzie, 1996a). Ways to increase setting sites for oysters include: increasing shell plantings for cultch, placing cultch in areas of moderate to high setting at the appropriate time, and using tow boards to blow off silt from setting surfaces. The latter is one of the less expensive ways to increase oyster seed production. Reducing mortality from predation is difficult (see Chapter 3), but exposure to heavy metals and organic contaminants, to which juvenile life stages are especially vulnerable, can be controlled. Seed oysters can be transplanted from setting areas to good growing areas from where they are marketed. Seed clusters are broken to produce double or single oysters which, when they have attained market size, are well-shaped for the half-shell trade. However, the movement of seed oysters is expensive and in moderate to high salinity waters disease may inhibit transplanted oysters from reaching market size. Cleaning shell and seed beds prior to replenishment can reduce the impact of resident endemic disease (Wesson *et al.*, 1999).

Restoration efforts are focusing more and more on the reconstruction of three-dimensional reef habitats with a view to restoring the ecological role of what are merely 'footprints' of former reefs. For example, in 1993 the Virginia

Marine Resources Commission Oyster Replenishment Program began reef restructuring in the Piankatank and St. James Rivers to investigate both the value of reef structures for the survival of the oyster as well as methods by which reefs could be constructed (Fig. 8.8). Reefs are built from shucked oyster, clam shells or stabilised coal ash, and are covered at high tide but exposed to some degree at low tide. All reefs are closed to oyster harvesting and for the foreseeable future will remain sanctuaries for broodstock restoration. Results to date are encouraging in that there are signs of a definite, albeit slow, increase in broodstock numbers at reef locations (Wesson *et al.*, 1999). A novel approach to restoration has been adopted throughout the Chesapeake Bay area where hatchery-produced mature oysters (>250 000) have been grown by school students and transplanted by them onto reefs. Surveys show an order of magnitude increase in the abundance of juvenile oysters on the reefs, and correspondingly high spat settlement rates on oyster grounds surrounding the reefs (Brumbaugh *et al.*, 2000). Similar reef reconstruction is underway in other parts of Virginia and also in Maryland, Alabama, Louisiana and North Carolina (Luckenbach *et al.*, 1999). Mann (2000) has argued that the primary objective in reef restoration programmes should be the re-establishment of this important cornerstone species in the ecosystem, with the added bonus of restoring the fishery, albeit not to former historical levels.

Table 8.12 presents some of the management strategies adopted to increase production of *C. virginica* in the main fisheries on the east and southern coasts of North America. It is clear from this table that the implementation of different management strategies does not invariably lead to an increase in production. Disease and the polluting effects of agricultural and urban runoff are still major impediments to oyster restoration. The imposition of restrictions to minimise sediment run-off, the development of MSX-resistant strains of oysters (see Chapters 10 and 11), and the increased production of hatchery seed for grow-out (see Castagna *et al.*, 1996 and Chapter 9 for details) will probably go some way to improving the situation. It is interesting that the one state where oyster production has remained steady over many years is Louisiana where there are large private plantings coupled with an abundant seed supply managed by the state. It is typical that private grounds out-produce public beds (Kennedy, 1989), and one of the repeated recommendations for the restoration of the Chesapeake Bay fishery has been to encourage more leasing of public beds to private groups or individuals (Andrews, 1991). However, fishermen in the area have consistently opposed such a move.

Mussel fisheries

There are hundreds of species of mussel but only about a dozen or so are fished commercially and most of these are in the genus *Mytilus*. Annual global landings from fisheries for the years 1997 to 1999 are presented in Table 8.1A. Three species, *Mytilus edulis*, *M. galloprovincialis* and *Aulacomya ater* are the main contributors to global mussel fisheries (Table 8.13) contributing 51%, 23% and 8%, respectively, to the fishery in 1999. As the vast majority of annual landings (>80%) come from aquaculture production (see Chapter 9) it is almost impossible to locate a natural fishery, i.e. one that is not subject to human

Table 8.12. Management methods employed on the Atlantic and Gulf coasts of North America from the 1960s to the present day to increase production of the oyster *Crassostrea virginica*. Adapted from the text of MacKenzie (1996a,b).

Location	Management method	Outcome
Prince Edward Island, Gulf of St. Lawrence, Canada	Transplantation of oysters plus shell to beds for harvest; removal of silt in high oyster setting areas.	More than a three-fold increase in production from 1972–1996.
Long Island Sound, Connecticut, USA	Increased shelling of beds; removal of gastropod predators by suction dredge, and starfish by mops or quicklime.	Forty five-fold increase in production from the late 1960s to 1995.
Chesapeake Bay, Maryland, USA	Shelling of grounds with good setting history.	Two-fold increase in oyster landings from the early 1960s to the early 1980s; however, disease has since reduced landings to less than 1% of 1980s landings.
James River, Virginia, USA	Shelling and, more recently, reef reconstruction for broodstock conservation (see text).	Much the same scenario as in Maryland.
Apalachicola Bay, Florida, USA	Rehabilitation of oyster reefs by spreading limestone or *Rangia cuneata* shells; relaying seed and juvenile oysters.	Annual production (10% of total USA landings) varies by a factor of 8 due to weather-related events.
Louisiana, USA	Shelling of beds with *Rangia cuneata* and oyster shells from old reefs.	Contributes 50% to annual USA landings; despite problems from oil pollution and occasional disease episodes production has remained steady, primarily due to the shelling programme and the fact that the majority of beds are privately owned.

Table 8.13. Global landings (live weight, tonnes) for the three most important contributing species to mussel, *Mytilus edulis*, fisheries. These figures do not include landings from the Netherlands and Germany treated as aquaculture production in FAO statistics, and amounting to 80 000–140 000 t year^{-1} (Table 8.14).

Species	1990	1991	1992	1993	1994	1995	1996	1997	1998	1999
Mytilus edulis	128 226	162 261	170 375	166 862	163 710	146 134	117 481	129 195	136 813	121 965
M. galloprovincialis	38 929	32 716	38 716	39 050	34 032	40 764	39 046	53 281	50 593	55 819
Aulacomya ater	23 521	9 214	15 269	13 541	16 843	17 580	13 428	16 078	22 831	19 738
Total	190 676	204 191	224 360	219 453	214 585	204 478	169 955	198 554	210 237	197 522

intervention at some stage of the mussel life cycle. Fisheries closest to being described as natural fisheries are those where seed is harvested from natural habitats and relayed on controlled beds for on-growing to market size. Other than collection and relaying there is no other human input. One such example is the Dutch, German and Danish mussel fishery in the Wadden Sea. Information on the Netherlands fishery has been principally gleaned from Dankers

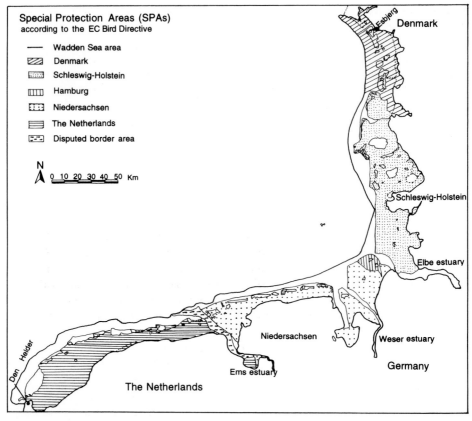

Fig. 8.9. The Wadden Sea and Special Protected Areas (SPA). Disputed border area according to Ems-Dollard treaty of 1960, the Supplementary Agreement of 1962 and the Environmental Protocol of 1996. From De Jong *et al.* (1999).

(1993), Dijkema (1997) and Dankers *et al.* (2001); for Germany from Seaman & Ruth (1997) and correspondence with both authors; and for Denmark from Kristensen (1997), Kristensen & Hoffmann (1999) and correspondence with P. Kristensen. The CWSS Mussel Fishery Report (CWSS, 1991), the Wadden Sea Quality Status Report (De Jong *et al.*, 1999) and Smaal & Lucas (2000) were very useful in providing information on the management and protection of the Wadden Sea.

The Mytilus edulis fishery in the Wadden Sea

The Wadden Sea proper (Fig. 8.9) is a shallow inshore sea of about 6000 km², extending along 450 km of coastline from Den Helder in the Netherlands across the coasts of Germany to Skallingen in Denmark. The sea has an average width of 10 km and is separated from the North Sea by a row of barrier islands. It is one of the most important wetlands of the world. Over 50 bird species (6–12 million individuals) use the Wadden Sea for feeding and breed-

Table 8.14. Annual landings (live weight, tonnes) of the mussel *Mytilus edulis* in the Netherlands, Germany and Denmark.

Country	1990	1991	1992	1993	1994	1995	1996	1997	1998	1999
Netherlands	98845	49254	51223	65981	104952	79772	94496	93244	113185	100800
Germany	20237	29977	50795	24666	4868	17782	38028	22330	31213	37912
Denmark	93348	125762	136271	136677	129317	107377	86002	90765	108329	96215
Total	212430	204993	238289	227324	239137	204931	218526	206339	252727	234927

ing, and it also serves as an important nursery area, spawning area or habitat for some 50 fish species, 300 species of macro invertebrates and >1000 micro- and meiobenthic species (Anon, 1999). The Wadden Sea is estuarine in character, with brackish water (20–34 psu), and high levels of turbidity and organic matter. This results in a natural nutrient enrichment, but the area is also heavily influenced by nutrients of anthropogenic origin (De Jonge & Van Raaphorst, 1995). It should be stressed that the Wadden Sea is subject to enormous variability (on a regional, local, daily, tidal, seasonal, inter-annual scale) with regards to physical/abiotic determinants such as temperature and salinity, and thus is also characterised by enormous biological variability in, for example, population sizes.

The first trilateral Danish-German-Netherlands governmental conference to protect and conserve the Wadden Sea was held in 1978. This led to a joint declaration, signed in 1982, to implement a number of directives regarding the protection of the flora and fauna of the Sea. The Common Wadden Sea Secretariat (CWSS) was established in 1987 to facilitate and support the co-operation. All policies, measures, projects and actions which have been agreed on by the three countries are contained in the Trilateral Wadden Sea Plan (1997), which is revised at regular intervals. Major parts of the sea have been designated as Wetlands of International Importance (Ramsar sites), Special Protected Areas (SPAs) for birds, and Man and Biosphere (MAB) reserves (De Jong *et al.*, 1999). All of these lie within the conservation areas of the three countries. The conservation areas extend along the coast from Den Helder in the Netherlands to Blåvands Huk, near Esbjerg in Denmark, and to three nautical miles offshore (Fig. 8.9).

The Netherlands

The Dutch part of the Wadden Sea covers about 2500 km² (Fig. 8.10). Most of the Netherlands annual landings of mussels, 50000 to 100000 t (Table 8.14) comes from the Wadden Sea. Until the beginning of the 20th century the Netherlands had a thriving wild fishery in the Zuiderzee and Zeeland (Fig. 8.10), serving both the home market, as well as Belgium, the UK, France and Germany. In 1918 a record 124000 t was fished but most of this was used as fertiliser because the potash trade had ceased during the First World War.

Mussel culture started in Zeeland in the middle of the 1800s. This new venture created a demand for seed mussels, which were supplied from the

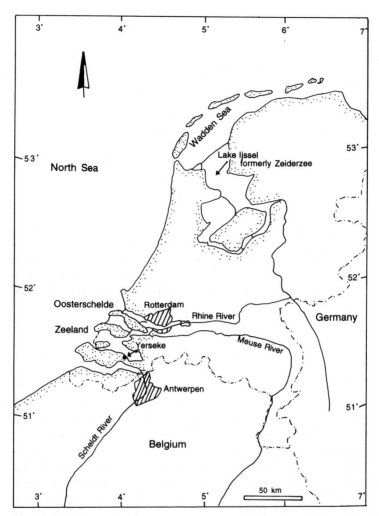

Fig. 8.10. Major fishing areas for the mussel *Mytilus edulis* in the Netherlands. From Dijkema (1997).

northern part of the Zuiderzee (now Western Wadden Sea) because little spat-fall takes place in the Zeeland estuaries. Consequently, a large seed mussel fishery developed in the Zeiderzee. By the end of the 1940s mussel production in Zeeland was about 50 000 t but had plummeted to 10 000 t by 1953. Heavy infestation of mussels with the copepod *Mytilicola intestinalis* has been blamed for this (see Chapter 11). This disaster gave the impetus to expand mussel culture to the Wadden Sea, and by the 1960s production was around 60 000 t in that area. After 1955 the Zeeland fishery recovered, principally because lower seeding densities were used in bottom culture.

Three types of mussel beds are used as a source of seed:

- Beds in the lower intertidal situated at the edge of tidal creeks and gullies and characterised by small hills or hummocks covered in brown algae.

Generally, these beds have a long life-span, although their size can vary depending on the strength of recruitment, fishing intensity and mortality from predation, ice or storms.

- Beds in the high intertidal away from creeks and only formed in years of strong recruitment. These beds have a life-span of just a few years, due to a combination of fishing and natural mortality.
- Beds in the subtidal. These are particularly common in the Dutch Wadden Sea due to larvae and whole mussel clumps (during storms) being swept onto them from nearby culture areas. The beds provide the majority of seed mussels for the Netherlands culture industry. If not fished, they are thought to have a life-span of only a few years because of predation from starfish, shrimp, crab, fish and birds.

During May and September seed mussels are dredged using four steel dredges 1.9 m wide operated by boats called cutters (35–40 m long) that can carry up to 100 t of seed. Most of the seed comes from the Wadden Sea and is about a year old, measuring 2–3 cm; older half-grown mussels 3–4 cm shell length, may also be relayed. In recent years younger seed (1–2.5 cm shell length) is dredged in the autumn after spatfall. Fishing is restricted to 5–6 weeks during May and June (4–5 days per week during daylight) and 1–5 days in September.

The majority of the collected seed is spread on culture plots in the western part of Wadden Sea. At the moment there are over 400 of these plots spread over 6000 ha but only 60% of this area has adequate current speed and bottom conditions. Seed is also spread in the Oosterschelde, Zeeland (Fig. 8.10) where there are over 300 plots covering 4000 ha, 50% of which is productive. The method of seeding is similar to that used in Germany (see below). Growers (~80 at present) rent their plots from the government and rent is calculated on the basis of culture performance of the plots and share of landings. The average area rented per firm is about 125 ha, which contains between 5 to 18 culture plots with water depths of 3–10 m. Seed fishing is restricted to fishermen who have rented culture plots, and government licences are valid for one year. Each boat must carry a black box that registers its movements, fishing activity and position.

Mussels reach market size (5–6 cm shell length) about 1.5–3.0 years after relaying depending on the productivity of the plot. When growers have productive plots with low storm risk and low predation rates they obtain at least 3 t of marketable mussels from about 1 t of seed. However, shallow areas in the Wadden Sea suffer severe storm damage. Such storms sweep away both cultured mussels and wild mussel banks, thus explaining the highly variable production in this area (Table 8.14). Predation is also a major problem in the Wadden Sea, e.g. about 2000–3000 t of mussels are consumed each year by a wintering population of about 150 000 eider ducks (Dijkema, 1997; see also Chapter 3). Because of these problems Netherlands growers generally need 1–2 t of seed to produce 1 t of marketable mussels.

Mussels are harvested between July and March from the culture beds and transported to Yerseke in Zeeland (Fig. 8.10), where the cargo of each boat is auctioned to the highest bidder among about 30 accredited mussel traders.

The traders and processors market the mussels themselves and quality standards and price are agreed before the start of the season, which runs from July to April. Mussels must have a minimum meat yield of 16%, and 35% of mussels must have a shell length greater than 50 mm. Mussels that do not meet these criteria are not auctioned, but are kept on separate plots and sold back to the growers at the end of the season.

The purchased mussels are re-laid on special re-watering plots rented by the traders. Here the mussels have a chance to cleanse themselves and recover from the stress of dredging and transport. Ultimately, the mussels are removed from these plots and transferred into containers and brought to the processing plant. There they are depurated (undergo controlled purification, see Chapter 12) for 4–7 h in UV-sterilised water, de-clumped, rinsed, de-byssused, cooled to 7–10°C in summer, graded and packed. About 70% of mussels are sold fresh on the home and export market, and the remainder is marinaded, canned or frozen. The home market consumes a small share of production (~20%), with the greater part destined for export to Belgium, Luxembourg and France (O'Sullivan, 1987).

There is a general decrease in the area of intertidal mussel beds in European estuaries, chiefly due to persistent fishing of beds when recruitment is poor or absent. Storm mortality and predation by birds are also important factors. The decrease has been particularly well documented for the Wadden Sea (Dankers *et al.*, 2001). In the late 1970s there were just over 4000 ha of mussel beds in the Dutch Wadden Sea but by 1987 this had decreased to 650 ha, and in 1998 less than 200 ha of mussel beds were present, thus leading to a serious shortage of seed and market-sized mussels.

In 1991 the Dutch government decided to close 25% of the intertidal fishery in the Wadden Sea to bivalve fishing, and this is valid until 2003 (Fig. 8.9 and De Jong *et al.*, 1999). In 1991 fishing was allowed in the remaining 75% once there was sufficient food for the protected bird population; about 60% of the average food demands of these birds is reserved (see cockle section below). In addition, in the unrestricted fishing area there were some protected beds, and there was a general policy to fish unstable beds first. This was the situation until 1999 when, on the basis of a Potential Mussel Habitat Map, 10% of the most suitable areas for mussel-bed development were also closed. This map is based on inventories of relatively young mussel beds between 1994 and 1998 (see Dankers *et al.*, 2001 for details). The aim is to achieve an area of 2000–4000 ha of stable mussel banks by 2003. But while it is possible to define, specify and compute the size of areas with potential to form mussel beds it is impossible to predict with any probability that they will actually form (M. Seaman & M. Ruth, personal communication 2001). Other management strategies enforced by the government include monitoring of the purity of the product and the water in the production areas, and preventing toxic algae introduction. Costs are shared between the government and the industry.

Germany

Germany, like the Netherlands, has extensive bottom culture with landings of 20 000–50 000 t annually (Table 8.14). The culture areas are on the west coast of Schleswig-Holstein and the coast of Niedersachsen (Fig. 8.9). The total culture area in Schleswig-Holstein increased to 3000 ha in the 1980s and 90s but is now being reduced to about 2000 ha by political decision. The size and location of the culture area does not depend on the state of the natural beds, or on the fishery, but exclusively on the granting of culture permits by the fisheries authority. All plots are situated north of the Eiderstedt peninsula in depths of up to 10 m. In Niedersachsen (Lower Saxony) the culture plots (~1000 ha) are in sheltered waters west of the Weser Estuary (Fig. 8.9).

In Schleswig-Holstein collection of seed for culture plots is restricted to the subtidal zone; seed fishing in the intertidal zone is illegal (Fig. 8.11A). In contrast, the Niedersachsen seed fishery is restricted to the intertidal zone simply because there are no seed beds in the subtidal zone. Fishing equipment is similar to that used in the Netherlands. Mussels go from the dredge onto a wooden platform where they are rinsed of empty shells, small animals, stones, mud etc. which fall through slits into a chamber below and are returned to the sea. The cleaned mussels are swept onto a conveyor belt which empties into the water-filled hold of the boat (Fig. 8.11B). The mussels are flushed out of the hold through hatches below the water line by means of strong injector pumps (Fig. 8.11C). The seeding density on the beds is regulated by the pump pressure and speed of the boat. Half-grown mussels (20–50 mm shell length) are seeded at densities of 100 t ha^{-1} and smaller mussels (5–20 mm) at densities of 30–40 t ha^{-1}. The success of a culture plot is strongly dependent on the care with which the seeding procedure is carried out. Small mussels reach the market-size of 50 mm in 1–2 years. When the mussels are harvested the plots are cleaned before re-seeding. The yield-to-seed ratio is 1:3 or 1:2 which is a slightly lower ratio compared to the Netherlands. This may be because of eider duck predation on mussels; in Germany, in contrast to the Netherlands, it is prohibited to frighten away eiders by noise or speed boats.

Most German mussels are sold fresh to Dutch wholesalers at Yerseke. The remainder are sold on the home market, or sold to wholesalers in Belgium, France and Italy. When there is a scarcity of seed mussels in the Netherlands undersized (see below) German mussels are sold to the Dutch growers. If the wholesale price of mussels is high at Yerseke then all of the German catch goes to the Netherlands, and Germany imports Danish mussels to satisfy the home market. Alternatively, when Dutch catches are high German growers suffer fierce competition on the home market from Yerseke traders. There are two processing plants in Schleswig-Holstein. One processes mussels for the fresh market (8000 t live weight annually), while the other produces cooked and frozen mussel meat (~30 000 t live weight annually, mainly from Denmark).

Since 1985 the majority of the Schleswig-Holstein Wadden Sea has been designated as a National Park, an area of 4410 km^2 (includes a marine mammal sanctuary and other areas outside of the Wadden Sea proper) that is divided into zones with different protection regimes (Fig. 8.12). In the core zones (Zone 1), which cover 36% of the National Park, no culture plots are allowed, and in more than 90% of the subtidal area of these zones seed fishing is also prohibited. Landing of wild mussels is prohibited. An area of 12 500 h (3% of the National Park), between the islands of Sylt and Föhr and the mainland, has been designated as a zero use zone where all activities are prohibited.

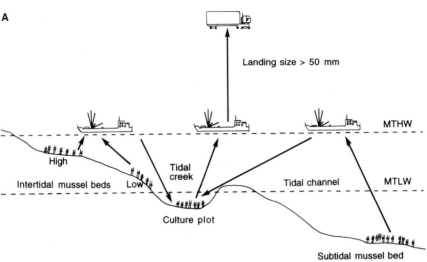

A

Landing size > 50 mm

MTHW

High

Intertidal mussel beds   Low

Tidal creek

Tidal channel   MTLW

Culture plot

Subtidal mussel bed

B

C

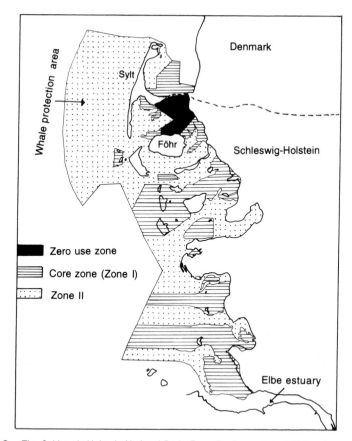

Fig. 8.12. The Schleswig-Holstein National Park. From De Jong *et al.*, 1999.

Regulations governing management of the culture lots in the German Wadden Sea from 1999 onwards (Anon, 1999, and M. Ruth, personal communication, 2000) include: gradual reduction of the culture lot area from 2400 ha in 1999 to 2000 ha by 2006; a closed season for mussel landings from 15 April to 30 June, and for seed fishing and half-grown mussels from 1 May to 30 June; and a minimum residence time of 14–19 months for seed planted on the culture lots. In addition, the number of licences has been set at eight and there is an annual governmental fee of €900 000. All vessels must have a satellite-

Fig. 8.11. Mussel (*Mytilus edulis*) production procedure in Germany. (A) Mussels are fished from wild beds (subtidal in Schleswig-Holstein, intertidal in Niedersachsen), relayed on culture plots, and later fished again for landing. After Seaman & Ruth (1997). (B) Mussel dredger: view of platform used to rinse the catch during the fishing procedure. The dredges are emptied onto the platform, which is flooded by water, and a jet of water creates a circular current that carries the mussels to the conveyor belt on the right, from which the catch drops into the hold. Small animals, dirt and debris flow through slits in the platform into a chamber beneath and are returned to the sea. This recent development reduces weight by 50%. (C) Mussel dredger: view of the hold. The pipe seen in the hold is used during the seeding procedure to pump water into the hold and flush out the mussels through lateral hatches below the water line. A dredge is visible in the background. Photographs courtesy of Maarten Ruth, Fisheries Agency, Kiel, Germany.

supported black-box system to allow the authorities to monitor daily fishing activities.

The Niedersachsen Wadden Sea National Park was established in 1986 and covers an area of about 2400 km². The core zone covers 54% of this area but, unlike the situation in Schleswig-Holstein, fishing is prohibited in only 39% of the core zone, most of which lies between the estuaries of the Elbe and the Weser (Fig. 8.9). Seed fishing and culture lots are permitted throughout the Niedersachsen Wadden Sea, but there are strict controls on the size of the lots, fishing areas and allowable catch. Seed fishing is only allowed between 1 October and the end of February, although mussels from culture lots may be landed year-round. Relaying is only permitted within the boundaries of the state, and there are frequent checks on minimum size in market mussels (50 mm shell length, with 10% by weight undersized mussels permitted) and maximum size in seed mussels (40 mm, with 10% oversized allowed), fishing areas and catch records.

After an outbreak of diarrhetic shellfish poisoning (DSP) in 1986 both states now test for bacteria, algal toxins, heavy metals and radioactive nuclides before the fishing season, and at weekly or biweekly intervals once the season starts. The coastal waters of both states are monitored routinely for noxious algae.

The main problems facing the industry are periodic shortages of seed and increasing pressure from environmental groups to increase the number of areas closed to fishing.

Denmark

About 100 000 t of mussels are landed annually in Denmark (Table 8.14) and most of this comes from the Limfjord (700 km²) in the north of the country (Fig. 8.13). Areas of minor importance are the Kattegat-Lille Bælt, the Isefjord and the Wadden Sea, contributing just over 10% to annual landings. The Danish fishery is closest to being a natural fishery as almost 100% of the mussels are wild-grown. Attempts to culture mussels on the bottom have met with little success, but commercial long-line culture is on-going at one location on the east coast of Jutland, and several long-line experiments are proceeding in the Limfjord (P. Kristensen, personal communication 2000).

In the Limfjord mussels are found at depths of 1–14 m and in shallower waters in the Isefjord and Kattegat-Lille Bælt areas. In the Wadden Sea mussels occur both in intertidal and subtidal areas. Fishing boats are generally old rebuilt fishing vessels with a holding capacity of 15–30 t, or in the Wadden Sea, old Dutch dredging vessels with capacities of 60–80 t are used. Old vessels in the Limfjord use either a single large dredge (1.5 t capacity; 80 mm mesh size), or two small dredges. In the Wadden Sea four small dredges (up to 0.5 t capacity) are used per boat because of the restrictions on fishing time, i.e. three hours before and three hours after high tide.

Seed mussels are dredged up along with adult mussels and, until relatively recently, these were discarded during the sorting process on land. This represented a significant loss to the industry because about one third of gross landings generally are under-sized mussels. Between 1990–1993 an experiment to relay these mussels was carried out in selected areas of the Limfjord. A small

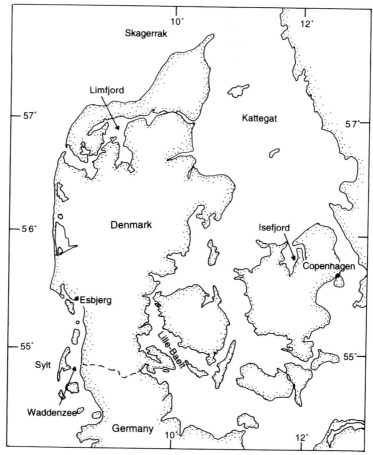

Fig. 8.13. Map of Denmark showing the Limfjord (~90% of landings), Kattegat, Lille Bælt, Isefjord and Wadden Sea areas.

percentage (3–8%) of mussels were damaged during the sorting process, but this loss was offset by good survival and growth rate in the individuals that were relayed. These attained commercial size within two years and had a 30% higher meat content than wild mussels. It has been estimated that returning 20 000–25 000 t of seed mussels to the sea each year may increase production of the Limfjord by about 40 000 t, and increase annual production by almost 40%.

About 90% of landings are processed (frozen, canned or bottled), while the remainder are sold fresh to the Netherlands and Germany. In years with a poor Dutch mussel harvest a major portion of Danish production is sold fresh on the Dutch market. The home market for fresh mussels is small (2000 t annually) but increasing. Mussels from the Limfjord are considered too small for the fresh market, so all of the catch is processed. Those from the Wadden Sea are deemed to be superior in quality to Limfjord mussels, and are sold fresh for export.

Fishery regulations vary depending on the region. The main fishing season is September–December in all areas. In the Limfjord the fishery is closed from the 24 June until the 11 August, and will eventually be closed in January and February also. Licences are issued for just one year and each licence covers a specified fishing area. Fishing is allowed only between sunrise and sunset, and not on Sundays. There is a minimum market size of 45 mm shell length, but up to 30% in the weight of a landing is allowed below this size, provided that the undersized mussels are later relayed on culture lots. The total allowable catch (TAC) is 85 t boat^{-1} week^{-1}, and the daily quota is only 30 t gross. Several areas have been closed to fishing and an additional 10% (150 km^2) of the fjord will be closed by 2001. Mussel dredging in a number of protected areas and at water depths <3 m will be prohibited in the future, leaving only about 60% of the Limfjord open to mussel fishing (P. Kristensen, personal communication 2000). Much the same regulations apply to the Kattegat-Lille Bælt and Wadden Sea fisheries, except the closed seasons are the month of July and the 1st May to 15th July, respectively, and the minimum market size is 50 mm shell length. In addition to daily and weekly quotas there is an annual quota for the Wadden Sea, which is based on allocating 50% of the annual production for birds, and 50% for the fishery. In all areas there is weekly monitoring of water samples for DSP and paralytic shellfish poisoning (PSP).

Kristensen (1997) predicts that the Danish mussel fishery will probably remain stable until at least 2005. However, one fear expressed by the industry is that the demand for mussels may decline since few young people eat mussels on a regular basis.

Clam fisheries

There are thousands of species of clams but only about two dozen are fished in commercial quantities. The surfclam, *Spisula solidissima*, and the ocean quahog, *Arctica islandica*, contribute about 20% each to annual global landings (Table 8.15; see also Fig. 3.3 for global distribution of the main commercial species). Global landings for clams are 800 000–900 000 t annually (Table 8.1A). This represents about 30% of total clam landings, the remainder being from aquaculture production (see Chapter 9).

The following sections will describe the wild fisheries for *Spisula solidissima* on the Atlantic coast of North America and the cockle, *Cerastoderma edule* in European waters. Information on the former is mainly from Murawski & Serchuk (1989), and Serchuk & Murawski (1997), and on the latter from Dankers (1993), Dijkema (1997), Edwards (1997) and Eric Edwards and Jaap Holstein (personal communications, 2000).

The Spisula solidissima fishery

The Atlantic surfclam is a large clam, maximum size 20–25 cm shell length, and has a life span of >30 years (Weinberg, 1999). The species is distributed from Nova Scotia to South Carolina on the eastern coast of North America, and is most abundant between New York and Virginia (Fig. 8.5A). The species is harvested from the subtidal zone to depths of about 60 m, although

Table 8.15. Annual global landings (live weight, tonnes) of the main commercially important clam species. †Includes *Mercenaria mercenaria*, *Mya arenaria*, *Protothaca thaca*, *Spisula polynyma*, *Mactra* and *Arca* spp.

Species	1990	1991	1992	1993	1994	1995	1996	1997	1998	1999
Spisula solidissima	174718	160608	183398	180239	168521	155908	155038	141417	131700	142370
Arctica islandica	177322	186390	192496	197559	177874	183777	173685	163799	157282	147933
Paphia spp.	86270	60500	71765	43089	34528	31207	53141	68790	36415	42282
Ruditapes philippinarum	84125	77985	72454	88584	61216	64523	56095	56514	51392	57051
Cerastoderma edule	76300	71976	91916	70939	63993	62980	35246	35762	86779	70400
Anadara spp.	37751	34301	46895	45558	48708	46595	48227	45446	46509	42627
Venus gallina	36758	41533	56348	61187	53071	46500	49522	37727	36798	45012
Others†	351570	318718	317002	358082	324853	359930	331537	281930	279823	264833
Total	1024814	952011	1032274	1045237	932764	951420	902491	831385	826698	812508

abundance declines sharply beyond depths of 40 m (see Fig. 3.3 for distribution and Table 3.4 for ecological preferences). In the United States it is the most important commercial clam, accounting for 45–50% of landings (meat weight). Despite encouraging results from earlier trials (Goldberg, 1989) there is no aquaculture production of *S. solidissima*. In Canada the Arctic clam, *Spisula polynyma*, supports a small fishery of about 25 000 t (live weight) per year (Jenkins *et al.*, 1997).

The fishery started in the 1870s when clams were initially fished for bait for the Atlantic cod fishery. In the 1930s a food fishery developed and expanded rapidly over the next four decades due to increased fishing effort, improved harvesting, shucking and processing methods, and to exploitation of new areas. Landings peaked in 1974 at 44 000 tonnes of meats, but by 1976 this figure had halved due to a combination of overfishing and wide-spread hypoxic water conditions off the coast of New Jersey during the summer of 1976. It was about this time that a fishery for the ocean quahog, *Arctica islandica*, developed in the offshore waters of New Jersey and Maryland. To conserve the surfclam fishery stringent regulations were implemented in 1977 (see below), and since then stocks have increased and annual landings are now about 150 000 t live weight (Table 8.15) which, using the conversion factor of 2.13 for marine molluscs (M. Perotti, FAO, personal communication, 2000), comes to about 70 000 t of meats. Most of the catch (75%) comes from northern New Jersey waters.

Surfclams are harvested by hydraulic dredges (up to 6 m wide) that use high pressure water jets at the cutting edge of the dredge to loosen the clams from the substrate. The dredge knife then lifts the clams into the rear portion of the dredge. The fishing vessels are large (50–150 gross registered tons) and typically operate two dredges, one off each side of the boat. Harvested clams are loaded into cages that weigh almost two tonnes when full. The cages are offloaded by crane and transported to the shucking and processing plants.

The average meat yield of market-sized (130 mm shell length) clams usually exceeds 100 g. All of the visceral mass is used to provide chopped flesh (fresh, frozen or canned) for chowder, clam cakes or stuffed clams. The mantle and foot are sliced into thin strips, breaded, fried and packaged dry. There is no market for live or whole clams.

Since 1977 the surfclam and ocean quahog fisheries in the US Exclusive Economic Zone (EEZ), 3 nautical miles offshore, have been managed under the Fisheries Management Plan of the Mid-Atlantic Fishery Management Council. Very restrictive measures were implemented under this plan in an attempt to rebuild and conserve surfclam stocks. These included annual and quarterly catch quotas, a moratorium on vessel entry into the fishery, mandatory logbook reporting for vessels and processing plants, fishing time limits and closed areas in order to protect small clams. A few years later a minimum size limit of 140 mm shell length and a target discard rate, no larger than 30% of the landed portion of the catch, were also implemented. The minimum size limit was later reduced to 127 mm. The dredge allows clams smaller than 80–90 mm shell length to escape. These measures, along with good recruitment, have been responsible for the recovery of the fishery; landings doubled between 1980 and 1986 and have remained fairly stable over the following decade (Table 8.15). Because the harvesting capacity of the fishing fleet is considerably greater than that necessary to catch the annual quota, an Individual Transfer Quota (ITQ) system was adopted in 1991 where percentages of the annual quota were allocated between vessels, based on performance history and vessel size. Since then the number of vessels have fallen by more than 50% as quota shares are bought, sold and combined on fewer vessels. Because of this, restrictions on vessel fishing time and new entries into the fishery have been lifted.

The *Cerastoderma edule* fishery

This species is distributed in north-west Europe from Norway to the Atlantic coasts of Spain and Portugal, and as far south as the west coast of Africa (Fig. 3.3). It occurs on sandy bottoms but can also inhabit a wide variety of substrates, ranging from soft mud to gravel. Cockles settle in densities of up to $100000\,m^{-2}$ mainly between mid-tide and low-water levels to depths of up to 10 m. Mortality from low winter temperature and bird predation rapidly reduces this to $<2000\,m^{-2}$ in one-year-old cockles. The maximum size attained by *C. edule* is about 65 mm shell length, but cockles are generally harvested at about 25 mm.

More than 90% of annual landings are from the Netherlands and the UK (Table 8.16). In 1988 total cockle landings were 106 258 t, with the Netherlands and UK contributing 70% and 23%, respectively. By 1997 annual landings had fallen to one third of the 1988 figure with a reduced contribution (30%) from the Netherlands but with UK landings remaining fairly stable. Since then the fishery has recovered in the Netherlands, but landings have started to fall in the UK. Fluctuations like these are primarily due to variable recruitment and weather conditions. For example, in the Netherlands there was excellent recruitment in 1987 and this resulted in good landings in 1988 and 1989. But poor recruitment, together with two severe winters in 1995/96 and 1996/97, resulted in very low landings in 1996 and 1997. Fortunately, there was an excellent spatfall in 1997, and landings in 1998 were about 68 000 t (Table 8.16).

In the Netherlands the majority of landings are from the Wadden Sea (50–80%), with the Zeeland estuaries (Figs. 8.9 & 8.10) and the southern part

Table 8.16. Annual landings (live weight, tonnes) of the cockle *Cerastoderma edule* from European waters. * FAO estimate.

Country	1990	1991	1992	1993	1994	1995	1996	1997	1998	1999
Netherlands	5 312	12 870	47 060	43 635	38 350	39 594	6 300	10 923	68 133	50 888
United Kingdom	19 593	46 612	32 051	21 360	22 330	21 796	24 176	19 493	12 035	14 123
France	4 183	7 849	7 843	4 565	2 152	379	583	694	447	482
Germany	3 721	2 615	495	0	0	0	0	0	0	0
Spain	130	632	903	331	573	600*	650*	700*	2 472	3 104
Denmark	3 042	321	2 423	543	31	0	5	2 603	1 993	246
Portugal	280	1 044	1 116	488	531	591	3 522	1 285	1 264	1 409
Ireland	39	33	25	17	26	20	10	64	296	1
Total	76 300	71 976	91 916	70 939	63 993	62 980	35 246	35 762	86 640	70 253

of the North Sea coast supplying the remainder. Up to the 1960s cockles were collected manually using a long-handled rake, but in the early 1960s the hydraulic dredge was developed. This used a water jet to dislodge cockles from the substrate in front of a steel blade that cut about 4 cm deep and scooped the cockles into the body of the dredge. After 1970 this dredge was replaced with the present-day version which has the added feature of a suction pump attached to the top of the dredge bag, through which the catch is continuously pumped aboard the fishing vessel. This development means that the dredge no longer has to be hauled aboard.

On deck under-sized cockles and by-catch are sieved from the catch and washed overboard. Most vessels are equipped with conveyor-belt cookers that are operated during low-tide and in areas where water quality can be monitored, and where shell dumping is permitted. Dumped shells are eventually removed by commercial operators and are ground to provide grit for poultry farms. On deck, cockles are de-sanded in seawater for a few hours, then cooked, shucked and brought to processing plants. There they are canned, block-frozen or individually quick frozen for the export market. The processing plants are mostly located in Zeeland and these also process mussels, as well as fish and vegetables in the off-season.

Regulatory measures to control the fishery include: a limit on the number of licences issued (in 2000 the number was 37), only one suction dredge (1 m wide) allowed per boat, a minimum mesh width of 15 mm in the sorting sieves and dredge cage, and agreed closed areas and seasons prior to the start of the fishing season. Substantial sections of the intertidal flats in the Wadden Sea and the Oosterschelde are nature reserves and closed to cockle (and mussel) fishing. A limit of 65 000 t (live weight) has been fixed for total yearly landings from the Wadden Sea and 78 000 t for the whole of the Netherlands. In view of the large amount of cockles (~20% of the standing stock) consumed by birds each year, it has been government policy since 1993 to reserve a certain quantity based on 60% of what the birds need, 49 000 t (live weight) for the Wadden Sea and 32 500 t for the Oosterschelde (Fig. 8.10). Fishing is prohibited if predicted landings are below these levels. If there is more than 60% but less than 100%, fishermen are allowed to fish the quantity above 60%. If there is more than 100% there are no restrictions, once landings are below the maximum allowable amounts given above.

Since the mid 1980s manual cockle picking has come back in vogue, and there are 80 individuals licensed to fish on a part-time basis, although this number will probably be reduced to 25 in due course. The main problems facing the fishery in the Netherlands are annual variations in recruitment and the ever-increasing pressure by the government to reduce the impact of the fishery on the environment and wildlife.

In the UK the major cockle beds are in The Wash and outer Thames Estuary on the east coast of England, the Dee and Ribble Estuaries in north-west England, Solway Firth in Scotland, and the Burry Inlet in South Wales. More cockles are landed in the UK than any other mollusc, and in 1999 20% of all cockles landed in Europe were from the UK (Table 8.16). The Wash, Thames, Burry Inlet and Dee Estuary account for 95% of the country's catch.

Landings range between 12000 and 46000t annually (Table 8.16), and this variability is primarily due to erratic recruitment.

Methods of harvesting vary depending upon location. In the Thames Estuary, The Wash and the Solway Firth hydraulic suction dredges (as described above) are used. In the Burry Inlet and Dee Estuary suction dredges are banned and cockles are hand-raked and gathered into piles, which are then sieved to allow small cockles to be returned to the sea. The market-size cockles are packed into sacks for transport to the processing plant.

Cockles are heat-processed and sold either freshly cooked, preserved in vinegar or brine, or individually quick frozen (IQF). Because some of the cockle beds are situated in sewage-polluted areas sufficient heat must be provided to raise the core temperature of the meat to 90°C for 90 seconds. In The Wash there is a growing trade in IQF cockle meats to the Netherlands.

There are no national regulations governing the fishery. Instead, the fishery is regulated through by-laws enforced by local sea fishery committees (SFC). In The Wash (Eastern SFC) spring surveys are carried out to estimate stock size; a total allowable catch (TAC) is set which in 1999 was set at 30% of the fishable stock. Each vessel must have a permit and there is an 8t daily quota per vessel. The fishing season is short (21 June–22 July), and the minimum landing size is 16mm shell length. Additional regulations in the Thames fishery (Kent & Essex SFC) include a limit on the number of licences (14 in 1998), a licence fee, and vessels are only allowed to fish for 2 days a week in June, and for 3 days from July to October. In areas where only hand-gathering is permitted e.g. Burry Inlet (South Wales SFC), there are only 50 licence holders and each has a daily quota of 300–400kg. No fishing is allowed on Sundays.

In Germany the cockle fishery was in the same areas as the present-day mussel fishery, namely Schleswig-Holstein and Niedersachsen. The fishery was always small but after 1992 there were no landings (Table 8.16). The creation of the national parks drastically reduced the area available to the cockle fishery, and there was increasing environmental concern about the effects of hydraulic dredging on the tidal flats. The fishery was banned in 1989 in Schleswig-Holstein, and in 1992 in Niedersachsen.

In Spain the fishery is concentrated in the Galician rias of north-west Spain (Caceres-Martinez & Figueras, 1997). Cockles are harvested either by shovel or rake, or by a boat with a hand dredge. They are either sold fresh, after depuration, or canned. Despite little mechanisation there has been a drastic reduction in landings since 1990 (Table 8.16). Overfishing and poor management are the main reasons for the decline. Culture of *C. edule* started in the early 1990s and production for the years 1993 to 1999 was 2000–4600t (FAO, 2001), higher than landings from natural populations over the same period (300–3000t; Table 8.16). The method of 'culture' is similar to that employed in the Wadden Sea mussel fishery. Seed is collected from natural cockle beds and taken to parks, protected areas with clean, fine sand, and distributed at densities up to $400\,m^{-2}$. Commercial reseeding is also carried out in France (Goulletquer & Héral, 1997), and annual production is around 2000t (FAO 2001), again greater than production from wild populations (Table 8.16). The

UK, Holland and Spain are the only other European countries where cockle reseeding is being carried out.

References

Andrews, J.D. (1991) The oyster fishery of eastern North America based on *Crassostrea virginica*. In: *Estuarine and Marine Bivalve Mollusk Culture* (ed. by W. Menzel), pp. 11–23. CRC Press, Boca Raton, Florida.

Anon. (1999) Scientists seriously worried about exploitation of the Wadden Sea. *Wadden Sea Newsletter*, **1**, 39–40.

Ansell, A.D., Dao, J.C. & Mason, J. (1991) Three European scallops: *Pecten maximus, Chlamys (Aequipecten) opercularis* and *C. (Chlamys) varia*. In: *Scallops: Biology, Ecology and Aquaculture* (ed. S.E. Shumway), pp. 715–51. Elsevier Science Publishers B.V., Amsterdam.

Beukers-Stewart, B.D., Jenkins, S.R. & Brand, A.R. (2001) The efficiency and selectivity of spring-toothed scallop dredges: a comparison of direct and indirect methods of assessment. *J. Shellfish Res.*, **20**, 121–26.

Beverton, R.J.H. & Holt, S.J. (1957) *On the dynamics of exploited fish populations. MAFF Fish. Invest. Ser. 2*, **19**, 1–533. HMSO, London.

Blake, N.J. & Moyer, M.A. (1991) The calico scallop, *Argopecten gibbus*, fishery of Cape Canaveral, Florida. In: *Scallops: Biology, Ecology and Aquaculture* (ed. S.E. Shumway), pp. 899–911. Elsevier Science Publishers B.V., Amsterdam.

Bourne, N. (1964) Scallops and the offshore fishery of the Maritimes. *Fish. Res. Board Can. Bull.*, **145**, 60p.

Bourne, N. (1991) Fisheries and aquaculture: west coast of North America. In: *Scallops: Biology, Ecology and Aquaculture* (ed. S.E. Shumway), pp. 925–42. Elsevier Science Publishers B.V., Amsterdam.

Bradshaw, C., Veale, L.O., Hill, A.S. & Brand, A.R. (2000) The effects of scallop dredging on gravely seabed communities. In: *Effects of Fishing on Non-Target Species and Habitats* (eds M.J. Kaiser & S.J. de Groot), pp. 83–104. Blackwell Science Ltd., Oxford.

Brand, A.R. (1991) Scallop ecology: distribution and behaviour. In: *Scallops: Biology, Ecology and Aquaculture* (ed. S.E. Shumway), pp. 517–84. Elsevier Science Publishers B.V., Amsterdam.

Brumbaugh, R.D., Sorabella, L.A., Johnson, C. & Goldsborough, W.J. (2000) Small-scale aquaculture as a tool for oyster restoration in Chesapeake Bay. *Mar. Technol. Soc. J.*, **34**, 79–86.

Bull, M.F. (1991) Fisheries and aquaculture: New Zealand. In: *Scallops: Biology, Ecology and Aquaculture* (ed. S.E. Shumway), pp. 853–59. Elsevier Science Publishers B.V., Amsterdam.

Caceres-Martinez, J. & Figueras, A. (1997) The mussel, oyster, clam and pectinid fisheries of Spain. In: *The History, Present Condition, and Future of the Molluscan Fisheries of North and Central America and Europe* (eds C.L. MacKenzie Jr., V.G. Burrell Jr., A. Rosenfield, & W.L. Hobart), Vol. 3, pp. 165–90. US Department of Commerce, NOAA Technical Report 129.

Caddy, J.F. (1971) Efficiency and selectivity of the Canadian offshore scallop dredge. *ICES, CM* 1971/**K:** 25.

Caddy, J.F. (1975) Spatial model for an exploited shellfish population, and its application to the Georges Bank scallop fishery. *J. Fish. Res. Board Can.*, **32**, 1305–28.

Caddy, J.F. (ed.) (1989a) *Marine Invertebrate Fisheries: their Assessment and Management.* Wiley Interscience, New York.

Caddy, J.F. (1989b) A perspective on the population dynamics and assessment of scallop fisheries, with special reference to the sea scallop, *Placopecten magellanicus* Gmelin. In: *Marine Invertebrate Fisheries: their Assessment and Management* (ed. J.F. Caddy), pp. 559–89. Wiley Interscience, New York.

Campbell, A. & Mohn, R.H. (1983) Definition of American lobster stocks for the Canadian Maritimes by analysis of fishery-landing trends. *Trans. Am. Fish. Soc.*, **112**, 744–59.

Carlton, J.T. & Mann, R. (1996) Transfers and world-wide introductions. In: *The Eastern Oyster* Crassostrea virginica (eds V.S. Kennedy, R.I.E. Newell & A.F. Eble), pp. 691–706. Maryland Sea Grant, College Park, Maryland.

Carvahlo, G.R. & Hauser, L. (1995) Molecular genetics and the stock concept in fisheries. In: *Molecular Genetics in Fisheries* (eds G. R. Carvahlo & T.J. Pitcher), pp. 55–79. Chapman and Hall, London.

Castagna, M., Gibbons, M.C. & Kurkowski, K. (1996) Culture application. In: *The Eastern Oyster* Crassostrea virginica (eds V.S. Kennedy, R.I.E. Newell & A.F. Eble), pp. 675–90. Maryland Sea Grant, College Park, Maryland.

Cobb, J.S. & Caddy, J.F. (1989) The population biology of decapods. In: *Marine Invertebrate Fisheries: their Assessment and Management* (ed J.F. Caddy), pp. 327–74. Wiley Interscience, New York.

CWSS (1991) *Mussel Fishery in the Wadden Sea* (Working Document 1991–2, 55p). Common Wadden Sea Secretariat, Wilhelmshaven, Germany.

Dankers, N. (1993) Integrated estuarine management – obtaining a sustainable yield of bivalve resources while maintaining environmental quality. In: *Bivalve Filter Feeders in Estuarine and Coastal Ecosystem Processes* (ed. R. Dame), Vol. G 33, pp. 479–511. Springer-Verlag, Berlin.

Dankers, N., Brinkman, A.G., Meijboom, A. & Ruiter-Dijkman, E. de (2001) Recovery of intertidal mussel beds in the Wadden Sea: use of habitat maps in the management of the fishery. *Hydrobiologia*, **465**, 21–30.

Dao, J.C., Fleury, P.G., Norman, M., Lake, N., Mikolajunas, J.P. & Strand, O. (1993) *Concerted action scallop seabed cultivation in Europe*. Report to Specific Community Programme for Research, Technological Development and Demonstration in the Field of Agriculture and Agro-industry, inclusive Fisheries (AIR 2–CT93–1647). European Commission, Brussels.

De Jonge, V.N. & Van Raaphorst, W. (1995) Eutrophication of the Dutch Wadden Sea (western Europe), an estuarine area controlled by the River Rhine. In: *Eutrophic Shallow Estuaries and Lagoons* (ed. A.J. McComb), pp. 129–49. CRC Press, Boca Raton, Florida.

De Jong, F., Baaker, J.F., van Berkel, *et al.* (eds) (1999) *Wadden Sea Quality Status Report 1999*. Wadden Sea Ecosystem No.9. Common Wadden Sea Secretariat, Trilateral Monitoring and Assessment Group and Quality Status Report Group, Wilhelmshaven, Germany.

Del Norte, A.G.C. (1991) Fisheries and aquaculture: Philippines. In: *Scallops: Biology, Ecology and Aquaculture* (ed. S.E. Shumway), pp. 825–34. Elsevier Science Publishers B.V., Amsterdam.

Dickie, L.M. (1955) Fluctuations in abundance of the giant scallop, *Pecten magellanicus* (Gmelin), in the Digby area of the Bay of Fundy. *J. Fish. Res. Board Can.*, **12**, 797–856.

Dijkema, R. (1997) Molluscan fisheries and culture in the Netherlands. In: *The History, Present Condition, and Future of the Molluscan Fisheries of North and Central America and Europe* (eds C.L. MacKenzie Jr., V.G. Burrell Jr., A. Rosenfield & W.L. Hobart), Vol. 3, pp. 115–35. US Department of Commerce, NOAA Technical Report 129.

Dore, I. (1991) *Shellfish: a Guide to Oysters, Mussels, Scallops, Clams and Similar Products for the Commercial User.* Van Nostrand Reinhold, New York.

Edwards, E. (1997) Molluscan fisheries in Britain. In: *The History, Present Condition, and Future of the Molluscan Fisheries of North and Central America and Europe* (eds C.L. MacKenzie Jr., V.G. Burrell Jr., A. Rosenfield & W.L. Hobart), Vol. 3, pp. 85–99. US Department of Commerce, NOAA Technical Report 129.

Eiríksson, H. (1997) The molluscan fisheries of Iceland. In: *The History, Present Condition, and Future of the Molluscan Fisheries of North and Central America and Europe* (eds C.L. MacKenzie Jr., V.G. Burrell Jr., A. Rosenfield & W.L. Hobart), Vol. 3, pp. 39–47. US Department of Commerce, NOAA Technical Report 129.

FAO (1999) *Yearbook of Fishery Statistics: Commodities 1993–1999.* Food and Agricultural Organisation, Rome.

FAO (2001) *Capture and Aquaculture Production 1970–1999.* Fishstat Plus (v. 2.30). Food and Agricultural Organisation, Rome.

Felix-Pico, E.F. (1991) Fisheries and aquaculture: Mexico. In: *Scallops: Biology, Ecology and Aquaculture* (ed. S.E. Shumway), pp. 943–80. Elsevier Science Publishers B.V., Amsterdam.

Goldberg, R. (1989) Biology and culture of the surf clam. In: *Clam Culture in North America* (eds J.J. Manzi & M. Castagna), pp. 263–76. Elsevier Science Publishing Company Inc., New York.

Goulletquer, P. & Héral, M. (1997) Marine molluscan production trends in France: from fisheries to aquaculture. In: *The History, Present Condition, and Future of the Molluscan Fisheries of North and Central America and Europe* (eds C.L. MacKenzie Jr., V.G. Burrell Jr., A. Rosenfield & W.L. Hobart), Vol. 3, pp. 137–64. US Department of Commerce, NOAA Technical Report 129.

Grant, W.S., García-Marín, J.L. & Utter, F.M. (1999) Defining population boundaries for fishery management. In: *Genetics in Sustainable Fisheries* (ed. S. Mustafa), pp. 27–72. Blackwell Science, Oxford, UK.

Gulland, J.A. (1983) *Fish Stock Assessment: a Manual of Basic Methods.* A Wiley-Interscience Publication, New York.

Hardy, D. (1991) *Scallop Farming.* Fishing News Books, Oxford, UK.

Hawkins, S.J. & Jones, H.D. (1992) *Marine Field Course Guide. 1. Rocky Shores.* IMMEL Publishing, London.

Hilborn, R. & Walters, C.J. (1992) *Quantitative Fisheries Stock Assessment: Choice, Dynamics and Uncertainty.* Chapman & Hall, New York.

Holme, N.A. & McIntyre, A.D. (eds) (1984) *Methods for the Study of Marine Benthos.* Blackwell Scientific Publications, Oxford.

Jenkins, J.B., Morrison, A. & MacKenzie C. Jr. (1997) The molluscan fisheries of the Canadian Maritimes. In: *The History, Present Condition, and Future of the Molluscan Fisheries of North and Central America and Europe* (eds C.L. MacKenzie Jr., V.G. Burrell Jr., A. Rosenfield & W.L. Hobart), Vol. 1, pp. 15–44. US Department of Commerce, NOAA Technical Report 127.

Jenkins, S.R., Beukers-Stewart, B.D. & Brand, A.R. (2001) Impact of scallop dredging on benthic megafauna: a comparison of damage levels in captured and non-captured organisms. *Mar. Ecol. Prog. Ser.*, **215**, 297–301.

Kennedy, V.S. (1989) The Chesapeake Bay oyster fishery: traditional management practices. In: *Marine Invertebrate Fisheries: their Assessment and Management* (ed. J.F. Caddy), pp. 455–77. Wiley Interscience, New York.

King, M. (1995) *Fisheries Biology, Assessment and Management.* Fishing News Books, Oxford.

Kristensen, P.S. (1997) Oyster and mussel fisheries in Denmark. In: *The History, Present Condition, and Future of the Molluscan Fisheries of North and Central America and Europe* (eds C.L. MacKenzie Jr., V.G. Burrell Jr., A. Rosenfield & W.L. Hobart), Vol. 3, pp. 25–38. US Department of Commerce, NOAA Technical Report 129.

Kristensen, P.S. & Hoffmann, E. (1999) *Fiskeri efter blåmuslinger i Danmark 1989–1999* (English summary) (DFU-Rapport 72–00): Danmarks Fiskeriundersøgelser, Charlottenlund, Denmark.

Luckenbach, M.W., Mann, R. & Wesson, J.A. (eds) (1999) *Oyster Reef Habitat Restoration: a Synopsis and Synthesis of Approaches.* VIMS Press, Virginia.

MacKenzie, C.L., Jr. (1996a) Management of natural populations. In: *The Eastern Oyster Crassostrea virginica* (eds V.S. Kennedy, R.I.E. Newell & A.F. Eble), pp. 707–21. Maryland Sea Grant, College Park, Maryland.

MacKenzie, C.L., Jr. (1996b) History of oystering in the United States and Canada, featuring the eight greatest oyster estuaries. *Mar. Fish. Rev.*, **58**, 1–78.

MacKenzie, C.L. Jr. (1997a) The US molluscan fisheries for Massachusetts Bay through Raritan Bay, N.Y. and N.J. In: *The History, Present Condition, and Future of the Molluscan Fisheries of North and Central America and Europe* (eds C.L. MacKenzie Jr., V.G. Burrell Jr., A. Rosenfield & W.L. Hobart), Vol. 1, pp. 87–117. US Department of Commerce, NOAA Technical Report 127.

MacKenzie, C.L. Jr. (1997b) The molluscan fisheries of Chesapeake Bay. In: *The History, Present Condition, and Future of the Molluscan Fisheries of North and Central America and Europe* (eds C.L. MacKenzie Jr., V.G. Burrell Jr., A. Rosenfield & W.L. Hobart), Vol. 1, pp. 141–69. US Department of Commerce, NOAA Technical Report 127.

Mann, R. (2000) Restoring the oyster-reef communities in the Chesapeake Bay: a commnentary. *J. Shellfish Res.*, **19**, 335–39.

Mason, J. (1983) *Scallop and Queen Fisheries in the British Isles.* Fishing News Books, Farnham, England.

Mason, J., Shanks, A.M. & Fraser, D.I. (1981) An assessment of scallop, *Pecten maximus* (L.) stocks at Shetland. *ICES, CM 1981/K:19*, 4pp.

Mason, J., Drinkwater, J., Howell, T.R. & Fraser, D.I. (1982) A comparison of methods of determining the distribution and density of the scallop, *Pecten maximus* (L.). *ICES, CM 1982/K:24*, 5pp.

Murawski, S.A. & Serchuk, F.M. (1989) Mechanized shellfish harvesting and its management: the offshore clam fishery of the eastern United States. In: *Marine Invertebrate Fisheries: their Assessment and Management* (ed. J.F. Caddy), pp.479–506. Wiley Interscience, New York.

Murawski, S.A., Brown, R., Lai, H.-L., Rago, P.J. & Hendrickson, L. (2000) Large-scale closed areas as a fishery management tool in temperate marine systems: the Georges Bank experience. *Bull. Mar. Sci.*, **66**, 775–98.

Naidu, K.S. (1984) *An analysis of the scallop meat count regulation.* Can. Atl. Fish. Sc. Advisory Comm. (CAFAC) Res. Doc., 84/73, 18p.

Naidu, K.S. (1991) Fisheries and aquaculture: sea scallop, *Placopecten magellanicus.* In: *Scallops: Biology, Ecology and Aquaculture* (ed. S.E. Shumway), pp. 861–97. Elsevier Science Publishers B.V., Amsterdam.

Navarro Piquimel, R., Figueroa Sturla, L. & Contreras Cordero, O. (1991) Fisheries and aquaculture: Chile. In: *Scallops: Biology, Ecology and Aquaculture* (ed. S.E. Shumway), pp. 1001–15. Elsevier Science Publishers B.V., Amsterdam.

NEFSC (2000) *Report of the 30th Northeast Regional Stock Workshop (30th SAW). E. Surfclams.* pp. 311–477, NEFSC, Woods Hole, Massachusetts.

O'Beirn, F.X., Luckenbach, M.W., Nestlerode, J.A. & Coates, G.M. (2000) Towards design criteria in constructed oyster reefs: oyster recruitment as a function of substrate type and tidal height. *J. Shellfish Res.*, **19**, 387–95.

Orensanz, J.M. (1986) Size, environment and density: the regulation of a scallop stock and its management implications. *Can. J. Fish. Aquat. Sci. Spec. Pub.*, **92**, 195–227.

Orensanz, J.M., Parma, A.M. & Iribarne, O.O. (1991a) Population dynamics and management of natural stocks. In: *Scallops: Biology, Ecology and Aquaculture* (ed. S.E. Shumway), pp. 625–713. Elsevier Science Publishers B.V., Amsterdam.

Orensanz, J.M., Pascual, M. & Fernández, M. (1991b) Fisheries and aquaculture: Argentina. In: *Scallops: Biology, Ecology and Aquaculture* (ed. S.E. Shumway), pp. 981–99. Elsevier Science Publishers B.V., Amsterdam.

O'Sullivan, G. (1987) *The production and market situation for mussels in the Netherlands, Denmark and West Germany* (Market Research Series). Bord Iascaigh Mhara (BIM), Ireland.

Parsons, G.J., Dadswel, M.J. & Rodtrom, E.M. (1991) Fisheries and aquaculture: Scandinavia. In: *Scallops: Biology, Ecology and Aquaculture* (ed. S.E. Shumway), pp. 763–75. Elsevier Science Publishers B.V., Amsterdam.

Renzoni, A. (1991) Fisheries and aquaculture: Italy. In: *Scallops: Biology, Ecology and Aquaculture* (ed. S.E. Shumway), pp. 777–88. Elsevier Science Publishers B.V., Amsterdam.

Rhodes, E.W. (1991) Fisheries and aquaculture: the bay scallop *Argopecten irradians*, in the eastern United states. In: *Scallops: Biology, Ecology and Aquaculture* (ed. S.E. Shumway), pp. 913–24. Elsevier Science Publishers B.V., Amsterdam.

Ricker, W.E. (1975) Computation and interpretation of biological statistics of fish populations. *Fish. Res. Board Can. Bull.*, **191**, 1–382.

Rodhouse, P.G. (1979) A note on the energy budget for an oyster population in a temperate estuary. *J. Exp. Mar. Biol. Ecol.*, **37**, 205–12.

Seaman, M.N.L. & Ruth, M. (1997) The molluscan fisheries of Germany. In: *The History, Present Condition, and Future of the Molluscan Fisheries of North and Central America and Europe* (eds C.L. MacKenzie Jr., V.G. Burrell Jr., A. Rosenfield & W.L. Hobart), Vol. 3, pp. 57–84. US Department of Commerce, NOAA Technical Report 129.

Serchuk, F.M. & Murawski, S.A. (1997) The offshore molluscan resources of the northeastern coast of the United States: surfclams, ocean quahogs, and sea scallops. In: *The History, Present Condition, and Future of the Molluscan Fisheries of North and Central America and Europe* (eds C.L. MacKenzie Jr., V.G. Burrell Jr., A. Rosenfield & W.L. Hobart), Vol. 1, pp. 45–62. US Department of Commerce, NOAA Technical Report 127.

Sinclair, M., Mohn, R.K., Robert, G. & Roddick, D.L. (1985) Considerations for the effective management of Atlantic scallops. *Can. J. Fish. Aquat. Sci. Tech. Rep.*, **1382**, 1–113.

Smaal, A. & Lucas, L. (2000) Regulation and monitoring of marine aquaculture in the Netherlands. *J. Appl. Ichthyol.*, **16**, 187–91.

SOFIA (2000) *The State of World Fisheries and Aquaculture. 1. World review of fisheries and aquaculture.* FAO, Rome.

Strand, O. & Vølstad, J.H. (1997) The molluscan fisheries of Norway. In: *The History, Present Condition, and Future of the Molluscan Fisheries of North and Central America and Europe* (eds C.L. MacKenzie Jr., V.G. Burrell Jr., A. Rosenfield & W.L. Hobart), Vol. 3, pp. 7–24. US Department of Commerce, NOAA Technical Report 129.

Sutherland, W.J. (1996) *Ecological Census Techniques: a Handbook.* Cambridge University Press, Cambridge.

Thompson, D.S. (1995) Substrate additive studies for the development of hardshell clam habitat in waters of Puget Sound in Washington State: an analysis of effects on recruitment, growth, and survival of the Manila clam, *Tapes philippinarum*, and on the species diversity and abundance of existing benthic organisms. *Estuaries*, **18**, 91–107.

Thouzeau, G. & Lehay, D. (1988) Variabilité spatio-temporelle de la distribution, de la crosisance et de la survie des juvéniles de *Pecten maximus* (L.) issus des pontes 1985, en baie de Saint-Brieuc. *Oceanol. Acta*, **11**, 267–83.

Trilateral Wadden Sea Plan (1997) In: *Ministerial Declaration of the Eighth Trilateral Governmental Conference on the Protection of the Wadden Sea*, Stade, 1997. Annex 1 (English version). Common Wadden Sea Secretariat, Wilhelmshaven, Germany.

Veale, L.O., Hill, A.S., Hawkins, S.J. & Brand, A.R. (2000) Effects of long-term physical disturbance by commercial scallop fishing on subtidal epifaunal assemblages and habitats. *Mar. Biol.*, **137**, 325–37.

Veale, L.O., Hill, A.S., Hawkins, S.J. & Brand, A.R. (2001) Distribution and damage to the by-catch assemblages of the northern Irish Sea scallop dredge fisheries. *J. mar. biol. Ass. U.K.*, **81**, 85–96.

Venvik, T. & Vahl, O. (1985) Opportunities and limitations for harvesting and processing Iceland scallops. *Can. Transl. Fish. Aquat. Sci.*, No. 5191, 76p.

Weinberg, J.R. (1999) Age structure, recruitment, and adult mortality in populations of the Atlantic surfclam, *Spisula solidissima*, from 1978 to 1997. *Mar. Biol.*, **134**, 113–25.

Weinberg, J.R. & Hesler, T.E. (1996) Growth of the Atlantic surfclam, *Spisula solidissima*, from Georges Bank to the Delmarva Peninsula, USA. *Mar. Biol.*, **126**, 663–74.

Wesson, J., Mann, R. & Lukenbach, M. (1999) Oyster restoration efforts in Virginia. In: *Oyster Reef Habitat Restoration: a Synopsis and Synthesis of Approaches* (eds by M.W. Luckenbach, R. Mann & J.A. Wesson), pp. 117–29. VIMS Press, Virginia.

Woolff, M. (1987) Population dynamics of the Peruvian scallop *Argopecten purpuratus* during the El Niño phenomenon of 1983. *Can. J. Fish. Aquat. Sci.*, **44**, 1684–91.

9 Bivalve Culture

Introduction

Aquaculture dates as far back as 500 BC when Fan Lei, a Chinese politician wrote the *Classic of Fish Culture*, which describes the rearing of carp in fishponds as a commercial venture. Bivalve culture has almost as long a history. About 350 BC Aristotle mentions the cultivation of oysters in Greece, and Pliny describes commercial holding ponds for oysters near Naples, Italy, around 100 BC. However, farming of oysters, as we know it today, did not begin until 1624 in Hiroshima Bay, Japan (Fujiya, 1970). Mussel culture is said to have started when a resourceful Irish sailor, shipwrecked off the Atlantic coast of France in 1235, used poles and netting on tidal mudflats to trap birds. He quickly discovered that the poles became covered with mussels and that the mussels grew better and were of a superior quality to mussels living on the bottom. This apparently was the start of the bouchot method of culture, still practised today in France and parts of Asia. However, it is only in the last three decades that the full potential of the mussel as a culture organism has been realised. Simple methods of clam culture have been practised for several centuries in China (Pillay, 1993) but there is little information on the actual methods used. Culture of scallops started in Japan about 1965 when a fisherman used an onion bag to collect seed (newly settled larvae) in the open sea (Ito, S., 1991). This novel method of seed collection formed the basis of what was later to become a thriving scallop culture industry in Japan.

Towards the end of the 19th century reliable techniques for culturing bivalves were developed. This allowed a rapid growth in production that was greatly augmented about a century later by the development of hatchery techniques for seed production. Overexploitation of wild stocks, deteriorating water quality and debilitating disease have compelled fishermen to actively manage the resource and turn to restocking and culture in deep water using juveniles originating in the hatchery.

An examination of the FAO marine aquaculture production data for 1970–1999 (FAO, 2001) shows that the number of bivalve species being cultured is increasing annually, and that more and more developing countries are becoming involved. Factors that have contributed to the continuing growth in bivalve culture include: the acknowledged need to achieve greater self-reliance in food production, especially by developing countries; the recognition that it is an efficient way to produce animal protein and that it improves the income and nutrition of rural populations; improved technology transfer in terms of hatchery and nursery techniques; and the availability of better transportation by road and air.

In 1999 over 8.8 million tonnes (t) of cultured bivalves were landed, with a value of over US$7500 million. This represents more than 80% of the combined yield from fisheries and culture, in contrast to the situation for finfish where less than 10% of the yield is cultured (Barnabé, 1994; SOFIA, 2000).

284

Table 9.1. Annual yields (live weight, tonnes) from aquaculture production (1989, 1998 and 1999) and capture fisheries (1999). Data from FAO (2001) Fishstat Plus Aquaculture production database 1970–1999. Figures in parentheses in aquaculture 1999 column represent production as a percentage of the combined annual yields from aquaculture and fisheries. Figures in parentheses in China 1999 column are the percentage contributions by China to global aquaculture.

Species	Aquaculture Global production			China	Fisheries Global
	1989	1998	1999	1999	1999
Mussels	1 078 603	1 377 831	1 452 222 (86%)	608 115 (42%)	237 816
Oysters	1 256 638	3 537 831	3 657 443 (96%)	2 988 613 (82%)	155 599
Scallops	310 863	874 226	951 810 (63%)	712 330 (75%)	567 263
Clams	575 419	2 226 030	2 753 931 (77%)	2 464 854 (90%)	812 508
Total	3 221 523	8 015 918	8 815 406 (83%)	6 773 912 (77%)	1 773 816

In 1999 over 95% of all oysters landed were from culture operations whereas the corresponding figures for mussels, scallops and clams were 86%, 63% and 77%, respectively (Table 9.1). China is by far the biggest contributor, producing almost 6.8 million of the total 8.8 million tonnes cultured in 1999. An examination of the data in Table 9.1 shows that over a ten-year period, harvest increased by 35% in the case of mussels, trebled for oysters and scallops, and increased more than four-fold for clams. Even when the data for two consecutive years are compared there was increased production in 1999 over the previous year, ranging from 3% in the case of oysters to 24% for clams (Table 9.1).

This chapter will first deal with some basic aspects of bivalve culture and will focus then on a number of key species for detailed treatment. Information on choosing a species for culture, selecting culture sites, and designing and constructing hatcheries and nurseries will not be included. This chapter cannot hope to provide exhaustive coverage of all aspects of bivalve culture. For more detailed information interested readers may wish to consult the various texts cited in this chapter.

Fundamentals of bivalve culture

In bivalve culture the supply of seed, also known as spat, is a critical element. Seed may be produced by wild or cultured stocks in the field, or alternatively from broodstock maintained in a hatchery. In the former case reproduction is left to nature, whereas in the latter situation it is controlled, albeit to varying degrees. Generally speaking, mussel culture relies on wild-caught seed while scallop, clam and oyster culture use either wild or hatchery-produced seed, depending on the species being cultured (see later).

Wild seed collection

The basic requirements for successful seed collection in the wild are information on the reproductive cycle of the population and knowledge of the spawning, larval development and settlement process of the species concerned. With this kind of information it is possible to forecast when and where to collect spat.

Thin sections of gonad or gonad squashes are examined to assess seasonal changes in the reproductive tissue of bivalves (Chapter 5). To take account of asynchronous development a sample size of about 25 females per month is recommended for squashes, and at least 25 individuals per month for gonad sections (Quayle & Newkirk, 1989). In scallops, where the gonad is a discrete structure, gamete development is often assessed through gross visual examination of gonad size, colour, and shape (Barber & Blake, 1991). In oysters, condition factor is sometimes used as an indirect way of assessing reproductive stage (see Chapter 5).

Temperature and food supply are the main exogenous factors influencing the reproductive cycle in bivalves (see Chapter 5). Temperature is also the main environmental trigger for spawning in temperate water species. In the tropics, where water temperatures are relatively stable, changing salinity, rather than temperature, may be the chief spawning stimulus. Endogenous factors, e.g. neurohormones, and neurotransmitters also play an important role in gametogenesis and spawning, and interact in a complex fashion with exogenous agents (see Chapter 5). As spawning triggers are species- and site-specific it is important that careful records of spatfall, temperature, salinity, lunar phase, tidal fluctuations and weather conditions are kept in order to establish relationships with spawning activity (Quayle & Newkirk, 1989).

After the gametes are released into the water column fertilisation occurs and the fertilised egg rapidly divides and goes through the trochophore and veliger larval stages. As soon as the veliger reaches 250–300 µm shell length it is ready to settle and metamorphose. The length of time that larvae are planktonic depends on temperature, salinity and food supply. Generally speaking, the larval period for species in temperate waters is between 15 and 30 days, but is less than 15 days for tropical species. The spat settle on a wide variety of substrates and attach themselves using either byssus threads (mussels, clams and scallops), or cement (oysters). Attachment by byssus or cement persists throughout the life of mussels and oysters, respectively. In scallops and clams the byssus is lost some time after settlement, although there are some species where it persists into adult life.

Accurate spatfall forecasting is a prerequisite for successful seed collection. This is particularly important when collection involves setting out poles, ropes or cultch, that require prior conditioning in seawater before spat will settle. If collectors are put out too early they may become covered with fouling organisms and hinder spat settlement. Alternatively, late placement of collectors, or not placing them at the proper place and depth, can result in low setting success. At the best of times forecasting is a difficult task but knowledge of the life cycle of the bivalve and of the local hydrographic conditions will help to improve accuracy. In temperate waters forecasting is based primarily on plankton sampling but in tropical regions forecasting is often not necessary because of the continuous nature of settlement. Details on the methodology of sampling, counting and identifying larvae are in Quayle & Newkirk (1989).

A wide variety of physical and chemical factors influence settlement in the wild (see Chapter 5). Generally speaking, larvae settle on clean, silt-free, irregular surfaces that can range all the way from filamentous algae through stone, wood, dead shell, and shells of their own species. Therefore, culture operations use many different materials to collect wild seed. For on-bottom oyster culture the preferred material is empty oyster shell, chiefly because it is generally available, cheap and attractive to the larvae. However, a myriad of other materials can also be used to good effect, such as rope, bamboo, rubber tyres, roof tiles, wood laths, tree branches and coconut shell. Mussel seed for off-bottom culture is collected on poles or stakes set into the seabed, or on ropes or mesh stockings suspended from rafts or longlines. For on-bottom culture the seed is dredged from natural beds and transplanted to culture plots (see Chapter 8). Newly settled clam spat are collected on oyster cultch or brush fences, or, because the byssal attachment phase is short, spat are more often sieved from the substrate using a series of graded sieves. This method is labour-intensive and also requires expertise in clam identification. Scallop spat are typically collected on net bags filled with monofilament gillnetting that are attached to lines suspended from a raft or longline. In Japan bundles of live kelp seaweed can be used in addition to net bags. In scallop species where the attachment phase is short the spat are removed from collectors to pearl nets or lantern nets.

Hatchery production of seed and juveniles

Hatchery production of seed began in response to declining stocks and the consequent seed shortage in the 1960s. Hatcheries guarantee the supply and maintenance of broodstock, provide the means to control maturation and spawning, and supply food for the early vulnerable stages in the life cycle. In addition, breeding techniques can be used to produce seed for desired characteristics such as fast growth or disease resistance. However, the costs involved in setting up and running hatcheries are high and very little of the seed used in culture operations actually comes from hatcheries; for oyster culture the figure is probably less than 1% (Quayle & Newkirk, 1989).

The basic techniques for the production of seed in hatcheries are very similar for all bivalves, although each hatchery will invariably modify these depending on local conditions and the species being cultured. The procedures include the holding, conditioning and spawning of broodstock, the rearing of larvae and juveniles, and the production of large quantities of micro-algae as food (Fig. 9.1). The latter will be dealt with first because of the crucial role it plays in the production of hatchery seed.

Culture of micro-algae

In most hatcheries the incoming seawater is filtered and sterilised to remove contaminants (see Petit, 1990 for details) but in doing so potential food is also removed, thus making it necessary to produce micro-algae cultures for the larvae, spat and breeding stock in the hatchery. The species chosen for culture chiefly depends on their nutritional value and ease of culture. Cell size, cell

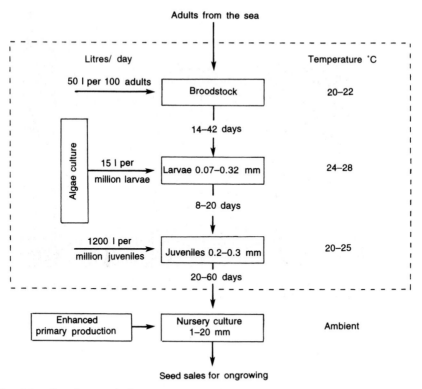

Fig. 9.1. Flow diagram of a hatchery operation. From Utting & Spencer (1991).

wall structure and chemical composition of the species are also important. The most widely used species for bivalve culture are the diatoms: *Phaeodactylum tricornutum, Chaetoceros calcitrans, Thalassiosira pseudonana, Skeletonema costatum,* and the flagellates, *Isochrysis galbana* (Tahitian strain) and *Tetraselmis suecica*. Essentially, a pure culture of the species is grown in a well-aerated enriched medium under temperature and lighting conditions that will ensure rapid growth. To avoid contamination the media, containers and tubing are sterilised and frequent examination of cultures under the microscope are carried out to check their purity. Pure cultures are usually started off in small 500 ml flasks and a small volume (e.g. 20 ml) of algae from these is aseptically transferred into 5 l flasks and then later into larger 20 l containers. These are used to inoculate transparent plastic tubes (80–120 l volume), which are the actual production units (Fig. 9.2). For mass culture, large (e.g. 1600 l) circular or rectangular fibreglass tanks are used and the seawater is not sterilised due to the large volume, but is filtered, generally at 0.2–1.0 µm. For further information on algal culture methods see Walne (1974), Castagna & Manzi (1989), Le Borgne (1990) and Lavens & Sorgeloos (1996).

Typical concentrations of cells in mass cultures are between 1 and 5 million cells ml^{-1}. The cells may be fed directly to the larvae at concentrations of 30–100 µl^{-1}, depending on larval density and size. Alternatively, the cells can be concentrated by centrifugation into a paste that is then stored at 0–6°C

Fig. 9.2. Large, transparent plastic tubes used in the culture of microalgae. Photo courtesy of Kevin O Kelly, Lissadell Shellfish Co. Ltd., Sligo, Ireland.

for several months, or frozen or freeze-dried. The number of cells eaten per day will vary depending on size of the larvae and the species of micro-algae. Daily consumption rates for *Ostrea edulis* larvae fed on *Isochrysis galbana* vary from 20 000 cells ml^{-1}, when the larvae are first liberated at 180 μm shell length, to 60 000 cells ml^{-1} as they approach metamorphosis at 250–300 μm (Walne, 1974).

In the past decade there has been increasing interest in screening new algal species for culture, and testing single or mixtures of algae on larval growth and mortality. An important factor in selecting algae is their essential fatty acid (EFA) composition. A significant positive correlation has been observed between growth of oyster spat (*C. virginica*) and strains of *Tetraselmis* that were high in EFAs and sterols (Wikfors *et al.*, 1996). In addition, combining algae with different EFA profiles can produce significantly faster growth than single-species diets (Laing & Psimopoulos, 1998). Interestingly, it would appear that the EFA composition of a species can change, depending on environmental conditions. For example, *Thalassiosira pseudonana* cultured under high light was a superior diet for oyster larvae than the same species grown under low light (Thompson *et al.*, 1996). Recently, the addition of lipid emulsions has been shown to be an acceptable lipid supplement for broodstock, larvae and spat of various species of bivalves (Caers *et al.*, 2000 and references therein).

Micro-algae culture is an expensive process, representing on average 30% of hatchery operating costs (Coutteau & Sorgeloos, 1992). Therefore, there is considerable interest in the development of suitable artificial diets to replace or supplement micro-algae. Southgate *et al.* (1992) have shown that when protein-walled microcapsules, containing a mixture of animal proteins, were fed to oyster larvae they were able to support a growth rate that was in excess of 80% that of algae-fed larvae. A major advantage of microcapsules is that they are made when required and, unlike micro-algae, require no additional maintenance. However, possible problems include settling, clumping and bacterial degradation of the particles, leaching of nutrients and low digestibility of the cell-wall material (Lavens & Sorgeloos, 1996). Another algal substitute that has been tried with good results is the use of yeast-based diets. Coutteau *et al.* (1994) have found that when baker's yeast mixed with algae (50:50 on dry weight basis) was fed to clam seed (*Mercenaria mercenaria*) the growth rate was similar to algal-fed controls. The high lipid content of the yeast formulation probably compensated for the deficiency in calories of the algae-yeast mix. Cornmeal or cornstarch also serve as good complements to live micro-algal diets (Pérez-Camacho *et al.*, 1998). However, a complete replacement for living algal diets has not yet been achieved.

An alternative less costly source of micro-algae that may be exploited in the future by the aquaculture industry is domestic wastewater treatment plants. Algae are a by-product of the water purification process, their growth aided by natural light and the high mineral and organic content of the water. Several techniques are used to separate the algae from the water. The main ones are industrial centrifugation and coagulation-flocculation, both of which have been used to provide algal food for humans and animals, but to date have not been used in the hatchery (see Barnabé, 1990 for details on these and alternative ways of harvesting micro-algae).

Broodstock conditioning and spawning

Broodstock are normally picked for the hatchery on the basis of desirable traits such as fast growth, disease resistance and meat weight. Genetic improvements can potentially be achieved through selective breeding, crossbreeding, hybridisation or chromosome manipulation (Chapter 10).

Broodstock may either be maintained in trays at ambient temperature or, if out-of-season reproduction is required, can be conditioned in indoor tanks with heated seawater and added algal food. For *Crassostrea virginica* the conditioning period is 6–10 weeks and the feeding regime is 0.5–1.01 algal culture per oyster per day (Castagna *et al.*, 1996). The length of the conditioning period is seasonally dependent. For example, in Italy the clam *Ruditapes philippinarum* needs 6–8 weeks in winter and early spring, but this period becomes progressively shorter as the natural spawning season approaches (Helm & Pellizzato, 1990). It is important that attention is paid to the fatty acid profile of the broodstock diet since it has been shown that it affects gonad composition, and hatching rate and quality of larvae (Utting & Millican, 1998 for review; Soudant *et al.*, 1996; Wilson *et al.*, 1996; Martinez *et al.*, 2000).

The easiest way to assess broodstock maturity is to examine a small piece of gonad from a few individuals under the microscope. Alternatively, a hole can be drilled in the shell and a piece of gonad extracted using a syringe, a method that does not involve sacrificing broodstock. Ripe oysters can be maintained without spawning for several months; in the case of clams this holding period can be as long as 8 months (Castagna & Manzi, 1989). However, holding broodstock for lengthy periods invariably entails extra costs.

When gametes are needed oysters are induced to spawn by raising the water temperature. If no spawning occurs within 30–45 min then the water temperature is lowered to 24°C for 30 min. This temperature cycle is repeated until spawning is achieved. Other stimuli that are used to trigger spawning include cold shock, addition of stripped gametes or a weak solution of ammonium hydroxide to the water, or injection of the neurotransmitter serotonin into the adductor muscle. If spawning does not occur gonad tissue may be dissected from a number of males and females. The tissue is blended with seawater for 5–10 sec and the eggs are collected on a screen, washed, and a sample is then examined under the microscope to check for fertilisation. Although this method produces fewer larvae than from a natural spawn it is a common way of obtaining gametes from *Crassostrea gigas*, one of the major oyster species being cultured on a global scale.

It is recommended to have an equal sex ratio and about 50 pairs of spawners in order to ensure a good supply of eggs and to keep inbreeding to a minimum (see Chapter 10). Spawners are segregated by sex as soon as spawning is observed and the eggs and sperm are collected separately. In hermaphrodites, e.g. many scallop species, the sperm is shed first followed by the eggs. In order to avoid self-fertilisation the sperm is collected and kept separate from the eggs. As soon as spawning stops the spawners are removed and the eggs are pooled and then sieved to remove extraneous material. In oysters (*Crassostrea* sp.) the egg concentration for optimum fertilisation success is between 30 000 and 100 000 l^{-1}, and to avoid polyspermy – the penetration of the egg by more than one sperm – the recommended sperm concentration is 500–5000 per egg (references in Thompson *et al.*, 1996). Polyspermy results in abnormal embryonic development. After fertilisation the eggs are then washed and sieved at 25 μm to remove excess sperm.

Investigations on the cryopreservation of oyster sperm have been ongoing since the early 1970s. After an oyster changes sex from male to female its eggs can be fertilised with its own cryopreserved sperm, to produce inbred lines for crossbreeding (see Chapter 10). Alternatively, the sperm could be used for shipment to other locations or to cross populations that breed asynchronously. Fertilisation success is about 70–90% of the rate for natural sperm, but larval yields are often as low as 20% (Yankson & Moyse, 1991). Cryopreservation of embryos is also being investigated but technical problems have prevented it being taken up to any great extent by the aquaculture industry (Renard, 1991; Gwo, 1995).

Larval rearing

The fertilised eggs are moved to flat-bottomed cylindrical tanks (500–50 000 l capacity) with high quality, filtered, well-aerated seawater (23–26°C), and are initially held at densities as high as 60 000 l^{-1}. In *C. virginica* fertilised eggs at 24°C reach the trochophore stage within 12 h of fertilisation, and the shelled larval stage within 24 h (Castagna *et al.*, 1996). There is no need for food during the first 24 h as the early larval stages do not feed. The first water change is at 24–48 h post-fertilisation and the water is changed about thrice weekly after that. When the tanks are being emptied the cultures are passed through a series of sieves to size-grade the larvae. In some hatcheries only the best-growing larvae are kept, but in others several size classes are segregated for rearing with only the slowest growers being discarded. At regular intervals samples of larvae are counted, staged and checked for disease under the microscope.

Shelled larvae are held at densities of $4000–15\,000\,l^{-1}$ and are fed once or twice daily at concentrations of 30–100 algal cells per micro-litre depending on development stage. The time it takes for the larvae to reach settlement stage depends on temperature and food quality. Generally, this stage is reached 14–21 days after fertilisation.

Some oyster hatcheries do not allow the larvae to settle but concentrate them on a screen that is then wrapped in moist paper and placed in a plastic cooler at 5°C. The larvae can be kept like this for about 5 days without ill effects. They may either be kept in the hatchery for subsequent setting or they may be transported to another location for setting. This technique has made it possible for oyster farmers working in locations remote from a hatchery to receive and set viable larvae. This process – called remote setting – has revolutionised oyster production on the west coast of the United States and Canada (references in Castagna *et al.*, 1996). Its success depends on producing very large numbers of larvae at a very low cost. One large American hatchery in Washington State produces many billions (10^9) of eyed larvae at a cost of $200 per million (Helm, 1990).

Setting

Cultch is the term used for the substrate provided for settling larva. The most popular cultch that hatcheries use is mollusc shells. For oyster setting the best cultch is clean dry oyster shells in plastic mesh bags that have been pre-soaked in filtered seawater for 12h so that the shell surfaces have a bacterial film to attract settling oysters (see Chapter 5). A few examples of other oyster cultch are grooved plastic PVC tubes, plastic cones, lime-coated strips of wood veneer, and rubber tyre chips. Some hatcheries use cultchless techniques for setting oysters that are to be grown off the bottom in trays. For example, the larvae may be allowed to settle on a flexible plastic sheet from which they can be easily removed soon after setting, or alternatively, finely crushed shell chips small enough to accommodate only 1–2 larvae at most can be used. Scallop larvae settle on fibreglass panels or sieves suspended in the setting tanks and the spat can be easily removed at a later stage. They also settle on pre-soaked (see above) cultch such as monofilament, shell, pebbles, polypropylene line, artificial turf, jute and sisal; maximum settlement occurs on filamentous material and kinran, an artificial material made in Japan. Clam larvae generally attach by a byssal thread on the bottom or to the sides of the rearing vessel, not needing any special substrate for setting. Sometimes the chemicals epinephrine and nor-epinephrine are used to induce larvae to settle and metamorphose without the provision of a settlement surface (Coon *et al.*, 1986).

Nursery culture of spat

If newly settled spat are placed out directly in the natural environment for on-growing they can suffer high mortality from smothering, predation, and competition. Therefore, maintaining the spat in a nursery system until they have reached a large enough size to withstand these factors often makes economic sense. Generally speaking, spat are held in the nursery until they have reached a few centimetres in size.

Many hatcheries initially maintain the newly settled spat in their larval rearing containers until they have reached about 1 mm shell length. The water is changed daily and the spat are fed large amounts of algae, starting at 2.5×10^6 cells ml^{-1} day^{-1} but increasing significantly as the spat grow (Fig. 9.1). As this is a costly process the spat are transferred to a number of different rearing systems, the type depending on the bivalve in question and the hatchery's production system and location. Nursery systems are either land-based (indoors or outdoors) or field-based. One type of land-based system uses raceways – long, shallow tanks or troughs made of fibreglass or concrete – that are continuously supplied with pumped seawater. The spat are layered on the bottom with enough water to barely cover them, or in deeper raceways they may be held on tiers of plastic mesh or tiers of shallow trays. Raceways can be either indoor or outdoor nursery systems. Alternatively, the spat may be held in tiers of rectangular, shallow fibreglass trays. A more common nursery system is the upflow or upweller unit that consists of screen-bottomed cylinders of spat held inside a larger container; seawater flows upward through the cylinders partially fluidising the seed mass, and discharges through a side exit (Fig. 9.3). The

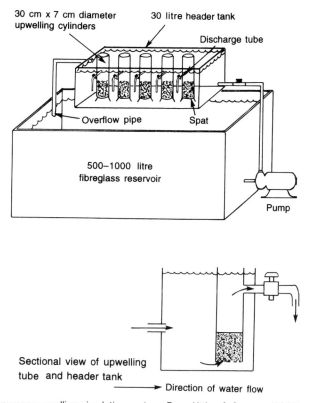

Fig. 9.3. A nursery upwelling circulation system. From Utting & Spencer (1991).

water is re-circulated by means of an electrically driven pump or airlift. If up-wellers are used indoors there can be significant costs in heating the seawater (22–25°C) and providing the micro-algae, although the latter can be supplemented with naturally occurring algae if the incoming seawater is only coarsely filtered (Utting & Spencer, 1991). With an indoor system, however, growth is rapid and the total time required to bring post-set spat to sizes large enough for flow-through nursery systems is minimised (Castagna & Manzi, 1989). Spat held in raceway or up-flow systems show similar rates of growth and survival. However, up-flow systems allow much higher stocking densities and also are easier to maintain than raceways (see Manzi & Castagna 1989 for a detailed discussion on the pros and cons of the two systems).

Instead of land-based nurseries spat may be transferred direct from the setting tanks to protected areas in the subtidal or low intertidal zones, or even to saltwater ponds. When spat are being transferred from the hatchery into the field the period out of water should be kept to a minimum and there should be as little change in water temperature as possible. Temperature shock can cause scallop larvae to release their attachment from cultch and become lost (Bourne & Hodgson, 1991). In the oyster *Crassostrea virginica*, if cultch has been used for setting, the cultch with attached spat is enclosed in mesh bags or trays, and these are placed on pallets or poles for on-bottom rearing, or hung from docks, rafts or longlines for suspended culture. In the case of single spat several layers are placed in plastic mesh trays and suspended as for spatted cultch. Alternatively, the trays may be placed on trestles at a level where they are exposed on each low tide. This hardens the oysters and apparently reduces competition, fouling and predation and increases survival (references in Castagna *et al.*, 1996). The field-based, subtidal nursery systems for scallops and clams are similar to those described for oysters.

Whatever system is used, the spat must be cleaned routinely of fouling organisms and silt, and thinned to reduce competition. Fouling organisms such as sponges, bryozoans, ascidians, barnacles, mussels and algae can settle and grow on the shell of spat, and will smother the spat eventually if they are not removed. Fouling is more of a problem in field-based than in land-based nursery systems; intertidal systems have fewer problems than suspended systems (see later section on mussel culture). Various removal methods include: high-pressure water hoses, air-drying in the sun or submergence in a bath of fresh water, brine or dilute copper sulphate (details in Quayle & Newkirk, 1989). A very effective biological control method uses grazing periwinkles to reduce algal growth on oyster bags (Cigarria *et al.*, 1998). Knowledge of the breeding cycle, settlement time and vertical distribution of the fouling organisms can help in deciding when and where to place spat (and larger individuals), especially for suspended culture.

Predation can also be a significant problem in a nursery system as the young stages are particularly vulnerable to major predators such as crabs, boring gastropods and starfish (see Chapter 3). Disease can also be a major problem in larval and nursery systems (see below). By the time the spat have reached 2–3 cm they are less vulnerable to predators and are ready for the grow-out phase. It is virtually impossible to give an estimate as to how long it takes to reach this size as so many different factors affect growth rate, e.g. type and siting of the nursery system, water temperature, quantity and quality of food, density of rearing, fouling, predators, pests, pollutants and disease.

Disease in the hatchery

The artificial, high density and sometimes stressful conditions under which hatchery larvae and juveniles are reared is an environment conducive to the spread of bacterial disease. The most important bacterial pathogens in hatchery cultures belong to the Genus *Vibrio*. *Vibrio anguillarum* (VAR) and *V. alginolyticus* cause frequent, massive mortality of cultured larvae. Symptoms of infection include: abnormal swimming behaviour, reduced feeding and general inactivity, due to proliferation of the bacteria through the soft issues. The spread of vibriosis is rapid; in experimental infections signs of the disease appear 4–5 hours after exposure and death begins at 8 hours, and complete mortality of the cultures occurs by 18 hours (Sindermann & Lightner, 1988). Larvae probably die as a result of starvation (Ford & Tripp, 1996). Because the spread of infection is so rapid antibiotics are often ineffective, may cause larvae to stop feeding and can lead to the development of resistant strains (Sindermann & Lightner, 1988). An alternative to antibiotic treatment is the recent use of probiotics, beneficial bacteria that override pathogens by producing inhibitory substances or preventing pathogenic colonisation in the host (see Gatesoupe, 1999 for review; also Gibson *et al.*, 1998; Riquelme *et al.*, 2000 & 2001). Outbreaks of vibriosis often coincide with high concentrations of *Vibrio* in the water supply to the hatchery, but continue well after field concentrations have declined. The bacteria may also enter larval cultures through the algal food supply or vertically through the broodstock (Sindermann, 1990). Once the disease has been diagnosed all equipment, containers, and water must be sterilised and infected broodstock and larval cultures should be destroyed. Preventative measures include: good hatchery hygiene; reduction of stress factors such as overcrowding, high temperatures and insufficient food; and routine water monitoring to provide advance warning of problems (Sindermann & Lightner, 1988).

Another disease that only affects hatchery-reared oysters, *Crassostrea virginica*, is juvenile oyster disease (JOD), caused by a newly described species of α-proteobacteria – CVSP (see Chapter 11). The disease strikes suddenly during the summer months in apparently healthy, fast growing oysters (Ford & Tripp, 1996). The first signs are growth inhibition and mantle lesions that appear about one week before the onset of mortality. Severe mantle retraction occurs, followed by deposition of conchiolin deposits around the shrunken tissue and mantle lesions. Bacteria are present in both lesions and shell deposits. Other tissues develop lesions, and additional deposits between the shell and mantle sometimes cause the adductor muscle to detach from the shell. This in turn causes the soft tissue to fall out of the shell. The disease outbreak lasts between 4 and 6 weeks. Highest mortality (60–90%) occurs in juveniles averaging 5–20 mm shell height; those larger than 25 mm suffer much lower mortality (0–30%) but a similar disease incidence (95%) to smaller juveniles. One of the most effective ways to avoid JOD is to transfer oysters into the field as early in the growing season as possible so that they will have reached the 25 mm size refuge before JOD appears. In the hatchery reducing density, and increasing water flow through up-welling systems can reduce mortality.

Serious diseases such as bonamiasis, dermo, brown ring disease (BRD) and MSX (see Chapter 11) do not tend to be a problem in spat or juvenile nursery systems. This is mainly because small bivalves filter less water and thus encounter fewer infective stages than older individuals.

See Sindermann (1990) for information on principles and mechanism of disease control in bivalve culture, and the economic effects of disease on producers and consumers.

Grow-out

The grow-out phase brings the small bivalve up to market-size and is therefore the longest phase in the culture process. The choice of site is of paramount importance not just for good growth but also for management and economic reasons (for more information see Quayle & Newkirk, 1989, and Pillay, 1993).

Oysters are either on-grown in trays or in nylon rearing nets in suspended culture, or in mesh bags or trays placed on trestles in the low intertidal area. They may also be grown directly on the bottom within protected crab-proof plots, or on unprotected plots if the oysters are large enough to resist crab attack (Spencer, 1990). Scallops are mainly grown in suspended culture in pearl nets, lantern nets or plastic trays. Clams are ongrown in the ground in parcs or plots in the intertidal area of the shore.

During grow-out the culture set-ups are regularly cleaned of fouling organisms and silt, samples are taken for measurement of growth rate (see Chapter 6), densities are regulated to optimise growth, and the animals are checked for signs of disease and predation. As the animals mature samples are taken to assess reproductive condition (see Chapter 5).

Rather than present a general account of the techniques used to on-grow bivalves, five key species have been chosen for a detailed account: culture of the Pacific oyster, *Crassostrea gigas* in France, the Mediterranean mussel, *Mytilus galloprovincialis* in Spain, the yesso scallop, *Patinopecten yessoensis* in Japan, the zhikong scallop, *Chlamys farreri* in China, and the Manila clam, *Ruditapes philippinarum* in Europe.

Mussel culture

Culture of Mytilus galloprovincialis in Spain

The following references have been the main sources of information: Lutz (1980); Figueras (1989); Quayle & Newkirk (1989); Dardignac–Corbeil (1990); Lutz *et al.* (1991); Hickman (1992); Caceres-Martinez & Figueras, (1997).

In 1999 Spain produced nearly 262 000 tonnes (t) of *Mytilus galloprovincialis* making it the second largest producer of cultured mussels in the world (Table 9.2; see also Table 9.1). The industry is mainly based in the rías of Galicia in north-western Spain (Figs. 9.4 & 9.5). Rías are deep sunken river valleys up to 25 km in length, 2–25 km wide and 40–60 m deep. Their mean surface water temperature ranges from 10 to 20°C, salinity is about 34 psu and tidal range

Table 9.2. Yields (live weight, tonnes) from the main producers of cultured mussels 1997–1999 (FAO, 2001). *species not specified, but largely *Mytilus galloprovincialis* with small amounts of *M. coruscus*, *Musculus senhousei* and *Perna viridis* (Tseng, 1993; Guo *et al.*, 1999). Data for the Netherlands and Germany have been excluded because mussel production resembles a fishery more than a pure culture operation (see Chapter 8). Method of culture: R, raft; L, longline; HP, hanging park; B, bouchot; P, bamboo pole; OB, on bottom.

Country	Method	Species	1997	1998	1999
China	L	Sea mussels*	398192	540901	608115
Spain	R	*Mytilus galloprovincialis*	188793	261062	261969
New Zealand	L	*Perna canaliculus*	65000	75000	71000
Italy	L, HP	*M. galloprovincialis*	103000	130000	130000
France	B, L	*M. edulis*	52350	50000	51600
		M. galloprovincialis	11000	10000	10900
Thailand	P	*P. viridis*	43087	43087	40300
Chile	R, L	*M. chilensis*	8635	11911	16203
Greece	R, L	*M. galloprovincialis*	11049	14535	16912
Canada	R, L	*M. edulis*	11449	14920	17339
Ireland	OB, R, L	*M. edulis*	16094	18317	16111
Korea	L	*M. crassistesta*	63573	17785	15042
Philippines	P	*P. viridis*	11658	15533	15478

Fig. 9.4. Rías in Galiciá NW Spain, the major raft culture region for *Mytilus galloprovincialis*. From Caceres-Martinez & Figueras (1997).

Fig. 9.5. Mussel rafts in the Ría de Vigo with Vigo in the background. Photo courtesy of N. Herriott, Community Environmental Services Ltd., Westport, Mayo, Ireland.

averages 3–4 m. Continuous upwelling of cold nutrient-rich waters, and run-off from the surrounding hills support a high primary production of 10.5 μg carbon $l^{-1}h^{-1}$. Protected from the full force of the Atlantic Ocean by islands at their mouth, these sheltered, nutrient-rich rías provide the perfect environment for suspended mussel culture.

The first culture started at the beginning of the 20th century at Tarragona and Barcelona in north-eastern Spain. Initially poles, similar to French bouchots (see below), were used but these were soon replaced by floating structures. These consisted of a central flotation chamber that supported a square wooden framework (raft) from which ropes were hung. In 1946 this system was introduced to Galicia and production increased rapidly from 22 460 t in 1956 to the present day figure of more than 260 000 t. About 60% of production comes from the ría de Arosa (Fig. 9.4).

Today the rafts consist of a framework of eucalyptus wood (100–500 m²) supported by four to six cylindrical fibreglass or steel floats (Fig. 9.6). The rafts are secured by mooring chains and a heavy concrete anchor in about 11 m of water at low tide. A raft has from 200 to 700 ropes, each about 9 m long. The rafts are spaced at a distance of 80–100 m from each other and located together in groups called parks. Currently, in the Galician rías there are over 3000 rafts. Most of these are family-owned and managed, a factor that contributes significantly to low operational costs and ultimately to the great success of mussel culture in the region.

The culture procedure is divided into five phases: collecting the seed, attaching the seed to the ropes, thinning the ropes, ongrowing, and harvesting. Most of the seed (70%) is collected from November to March from natural settlement on exposed rocky shores at the mouth of the rías. The remainder is collected on ropes hung from rafts during March and April. Farmers can

Fig. 9.6. Spanish raft with polyurethane floats. © A. Figueras.

collect up to 1500 kg of seed on the low tide over a four-hour period. The seed (6–8 mm shell length) is wrapped by hand or by machine onto the ropes using water-soluble rayon netting that dissolves within a few days (Fig. 9.7). By that time the small mussels will have attached themselves by byssus to the ropes suspended from the rafts. The average weight of seed per metre of rope is 1.5 to 1.7 kg, about 14 kg per rope. Wooden or plastic pegs are inserted in the rope every 30–40 cm to prevent subsequent slippage of mussel clumps. However, it is claimed that plastic 'mussel discs' inserted at similar intervals down the length of the rayon netting or rope are more effective (Jefferds, 2000). Ropes are moved to the raft as quickly as possible and attached at a density of 2–3 ropes per m² of raft. So that the grower maintains continuous production there are normally ropes for collecting seed, ropes for producing half-grown mussels, and ropes on which marketable mussels are growing, all on the same raft.

After 5–6 months when mussels are half-grown (4–5 cm shell length) and the average weight per rope is 46 kg, the mussels are removed from each rope, cleaned, and redistributed onto two or three new ropes. Thinning accomplishes two things: it decreases the risk of clumps of mussels falling off the ropes and it increases growth rate. In general, thinning is automated which means that it requires less than 14 hours for 500 ropes of 10 m length (Fig. 9.8). The process is repeated once more before harvesting. Mussels can reach market size (8–10 cm shell length) in 13–16 months. Apparently, in the past this size was reached in 8–9 months when mussel rafts were at a lower density than they are today. Other factors that affect growth rate are: season, location of raft and depth of cultivation. Growth is slowest during summer and highest during winter because water stratification in summer causes a scarcity of phytoplankton feed for the mussels. Mussels situated on rafts at the ocean end of the ría have faster growth than those located further in. Fuentes *et al.* (2000) have shown that depth of cultivation is a more important factor in

Fig. 9.7. Binding the mussel seed on the ropes in the Ría de Vigo. Wooden pegs to prevent slippage of mussel clumps can be seen on the nearest rope. Photo courtesy of N. Herriott, Community Environmental Services Ltd., Westport, Mayo, Ireland.

influencing growth than position of the culture rope on the raft, or stocking density; mussels in the upper part of the water column, above the thermocline (2.5 m) were significantly larger and heavier than those cultivated in deeper water (7.5 m).

In general, mortality of mussels is low on the rafts. Occasionally, extremely heavy rainfall may lower salinity to such an extent that mussels on the upper 0.5–1.0 m of ropes die. Predation from starfish or crabs is minimal as ropes are off the bottom, and disease or parasites do not appear to be a problem. Fouling by algae, tunicates and barnacles is more of a problem as these organisms must be removed manually during ongrowing and at harvesting before the mussels are sent for processing or depuration (see below).

At harvest time, usually October to March when mussel condition is best, the ropes of mature mussels are lifted by crane and a large wire basket is lowered under the rope before both are winched onto the boat. A vigorous shake of the rope removes the mussels, which are then cleaned and graded. Those less than market size (8–10 cm shell length) are wrapped onto new ropes for re-hanging, while marketable mussels are packed in nylon bags and transported either to processing plants for canning or freezing, or, in the case of

Fig. 9.8. Crane and basket used to lift and move ropes on the raft. © A. Figueras.

mussels for the fresh market, to depuration stations. Production is about 10 kg per metre of rope, which works out at about 60 t for a raft with 700 nine-metre ropes. Losses due to natural mortality and handling are estimated to be about 15% per year. About 60% of harvested mussels are for the fresh market with over 70% of this for local consumption; most of the remaining 40% is canned although the percentage being frozen is increasing annually. Exports are mostly to Italy, France and Germany.

Mussels for depuration are placed in trays or baskets in large concrete tanks. Chlorinated seawater is slowly pumped over and through the mussels for 48 hours. The mussels are then rinsed, bagged (1–15 kg quantities), placed in yellow net bags with health certification, and transported to national markets, or to markets outside of Spain. Government regulations stipulate that growers must have a licence to market mussels, and that all mussels for the fresh market must be depurated before sale. The Department of Health constantly screens mussels for red-tide toxins such as PSP and DSP (see Chapter 12). If toxins are detected the entire ría, or a section of it, is immediately closed and harvesting is not permitted until the area has been given the all clear.

About 4600 t of seed per year are needed to maintain the present level of mussel production in Galicia and most of the seed for seeding ropes comes from the shore. Because of increasing pressure on this source and the lack of any specific government regulations governing its removal it is important that

Fig. 9.9. Bouchot mussel culture in France. Photo courtesy of P. Goulletquer, IFREMER, France.

the alternative method of seed collection, namely on collector ropes, be improved. For the Ría de Arosa, Fuentes & Molares (1994) have recommended that to maximise settlement, collectors be deployed before July, that collectors be placed in the outer and middle parts of the ría, and that ropes be hung within the first 5 m of the water column where settlement is highest.

Alternative methods of cultivation

While mussel culture in Galicia is based solely on rafts there are alternative methods of production (see Table 9.2). These fall into two categories: on-bottom cultivation (see Chapter 8) and off-bottom cultivation. Compared to on-bottom culture, off-bottom cultivation makes better use of the water column and mussels are less accessible to bottom predators. Off-bottom cultivation includes raft culture described above, and methods that employ sticks or poles driven into the ground, or longline systems.

In France mussel culture (*Mytilus edulis*) on the Atlantic and English Channel coasts is based on the bouchot method – a line of poles, usually oak tree trunks, 12–25 cm in diameter and 4–7 m long with half their length embedded in the seabed (Fig. 9.9). Spat bouchots are situated offshore and consist of a parallel row of poles with horizontal ropes for seed collection strung between the poles (Fig. 9.10). When the seed is a few months old it is removed from the ropes, placed in mesh tubes and transferred to bouchots in the intertidal zone for ongrowing. The mesh tubes are wound around the poles and secured by nailing each end (Fig. 9.11). These eventually disintegrate but by that time the mussel seed has spread to cover the entire post. A typical bouchot operation is placed perpendicular to the shoreline and consists of a single line of 125 poles running for 50–60 m and spaced 15–25 m from the next line. Most are family-run, each with 15 000–20 000 poles on average. Marketable

Fig. 9.10. Bouchot mussel culture: spat-collecting ropes (coconut fibres) covered with *Mytilus edulis*. Photo courtesy of P. Goulletquer, IFREMER, France.

mussels (>4 cm shell length) are harvested when they are 12–18 months old; an average of 25 kg of mussels is harvested from each pole annually. All mussels are sold fresh but production from bouchots and suspended culture is insufficient to satisfy the home market (see Goulletquer & Héral, 1997 and Dardignac–Corbeil, 1990 for details).

In the Philippines and Thailand bamboo poles (6–8 m long) are embedded in the seabed either singly, or grouped in a circle and tied at the top to form a wigwam structure. The poles are used both for spat collection and grow-out of *Perna viridis*. Growth is rapid with a harvest size of 5–10 cm shell length attained in 6–10 months. Each pole yields 8–12 kg of mussels. Alternatively, fixed suspended systems can be used such as the rope-web system in the Philippines, or the hanging parks in the Mediterranean. The former consists of rope strung in a zig-zag fashion between two bamboo poles in shallow water. Hanging parks consist of metal posts with wooden horizontal beams on top from which ropes are hung. Both systems use the area between the poles or posts to grow the mussels (Dardignac–Corbeil, 1990 and Vakily, 1989). Elsewhere, countries such as Ireland, New Zealand and the UK, use longline systems. Each longline consists of a series of floats connected by horizontal lines from which rope or mesh-stocking collectors are hung (Figs. 9.12 &

Fig. 9.11. Bouchot mussel culture: wrapping tubular nets filled with mussel seed onto wooden poles. Photo courtesy of P. Goulletquer, IFREMER, France.

9.13). Seed collection, thinning and grow-out practice is similar to that for raft culture. Compared to rafts, longlines are cheaper, easy to construct and maintain, and are better suited to harsh winters.

Oyster culture

Culture of the Pacific oyster, *Crassostrea gigas, in France*

More than 3.6 million tonnes of oysters are cultured annually (Table 9.1), a third of which is the Pacific oyster, *Crassostrea gigas*. According to official statistics the main producer of *C. gigas* is China, followed by Japan, Korea and France (Table 9.3) but the FAO statistics for *C. gigas* production in China must be treated with caution. A paper published by Guo *et al.* (1999) after two visits to China in 1996 and 1997, reports that it is another species, the zhe oyster, *Crassostrea (Alectryonella) plicatula*, that accounts for 50–60% of national production; *C. rivularis* (or *C. ariakensis*) accounts for 20–30%, and *C. gigas* only accounts for 10–20%. For this reason, harvest for the latter in China is taken as 15% of the FAO figure (Table 9.3). The main species that make up the remaining world production of oysters are: the slipper oyster, *Crassostrea*

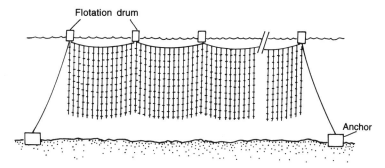

Fig. 9.12. A longline unit with anchoring and flotation systems. Typically the horizontal line is 60 m long with large plastic 200 litre air-filled flotation drums attached at 6 m intervals. The vertical substrates are about 6 m long and are spaced at intervals of ~50 cm. Thirteen or fourteen vertical substrates are hung between each flotation drum. Each end of the longline is anchored securely. There is usually about 3 m distance between each horizontal longline system. From Hurlburt & Hurlburt (1980).

Fig. 9.13. Longline system of mussel culture showing mussel-covered ropes with pegs to prevent slippage. Photo courtesy of P. Goulletquer, IFREMER, France.

iredalei (Philippines), the eastern oyster, *Crassostrea virginica* (United States and Canada), the European flat oyster, *Ostrea edulis* (France, Spain, UK, Ireland, Netherlands) and the Sydney rock oyster, *Saccostrea glomerata* (Australia).

Despite China's high production of *Crassostrea* spp. information on culture methods is difficult to access, simply because almost all publications are in Chinese. In contrast, there is a wealth of published information in English on the culture of *C. gigas* in France. Consequently, production will be dealt with from the French rather than the Chinese perspective. Information has been obtained primarily from Héral (1990), Héral & Deslous-Paoli (1991) and Goulletquer & Héral (1997).

Table 9.3. Yields (live weight, tonnes) from the main producers of cultured Pacific oyster, *Crassostrea gigas*, 1997–1999 (FAO, 2001). *includes yields from Hong Kong and Taiwan; figures for China are 15% of the figure given in FAO data (for explanation see text). Each country's percentage of *C. gigas* global yields for 1999 is in parentheses.

Country	1997	1998	1999
China*	352949	427935	451132 (43%)
Japan	218056	199460	205345 (20%)
Korea	200913	175926	177259 (17%)
France	147150	143000	134800 (13%)
United States	24796	31715	33259 (3%)
New Zealand	7000	13000	14950 (1%)
Others	23553	22826	28297 (3%)
Total	974417	1013862	1045042

Oyster culture started in France in the 17th century with the indigenous species *Ostrea edulis*. Oyster seed was collected from natural beds and grown for four or five years in rearing ponds or 'claires' at Atlantic coast sites. However, progressive over-fishing of the natural beds led to a shortage of oysters, and in 1860 the cupped oyster, *Crassostrea angulata*, was imported from Portugal. This hardy species quickly colonised the west coast of France and by the 1960s the yield was more than three-fold higher than that from flat oysters. However, between 1970 and 1973 massive mortalities from gill disease effectively wiped out the cupped oyster from the French coast. Faced with this crisis France imported commercial quantities of the Pacific oyster, *Crassostrea gigas*, from Canada. These were directly planted into the main cupped oyster bays between 1971 and 1975. In addition, between 1971 and 1977 *C. gigas* spat was imported from Japan to reseed the oyster grounds (Grizel & Héral, 1991) and to build oyster reef sanctuaries. The introduction of *C. gigas* was extremely successful, but disaster struck once again in the late 1970s with the appearance of two disease organisms, *Marteilia refringens* and *Bonamia ostreae* (see Chapter 11) that decimated production of *O. edulis* in almost all rearing areas. Production has never recovered and today harvests of *O. edulis* represent only a fraction (2%) of those of *C. gigas*. Fortunately, the diseases that have destroyed *C. angulata* and *O. edulis* populations have not so far affected *C. gigas*.

In France seed supply is from natural settlement onto artificial collectors. Spatfall is reliable and regular on the south-west Atlantic coast; the Gironde Estuary, and Arcachon and Marennes-Oléron Bays furnish the entire French production for natural spat settlement, representing about 90% of the total spat production. From here seed is transplanted to sites in Brittany, Normandy and the Mediterranean. Although oysters settle on a variety of substrates the most widely used collectors nowadays are PVC tubes and PVC dishes that have been pre-soaked and sun-dried to remove hazardous chemicals (Fig. 9.14).

Fig. 9.14. PVC collectors used for natural settlement of the oyster *Crassostrea gigas*. The tubes are placed at high density for catching the spat, but the density is subsequently reduced to improve growth of settled spat. Photo courtesy of P. Goulletquer, IFREMER, France.

Recruitment is around 5 trillion (10^{12}) spat in Arcachon Bay and double this in Marennes-Oléron Bay, although these figures can vary widely from one year to the next. In recent years French hatcheries have begun to produce cultchless spat and larvae for remote setting. This is good news for farmers in areas with no natural recruitment because of low summer water temperatures (<18°C). The share of spat produced by hatcheries is increasing significantly each year, mainly due to the interest in triploid oysters (see Chapter 10).

The three main methods of ongrowing are bottom culture, rack culture and hanging-rope culture. Bottom culture is carried out either in 'parcs' or 'viviers' in the intertidal zone, or in deep water. The ground is first hardened and the spat (6–10 months old) are then either sown directly on the ground, placed in small mesh oyster bags, or if they are retained on the collectors these are placed on the bottom. In the intertidal area the spat are protected against crab predation by plastic mesh fences (40 cm high) or hedges of twigs or stones. The spat are left for a 'pre-growing' period of 1–2 years after which time if they are still on collectors they are scraped off, graded and put back on the bottom for a further period of 1–2 years. Densities for the pre-growing and maturing phases are on average 5 and 7 kg (total weight) per square metre, respectively and one tonne of spat produces about 20 tonne of market-size oysters (>70 g; note that for oysters it is weight rather than size that is nor-

Fig. 9.15. Oysters (*Crassostrea gigas*) growing in bags on metal trestles in the intertidal area of the shore. Photo courtesy of P. Goulletquer, IFREMER, France.

mally used). The oysters are harvested at low water with a rake, put into baskets and at flood tide loaded into boats. Because of high labour costs, this method is being replaced more and more by mechanical harvesting methods.

For deep-water culture large areas of several hundred hectares are dredged clean and marked out by buoys. Densities of planting are either 50 kg per 100 m² for spat, or 70–90 kg per 100 m² for 2-year-old oysters. Growth is rapid but there is high mortality and a high investment due to dredging costs. In addition, submerged oysters have a weak adductor muscle and do not keep well out of water. They are 'trained' to keep the shell tightly closed by being held for a time in the intertidal zone, where they are alternately exposed and submerged as the tide ebbs and flows.

Rack culture is the most common technique used on the Atlantic and English Channel coasts. Oyster spat are placed in plastic mesh bags (0.5 × 1.0 m) and tied to metal trestles (3 m long, 0.5 m off bottom) that are arranged in parallel rows in the intertidal zone (Fig. 9.15). As the oysters grow they are thinned and bag mesh size is enlarged to encourage good growth. Biomass per bag varies between 5 and 15 kg depending on the size of the oysters. The bags are turned regularly, often by machine, to produce a well-shaped oyster and to control fouling. Oysters stay in the bags for 1–3 years depending on growth conditions. Advantages of rack culture include: good growth and quality, low mortality and ease of access; but overcrowding, fouling and silting under the trestles are some of the disadvantages. These have led to strict controls being imposed e.g. restrictions on the number of trestles per leased area, and the removal of trestles during winter to improve silt transport. Oysters are harvested by flatboat at high tide, or by tractor at low tide.

Suspended culture is mainly used in Mediterranean lagoons, where there is good water depth (10 m) and no tide. Spat collectors transported from the

Fig. 9.16. Aerial view of oyster claires in Marennes-Oléron Bay, W France. Photo courtesy of P. Goulletquer, IFREMER, France.

Atlantic coast are suspended under metal tables (11 m × 50 m) driven into the seabed. Some of the oysters are marketed after 12–18 months, while the rest may be individually cemented onto wooden bars or ropes and hung for an additional year to give an oyster of very fine quality. Dredgers harvest the oysters and average yield per table is 5–7 tonnes. Along the Atlantic coastline several companies are now using longlines for pre-growing oysters, while a number of such projects are currently seeking state authorisation (P. Goulletquer personal communication, 2001).

In salt marsh areas e.g. the Marennes-Oléron Bay on the west coast, oysters are often placed in claires for fattening. These shallow (0.4 m) claires are rich in nutrients with consequent high phytoplankton productivity (Fig. 9.16). One species of particular interest is the diatom *Haslea ostrearia*. After death the green pigment from the diatom diffuses into the water and is absorbed by the oyster gills, giving them an attractive dark–green colour, much in demand by the consumer. Oysters fattened for three weeks at a stocking density of $3 \, \mathrm{kg \, m^{-2}}$ are called *fines de claires* with a <10.5 fattening index, while those with a >10.5 are *speciales de claires*. Moreover, oyster spending more than four months at a density below five individuals $\mathrm{m^{-2}}$ and with a >12 fattening index are called *pousses en claires*. About 50% of French production is *fines de claires* while <10% is *speciales de claires*, *speciales de claires label rouge* and *pousse en claires label rouge* (higher and certified quality oyster). This fattening technique however, does not always provide the appropriate quality oyster. The addition of algae (380 000 cells $\mathrm{oyster^{-1}}$) to the ponds during autumnal neap tides (when no seawater replenishes the water) has recently been shown to greatly improve the fattening process (Soletchnik *et al.*, 2001). When the oysters are harvested they are transported to a processing plant where they are washed and mechanically sorted by weight, packed and brought to market. Almost all oysters are sold fresh either to local markets, supermarkets or restaurants. Production is solely for the home market with more than 50% marketed for the Christmas and New Year periods.

Local authorities lease oyster beds for a period of 30 years at a set annual cost per hectare. The state and local agencies manage the dredging of the bays and set regulations governing type of culture, densities, harvesting times, and quotas based on yearly stock assessments. Areas are classified into sanitary and unsanitary sectors as stipulated by the EU Shellfish Directive (EEC, 1991) [see p. 324], and no production from the latter is permitted. Even when oysters are grown in sanitary waters they must spend 2–3 days in clean, food-free water, mainly to discharge silt particles.

While *C. gigas* culture appears to be free of disease it suffers similar problems to the mussel culture industry, i.e. fouling, predation, competition, and toxic blooms. One competitor that is particularly damaging is the slipper limpet, *Crepidula fornicata*, which attaches to the shell, sometimes completely covering and suffocating the oyster. Another problem is the annelid *Polydora* spp. that burrows into the shell and weakens it, making the oyster more vulnerable to predators, and also less suitable for market. Pollution can also be a problem for the industry because oysters tend to be cultivated in estuarine areas that are subject to changing bacteriological and chemical input.

Before leaving this section it is probably opportune to give a brief description of some of the methods used in China to culture species of *Crassostrea*. The sole source for this account is Guo *et al.* (1999). Culture of *C. plicatula* depends entirely on natural seed that settles on stone pilings, vertical stone strips (>1 m in height), and bamboo or wooden stakes placed in the intertidal zone. About 18 500 ha of these structures cover miles of coastline, mainly in Fujian province (see Fig. 9.19). Longlines and rafts are also used in deeper waters. The culture period is under a year with settlement in May and September, and harvesting of 6–7 cm oysters between December and March. The Suminoe oyster *C. rivularis* is cultured mainly in southern China (Fujian and Guangdong). The species is euryhaline but prefers low salinity estuaries and riverbeds. Spat settle on concrete stakes (50 cm long and 6 cm × 6 cm cross section) placed at upper river sites. The stakes are subsequently moved to lower river sites for grow-out. About 30 000 stakes are planted per hectare and each stake carries about a dozen oysters. Longline and raft culture methods are also used. Oysters are harvested at 2–3 years when shell length is 10–15 cm. Culture of *C. gigas* occurs all along the Chinese coast but main producers are in the north (Liaoning and Shandong) and south (Guangdong) of the country. Culture depends entirely on hatchery-produced seed. The spat settle on strings of scallop or oyster shells that are subsequently inserted on ropes that are then suspended on longlines or from rafts. Oysters in suspended culture grow rapidly reaching a shell length of 8–10 cm after one growing season.

Scallop culture

Culture of the yesso scallop, Patinopecten yessoensis, in Japan

Currently, nearly one million tonnes of scallops are cultured annually (Table 9.1) and the yesso scallop, *Patinopecten yessoensis*, makes up about 20% of this

Table 9.4. Yields (live weight, tonnes) for the main cultured scallop species and their principal producers 1997–1999 (FAO, 2001). Values for China are approximate figures (Guo *et al.* 1999; Guo & Luo, in press).

Species	Country	1997	1998	1999
Patinopecten	Japan	254086	226134	216017
yessoensis	China	~10000	6500	~7500
	Korea	637	360	377
	Russian Federation	600	350	0
Chlamys farreri	China	800000	505000	570000
C. nobilis	China	10000	6500	7500
Aequipecten	UK	46	147	114
opercularis				
Argopecten irradians	China	165000	105000	120000
A. purpuratus	Chile	11482	16474	20668
	Peru	311	1021	1585
Pecten maximus	France	150	150	500
	Spain	206	149	156
	UK	27	41	27
	Ireland	30	20	33
	Total	1252575	867846	944477

figure, virtually all of which is cultured in Japan (~250000t year^{-1}). Small amounts are cultured in China, Korea and the Russian Federation (Table 9.4). The main sources of information on the culture of *P. yessoensis* in Japan have been: H. Ito (1991), S. Ito (1991) and Dao *et al.* (1993). At the time of writing no more recent reviews or reports than these were available, at least not in the English language.

Culture of *P. yessoensis* started because of the decline at the end of the 1960s in the wild fishery for this species in Japan. The first attempt at spat collection actually started well before this in 1934 (Kinoshita 1935), and after many years of trials a successful spat collector was invented by a fisherman in Mutsu Bay in 1964. Subsequently, production increased rapidly from a figure of 6000t in 1965, to 100000t by 1975, to more than 200000t by 1985 where it has remained fairly stable to this day (Table 9.4).

The culture of *P. yessoensis* is confined mostly to northern areas of Japan where water temperatures are ideal (10–17°C) for this cold–water species (Fig. 9.17). Culture is equally divided between suspended and bottom production. The industry is based on natural spat that settle on collectors hung from long-lines. Various types of collectors are used but the most common type is a net onion bag loosely filled with plastic netting or used gill netting. The mesh size must be large enough (usually 0.5mm) to allow the water with settling larvae through but small enough to prevent the spat from escaping when they eventually detach from the netting. Each bag measures ~60cm long by 35cm wide and about 12 of these are attached at intervals to each 8–12m longline (Fig. 9.18A). Numbers of spat range from over 10000 to a low of 300 per collector. In Mutsu Bay (Fig. 9.17) spat collectors are in place from mid April to the beginning of August, but where water temperatures are lower, e.g. Hokkaido, spat collection starts and ends later.

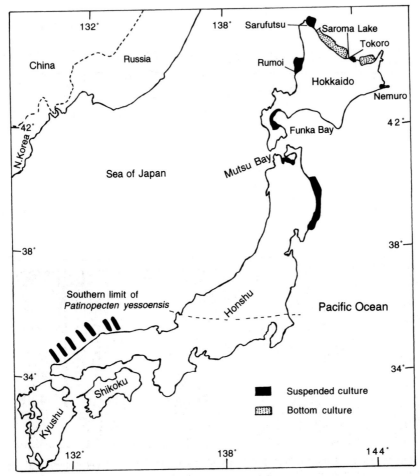

Fig. 9.17. Cultivation areas for the scallop *Patinopecten yessoensis* in Japan. From text of Dao *et al.* (1993).

Between mid July to mid August, when the scallops are 5–10 mm shell height, they are removed from the collectors and placed in pearl nets (4–9 mm mesh) suspended from longlines that are 5–10 m below the water surface (Fig. 9.18B). This stage is referred to as intermediate or nursery culture. The initial stocking density is 100 spat per net but by September–October this is reduced to 20 spat per net if adults are for suspended culture, and 50 spat per net for bottom culture. Scallops do not tolerate as high a stocking density as other bivalves. Growth is fast and juveniles have a shell height of about 4 cm by December and about 5.5 cm by mid March. The survival rate is generally higher than 90%. Juveniles from intermediate culture are used for both suspended and bottom culture. In suspended culture scallops are either grown by 'ear hanging' or in different types of nets. Scallops need to be contained like this because they can break their byssus and swim off. In the ear-hanging method a small hole (1.3–1.5 mm) is drilled in the anterior ear of the left shell

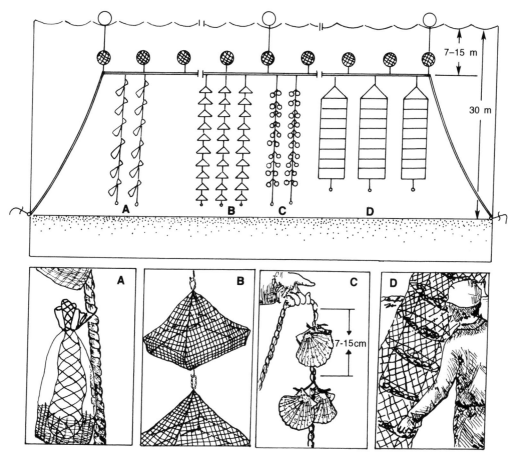

Fig. 9.18. Longline system used in the culture of scallops: (A) onion bags for spat collection; (B) pearl nets for nursery phase; (C) ear hanging and (D) lantern nets for on-growing phase. Redrawn and modified from K. Freeman (1988).

valve, or through both the right and left anterior ears. The scallops are tied by nylon thread to a rope (6–8 mm diameter) that is suspended from a long-line (Fig. 9.18C). Scallops are usually attached in pairs at intervals of 10–20 cm depending on their size. One rope contains about 130 individuals. This method gives a high meat yield, high survival and a well-shaped shell. Although attaching the scallops to a line is labour-intensive, once hung the ropes need minimal attention. Fouling can be a problem and if severe can cause the scallops to fall off the ropes. This method is not suitable in areas subject to excessive wave action.

Various types of nets are used in suspended culture, the most common one being the lantern net. This consists of 10 wire hoops, 50 cm in diameter and 15–20 cm apart, enclosed by monofilament netting 20 mm mesh size (Fig. 9.18D). Each compartment has an opening through which scallops can be inserted or removed. Several lantern nets can be strung one below the other if there is sufficient water depth. The stocking density is between 7 and 10 scallops per compartment, depending on size. It usually takes 10–15 months for scallops to reach the market size of 10–11 cm shell height. Although all suspended culture methods are subject to heavy fouling, cleaning must be kept to a minimum and executed rapidly as scallops are very sensitive to being exposed to air and direct sunlight. Fouling can be minimised by deep suspension, or by planting the scallops at a lower density in the nets to lessen time before harvest.

The other grow-out method is bottom cultivation, which is mainly concentrated on the north coast of Hokkaido and Mutsu Bay in Honshu (Fig. 9.17). Prior to seeding predators such as starfish and sea urchins are dredged from the bottom. Seed measuring about 3 cm shell length is sown between December and April at a density of 5–10 individuals m^{-2}, a value based on natural stock density. The seed is obtained either from natural settlement or from hanging culture. It takes between 3 and 4 years for the seed to reach market size. The culture beds are rotated on a regular basis and the Tokoro cooperative will serve as a good illustration of how this operates. The co-operative situated on the north coast of Hokkaido have divided their coastal area into four units, each averaging 25–30 km^2 with depths ranging from 30 to 60 m. A smaller unit is set aside as a reserve in case of under-production. Each unit is harvested every four years and is followed by a new seeding (density 10 m^{-2}). The spat comes from natural settlement in either the nearby Saroma Lake, or from Funka Bay or Mutsu Bay (Fig. 9.17). For 300 million spat the cooperative harvests about 30 000 tonnes live weight of scallop three years later. When the co-operative started in 1976 the ratio of adults fished to the number of juveniles seeded was about 1:4, but this ratio has improved to 1:2 due to better management of the beds in terms of stocking, predator control etc. The scallops are harvested by dredge from June–July to November and each boat has a daily catch quota of 6 t, usually caught over a period of 3–4 hours. In Japan both the adductor muscle and soft body parts are processed. The following are processing yield data for Hokkaido in 1987–88: fresh scallop 20%; boiled whole body 32%; frozen muscle 27%; dried muscle 13%; canned and other 8%. These proportions, however, can vary considerably from one year to the next.

The continuing success of scallop culture in Japan will be guaranteed if there is a reliable seed supply each year and if the carrying capacity of bays used for scallop culture are not exceeded. To date there is no hatchery production of seed scallops, but this situation must change since seed production from Saroma Lake and Funka Bay, two of the main supply areas, will not be sufficient to meet future demands. Many bays used for scallop culture are now overcrowded and offshore culture is not successful. Therefore, density in both suspended and bottom culture must be carefully regulated. This will increase food availability and consequently improve growth rate. Other areas for attention might include: location of additional sites for bottom seeding culture, routine bed rotation, better predator control on seeding plots, and improved methods in dealing with fouling organisms in suspended culture.

Outside of Japan other species that are cultured worldwide are the zhikong scallop, *Chlamys farreri*, the huagui scallop, *Chlamys nobilis*, the bay scallop, *Argopecten irradians*, the Peruvian scallop, *Argopecten purpuratus*, the great Atlantic scallop, *Pecten maximus*, and the queen scallop, *Aequipecten (Chlamys) opercularis* (Table 9.4 and Fig. 3.4 for distribution). Moves are afoot to increase production of some of these species, and to initiate new scallop species into the industry. Hatchery-reared *Pecten maximus* spat are now produced on a commercial scale in France (Dao *et al.*, 1996), Norway (M. Norman, personal communication 2001), and the UK (Laing & Psimopoulous, 1998; Utting & Millican, 1998) and the process is at the development stage for *Patinopecten yessoensis* (Bourne, 2000) and *Placopecten magellanicus* (Harvey *et al.*, 1997) in Canada, *Pecten fumatus* in Australia (Heasman *et al.*, 1995), *Argopecten purpuratus* in Chile (Martinez *et al.*, 1995) and *Argopecten ventricosus* (= *circularis*) in Mexico (Ibarra *et al.*, 1995).

Culture of the zhikong scallop, Chlamys farreri, in China

China has a history of mollusc culture dating back several thousand years but it is only in the last three decades that the industry has expanded exponentially. China is now the top producer of cultured oysters, scallops and clams in the world (Table 9.1). In 1999 China produced 6.7 million tonnes of cultured bivalves of which 0.7 million t was scallops. The zhikong scallop *Chlamys farreri* accounts for about 80% of this figure, *Argopecten irradians* for 15–18%, and *Patinopecten yessoensis* and *Chlamys nobilis* together account for ~2% (Table 9.4). Most publications on culture methods in China are in Chinese so the following description of *C. farreri* culture has been heavily dependent on three reports in English: Luo (1991), Guo *et al.* (1999) and Guo & Luo (2003).

The zhikong scallop, *C. farreri*, is naturally found in north China (40°N) to Fujian (25°N) where water temperatures do not exceed 25°C (Fig. 9.19). It also inhabits the waters of Japan and Korea. The distribution of the species is patchy and the total natural resource is only about 2000 t. Experimental culture of scallops started in the late 1960s and expanded rapidly through the 1980s and 1990s to become the very impressive industry it is today. The most important culture areas for *C. farreri* are Shandong and Liaoning provinces, with the former producing about 80% of the national total. Seed is collected in early summer and in the autumn on bags (30 cm × 40 cm, 1.2–1.5 mm mesh) stuffed with about 100 g of nylon screens. The bags are strung together, about 10–12 per string, and these are hung from a raft that carries about 500–600 strings. Each bag collects about 100–1000 spat depending on site, season and year. One of the bays in north Shandong produced about 130 billion seed in 1996, all from cultured populations. Most of the seed is collected from the first spawning season that set between late June and mid July. Spat are left on the collectors until early October when they have reached about 5–10 mm shell height. The seed is put into lantern nets (mesh size 4–8 mm) at a stocking density of 200–300 per layer, giving a total of 2000–3000 per lantern net. The nets are transferred into a nursery area and hung on longlines (80–100 m long) until the following March when shell height is ~30 mm. The scallops are then thinned to 50–80 per layer for the grow-out phase. By December the scallops have reached the market size of 60–70 mm shell height. In some areas seaweed and scallop culture take place on the same longline system, and sometimes sea cucumbers are cultured in

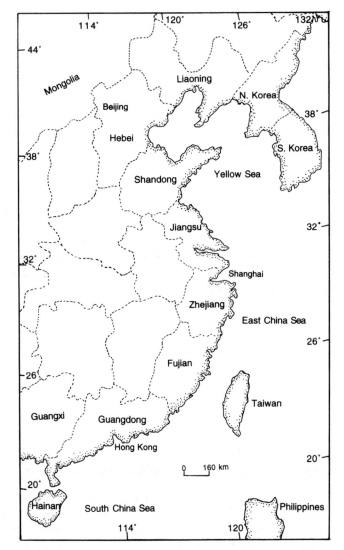

Fig. 9.19. The coastal provinces in China where scallops are cultured: *Chlamys farreri* and *Patinopecten yessoensis*, Liaoning and Shandong; *Argopecten irradians*, Liaoning, Shandong and Fujian; *Chlamys nobilis*, Guangdong and Fujian. From the text of Guo *et al.* (1999).

the same lantern nets to keep down fouling, a severe problem particularly near the water surface.

Since 1996 there has been massive summer mortality of zhikong scallops. In 1998 this was so severe that production was reduced by 37% with an estimated loss of revenue of around $360 million. The cause was probably a combination of over-crowding, high summer temperatures and deteriorating water quality. The recommended stocking density for grow-out is 30–35 scallops per layer, but farmers often stock at 2–3 times this density (see above).

Also, the huge expansion in the number of longlines means that the carrying capacity of many areas has been exceeded. Both of these factors result in a low-quality environment for the culture of zhigong scallops, and indeed for other bivalves.

The other scallop species that accounts for 5–18% of scallop production is *Argopecten irradians*, introduced from the United States in 1982. Seed is entirely hatchery-produced and the spat are ongrown in lantern nets suspended from longlines, mainly in the Shandong and Fujian provinces (Fig. 9.19). Growth of this scallop is fast and market size (50–60 mm shell height) is reached within a year, a definite advantage over zhikong scallops that take 1.5–2 years to reach market size. The huagui scallop *Chlamys nobilis* is cultured to a very limited extent in southern China, while *Patinopecten yessoensis*, introduced from Japan in the late 1970s, is grown in the northern provinces of Liaoning and Shandong.

Very few scallops are for the live market. Traditionally, most were processed by simply washing the adductor muscles in seawater, and then drying them in the sun. Alternatively, the muscles were first cooked and then dried in the sun. These were marketed as 'dry shellfish' in China and other parts of south-east Asia. However, since the late 1980s most of the harvest is processed into individually quick frozen (IQF) adductor muscles. In the early 1990s more than 60% of these were exported to North America and other regions but because of the later summer mortality problem, and because the home market is growing rapidly, only a small percentage of scallops are now exported. In fact, by 2000 China was importing sea scallops, *Placopecten magellanicus*, from North America.

Clam culture

Culture of the Manila clam, Ruditapes philippinarum, *in Europe*

More than 2.75 million tonnes of clams (live weight) are cultured each year on a global scale (Table 9.1), with China responsible for about 90% of this figure. Worldwide, there has been more than a four-fold increase in production since 1989, mostly due to the expansion of culture operations in China, Malaysia, Italy and the United States. A wider variety of clam species are cultivated compared to the other bivalve groups. The main cultured species are: the blood cockle, *Anadara granosa*, various ark shells, *Arca* spp., various razor clam species, the Manila clam (Japanese carpet clam), *Ruditapes philippinarum*, and the hard clam, *Mercenaria mercenaria* (Table 9.5). These make up about 96% of annual global yields of clams.

The Manila clam, *Ruditapes philippinarum*, contributes 67% to annual global yields and most of this comes from China (Table 9.6). The constraints of little or no published information, difficulty in accessing information in the 'grey literature' and language barriers, mean that culture of this species will be described from minor producing countries, Italy, France and the British Isles, rather than from China, the major player in the field (Table 9.6). The main sources of information were De Valence & Peyre (1990), papers in Allessandra

Table 9.5. Yields (live weight, tonnes) for the main cultured clam species and their principal producers 1997–1999 (FAO, 2001); [†] includes Taiwan. * species not specified.

Species	Country	1997	1998	1999
Anadara granosa	China	129488	157525	188355
	Malaysia	58400	81717	79912
Arca sp.	Korea	13156	23029	8550
Razor clams*	China	354152	415032	479252
Mercenaria mercenaria	Taiwan	25761	25874	26533
	United States	17992	19943	26517
Ruditapes philippinarum	China[†]	1258207	1405366	1797505
	Italy	40000	48000	50000
	Total	1897156	2176486	2656624

Table 9.6. Yields (live weight, tonnes) from all producers of cultured Manila clam, *Ruditapes philippinarum* 1997–1999 (FAO, 2001). China includes Taiwan. Figures in parentheses in 1999 column represent country's percentage (rounded up) of total yields.

Country	1997	1998	1999
China	1258207	1405366	1797505 (96%)
Italy	40000	48000	50000 (3%)
Korea	13958	17178	16135 (1%)
United States	2363	1896	3997
France	400	400	800
Ireland	200	178	121
Spain	140	1630	1826
UK	38	34	29
Total	1318906	1474682	1870413

(1990), Britton (1991), Spencer *et al.* (1991), Rossi & Paesanti (1992) and unpublished reports from an international clam workshop on husbandry and quality held in Ireland in May, 1997.

Overfishing and irregular yields of the native clam, *Ruditapes (Tapes) decussatus*, led to the importation of the closely related species *R. philippinarum* into north-west Europe in the 1970s and 1980s. The species, a native of Japan, Korea and the Philippines, was introduced accidentally along with oysters into North America during the 1930s. From there it was deliberately introduced as hatchery broodstock into France in 1972, and into the UK and Ireland in 1980 and 1982, respectively. In contrast, between 1983 and 1984 Italy imported large quantities of seed from a UK hatchery for direct planting into the ground. In all regions *R. philippinarum* proved to be hardier and faster growing than *R. decussatus* and today contributes 91% to European yields of the two species. The techniques used to culture *R. philippinarum* work equally well with *R. decussatus*.

The main culture areas in Europe are the coasts of the northern parts of the Adriatic Sea, Italy; the Atlantic coasts of France from Normandy to Arcachon; the Atlantic coasts of Ireland, Galicia and the Basque region of Spain; and Poole Harbour on the south coast of England. The Manila clam does not breed successfully in northern European waters, therefore the industry in these regions is dependent on hatchery-produced seed. Broodstock are conditioned for 30–40 days at a temperature of 20°C. Each hatchery has its own tried and tested method to induce spawning (see above). In France clams are placed in individual dishes, and spawning is induced by raising the temperature of the water to 26–28°C, followed by the addition of a small amount of sperm from a sacrificed male. When eggs are released a few more drops of sperm are added for fertilisation. If gamete release has not occurred within 30 min the water is replaced by water at 14–15°C to provide a cold shock stimulus. This procedure is repeated until the clams have spawned. The eggs are filtered through a 40 µm sieve and maintained in 10 l containers until after the veliger stage two days later. Larvae are then collected on a 40 µm sieve and distributed into containers at a density of 3000 larvae l^{-1}. For the first week the larvae are fed every day with unicellular algae at a density of 20–40 cells µl^{-1}, depending on clam size, and then every second day until metamorphosis.

In the spring, when the clams have reached about 2–3 mm shell length, they are transferred outdoors to the nursery. In France this can be an up-welling system (Fig. 9.3) where the clams remain until they have reached 10–15 mm shell length, a period of 3–4 months, depending on the season. Alternatively, slightly larger clams (6–7 mm) may be enclosed in 4 mm mesh bags (1.5 m × 2 m) at a density of 3000 m^{-2}. The bags are placed on the seabed in spring and by summer the seed measures 13–15 mm. In some areas of France small clams are reared all the way to market-size (35–50 mm shell length) in 400 m^2 coastal ponds, and a sprinkling system has been developed that can deliver seawater and live algal food (Sauriau *et al.*, 1997).

In 1987 SATMAR, a French hatchery consortium, first noted brown ring disease (BRD) in 20–30% of their clam stocks (see Chapter 11 for details on this disease). Since then various procedures have been adopted by the industry to reduce mortality. For example, growers no longer transfer seed between 5–10 g (~10 mm shell length) or nurse seed at ongrowing sites. As small seed is apparently more resistant to the disease, growers only stock seed that is less than one year old. In high-risk areas the length of the cultivation cycle is limited to 20 months maximum. BRD has also been reported in Irish, UK, Spanish and Italian stocks.

Italy has several major clam-growing areas in the upper Adriatic Sea, one of which, the Marano Lagoon to the east of Venice, is fairly typical of the industry. It is interesting that in Italy there are no commercial hatcheries, so culture of *R. phillipinarum* depends on bought-in seed from hatcheries in the UK, France, Spain and even the United States. In the Marano Lagoon the seed (4–5 mm shell length) is placed at a density of 10 000 m^{-2} in wooden frames (1 m × 5 m) that are covered in plastic netting (2–4 mm mesh), and stacked underwater until the seed is 10–12 mm, a period of 3–4 months. In Ireland seed (2 mm) is placed in mesh-covered wooden frames (3 m × 1 m) for 12 months, or in mesh bags on trestles, around the low spring tide area of the

Fig. 9.20. Mobile seed trays for nursing clams (*Ruditapes philippinarum*). Photo courtesy of Kevin O Kelly, Lissadell Shellfish Co. Ltd., Sligo, Ireland.

shore (Fig. 9.20). Stocking density is initially high at 300 000 per frame but once the seed has reached 6 mm it is thinned to 30 000 per frame for over-wintering. Due to Ireland's northerly latitude it takes until the spring of the following year for the seed to reach 9–10 mm shell length.

During the nursery stage clams must be regularly graded and cleaned. In the field a high-pressure hose is used to clean the meshes and frames of silt and fouling organisms but careful hand cleaning is necessary to remove filamentous weed from clumped seed. Care must be taken to ensure that small crabs and mussel spat are removed from the bags or frames and that there are no tears or openings through which crabs can enter. Even very small crabs can do a lot of damage by nipping off the siphons of feeding clams. Crabs are usually only a problem from March to October as they migrate offshore during the winter. Another problem is mussel spat that compete with clams for food, bind them in byssus clumps, very quickly outgrow them, and thus retard clam growth rate.

When the clams have reached about 10 mm shell length they are ready for seeding in the substrate. Prior to the 1990s two main methods were in use, the French parc system and the plot system. Parcs consisted of a fenced off area of the shore. The fence was made from wooden posts about 80 cm high and was designed to prevent crabs from entering the parc. Baited crab traps were used inside to remove the crabs that did manage to breach the fence. The floor of the parc was covered by mesh to protect against bird predators such as oystercatchers. This system was expensive and difficult to maintain and has since been superseded by the plot system of cultivation. In the plot system strips of mesh are laid over the seeded clams and ploughed in along the edges of the plot to make it crab and bird-proof (see below). Ideally the seeding site should be sheltered, with a salinity of 20–34 psu, the substrate should be free

Fig. 9.21. Seeding clams (*Ruditapes philippinarum*) with a mechanical net-layer. The roll of netting has just run out. Photo courtesy of Kevin O Kelly, Lissadell Shellfish Co. Ltd., Sligo, Ireland.

from rocks, boulders and gravel and firm enough for machinery, there should be good access to the site, and trials should show evidence of good clam growth and survival. The chosen site should be monitored over a period of one year for changes in substrate, shifting of estuarine channels, water coverage on different tides and under different weather conditions, and settlement patterns of potential competitors and fouling organisms.

In Ireland, before constructing a plot, all large stones, weeds, mussels and crabs are removed from the surface of the seeding area. Crabs in the substrate are killed by a crab killer, a device with vertical tines that penetrates to a 75-mm depth, killing all crabs, and in the process loosening the sand to help the young clams bury themselves. Clams of 10–14 mm shell length are seeded at a density of ~300 m^{-2} and covered with 4 mm aperture netting (1–1.5 m wide, and up to 300 m long). A 0.5 m gap is left between each net to allow a tractor to pass over the plot without harming the clams. A planting machine ploughs in the netting and sows the seed simultaneously. Deflector plates backfill the trenches, burying the edges of the net up to 100 mm (Figs. 9.21 & 9. 22). Ideally, the net strips should be laid parallel to the prevailing wind direction and also parallel to the main run off. Seeding is generally carried out in the spring to maximise growth and to avoid the main period of crab predation (July–September). The UK and France use much the same method as in Ireland with slight variations in terms of plot size, type of netting etc. In Italy the plots are sown by manually scattering seed (~200 individuals m^{-2}) over the prepared area at low tide. The nets should be cleaned routinely of fouling organisms and sediment and checked for holes and for predators under the mesh. In Ireland a ganged static brush assembly mounted on a tractor is drawn over the nets to clean them. This is done once or twice every spring tide depending on the season.

Fig. 9.22. Clams (*Ruditapes philippinarum*) planted under nets. Photo courtesy of Kevin O Kelly, Lissadell Shellfish Co. Ltd., Sligo, Ireland.

Estimated growth of Manila clams at yearly intervals in UK plots is shown in Fig. 9.23. A clam placed in a ground plot at 10 mm shell length should grow to 30 mm, 42 mm and 51 mm after one, two and three years, respectively. Similar growth rates are observed in Ireland. Bearing in mind that it takes one year of hatchery and nursery rearing before clams are seeded in the ground, the length of time it takes to produce minimum market-size (35–40 mm shell length) clams in the UK and Ireland is about three years. This size is reached in Italian plots in 16–18 months (Mattei & Pellizzato, 1997), and in France is less than two years. However, in all countries the optimum market size is 45–50 mm as demand and price is better for these larger clams.

Various factors influence survival rate of clams in ground culture systems. Data from the UK show that small clams survive less well than larger clams in ground plots. Clams put in the ground at 3 mm had only a 34% survival rate compared to 10 mm clams with a 60% survival rate. Larger clams of 24 mm had a 77% survival rate, but the gain in survival is outweighed by the higher cost of rearing clams to this size. Various characteristics of the plot, e.g. silt content of the water and water coverage, are important to survival. In UK trials Manila clams had a 70% survival in plots with up to 30% exposure (low water neap tide) to air, but at 50% exposure higher on the shore no clams survived due to smothering by moving sediment. Winter storms can also be a problem in that clams fill up with sand particles during turbulent weather when there is a lot of sand in suspension. The clams cannot expel the sand so they come to the surface for oxygen, probably use up their energy reserves, gape, and eventually die. Predation and BRD are additional factors that cause high mortality in cultured clams.

Before harvesting the plots must be sampled to make sure that the clams are of marketable size. In small operations, and in most Italian regions, clams

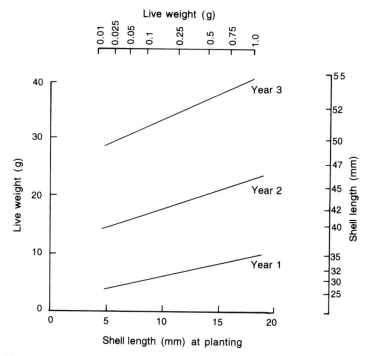

Fig. 9.23. Estimated growth of *Ruditapes philippinarum* spat after one, two and three years in the ground. After Spencer *et al.* (1991).

are harvested manually by raking them out of the substrate. In France where net envelopes are used the sides of the envelope are manually lifted out of the substrate and the net is cleaned in the water with all the clams inside. However, most farms use a mechanical harvester operated off the back of a tractor (Fig. 9.24). In England where the plots are under water growers use either suction or elevator dredges to harvest the clams. Once harvested the clams are stored in boxes or bags and transported for mechanical grading. In France and the UK all cultured clams are sold fresh to local markets and restaurants. In Ireland there is little demand for clams on the home market and so almost all of the harvested clams are exported fresh to France and Spain. In Italy clams are sold on the home market and large quantities are also exported to Spain.

To illustrate the management and regulation of clam culture in Europe the UK will serve as a good example. In the UK the culture of clams and other bivalves is governed by the Sea Fisheries (Shellfish) Act 1967. Licensed farms must register in England with the Department of Environment, Food and Rural Affairs (DEFRA), or in Wales or Scotland with the National Assembly of Wales Agriculture Department (NAWAD) or Scottish Executive Environment and Rural Affairs Department (SERAD) respectively, to assist in dealing with disease outbreaks should they occur (Anon, 1999). There are strict controls on the movement of all live bivalves, eggs and seed stock within and between coastal zones in the UK, and also from other parts of the European Union (EU) and from non-EU countries. Within the UK permission must be

Fig. 9.24. Harvesting clams (*Ruditapes philippinarum*) sown under nets. Photo courtesy of Kevin O Kelly, Lissadell Shellfish Co. Ltd., Sligo, Ireland.

obtained to move bivalves from restricted areas, i.e. where disease has been found, to clean areas (see Chapter 12). There are separate restrictions under the Wildlife and Countryside Act 1981 regarding the release of non-native bivalves, e.g. *Ruditapes philippinarum* or *Crassostrea gigas*, into the wild. For trade within the EU controls operate through a system of approved zones or farms that are free of disease. However, the only listed diseases are bonamiasis and Aber's disease of oysters (*Ostrea edulis*); BRD of clams is not listed. Up to 2002 there were no approved zones or farms within the EU but the UK, Guernsey, Isle of Man, and parts of France were undergoing testing programmes for disease and were being treated as though they were approved zones. Bivalves from non-EU countries may only be deposited in the UK provided they are certified as disease-free by a testing programme as stringent as that which applies to trade within the EU. All bivalve beds are classified according to faecal *E. coli* levels in bivalve flesh as stipulated by the EU Shellfish Directive (EEC, 1991) implemented in England, Wales and Scotland by the Food Safety (Fishery Products and Live Shellfish) (Hygiene) Regulations 1998. Bivalves from Category A beds (<300 of *E. coli* per 100 g flesh) can be marketed without treatment, while those from Category B areas (<4600 *E. coli* per 100 g flesh in 90% of samples) need depuration or relaying in Category A areas or cooking by an EU approved method. Those in Category C areas (<46 000 *E. coli* per 100 g flesh in 100% of samples) need relaying for a minimum period of 2 months in Category A or B areas, or cooking by an EU approved method. Bivalves cannot be marketed from areas where the bacterial count is higher than in Category C areas. The transportation and marketing of live shellfish in the UK is also governed by fishery department regulations. A consignment of bivalves must be accompanied by sufficient information to enable place of origin to be traced. To ensure the health and welfare of bivalves journeys must be kept as short as possible and the vehicle must be cleaned and disinfected at the start of the journey. Bivalves for the live market must be disease-free at time of dispatch and comply with regulations laid down by the local health authorities.

Bivalve culture and the environment

Bivalve production from aquaculture has nearly trebled in the ten years between 1989 and 1999 (Table 9.1). This increased production has been achieved through the expansion of areas of land and water under culture. Fortunately, bivalve culture does not affect the environment in the same way that intensive fish-culture operations do. Indeed, fish farms with their high inputs of waste-water, feeds, fertiliser and chemicals are now counted as potential polluters of the aquatic environment, and the cause of degradation of wetland areas (Pillay, 1992). As a result, more and more restrictions are being imposed on fish-farming ventures. For example, in most countries environmental impact assessments (EIA) are essential before permission is granted from designated authorities to start a fish farm. To date, there is no such requirement for bivalve culture.

If anything, the environment has a more adverse effect on bivalve culture than *vice versa*. For example, blooms of toxic algae blooms are a naturally occurring phenomenon and are not associated with farm waste input into coastal waters. Filter-feeding bivalves accumulate the algal toxins and if the bivalves are consumed can cause paralytic (PSP), amnesic (ASP), diarrhetic (DSP), and neurotoxic (NSP) shellfish poisoning in humans (see Chapter 12 for details). Monitoring programmes for biotoxins are well established in countries such as the United States, Canada, Japan, Tasmania and most of the maritime European countries, but to date there is little, if any, monitoring carried out in developing countries. When bivalves are declared toxic harvesting is prohibited until such time as the bivalves are pronounced safe to eat. Long-term closures obviously have serious repercussions on the livelihood of the growers. For example, in the south-west of Ireland a nine month closure from mid May 1994 to mid February 1995 led to a reduction of nearly 50% of the expected production total of rope grown mussels for that period, a financial loss of about 1.7 million Euro (unpublished report, Task Force on Biotoxin Contamination and Monitoring, 1995). Toxic blooms can have even more drastic, long-term effects. Repeated occurrences of DSP outbreaks forced Norway and Sweden to abandon mussel rearing as a commercial venture for a number of years (Stewart, 1997).

Another adverse impact of the environment on aquaculture is when bivalves are grown in water that is contaminated with pollutants, e.g. trace metals and organochlorines, or with high bacterial or viral concentrations. In the case of bacterial and viral contamination bivalves must be purified before they are marketed, either by relaying them in clean water or by depuration (see Chapter 12 for details). Complete elimination of most bacterial types normally occurs within 48–72 hours. In contrast, viruses are released slowly and at different rates depending on the virus. For example, elimination of hepatitis A virus in mussels is slower than elimination of polio virus, taking about seven days for complete elimination (Enriquez *et al.*, 1992). Relaying or depuration are not used for the elimination of pollutants. Clearance of pollutants is markedly affected by the duration of exposure to the chemical. Elimination is rapid and complete following short-term exposure, but long-term exposure results in slower and often incomplete clearance of the chemical. Elimination rate also depends on the molecular weight and water solubility of the chemical, as well as on environmental factors such as temperature.

Bivalve culture can also be adversely affected by biological waste from the culture operation. For example, after many years of production at the same site Japanese pearl culture operations in Mie province experienced a decline in the size of the average pearl and the inclusion of a pigment that gave a yellowish hue to the pearls, thus reducing their value (Stewart, 1997). Apparently, the deteriorating quality of the pearls was due to feedback from the accumulation on the seabed of biological waste from the pearl oysters themselves. To eliminate the problem culture farms had to be relocated at considerable cost.

The negative effect of fish farming on the environment (see above) is often further compounded by the frequent use of chemicals, which in turn can have serious repercussions when bivalves are cultured in the vicinity of fish cages or pens. A case in point has been the use of tributyl tin (TBT) used as an anti-fouling paint on sea cages, boats, ships etc. At concentrations as low as 1 ppm TBT is lethal to the larvae of many commercially important bivalve species, and at lower concentrations causes shell malformation and reduced shell and tissue growth in larvae and juveniles (Widdows & Donkin, 1992). In 1999, following concern about the damage caused by this toxic chemical, the International Maritime Organisation passed a resolution calling for a complete phase-out of the use of organotins, especially TBT, in ship paints by 2003 and for TBT-free alternatives to be used. At the same time the European Commission rejected draft proposals by Belgium for a national ban on all organotin anti-fouling paints for use on ships. Another chemical, dichlorvos, the active ingredient in the delousing agent Aquagard SLT®, kills bivalve larvae (and adult shrimp and crab) at concentrations that are several orders of magnitude lower than the dosing concentration used on salmon farms (Duggan, 1990). The chemical has not been banned although its use is controlled in EU countries.

Deliberate introductions of non–native (exotic) bivalves may have detrimental effects on local animal and plant species. Species such as *Crassostrea gigas* and *Ruditapes philippinarum*, have proved remarkably successful from the culture viewpoint. However, in Portugal natural populations of introduced *R. philippinarum* have significantly reduced native *R. decussatus* populations (B. Ottway, personal communication 2001). The introduction of *C. gigas* into Europe in the mid 1960s was accompanied by the inadvertent introduction of the seaweed *Sargassum muticum* which has spread along the coasts of France and England forming dense obstructive beds. In the 19th century two predatory gastropods, the oyster drill, *Urosalpinx cinerea*, and the slipper limpet, *Crepidula fornicata*, accompanied the introduction of the oyster *Crassostrea virginica* into Europe (see Andrews, 1980 for details). To avoid problems stemming from indiscriminate introductions strict criteria for selecting species have been drawn up by the International Council for the Exploration of the Sea (ICES), the European Inland Fisheries Advisory Commission (EIFAC) and the American Fisheries Society (AFS). Their procedure for evaluating proposed introductions was published by ICES (1988) and has been updated in ICES (1995).

Another concern regarding the introduction of exotics is the possibility that they will interbreed with or replace indigenous species. In the case of *Crassostrea gigas* there is little danger of this in Europe because the species was either introduced to replace the indigenous specie(s), as in France, or the species does not breed successfully, e.g. the UK and Ireland. In exceptionally warm summers spatfalls of *C. gigas* have been reported in British waters but there is no evidence that self-sustaining populations have developed (Spencer *et al.*, 1994). When *C. gigas* was introduced into the cool southern waters of Australia a successful fishery developed based on hatchery-produced seed. However, the species has spread through deliberate transfers of adults into the warmer waters of Victoria and New South Wales where it has become established and is displacing the native rock oyster, *Saccostrea glomerata* (Holliday & Nell, 1985). This has been exacerbated by the fact that large numbers are killed each year by an undefined disease to which *C. gigas* is totally resistant (J. Nell, personal communication 2001). As for *Ruditapes philippinarum*, the situation is similar to *C. gigas* in that, with the exception of Italy, the species does not breed successfully in Europe and so all seed is hatchery-produced. To avoid possible interbreeding with the native species, *Ruditapes desussatus*, additional precautions have included producing sterile *R. philippinarum* triploids, or growing the species in areas where the native clam is not cultivated (Spencer *et al.*, 1991). With the growth in aquaculture production in the last decade (Table 9.1) more and more countries are adopting control measures to minimise the impact of the industry on the environment. Far from being a threat to the aquaculture industry properly implemented and scientifically based legislation will encourage a healthy, productive and profitable industry, living in harmony with its natural surroundings.

References

Allessandra, G. (ed.) (1990) Tapes philippinarum *Biologia e Sperimentazione*. Ente Sviluppo Agricola Veneto, Venice.

Andrews, J. (1980) A review of introductions of exotic oysters and biological planning for new importations. *Mar. Fish. Rev.*, **42**, 1–11.

Anon. (1999) *A guide to shellfish health controls: an explanation of the controls governing the movement of shellfish, their eggs and gametes into, from and within Great Britain*. Joint publication of Ministry of Agriculture, Fisheries and Food (MAFF), the Welsh Office Agriculture Department (WOAD) and Scottish Office Agriculture, Environment and Fisheries Department (SOAEFD), 26pp.

Barber, B.J. & Blake, N.J. (1991) Reproductive physiology. In: *Scallops: Biology, Ecology and Aquaculture* (ed. S.E. Shumway), pp. 377–428. Elsevier Science Publishers B.V., Amsterdam.

Barnabé, G. (1990) Harvesting micro-algae. In: *Aquaculture*, 2nd edn. Vol. I (ed. G. Barnabé), pp. 207–12. Ellis Horwood, Chichester, UK.

Barnabé, G. (ed) (1994) *Aquaculture: Biology and Ecology of Cultured Species*. Ellis Horwood, Hemel Hempstead, UK.

Bourne, N.F. (2000) The potential for scallop culture – the next millenium. *Aquacul. Internatl.*, **8**, 113–22.

Bourne, N.F. & Hodgson, C.A. (1991) Development of a viable nursery system for scallop culture. In: *An International Compendium of Scallop Biology and Culture* (eds S.E. Shumway & P.A. Sandifer), pp. 273–80. World Aquaculture Society, Louisiana State University, Baton Rouge, Louisiana.

Britton, W. (1991) *Clam Cultivation Manual*, Vol. 8. Taighde Mara Teo and Bord Iascaigh Mhara, Dublin.

Caceres-Martinez, J. & Figueras, A. (1997) The mussel, oyster, clam and pectinid fisheries of Spain. In: *The History, Present Condition, and Future of the Molluscan Fisheries of North and Central America and Europe* (eds C.L. MacKenzie Jr., V.G. Burrell Jr., A. Rosenfield & W.L. Hobart), Vol. 3, pp. 165–90. US Department of Commerce, NOAA Technical Report 129.

Caers, M., Coutteau, P. & Sorgeloos, P. (2000) Incorporation of different fatty acids, supplied as mulsions or liposomes, in the polar and neutral lipids of *Crassostrea gigas* diets. *Aquaculture*, **186**, 157–71.

Castagna, M. & Manzi, J.J. (1989) Clam culture in North America: hatchery production of nursery stock clams. In: *Clam Mariculture in North America* (eds J.J. Manzi & M. Castagna), pp. 111–25. Elsevier Science Publishing Company Inc., New York.

Castagna, M., Gibbons, M.C. & Kurkowski, K. (1996) Culture: application. In: *The Eastern Oyster Crassostrea virginica* (eds V.S. Kennedy, Eble A.F. & Newell R.I.E.), pp. 675–90. Maryland Sea Grant, College Park, Maryland.

Cigarria, J., Fernandez, J. & Magadan, L.P. (1998) Feasibility of biological control of algal fouling in intertidal oyster culture using periwinkles. *J. Shellfish Res.*, **17**, 1167–69.

Coon, S.L., Bonar, D.B. & Weiner, R.M. (1986) Chemical production of cultchless oyster spat using epinephrine and nor-epinephrine. *Aquaculture*, **58**, 255–62.

Coutteau, P. & Sorgeloos, P. (1992) The requirement for live algae and the use of algal substitutes in the hatchery and nursery rearing of bivalve molluscs: an international survey. *J. Shellfish Res.*, **11**, 467–76.

Coutteau, P., Hadley, N., Manzi, J.J. & Sorgeloos, P. (1994) Effect of algal ration and substitution of algae by manipulated yeast diets on the growth of juvenile *Mercenaria mercenaria*. *Aquaculture*, **120**, 135–50.

Dao, J.C., Fleury, P.G., Norman, M., Lake, N., Mikolajunas, J.P. & Strand, O. (1993) *Concerted action scallop sea bed cultivation in Europe*. Report to Specific Community Programme for Research, Technological Development and Demonstration in the Field of Agriculture and Agro-industry, inclusive Fisheries (AIR 2–CT93–1647).

Dao, J.C., Barret, J., Devauchelle, N., Fleury, P.G. & Robert, R. (1996) Rearing of scallops (*Pecten maximus*) in France, from hatchery to intermediate culture: results of a 10-year programme (1983–1993). In: *Improvement of the Commercial Production of Marine Aquaculture Species* (eds G. Gajardo & P. Coutteau), pp. 121–34. Impresora Creces, Santiago, Chile.

Dardignac-Corbeil, M.J. (1990) Traditional mussel culture. In: *Aquaculture*, 2nd edn. Vol. I (ed. G. Barnabé), pp. 285–341. Ellis Horwood, Chichester, UK.

De Valence, P. & Peyre, R. (1990) Clam culture. In: *Aquaculture*, 2nd edn. Vol. I (ed. G. Barnabé), pp. 388–415. Ellis Horwood, Chichester, UK.

Duggan, C. (1990) Chemical usage in aquaculture. In: *Interactions between Aquaculture and the Environment.* (eds P. Oliver & E. Colleran), pp. 23–26. Print World, Dublin, Ireland.

EEC (1991) Council Directive 91/492/EEC: Laying down the health conditions for the production and the placing on the market of live bivalve molluscs. *Official Journal of the European Communities*, No. **L 268/1**, 24. 09.91.

Enriquez, R., Frosner, G.G., Hochsteinmintzel, V., Riedemann, S. & Reinhardt, G. (1992) Accumulation and persistence of hepatitis A virus in mussels. *J. Med. Virol.*, **37**, 174–79.

FAO (2001) *Aquaculture production 1970–1999.* Fishstat Plus (v. 2.30). Food and Agricultural Organisation, United Nations, Rome.

Figueras, A. (1989) Mussel aquculture in Spain and France. *World Aquacul.*, **20**, 8–17.

Ford, S.E. & Tripp, M.R. (1996) Diseases and defense mechanisms. In: *The Eastern Oyster Crassostrea virginica* (eds V.S. Kennedy, Eble, A.F. & Newell R.I.E.), pp. 581–660. Maryland Sea Grant, College Park, Maryland.

Freeman, K. (1988) The hanging gardens of Mutsu Wan. *Seafood Leader*, Spring volume, 122–36.

Fuentes, J. & Molares, J. (1994) Settlement of the mussel *Mytilus galloprovincialis* on collectors suspended from rafts in the Ria de Arousa (NW of Spain): annual pattern and spatial variability. *Aquaculture*, **122**, 55–62.

Fuentes, J., Gregorio, V., Giráldez, R. & Molares, J. (2000) Within-raft variability of the growth rate of mussels, *Mytilus galloprovincialis*, cultivated in the Ría de Arousa (NW Spain). *Aquaculture*, **189**, 39–52.

Fujiya, M. (1970) Oyster farming in Japan. *Helgo. Wiss. Meeresunters*, **20**, 464–79.

Gatesoupe, F.J. (1999) The use of probiotics in aquaculture. *Aquaculture*, **180**, 147–65.

Gibson, L.F., Woodworth, J. & George, A.M. (1998) Probiotic activity of *Aeromonas media* on the Pacific oyster, *Crassostrea gigas*, when challenged with *Vibrio tubiashii*. *Aquaculture*, **169**, 111–20.

Goulletquer, P. & Héral, M. (1997) Marine molluscan production trends in France: from fisheries to aquaculture. In: *The History, Present Condition, and Future of the Molluscan Fisheries of North and Central America and Europe* (eds C.L. MacKenzie Jr., V.G. Burrell Jr., A. Rosenfield & W.L. Hobart), Vol. 3, pp. 137–64. US Department of Commerce, NOAA Technical Report 129.

Grizel, H. & Héral, M. (1991) Introduction into France of the Japanese oyster (*Crassostrea gigas*). *J. Cons. int. explor. Mer*, **47**, 399–403.

Guo, X. & Luo, Y. (2003) Scallop culture in China, in press.

Guo, X., Ford, S.E. & Zhang, F. (1999) Molluscan aquaculture in China. *J. Shellfish Res.*, **18**, 19–31.

Gwo, J.C. (1995) Cryopreservation of oyster (*Crassostrea gigas*) embryos. *Theriogenology*, **43**, 1163–74.

Harvey, M., Bourget, E. & Gagné, N. (1997) Spat settlement of the giant scallop, *Placopecten magellanicus* (Gmelin, 1791), and other bivalve species on artificial filamentous collectors coated with chitinous material. *Aquaculture*, **148**, 277–98.

Heasman, M.P., O Connor, W.A. & Frazer, A.W. (1995) *Evaluation of hatchery production of scallops.* Final report to Fisheries Research and Development, Deakin, ACT, Australia. 191pp.

Helm, M.M. (1990) Hatchery design and general principles of operation and management and new developments. In: Tapes philippinarum *Biologia e Sperimentazione* (ed. G. Allessandra), pp. 63–87. Ente Sviluppo Agricola Veneto, Venice.

Helm, M.M. & Pellizzato, M. (1990) Hatchery breeding and rearing of the *Tapes philippinarum* species. In: Tapes philippinarum *Biologia e Sperimentazione* (ed. G. Allessandra), pp. 115–40. Ente Sviluppo Agricola Veneto, Venice.

Héral, M. (1990) Traditional oyster culture in France. In: *Aquaculture*, 2nd edn. Vol. I (ed. G. Barnabé), pp. 342–87. Ellis Horwood, Chichester, UK.

Héral, M. & Deslous-Paoli, J.M. (1991) Oyster culture in European countries. In: *Estuarine and Marine Bivalve Mollusk Culture* (ed. W. Menzel), pp. 153–90. CRC Press Inc., Boca Raton, Florida.

Hickman, R.W. (1992) Mussel Cultivation. In: *The Mussel* Mytilus: *Ecology, Physiology, Genetics and Culture* (ed. E.M. Gosling), pp. 465–510. Elsevier Science Publishers, B.V., Amsterdam.

Holliday, J.E. & Nell, J.A. (1985) Concern over Pacific oysters in Port Stephens. *Aust. Fish.*, **44**, 29–31.

Hurlburt, C.G. & Hurlburt, S.W. (1980) European mussel culture technology and its adaptability to North American waters. In: *Mussel Culture and Harvest: a North American Perspective* (ed. R.A. Lutz), pp. 69–98. Elsevier Science Publishing Co. Inc., New York.

Ibarra, A.M., Cruz, P. & Romero, B.A. (1995) Effects of inbreeding on growth and survival of self-fertilized catarina scallop larvae, *Argopecten circularis. Aquaculture*, **134**, 37–47.

ICES (1988) *Codes of Practice and Manual of Procedures for Consideration of Introductions and Transfers of Marine and Freshwater Organisms.* ICES Co-operative Research Report, No.159, 44pp.

ICES (1995) *Code of Practice on the Introductions and Transfers of Marine Organisms.* Annex 3, Report of the Advisory Committee on the Marine Environment, 1994. ICES Co-operative Research Report, No. 204.

Ito, H. (1991) Fisheries and aquaculture: Japan. In: *Scallops: Biology, Ecology and Aquaculture* (ed. S.E. Shumway), pp. 1017–55. Elsevier Science Publishers B.V., Amsterdam.

Ito, S. (1991) *Patinopecten (Mizuhopecten) yessoensis* (Jay) in Japan. In: *Estuarine and Marine Bivalve Mollusk Culture* (ed. W. Menzel), pp. 211–25. CRC Press Inc., Boca Raton, Florida.

Jefferds, I. (2000) New 'Mussel Disc' is a great mussel saver. *Shellfish News*, **9**, 7–8.

Kinoshita, T. (1935) A test for natural spat collection of the Japanese scallop. *Rep. Hokkaido Fish. Res. Stn.*, **273**, 1–8. (In Japanese).

Laing, I. & Psimopoulos, A. (1998) Hatchery cultivation of king scallop (*Pecten maximus*) spat with cultured and bloomed algal diets. *Aquaculture*, **169**, 55–68.

Lavens, P. & Sorgeloos, P. (eds) (1996) *Manual on the production and use of live food for aquaculture.* FAO Fisheries Technical Paper. No. 361. Rome, FAO. 295pp.

Le Borgne, Y. (1990) Culture of micro-algae. In: *Aquaculture*, 2nd edn. Vol. I (ed. G. Barnabé), pp. 197–206. Ellis Horwood, Chichester, UK.

Luo, Y. (1991) Fisheries and aquaculture: China. In: *Scallops: Biology, Ecology and Aquaculture* (ed. S.E. Shumway), pp. 809–24. Elsevier Science Publishers B.V., Amsterdam.

Lutz, R.A. (1980) Pearl incidences: mussel culture and harvest implications. In: *Mussel Culture and Harvest: a North American Perspective* (ed. R.A. Lutz), pp. 193–222. Elsevier Science Publishing Co. Inc., New York.

Lutz, R.A., Chalermwat, K., Figueras, A.J., Gustafson, R.G. & Newell, C. (1991) Mussel aquaculture in marine and estuarine environments throughout the world. In: *Estuarine and Marine Bivalve Mollusk Culture* (ed. W. Menzel), pp. 57–97. CRC Press Inc., Boca Raton, Florida.

Manzi, J.J. & Castagna, M. (1989) Nursery culture of clams in North America. In: *Clam Mariculture in North America* (eds J.J. Manzi & M. Castagna), pp. 127–47. Elsevier Science Publishing Company Inc., New York.

Martinez, G., C, L., Uribe, E. & Diaz, M.A. (1995) Effects of different feeding regimens on larval growth and the energy budget of juvenile Chilean scallops, *Argopecten purpuratus* Lamarck. *Aquaculture*, **132**, 313–23.

Martinez, G., Aguilera, C. & Mettifogo, L. (2000) Interactive effects of diet and temperature on reproductive conditioning of *Argopecten purpuratus* broodstock. *Aquaculture*, **183**, 149–59.

Mattei, N. & Pellizzato, M. (1997) Mollusc fisheries and aquaculture in Italy. In: *The History, Present Condition, and Future of the Molluscan Fisheries of North and Central America and Europe* (eds C.L. MacKenzie Jr., V.G. Burrell Jr., A. Rosenfield & W.L. Hobart), Vol. 3, pp. 201–16. US Department of Commerce, NOAA Technical Report 129.

Pérez-Comacho, A., Albentosa, M., Fernández-Reiriz, M.J. & Labarta, U. (1998) Effect of microalgal and inert (cornmeal and cornstarch) diets on the growth performance and biochemical composition of *Ruditapes decussatus* seed. *Aquaculture*, **160**, 89–102.

Petit, J. (1990) Water supply, treatment, and recycling in aquaculture. In: *Aquaculture*, 2nd edn. Vol. I (ed. G. Barnabé), pp. 63–196. Ellis Horwood, Chichester, UK.

Pillay, T.V.R. (1992) *Aquaculture and the Environment*. Fishing News Books, Oxford.

Pillay, T.V.R. (1993) *Aquaculture: Principles and Practices*. Fishing News Books, Oxford.

Quayle, D.B. & Newkirk, G.F. (1989) *Farming Bivalve Molluscs: Methods for Study and Development*. The World Aquaculture Society and the International Development Research Centre, Baton Rouge, Louisiana.

Renard, P. (1991) Cooling and freezing tolerance in embryos of the Pacific oyster, *Crassostrea gigas*: methanol and sucrose effects. *Aquaculture*, **92**, 43–57.

Riquelme, C., Araya, R. & Escribano, R. (2000) Selective incorporation of bacteria by *Argopecten purpuratus* larvae: implications for the use of probiotics in culturing systems of the Chilean scallop. *Aquaculture*, **181**, 25–36.

Riquelme, C.E., Jorquera, M.A., Rojas, A.I., Avendano, R.E. & Reyes, N. (2001) Addition of inhibitor-producing bacteria to mass cultures of *Argopecten purpuratus* larvae (Lamarck, 1819). *Aquaculture*, **192**, 111–19.

Rossi, R. & Paesanti, F. (1992) Rearing grooved carpet shell clams (*Tapes decussates* and *T. philippinarum*): production and markets in Europe and in the Mediterranean basin. *Il Pesce*, **3**, 25–29.

Sauriau, P.G., Haure, J. & Baud, J.P. (1997) Sprinkling: a new method of distributing live algae food in marine coastal ponds used for Manila clam *Tapes philippinarum* (Adams & Reeve) intensive culture. *Aquacul. Res.*, **28**, 661–69.

Sindermann, C. (1990) *Principal Diseases of Marine Fish and Shellfish*, 2nd edn. Vol. II. Academic Press, San Diego, California.

Sindermann, C.J. & Lightner, D.V. (eds) (1988) *Disease Diagnosis and Control in North American Marine Aquaculture* 2nd edn. Vol. 17. Elsevier Science Publishers B.V., Amsterdam.

SOFIA (2000) *The State of World Fisheries and Aquaculture*. 1. World review of fisheries and aquaculture. Food and Agricultural Organisation, United Nations, Rome.

Soletchnik, P., Le Moine, O., Goulletquer, P. *et al.* (2001) Optimisation of the traditional Pacific cupped oyster (*Crassostrea gigas* Thunberg) culture on the French Atlantic coastline: autumnal fattening in semi-enclosed ponds. *Aquaculture*, **199**, 73–91.

Soudant, P., Marty, Y., Moal, J. *et al.* (1996) Effect of food fatty acid and sterol quality on *Pecten maximus* gonad composition and reproduction process. *Aquaculture*, **143**, 361–78.

Southgate, P.C., Lee, P.S. & Nell, J.A. (1992) Preliminary assessment of a microencapsulated diet for larval culture of the Sydney rock oyster, *Saccostrea commercialis* (Iredale & Roughley). *Aquaculture*, **105**, 345–52.

Spencer, B.E. (1990) *Cultivation of Pacific oysters*. Lab. Leafl., MAFF Direct. Fish. Res., Lowestoft (63): 47pp.

Spencer, B.E., Edwards, D.B. & Millican, P.F. (1991) *Cultivation of Manila clams*. Lab. Leafl., MAFF Direct. Fish. Res., Lowestoft (65): 29pp.

Spencer, B.E., Edwards, D.B., Kaiser, M.J. & Richardson, C.A. (1994) Spatfalls of the non-native Pacific oyster, *Crassostrea gigas*, in British waters. *Aquatic Conserv.: Mar. Freshw. Ecosyst.*, **4**, 203–17.

Stewart, J.E. (1997) Environmental impacts of aquaculture. *World Aquacul.*, **28**, 47–52.

Thompson, P.A., Guo, M.X. & Harrison, P.J. (1996) Nutritional value of diets that vary in fatty acid composition for larval Pacific oysters (*Crassostrea gigas*). *Aquaculture*, **143**, 379–91.

Tseng, C.K. (1993) Notes on mariculture in China. *Aquaculture*, **111**, 21–30.

Utting, S.D. & Spencer, B.E. (1991) *The hatchery culture of bivalve mollusc larvae and juveniles*. Lab. Leafl., MAFF Direct. Fish. Res., Lowestoft (68): 31pp.

Utting, S.D. & Millican, P.F. (1998) The role of diet in hatchery conditioning of *Pecten maximus* L: a review. *Aquaculture*, **165**, 167–78.

Vakily, J.M. (1989) The biology and culture of mussels of the genus *Perna*. ICLARM Stud. Rev., **17**, 1–63.

Walne, P.R. (1974) *Culture of Bivalve Molluscs: 50 Years' Experience at Conwy*. Fishing News Books, Oxford.

Widdows, J. & Donkin, P. (1992) Mussels and environmental contaminants: bioaccummulation and physiological aspects. In: *The Mussel* Mytilus: *Ecology, Physiology, Genetics and Culture* (ed. E.M. Gosling), pp. 383–424. Elsevier Science Publishers B.V., Amsterdam.

Wikfors, G.H., Patterson, G.W., Ghosh, P., Lewin, R.A., Smith, B.C. & Alix, J.H. (1996) Growth of post-set oysters, *Crassostrea virginica*, on high-lipid strains of algal flagellates *Tetraselmis* spp. *Aquaculture*, **143**, 411–19.

Wilson, J.A., Chaparro, O.R. & Thompson, R.J. (1996) The importance of broodstock nutrition on the viability of larvae and spat in the Chilean oyster *Ostrea chilensis*. *Aquaculture*, **139**, 63–75.

Yangson, K. & Moyse, J. (1991) Cryopreservation of the spermatozoa of *Crassostrea tulipa* and 3 other oysters. *Aquaculture*, **97**, 259–67.

10 Genetics in Aquaculture

Introduction

For about 10 000 years humans have modified useful species over many generations and as a result, through a process of artificial selection, we now have thousands of varieties of crop plants and domesticated livestock. The early attempts at domestication were achieved without any real knowledge of genetic principles and concepts. Nevertheless, the procedures employed then still form the basis of genetic management programmes today: breeding from selected parents, culling inferior offspring, crossbreeding with other species and producing and crossing different strains within a species (Wilkins, 1981).

The application of these procedures to aquatic organisms is still in its infancy. Up to a few decades ago, aquaculture management procedures were largely concerned with the physical constraints inherent in rearing organisms in restricted bodies of water. With husbandry problems largely sorted for a wide variety of fish, bivalve and crustacean species, attention is now increasingly focused on the application of genetic techniques, such as selective breeding, chromosome manipulation and, more recently, gene transfer, to the aquaculture industry. Several features of the life cycle make aquatic organisms more amenable than domesticated livestock to these procedures, e.g. shorter generation time, higher fecundity, external fertilisation and larval development, and more plastic sex determination. In addition, the amount of variation for genetic manipulation is greater in fish and bivalves than in domesticated livestock, which are already considerably improved by a long history of artificial selection. It should be pointed out at this stage that because of their higher market value genetic advances in aquaculture are much further ahead for fish than for bivalve species.

The first part of this chapter deals with quantitative genetics and the various types of breeding schemes used in the selective breeding of bivalves. Protein and DNA molecular markers and their use in quantifying genetic variability will then be discussed, as will the technology of chromosome manipulation in the production of triploids, tetraploids and gynogens. More detailed information on methods and applications can be found in Wilkins & Gosling, 1983; Gall & Busack, 1986; Gjedrem, 1990; Gall & Chen, 1993; Doyle et al., 1996; McAndrew & Penman, 1999; Benzie, 2002.

Quantitative genetics

Populations exhibit an enormous amount of variation and much of this is inherited. In a selective breeding programme the success of selection depends upon the extent to which the variation for a particular trait in the population from which the parents were chosen is inherited. The traits of most interest, from a production viewpoint, are growth rate, survival, meat yield and disease resistance. Quantitative traits, as they are called, exhibit continuous variation in populations and their variance usually reflects both genetic and environmental influences. Typically these traits are controlled by a large number of genes, each with a small effect.

Genetic analysis of quantitative traits involves working with means and variances, trying to partition the variance of the trait (V_P) into the variance due to genes (V_G), to the environment (V_E) and to interaction between the two ($V_G \times V_E$):

$$V_P = V_G + V_E + V_{G+E}$$

The genetic component can itself be partitioned into additive and non-additive components:

$$V_G = V_A + V_D + V_I$$

Additive effects result from the cumulative contribution of alleles at all the loci governing a quantitative trait, and as such are important because they contribute to the breeding value of individuals and are passed on to progeny in a predictable manner. Non-additive genetic effects are due to dominance and epistasis. Dominance effects (V_D) result from interactions among alleles at the same gene locus, while epistatic effects (V_I) are due to interactions among loci. Neither of these is passed on to progeny, due to segregation of parental alleles at meiosis.

The ratio of the additive genetic variance to the total phenotypic variance for a trait is called the heritability, denoted by h^2.

$$h^2 = V_A/V_P$$

One of the simplest methods for estimating the heritability of a trait is to compare the mean phenotypic value of full-sibs (individuals that share the same two parents) to the mean phenotypic value of their parents (mid-parent mean) in a regression analysis. In a randomly mating population the slope of the line obtained from a regression of full-sib means on mid-parent means is the heritability (Fig. 10.1). Values of h^2 may theoretically range between zero, where V_P is entirely due to environmental effects, and one, where all the variance is due to additive genetic effects. The response to selection of a particular trait can be predicted from the heritability estimate and the phenotypic variance, and selection methods are chosen on the basis of these values. Table 10.1 presents h^2 estimates for some quantitative traits in bivalves.

Various factors of which one must be aware when estimating heritability have been summarised in Beaumont (1994 & 2000): estimates are only valid for the population and specific environmental conditions in which measurements are made; there are large standard errors associated with h^2 estimates unless large numbers of individuals are used; frequently, different traits are correlated so that an increase in value for one trait will be accompanied by a decrease in value for another.

Selective breeding

The four objectives in a breeding plan are to: ascertain the breeding objectives, identify the broodstock, choose a mating scheme for the reproduction

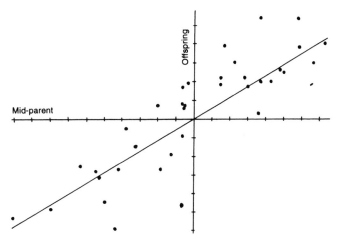

Fig. 10.1. Estimation of heritability by regression of the mean phenotypic value of offspring on the mean phenotypic value of their parents (mid-parent mean). Each point on the graph represents the mean value of one pair of parents (horizontal axis), and the mean value of their offspring (vertical axis). Heritability is estimated from the slope of the regression line and is 0.577 ± 0.07 in this example (*Drosophila* wing length). From Falconer (1960) and reprinted by permission of Pearson Education Ltd.

Table 10.1. Heritability estimates (h^2) for various quantitative traits in bivalves.

Species	Trait	h^2	Reference
Crassostrea gigas	Meat weight (18 mo) Larval survival	0.37 ± 0.06 0.31 ± 0.06	Lannan, 1972
C. virginica	Larval growth rate Spat growth rate	0.24 0.93	Longwell, 1976
Ostrea chilensis	Live weight (8–27 mo)	0.43 ± 0.18 – 0.69 ± 0.11	Toro *et al.*, 1995
O. edulis	Live weight (19 mo) Shell height (19 mo)	0.24 ± 0.20 0.19 ± 0.07	Toro & Newkirk, 1990
Pinctada fucata martensii	Shell width Shell concavity	0.13–0.47 0.13–0.37	Wada, 1984, 1986
Mercenaria mercenaria	Growth rate at 2 yr Juvenile growth rate	0.42 ± 0.01 0.37	Hadley *et al.*, 1991 Rawson & Hilbish, 1990
Mytilus edulis	Larval growth rate	0.12–0.62	Newkirk, unpublished
Argopecten ventricosus	Live weight at 1 yr Shell width at 1 yr	0.33 ± 0.08–0.59 ± 0.13 0.10 ± 0.07–0.18 ± 0.08	Ibarra *et al.*, 1999
A. irradians	Growth rate at 12 mo	0.21	Crenshaw *et al.*, 1991

of selected broodstock, and identify how the selected stock can be expanded for production purposes (Gall, 1990).

Breeding objectives

Breeding objectives are generally set by the industry and the consumer, and should be precisely defined at the start of a breeding programme. Frequently, the objectives differ from one species to the next, from one country to another, and even between different regions of a country. In a questionnaire on 22 prospective breeding goals in oysters Mahon (1983) found that all respondents gave a high rating for growth rate and survival from settlement to market size. Despite this consensus some countries gave a high rating to additional traits. For example, Irish and Spanish respondents working with the oyster *Ostrea edulis* gave a high score to proportion of larvae that settle, while those from north-west Europe who worked with the oyster *Crassostrea gigas* assigned a high score to resistance to low temperature. Not surprisingly, respondents from North America working with *Crassostrea virginica* gave particularly high scores to resistance to disease (see below). Respondents suggested additional prospective breeding goals, e.g. fast larval growth, good food conversion efficiency, ability to use non-algal feeds, shell hardiness to withstand handling, good appearance and flavour of meat. Almost 20 years later the main breeding objectives have not changed, as is evidenced from a recent report from a workshop on genetic improvement in the Australian aquaculture industry (Lymbery, 2000; Table 10.2).

Table 10.2. Biological traits included in the breeding objective for different aquaculture species groups (freshwater and marine crustaceans, finfish and edible molluscs (oysters *Crassostrea gigas*, *Saccostrea commercialis* and abalone *Haliotis* spp.). Participants (N =61) in the survey were split into four groups depending on their major commercial or research interest. Ranking refers to the number of groups (out of four) that placed that trait in the breeding objective. From Lymbery (2000).

Trait	Ranking
Size at harvest	4
Survival to harvest	4
Meat yield at market	3
Feed efficiency	3
Size uniformity	2
Disease resistance	2
Taste	2
Flesh colour	2
Reproductive output	1
Temperature tolerance	1
Survival to (live) market	1
Shell shape	1
Claw size	1
Peelability	1

Establishing broodstock

It is inevitable that some of the natural variability of the species is lost when individuals are chosen as broodstock, since they represent but a small fraction of the population. In addition, over the following generations the offspring of these parents will be bred together, with possible inadvertent loss of variation through inbreeding (see below). To avoid these pitfalls Newkirk (1993) has recommended that for each generation an equal sex ratio be used, and that there should be fifty pairs of adults from each stock or line. He estimated that this strategy would result in an inbreeding rate of 0.5% per generation and a total accumulation of inbreeding after 5, 10 and 20 generations of 1%, 3% and 5%, respectively. Inbreeding can be minimised within each stock or line by using a crossing system, such as rotational line crossing (see later). This involves crossing the females of one line with the males of another, using three or more lines, in a sequential manner each generation.

Breeding schemes

Several different breeding techniques will be covered in this section. Some of these, e.g. individual selection and family selection, make use of additive genetic variation for improvement of stocks, while others such as hybridisation make use of non-additive genetic variation. Techniques such as line crossing and out-crossing can be used to reduce inbreeding in broodstock.

Much of the information in the following section comes from Kapuscinski & Jacobson (1987) and readers are referred to this source, and to Gjedrem (1983, 1992), Gall (1990), Tave (1993) and Falconer & Mackay (1996) for additional information.

Individual selection

Individual selection (mass selection) is the simplest form of artificial selection and in many cases yields the most rapid response. The best individuals are selected from the population for breeding and the rest are discarded (Fig. 10.2A). This can be repeated in each generation until the desired phenotypic change is achieved. The difference between the mean phenotypic value of the population and the mean phenotypic value of those individuals selected for breeding is called the selection differential, S (Fig. 10.3). The response to selection (R) is the difference between the mean of the progeny from the selected parents and the mean of the original population, and is given by:

$$R = h^2 S$$

It is important that the heritability value for the trait under selection has been determined in the same environment and for the same population on which selection is being carried out. Otherwise, the predicted response to selection may be unreliable. Individual selection works best for traits with high heritability (>0.30), and where the population is large so that large selection differentials can be used.

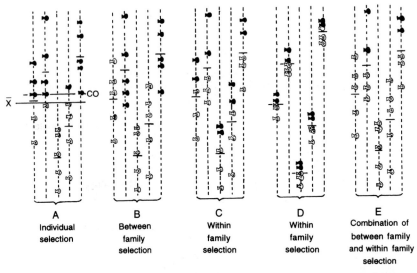

A	B	C	D	E
Individual selection	Between family selection	Within family selection	Within family selection	Combination of between family and within family selection

Fig. 10.2.   Schematic representation of different selection methods. There are 5 families (represented by vertical dashed lines) each with 5 individuals: A, B, C & E represent identical arrangements of the same 25 individuals. A small crossbar shows the mean of each family. Individuals are plotted on a vertical scale of merit; the black individuals, those with the best phenotypic values, are selected, while the rest are culled. In individual selection (A) family means are ignored and individuals are compared to the population mean ($\bar{X}$). A cut-off (CO) value is arrived at based on the population mean and all individuals equal to, or larger, than the CO value are selected, and the rest are culled. In between-family selection (B) the population mean is ignored and only families (two) with the highest mean phenotypic values are selected. The remaining families (three) are culled. In within-family selection (C & D) the population mean is ignored, and individuals within each family are compared only to their family mean. A proportion of the best individuals are selected (2/5 in C & D) and the rest are culled. This method of selection is most useful in situations where the variation between families is large compared to the variation within families, as in D. Between-family and within-family selection (E) may be combined; the best families are initially selected and then the best individuals within those families are selected. Modified from Tave (1993) based on Falconer (1960). Reprinted by permission of Pearson Education Ltd.

Family selection

When the heritability for a trait is low (<0.3) the phenotype of an individual is not an accurate estimate of the breeding value. Then it is more efficient in terms of gain per generation to base selection on family performance rather than on individual performance. Entire families, usually groups of full–sibs or half–sibs (individuals that share one parent) are selected for breeding, based on their mean phenotypic value (Fig. 10.2B). Environmental variance among family means is kept low by raising families in the same environment, and by averaging over a large number of individuals in calculating the family mean. If a reasonable intensity of selection is to be achieved, and at the same time inbreeding is to be avoided, then the number of families raised and measured has to be two to four times greater than the number of families selected for breeding (Falconer, 1960). Until recently, family selection was costly of space, because each family had to be maintained separately. The use of highly variable microsatellite DNA markers allows the genetic tagging of individuals

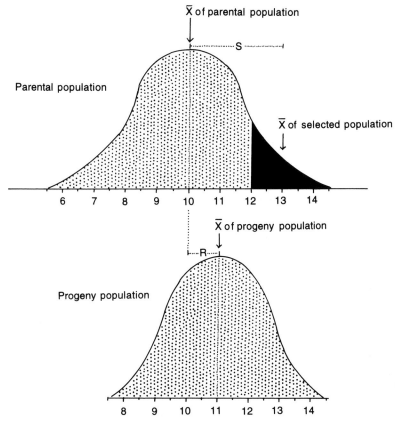

Fig. 10.3. Response to individual (mass) selection. A population of fish have a mean weight of 10 g at six months of age. The adults selected for breeding (dark section of parental distribution) have a mean weight of 13 g at six months of age. The selection differential $S = 13 - 10 = 3$ g. The heritability for weight at six months of age is 0.33, therefore the predicted response to one generation of selection is $R = h^2S = 3 \times 0.33 = 1$ g. From Kapuschinski & Jacobson (1987).

within a specific family (see later), which means that many families may now be reared communally.

Within-family selection

This type of selection is useful when phenotypic differences among families are mainly due to environmental factors. The best individuals, those that exceed the mean phenotypic value of the family to which they belong, are selected for breeding (Fig. 10.2C & D). If two members of every family are chosen to be parents of the next generation then the variance of family size is zero, and provided that the sexes are equal in numbers, $N_e = 2N$ (Falconer, 1960). This means that the effective population size N_e (the number of breeding individuals in the population) is twice the actual, and thus the rate of inbreeding is half that under individual selection.

It is possible to combine family selection with within-family selection by choosing the best individuals from the best families for breeding (Fig. 10.2E).

The advantage of combined selection is an increased response because the additive genetic variance among individuals, as well as within families is exploited. Gall (1990) has advised that 'combined selection should nearly always be the breeding plan of choice for aquacultural production'.

Progeny and sibling selection

Progeny selection is widely used in animal breeding but is less often used in fish and bivalve breeding. The method involves selecting individuals for breeding on the mean phenotypic value of their progeny. This is a form of family selection, the families in this case being groups of progeny, either full or half-sibs. The advantage of this selection method is that relatively small families can be used to determine which families should be used as broodstock. To avoid inbreeding the size of the best families can be increased by allowing the original parents to continue breeding. A serious drawback to progeny selection is a slower response to selection per unit of time because the parents cannot be selected until the progeny have been measured.

Inbreeding, crossbreeding and hybridisation

Inbreeding is defined as the mating of individuals who are more closely related to each other than individuals mating at random within a population (Kincaid, 1983). Pairs mating at random are more closely related to each other in a small population than in a large one. In fish and bivalve hatcheries the populations used are finite populations, and it is therefore not surprising that inbreeding can occur especially if too few parents are used as broodstock, or if the sex ratio departs from 1:1. The extent of inbreeding in a population is measured by the coefficient of inbreeding, F, which is the probability that uniting gametes contain identical alleles (alternative forms of a gene) derived from a common ancestor. The value of F increases, as alleles are lost from the population through inbreeding or genetic drift (random changes in allele frequencies through natural sampling errors that occur in each generation). Values of F range from 0 to 1; an F value of 0 indicates no inbreeding and therefore maximum heterozygosity (genetic variability) in the population, while a value of 1 indicates complete inbreeding and total homozygosity. When broodstock is first established in the hatchery it is, by convention, considered to have an inbreeding coefficient of zero. After that, the inbreeding coefficient should be calculated for subsequent generations to track inbreeding and drift. The rate of inbreeding (ΔF) depends on population size:

$$\Delta F = 1/(2N_e)$$

N_e is the effective population size and can often be considerably less than the actual population size. The calculated ΔF value for each generation is added to the inbreeding coefficient of the preceding generation to yield the new inbreeding coefficient.

The primary effect of inbreeding is an increase in the level of homozygosity in a population (Table 10.3). This is usually detrimental since many recessive deleterious alleles, which normally remain concealed by the presence of

Table 10.3. Effects of inbreeding on genotype frequencies at a single locus. Note that inbreeding increases homozygosity, but does not change allele frequencies. The following matings occur each generation: AA × aa, Aa × Aa and aa × aa. Due to rounding error genotype frequencies do not always sum to 1.0 in a given generation. From Tave (1993).

Generation	Genotype frequency			Allele frequency	
	AA	Aa	Aa	A	a
P_1	0.25	0.5	0.25	0.5	0.5
F_1	0.375	0.25	0.375	0.5	0.5
F_2	0.4375	0.125	0.4375	0.5	0.5
F_3	0.46875	0.0625	0.46875	0.5	0.5
F_4	0.48437	0.03125	0.48437	0.5	0.5
F_5	0.49218	0.015625	0.49218	0.5	0.5
F_6	0.49609	0.007812	0.49609	0.5	0.5
F_7	0.49804	0.003906	0.49804	0.5	0.5
F_8	0.49902	0.001953	0.49902	0.5	0.5
F_9	0.49951	0.000976	0.49951	0.5	0.5
F_∞	0.5	0.0	0.5	0.5	0.5

their dominant allele, are expressed. Inbreeding results in what is known as inbreeding depression, normally measured as the average performance difference between an inbred population and the base population for a particular trait(s). The deleterious effects of inbreeding depression in domesticated species have long been recognised, but it is only in the last two to three decades that its effects in fish, and to a much lesser extent in bivalves, have been documented. Inbreeding depression in fish species is characterised by increased fry abnormalities, reduced survival, reduced growth rate and lowered reproductive success (references in Kincaid, 1983). For example, one generation of brother-sister mating in rainbow trout (*Oncorhynchus mykiss*[= *Salmo gairdneri*]) produced an increase in fry deformities (38%), and decreased fry survival (19%), food conversion efficiency (6%) and fish weight (up to 20%). After two generations even greater differences were recorded: fry deformities increased to 191%, fry survival decreased to 30% and fish growth decreased by up to 33% (Kincaid, 1976a,b).

The data on inbreeding depression in bivalve species are sparse and results are not as clear-cut as for fish. In the oyster *Crassostrea virginica* Longwell & Stiles (1973) found that progeny from full-sib matings produced significantly lower survival of larvae to metamorphosis, and higher frequencies of larval abnormalities, than outbred controls. However, Lannan (1980) found no evidence of inbreeding depression in *Crassostrea gigas* through two generations of inbreeding. Mallet & Haley (1983) working with *C. virginica* actually found higher larval survival rates in inbred versus outbred families, although adults exhibited increased growth performance in the outbred families. Self-fertilisation, an extreme form of inbreeding, was also performed in this species, with no obvious ill effects. However, self-fertilisation in *Pecten maximus* did produce inbreeding depression in larval growth rate after only one generation of selfing (Beaumont & Budd, 1983). Wada & Komaru (1994) also found

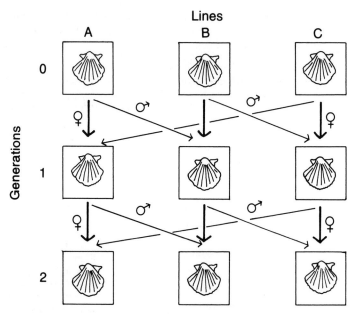

Fig. 10.4. Rotational line-crossing scheme using three separate lines A, B, and C. Each box represents a group of scallops from a separate line. Bold vertical arrows show the source of females used for breeding, while thin diagonal arrows show the source of males mated with those females. It is essential that a high level of genetic diversity is present in the starting broodstock. Modified from Kapuschinski & Jacobson (1987) based on Kincaid (1977).

evidence of inbreeding depression for growth and survival of inbred versus crossbred lines of the pearl oyster *Pinctada fucata martensii*.

Although the results are conflicting there is some evidence for inbreeding depression in bivalves. Therefore, until further information is available, the best management strategy is to adopt breeding methods that will minimise inbreeding. There are three general methods by which this can be achieved: the use of large random mating populations, the use of systematic line crossing schemes to eliminate the mating of close relatives, and strain crossing to produce hybrid populations. It is virtually impossible to have a truly random mating broodstock under hatchery conditions, but in order to keep the inbreeding rate to a low level of <1% a minimum number of 50 adult pairs is recommended. Gaffney *et al.* (1992) recommend multiple small spawnings with subsequent pooling of larvae, rather than a single mass spawning with many ripe adults.

Line crossing involves breeding individuals from one line or strain to superior individuals from unrelated lines or strains. This serves to minimise further inbreeding and genetic drift within lines, and can bring new genes into a selection programme. Rotational line crossing is one such mating scheme (Fig. 10.4). Sometimes intentional inbreeding is carried out to generate inbred, completely homozygous lines. When two such lines, homozygous for different alleles at every gene locus, are crossed, the progeny are completely heterozygous at all loci, and are genetically identical. Such progeny are genetically superior to the parents and display what is known as 'hybrid vigour' or heterosis. This method has been very successful with plants and is responsible for the production of many of the hybrid seed cereal crops grown world-wide today. Its success with animals has been less notable, primarily due to lines dying out from inbreeding depression before a reasonably high level of inbreeding has been attained (Falconer, 1960). Consequently, animal breeders make extensive use of crossbreeds, produced by cross-mating selected populations, breeds or, very infrequently, inbred lines. A large proportion of the poultry, pigs, sheep and beef cattle currently used for the commercial production of meat, eggs and wool are crossbreeds (Bowman, 1974). To date, there has been little interest by the aquaculture industry in the production of inbred lines for crossbreeding. However, Bayne *et al.*, (1999) have shown that when inbred lines of the oyster *Crassostrea gigas* were crossed the hybrid offspring had higher rates and efficiencies of feeding and growth than the inbred parents.

An alternative technique for producing offspring with hybrid vigour is to cross individuals from strains, or closely related species or subspecies. This method has been extensively employed by fish aquaculturists but has not, to date, been fully explored by bivalve breeders. Results from the few studies that have been published are, in general, not encouraging. When different geographically isolated populations of *Crassostrea virginica* were crossed, heterosis for early larval growth was observed but was not present in older larvae (Mallet & Haley, 1984). Kraeuter *et al.* (1984) crossed different subspecies of *Argopecten irradians* from the east coast of the United States, but found no evidence for heterosis in growth rate. Crosses between stocks of *Mercenaria mercenaria* do not produce superior offspring (Manzi *et al.*, 1991) although, surprisingly, the hybrid from the cross between *M. mercenaria* and *M. campechiensis* was reported to show improved growth rates and tolerance to a wider range of environmental variables than either species (Menzel, 1962, 1989). However, more recent evidence has shown that the hybrids show growth characteristics that are intermediate to the two parent species (Arnold *et al.*, 1998). The hybrid between *Crassostrea gigas* and *C. virginica* showed an increased growth rate over *C. virginica* (Menzel, 1968), even though there was no evidence for heterosis when different geographic strains of either species were crossed (Newkirk, 1978; Purdom, 1987). There is the suggestion, however, that these were not hybrids but merely the products of contaminated cultures (Gaffney & Allen, 1993). Allen *et al.* (1993) have shown that while hybrids of *C. virginica* and *C. gigas* can be readily produced they are inviable after 8–10 days. In contrast, when *C. gigas* and *C. rivularis* were crossed they did produce hybrid spat, although there was no evidence for hybrid superiority (Allen & Gaffney, 1993). Overall, there are no clear indications that crosses between different bivalve species produce superior hybrids.

Bivalve breeding programmes

Bivalves are prime candidates for selective breeding programmes for a number of reasons: their high economic value in temperate regions of the world; increasing control over the complete life cycle of many species, particularly oysters; high levels of genetic variability as evidenced from electrophoretic studies (see below); and high fecundity – of the order of 10^6 eggs per female per season. Newkirk (1983) found it surprising that, despite many reasons for placing a high priority on selective breeding, 'there are very few established selective breeding programmes for molluscs'. Twenty years later the situation is largely unchanged. Hershberger *et al.* (1984) have suggested that those involved in commercial bivalve production have been reluctant to put resources and time towards the development of genetically improved strains of molluscs. At present the bulk of the research is being conducted in universities and research institutes and it will probably take several decades more before commercial concerns take charge of their own research programmes. It is crucial, however, that universities and research institute personnel should maintain constant two-way relations with private industry to ensure that the goals of any programme are pertinent to the producers (Hershberger *et al.*, 1984). A national consortium of laboratories and/or companies involved in coordinated research programmes could be one solution (Allen, 1998).

There are several oyster breeding programmes in the United States. One such programme is concerned with mass mortality in the oyster *Crassostrea virginica*. The research is carried out at Rutgers University, New Jersey, United States, and is supported by the National Marine Fisheries Service and the Division of Fish, Game and Shellfisheries of the State of New Jersey. The causative agent of mortality is the parasite *Haplosporidium nelsoni* (MSX), which killed up to 95% of native oysters in Delaware Bay between 1957 and 1959. Within several years of that episode newly set native oysters had about three times as many survivors after a 2.5 year exposure period than imported oysters (Haskin & Ford, 1979). Since 1960 strains of oysters have been produced that are resistant to *H. nelsoni* mortality. This has been achieved by exposing lines (F_1 generation), developed from survivors of the 1957–59 epizootic, to MSX infected waters in Delaware Bay for a period of 33 months. Survivors of the F_1 line were then used to produce the F_2 line and these were again challenged as before. This procedure was followed for six generations and the progress of selection is illustrated in Fig. 10.5. Strains of oysters now exist that have up to ten times the survival of unselected oysters. A clear defence mechanism against *H. nelsoni* has not been demonstrated; selected oysters are infected but they have the ability to tolerate the parasite, restrict its development and thus delay mortality (Ford & Haskin, 1987; see also Chapter 11). A number of researchers from Rutgers University and from neighbouring states have organised a Co-operative Regional Oyster Selective Breeding programme (CROS-Breed) to cooperate in the design, analysis and execution of a brood stock programme based on Rutgers' disease resistant strains (Allen, 1998). Another programme is the Molluscan Broodstock Programme (MBP) based in the Pacific North-west. Academics from the region have formulated and initiated a *C. gigas* broodstock selection programme and are now looking for endorsement and cooperation among the growers (Allen, 1998).

In Australia there is a mass selection programme for weight gain in the Sydney rock oyster, *Saccostrea glomerata* (= *commercialis*). This was initiated in 1990 at the Port Stephens Research Centre, New South Wales (Nell *et al.*, 1996). After two generations of selection average improvement for four breed-

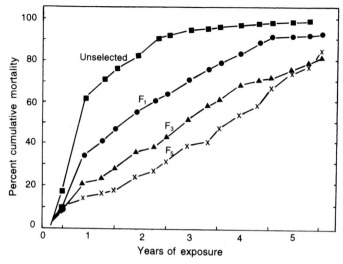

Fig. 10.5. Cumulative mortality of laboratory-reared strains of the oyster *Crassostrea virginica* over 5.5 years of continuous exposure to the protozoan parasite *Haplosporidium nelsoni*. The graphs were computed by averaging seasonal mortality for all strains with at least 100 oysters remaining, and then cumulating these figures. From Ford & Haskin (1987).

ing lines was an 18% (range 14–23%) increase in whole weight (Nell *et al.*, 1999). This effectively reduces time to market by three months. Combining selection with a 6-month reduction in grow-out time provided by triploidy (see below) should prove attractive to oyster growers (Nell *et al.*, 1999).

Other breeding programmes currently underway are: selection for resistance to the disease organism *Bonamia ostreae*, in the oyster *Ostrea edulis* (Naciri-Graven *et al.*, 1998); selection for growth in the Chilean oyster, *Tiostrea chilensis* (Toro *et al.*, 1996); and selection to improve the colour and weight of artificially produced pearls in the pearl oyster, *Pinctada fucata martensii* (Wada, 2000).

Protein and DNA markers

That large amounts of genetic variation are present in natural populations has been known since Darwin's time. But the ability to measure this variation only became possible in the 1950s with two developments: the elucidation of the structure of the DNA molecule, which ultimately clarified the direct relationship between genes and proteins; and the development of gel electrophoresis, an analytical method of protein separation, which permitted rapid and reliable identification of protein variations (technical details in Murphy *et al.*, 1996).

In gel electrophoresis a piece of tissue from an individual organism is ground up to disrupt cells, centrifuged to remove insoluble material, and the resulting protein solution is inserted into the gel (usually starch or polyacrylamide) on a piece of filter paper. An electric current is applied for a fixed time and different proteins migrate at different rates, depending on their charge and configuration. The gel is sliced and each slice is treated with appropriate chemical solutions to visualise the position of proteins or specific enzymes, which appear as discrete zones or bands on the gel. The usefulness of the method lies in the fact that the genotype of an individual with respect to the gene locus (loci) coding for a particular protein, can be inferred from the banding pattern on the gel: homozygotes are single-banded, while heterozygote are either double-, triple- or quintuple-banded phenotypes, depending on whether the enzyme is composed of one, two or four polypeptide chains, respectively. By counting all the different homozygotes and heterozygotes in a sample of individuals one can estimate the number of alternative genes (alleles) coding for a particular protein, and also the frequency of the different alleles. The genetic variability, or heterozygosity, of a population is calculated by first obtaining the frequency of heterozygotes at each locus, and then averaging these frequencies over all loci to get a mean heterozygosity value.

The electrophoretic technique makes it possible to compare allele frequencies and levels of heterozygosity within and between different populations of a species, between different species, and so on. The technique has a number of advantages: a large number of loci and individuals can be examined in a relatively short time, and at moderate expense; and environmental effects on banding patterns are usually unimportant. However, the fact that only the products of structural and not regulatory (non-protein coding) genes are detected means that the technique underestimates overall genomic variability. In addition, because protein expression is two steps away from coding DNA, a change in a DNA base will not necessarily result in an altered amino acid sequence of the protein (because of the degeneracy of the genetic code), or altered electrophoretic mobility. Also, fresh or freshly frozen tissue in relatively large amounts is needed, which invariably means killing the specimens.

Investigations of DNA polymorphisms can greatly enhance information already obtained from allozyme (protein product of an allele) studies. DNA is found in two locations within animal cells: mitochondrial DNA (mtDNA) which is about 15 to 20 kilo-bases in length, and nuclear DNA (nDNA) which makes up about 99.9% of all cellular DNA, most of which is non-coding – either single copy or repetitive sequences. There are several additional features which distinguish mtDNA from nDNA: it is haploid, i.e. each mitochondrion generally contains only one type of mtDNA; the mode of inheritance is primarily through the maternal line and therefore all individuals with a particular mtDNA genotype probably belong to the same maternal clone; most mutations are selectively neutral and mtDNA evolves about 5–10 times faster than nDNA (Brown *et al.*, 1979). Because of its higher mutation rate it is more likely to show differences between populations or species.

After extraction from fresh, frozen or alcohol-preserved tissue, both types of DNA can be cut at specific sites by special enzymes called restriction endonucleases. Cleavage results in a series of DNA fragments called 'restriction fragments' whose number and size are quantified by gel electrophoresis. DNA fragments can be detected by direct staining, usually with ethidium bromide, radio end-labelling, or southern blotting (see Dowling *et al.*, 1996 for details). Polymorphisms in the nucleotide sequence at a restriction enzyme recognition site result in varied numbers of DNA fragments, while polymorphisms

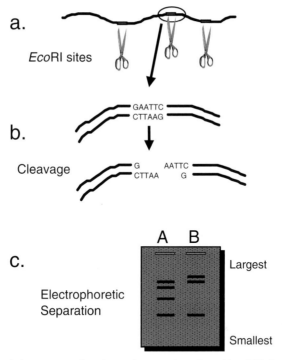

a.

EcoRI sites

b.

GAATTC
CTTAAG

Cleavage

G   AATTC
CTTAA   G

c.

A   B

Electrophoretic
Separation

Largest

Smallest

Fig. 10.6. Restriction enzyme digestion and separation of resulting DNA fragments by electrophoresis. There are three restriction sites in the DNA molecule, each marked with open rectangles in (a). The restriction enzyme *EcoRI* cuts the DNA molecule at every position where the sequence GAATTC occurs (b), thus producing four DNA fragments. These are separated on an agarose gel [lane A in (c)], with the smallest fragments migrating fastest. The loss of a restriction site, circled in (a) means that the two middle fragments in A profile will not be cleaved and will be seen as the single larger more slowly migrating fragment in the B profile. From Park & Moran (1995).

in the length of DNA fragments result in an altered electrophoretic mobility of a DNA fragment through a gel matrix (Fig. 10.6). These are called restriction-fragment-length polymorphisms or RFLPs. In practice, when individuals within or between populations are being compared a number of different restriction enzymes are used to increase the probability of detecting RFLPs. Digestion of nDNA with most enzymes results in too many fragments, but the technique works well on the smaller mtDNA genome. The most important advantage of RFLP analysis is that it provides additional genetic markers for population characterisation.

Most nDNA does not code for any gene product, and even within the DNA sequences that are recognised as genes there are non-coding regions, called introns. One class of non-coding DNA consists of a variable number of tandem repeat (VNTR) short nucleotide sequences. The repeat unit can be anywhere from one to several hundred nucleotides long; repeat units from ten to a hundred nucleotides long are referred to as minisatellites, while shorter repeat units of one to four nucleotides long are called microsatellites. Each VNTR locus (tandem repeat array) is flanked by a unique sequence and if

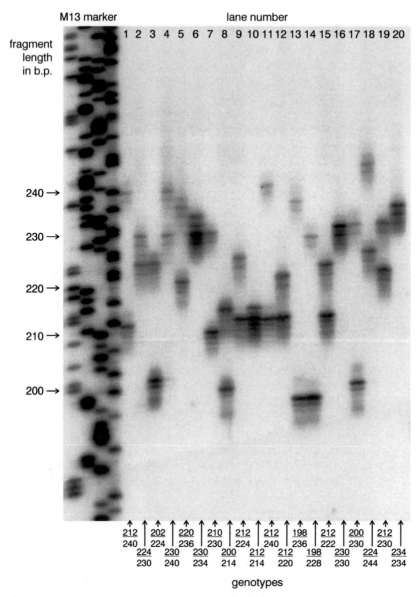

Fig. 10.7. Autoradiograph of microsatellite variation at the *Oedu* HA11a locus in the flat oyster *Ostrea edulis*. From Sobolewska *et al.* (2001). © A. Beaumont, University of Wales, Bangor, UK.

these sequences are known primers can be synthesised that are complementary to them. Using the primers, the tandem array of the locus can be amplified by the polymerase chain reaction (PCR). The high rate of mutation in these loci, detected as length variants on gels (Fig. 10.7), leads to extensive allelic variation and makes them particularly useful as genetic markers for numerous applications in aquaculture and fisheries (Wright & Bentzen, 1995). For example, they are very useful in detecting variability in situations where conventional markers such as allozymes or mtDNA markers have revealed low inter-population or species variability. Microsatellites have also been used to identify individuals and family groups so that they can be reared communally, or to quantify reproductive contributions of individual males and females in laboratory crosses (Boudry *et al.*, 2002). Many laboratories are now attempting to identify marker loci that are associated with nuclear loci that control economically important traits (quantitative trait loci, or QTLs). When such markers are found they can be used in marker-assisted selection (MAS), which has the potential to convert polygenic quantitative variation into individually defined Mendelian entities and provide direct monitoring of the genetic consequences of selection (Ferguson, 1995; Davis & Hetzel, 2000). The most complete linkage maps are for salmonid species and it is anticipated that the genome of several species will be mapped by 2004 (Høyheim *et al.*, 2002). A preliminary genetic linkage map is now available for *Crassostrea virginica* (Gaffney *et al.*, 2002) and *C. gigas* (McGoldrick *et al.*, 2002).

Other approaches to detect nDNA polymorphisms include: DNA fingerprinting where many loci are visualised simultaneously; random amplified polymorphic DNA (RAPD), where DNA polymorphisms are detected by randomly amplifying multiple regions of the genome using single arbitrary primers; amplified fragment-length polymorphisms (AFLPs), where multiple loci can be amplified using only one primer combination; and nucleic acid sequencing of a target DNA sequence, e.g. a gene, or sequencing of an RNA transcript of a gene. Further information can be found in Hadrys *et al.* (1992), Hoelzel (1992), Carvahlo & Pitcher (1995) and Hillis *et al.* (1996).

Molecular markers have been used to resolve a number of important problems in fisheries and aquaculture. Some of these will be outlined in the following section. Because the published literature in this area is so vast it is not feasible to mention all of the relevant papers. Therefore, only those examples that best illustrate the utility of allozyme and DNA markers will be cited.

Identification of populations and species

Fishery managers must know the composition of their particular fishery. Is it made up of a single panmictic population, or is it composed of a number of genetically differentiated populations or stocks (see Chapter 8)? If electrophoresis of samples from two geographically distinct areas reveals significant allele frequency differences at a number of loci, it suggests that the two samples are sufficiently differentiated to be regarded as two separate populations or stocks with little gene flow between them. Using data from a protein locus (*Pt-A*) Beaumont (1982) found significant differences in allele frequencies between Irish Sea and west coast of Ireland queen scallops, *Aequipecten* (*Chlamys*) *opercularis*, and proposed that these should be regarded as separate stocks. On the basis of data from five allozyme loci Meehan (1985) suggested that populations of the clam *Macoma balthica* from the eastern and western Atlantic are not conspecific but instead should be regarded as separate sibling species. In addition, allele frequency data indicated that *M. balthica* populations from San Francisco Bay, California were more closely related to east Atlantic coast populations than to Pacific coast populations (Meehan *et al.*, 1989). These authors suggested that San Francisco Bay clams are a relatively recent introduction from the east coast. Analysis of 29 allozyme loci on what were believed to be two separate oyster species, *Tiostrea chilensis* and *T. lutaria*, have confirmed that these are merely geographical populations of the one species, *T. chilensis* (Buroker *et al.*, 1983). RFLPs of mtDNA fragments have been used to separate the Portuguese oyster, *Crassostrea angulata*, from the Pacific oyster, *C. gigas*, (Boudry *et al.*, 1998). The taxonomic status of *C. angulata* was unclear since no morphological or genetic differences had ever been observed between the two taxa. Brown and Paynter (1991) have shown that *Crassostrea virginica* selectively inbred for rapid growth was characterised by mtDNA genotypes that were distinctly different from those of the progenitor population. Thus, mtDNA markers could be used to distinguish domesticated from native oysters in field comparisons of, e.g. growth, survival and reproductive performance. The mussel *Mytilus trossulus* has been identified solely from allozyme and nDNA marker studies (Gosling, 1992 for review; Geller *et al.*, 1994; Inoue *et al.*, 1995; Suchanek *et al.*, 1997).

Quantification of levels of genetic variability

The ability of a population to respond to selection depends on there being sufficient genetic variation present in the population for the traits being selected. Artificial production of bivalves in hatcheries can lead to a reduction in genetic variability through small effective population size (see above). Allozyme, and more recently mtDNA, analysis has proved useful in tracking the genetic consequences of current hatchery practices. By comparing the number and distribution of alleles and level of heterozygosity of derived and source populations one can get a measure of the loss, or otherwise, of genetic diversity as well as effective population size (N_e) of the hatchery stock. Data from 14 variable enzyme loci indicated that N_e of a commercial oyster (*Crassostrea gigas*) stock from Humboldt Bay, California was only 9 individuals, confirming that N_e in mass spawnings can be very much smaller than the apparent number of progenitors used (Hedgecock & Sly, 1990). A significant loss of genetic diversity following hatchery culture has been demonstrated in *Crassostrea gigas* (Gosling, 1982; Hedgecock & Sly, 1990; but see English *et al.*, 2000), *Crassostrea virginica* (Vrijenhoek *et al.*, 1990) and *Mercenaria mercenaria* (Dillon & Manzi, 1987). Similar results have been obtained from the more recent usage of mtDNA analysis (Brown & Paynter, 1991). Loss of diversity has been generally manifested by loss of rare alleles rather than by any overall reduction in heterozygosity. However, continuation of poor hatchery practice over many generations will eventually lead to substantially reduced heterozygosity.

Allozyme analysis has also been used to compare levels of genetic variability between accidentally introduced species, e.g. *Crassostrea gigas* to New Zealand from Japan (Smith *et al.*, 1986). Unexpectedly, although the actual number of founder oysters is not known, there was no evidence for loss of rare alleles or a reduction in heterozygosity in the New Zealand population compared to Japanese populations.

Identification of hybrids

Detection of hybrids by morphological means is dependent on the hybrids being intermediate to the parent species. This is not always the case, particularly beyond the F_1 generation. Allozyme analysis has proved useful, particularly in fish, in identifying hybrids and backcross individuals, and also in quantifying the extent of gene pool mixing (Wirgin & Waldman, 1994). The two clam species *Mercenaria mercenaria* and *M. campechiensis* hybridise in the wild, but the extent of hybridisation differs depending on locality. For example, Dillon & Manzi (1989) have reported extensive hybridisation in Florida, United States, but on the South Carolina coast hybrids are rare, indicating that reproductive isolation between the two species is clearly more complete at the latter location (Dillon, 1992). Allozyme analysis has been used to confirm extensive hybridisation and introgression between *Mytilus edulis* and *M. galloprovincialis* (see review in Gosling, 1992). Mitochondrial DNA analysis of hybrid mussels provided no evidence for unidirectional gene flow from one mussel type to the other (Gardner & Skibinski, 1991).

Forensic analysis and other applications

Two scallop species, *Placopecten magellanicus* and *Chlamys islandica*, are fished commercially on the Atlantic coasts of Canada (see Chapter 8). Most of the processing is carried out at sea, with only the adductor muscles commonly landed. To ensure that fishery regulations were not being infringed as to numbers of each species landed it was necessary to be able to positively identify the two species solely on the basis of adductor muscle tissue. Kenchington *et al.* (1993) have been successful in locating several allozyme and nDNA markers to discriminate between the species and they believe that time and costs involved are reasonable compared to the costs involved in successful prosecution of fishery regulation infractions. Other ways in which molecular markers have, or could prove useful are as tags, or in translocation experiments. For example, large holding facilities are required for the separate rearing of individual families in selective breeding programmes. The use of genetic markers, e.g. VNTRs (see above) to tag the parents means that identity of offspring can be unambiguously assigned. This has the advantage that many families can be reared *en masse* under identical environmental conditions. In addition, estimates of selective mortality, within and among families, may be made by comparing genotypic ratios of marker loci with Mendelian expectations (Gaffney, 1990). Both allozyme and DNA markers have been developed for fish (Jørstad *et al.*, 1994; see also Doyle *et al.*, 1996) and are mainly used to study genetic interactions between marked and wild stocks, in particular the potentially harmful effects of farmed escapees, and in large-scale ranching programmes. As far as I am aware there are no reports on the use of molecular tags in bivalves.

Reciprocal transplant experiments are used to determine the extent to which differences between populations occupying different habitats are environmentally induced or inherited. If mortality occurs in the transplanted samples allozyme analysis of survivors can determine whether random or selective mortality of specific genotypes has occurred. However, mortality of specific allozyme genotypes does not imply that selection is acting directly on these loci; it is more likely, since most allozyme loci act as neutral markers, that selection is acting at loci in linkage with the allozyme loci under study. Allozyme analysis of the survivors from a transplant experiment involving two genetically differentiated *Chlamys varia* populations demonstrated that transplanted scallops after one year were not significantly different from the indigenous population (Gosling & Burnell, 1988). Similar results have been obtained for other bivalves (Theisen, 1978; Blot *et al.*, 1989; Johannesson *et al.* 1990). These results are relevant in view of large-scale movements of bivalves both within and between countries.

Chromosomal genetics and ploidy manipulation

The genes of an organism are located on chromosomes that reside in the cell nucleus. In somatic cells (cells that are not eggs or sperm) chromosomes occur in homologous pairs, consisting of one chromosome from each parent bearing essentially the same genes in the same order. The total number of chromosomes in a somatic cell is called the diploid number (2N). This number is normally constant within a species but may vary between species; for instance, mussels have 2N = 28, oysters have 2N = 20, and most scallops and clams have 2N = 38. The number of chromosomes in the egg or sperm is half that in somatic cells and is called the haploid (N) number. When the egg is fertilised by a sperm the diploid number is restored. Within a species there can be variation in the ploidy of individuals, but this is usually associated with a decrease in fitness e.g. reduced viability and fertility, sterility and deformity. Individuals with more than two sets of homologous chromosomes are called polyploids; an organism is triploid if it contains three sets, tetraploid if it contains four sets, and so on. Polyploidy is very common in some groups, such as flowering plants and cereals, but is rare in animals. This is because it upsets complex gene interactions that are crucial in animal development.

For an organism to grow cells must divide to produce more cells. This type of cell division, called mitosis, involves duplication of whole chromosome sets in order to produce two daughter cells that are diploid and genetically identical to each other and to the cell that gave rise to them. The type of cell division that produces gametes is called meiosis. This process involves two meiotic cell divisions in a diploid germ cell: meiosis I which produces two new cells each with half the number of chromosomes as their precursor; and meiosis II, which is essentially a mitotic division, and results in four haploid cells. Male germ cells produce four functional gametes, whereas female germ cells divide unequally at meiosis I to produce one large haploid cell and a small, attached first polar body, PB1 (Fig. 10.8). Meiosis II produces the egg cell (oocyte) and a second attached polar body, PB2. Occasionally, in some species the first polar body undergoes meiosis II, giving three attached polar bodies, none of which contribute genetic material to the oocyte.

With the advent of a worldwide interest in fish and bivalve culture came the realisation that polyploids, produced through manipulation of chromosome sets, could have economic implications for the industry. The technique of

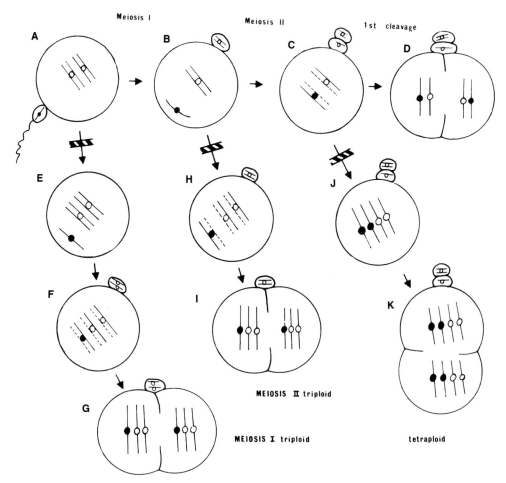

Fig. 10.8. Normal meiosis in the bivalve egg (A–D). For simplicity sake only one pair of chromosomes is shown. (A) Egg when released is at metaphase of meiosis I; activated by sperm. (B) Meiosis I completed, PB1 extruded, sperm nucleus enters egg. (C) Meiosis II completed, PB2 extruded, male and female pronuclei unite. (D) First cleavage mitotic division. Ploidy manipulation in the bivalve egg (E–K). (E) Shock administered during meiosis I, both chromosomes of the pair retained in the egg; PB1 not extruded. (F) Normal meiosis II allowed, PB2 extruded; 2N female pronucleus and N male pronucleus unite. (G) Triploid first cleavage. (H) Shock administered during meiosis II, PB2 not extruded; 2N female pronucleus and N male pronucleus unite. (I) Triploid first cleavage. (J) Shock administered during first cleavage. (K) Tetraploid chromosome complement in second cleavage. From Beaumont & Fairbrother (1991).

ploidy manipulation was first applied to fish in the early 1970s (Purdom, 1972), and a decade later to bivalve species (Stanley *et al.*, 1981).

Triploidy, tetraploidy and gynogenesis

In bivalves eggs and sperm are released directly into the sea, where fertilisation takes place. Although the sperm are haploid, the eggs are arrested in the early stage of meiosis I, and meiosis I is not completed until the egg is activated by the sperm at fertilisation. Triploidy can be induced in eggs by inhibit-

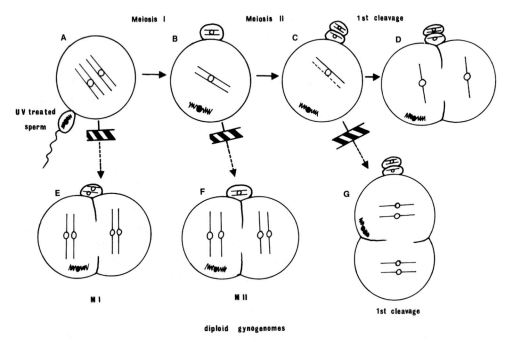

Fig. 10.9. Gynogenesis in the bivalve egg. A–D Maturation divisions and first cleavage of egg after activation by UV treated sperm. Diploid gynogenomes produced by shock treatment administered at meiosis I (E), meiosis II (F), or first cleavage (G). From Beaumont & Fairbrother (1991).

ing either meiosis I or II by chemical or physical shock treatment. If meiosis I is targeted PB1 is not extruded and homologous pairs of chromosomes are retained in the egg; meiosis II proceeds normally and the result is a diploid egg which, when fertilised by a haploid sperm, results in a triploid zygote. When meiosis II is targeted PB2 is not extruded and once again the result is a diploid egg (Fig. 10.8). If treatment is administered during the first cleavage division of the fertilised egg the result is a cell with twice the diploid chromosome number. Subsequent divisions should produce a tetraploid embryo (Fig. 10.8). Sperm can be UV or gamma-irradiated to destroy the DNA, and this genetically inactive sperm can then be used to stimulate the parthenogenetic development of the egg. The result is a haploid egg that contains genetic information only from the mother. If first cleavage is inhibited this results in retention of an extra functioning set of maternal chromosomes in the egg (Fig. 10.9). This compensates for the missing set from the sperm, and thus development can proceed normally. The result is a diploid gynogen that is homozygous at almost all its gene loci.

Benefits of ploidy manipulation

Adult triploids are commonly sterile due to the inability of homologous chromosomes to pair during meiosis; this results in uneven or aborted separation of chromosome triplets. From a commercial viewpoint sterility is desirable for several reasons. Metabolic energy normally utilised in gonad development may be available for increased somatic growth, thereby resulting in larger animals. Reproduction is sometimes accompanied by deterioration in the quality and taste of flesh. For example, in sexually mature *Crassostrea gigas* from the Pacific north-west United States the flesh softens and glyco- gen stores, which normally impart a sweet taste to the non-reproductive oyster, are converted to less flavoursome products. Sexually mature *C. gigas* are thus inferior products for marketing, but the high consumer acceptance of superior tasting triploids ensures their year-round marketability (Allen, 1988). Sexual maturity is often accompanied by high summer mortality, particularly in *C. gigas*. It is not surprising that about one third of all hatchery-produced *C. gigas* seed on the US west coast are triploids (Chew, 1994).

The value of tetraploid production is that tetraploid females should produce diploid eggs, which when fertilised by normal haploid sperm would be expected to give 100% triploid offspring. Artificial gynogenesis was originally used in fish as a method of sex control but later as a method for the rapid production of inbred lines that could be crossed to produce heterozygous F_1 hybrids.

Induction methods

Heat or cold shock treatment can be used to inhibit meiosis I, II, or first cleav- age. The usual temperature range employed is between 25°C and 32°C for heat shocking, and between 0 and 5°C for cold shocking. Choice of temperature will depend on there being sufficient differential between normal spawning temperature and thermal shock treatment temperature. Pressure shock treatment can also be used to produce triploids. Activated eggs are placed under a pressure of 6000–8000 psi (c. 410–550 bar) for 10 min (Chaiton & Allen, 1985). When pressure is reduced development proceeds as normal.

While pressure and thermal shock treatments are the methods of choice for fish, for bivalves chemical methods are most often used, primarily because a higher percentage triploidy is achieved. Cytochalasin B (CB) a fungal metabo- lite, is most commonly used, although the cheaper, less toxic, and equally effec- tive 6-dimethylaminopurine (6-DMAP) is an attractive alternative. The effects of temperature and pressure shock are similar in that both methods disrupt the formation of the spindle, thereby preventing replicated chromosomes from separating into daughters cells. In contrast, both CB and 6-DMAP inhibit polar body formation but chromosome movement is unaffected by the treat- ment. Table 10.4. gives details of chemical and physical methods of triploid induction in a variety of bivalve species. All methods, but in particular chemi- cal methods, are accompanied by high early larval mortality. This is particu- larly true for CB treatment, where Gérard *et al.* (1994a) reported a mortality of 64% for CB treatment, but a lower mortality of 36% when 6-DMAP was used. Attention to dosage, timing, duration and temperature of treatment, as well as to factors such as egg quality and larval rearing techniques can sub- stantially increase survival rates. Once the D-larva stage of development has been reached triploids and diploids exhibit similar survival rates (Downing & Allen, 1987).

Table 10.4. Triploid induction in bivalves. The % triploidy figure, estimated in early embryos, is the best obtained for that study. P = pressure method; T = temperature method; CB = cytochalasin B; 6-DMAP = 6-dimethylaminopurine. First figure after CB treatment is mg l^{-1} and after DMAP treatment is μM l^{-1}; second (†) and third (*) figures refer to duration of treatment and time after fertilisation, respectively.

Species	Method	% Triploidy	Reference
Mytilus edulis	T; 32°C; 10†; 20*	97	Yamamoto & Sugawara, 1988
	T; 1°C; 10; 35	85	
	T; 25°C; 10; 10	25	Beaumont & Kelly, 1989
	CB; 0.1 mg; 20; 30–50	67	
Ruditapes philippinarum	CB; 0.5 mg; 15; 15–30	82	Beaumont & Contaris, 1988
	CB; 1 mg; 15; 20–35	76	Dufy & Diter, 1990
	CB; 0.5 mg; 15; 20	50	Gosling & Nolan, 1989
	T; 32°C; 10; 20	56	
Placopecten magellanicus	6-DMAP; 400 μM; 15; 70	95	Desrosiers et al., 1993
Pecten maximus	CB; 0.5 mg; 20; 10	30	Beaumont, 1986
Chlamys varia	CB; 1 mg; 15; 20	78	Baron et al., 1989
Crassostrea gigas	CB; 0.5 mg; 20; 10	67	Yamamoto et al., 1988
	T; 0°C; 10; 15	67	
	T; 37°C; 15; 45	83	
	6-DMAP; 450 μM; 10;15	85	Gérard et al., 1994a
	6-DMAP; 300 μM; 20; 15	90	Desrosiers et al., 1993
	P; 7200 psi (~500 bar); 10; 10	60	Allen et al., 1986
C. virginica	CB; 1 mg; 20; 10	79	Allen & Bushek, 1992
Ostrea edulis	CB; 1 mg; 20; 30–50	70	Gendreau & Grizel, 1990
Saccostrea glomerata	CB; 0.5 mg; 20; 23	81	Nell et al., 1994
Pinctada martensii	P; ~3000 psi (~205 bar); 10; 5–7	76	Shen et al., 1993

Tetraploids are produced by using CB to block the release of PB1 in eggs from triploids fertilised by sperm from diploids. This method, pioneered by Guo & Allen (1994a) and refined by Eudeline *et al.* (2000), has been successfully used to produce tetraploid oysters (*Crassostrea gigas*), which in turn have produced viable second-generation tetraploids (Guo *et al.*, 1996). The sperm from tetraploid *C.gigas* has been used to fertilise eggs from diploids to produce 100% triploids (Chew, 2000). Tetraploid *C. virginica* (Supan, 2000) and *Pinctada martensii* (He *et al.*, 2000) have also been produced using this method, but to date the production of tetraploid *Saccostrea glomerata* has been unsuccessful (Nell *et al.*, 2002).

To produce gynogens sperm are irradiated with UV light (254 nm) for sufficient time to destroy the DNA. Dosage time ranges from 1.5 to 10 min at intensities of 640–1400 μW cm^{-2}s^{-1}, depending on the species (Guo *et al.*, 1993; Guo & Allen, 1994c; Scarpa *et al.*, 1994; Li *et al.*, 2000). Chromosome doubling of activated eggs is achieved by the use of CB to suppress either PB1 or PB2 to produce MI or MII gynogens respectively, or to suppress first cleavage of the haploid egg to produce mitogynes (Fig. 10.9).

Ploidy verification

Since no method of triploidy induction is 100% efficient it is important to have a quick, reliable and cost-effective means to check the ploidy level at every stage in the life cycle. The main methods for molluscs are: chromosome counts, polar body counts, cell nuclear measurements, flow cytometry, micro-fluorometry and image analysis. In this section each of these will be briefly described, and their main advantages and disadvantages highlighted. More details on the individual techniques are in Allen *et al.* (1989), Komaru *et al.* (1988), Beaumont & Fairbrother (1991) and Gérard *et al.* (1994b).

Chromosome counts can be carried out at the embryo as well as at later stages of development. Although the technique is time consuming and involves a certain degree of expertise it is a reliable and direct method of ploidy verification. Since triploidy induction involves the suppression of either the first or second polar body triploid zygotes should have only one polar body instead of the two normally present in a haploid egg. Polar body counts are a simple and fast method of ploidy verification and counting can start within a few hours of treatment. The use of DAPI (4-6-diamidino-2-phenylindole), a fluorescent dye that binds preferentially to the adenine-thymine base pairs of DNA, makes the task of locating polar bodies an easier one (Komaru *et al.*, 1990). One disadvantage of polar body counts is that the method can only be used in embryos, and therefore an alternative method is required to confirm the ploidy of juveniles and adults.

Due to the extra DNA content the nuclei of triploid cells are 1.5 times greater in volume and therefore have a larger diameter than diploid nuclei. The difference in diameter between diploid and triploid cells can be measured microscopically (Child & Watkins, 1994). Generally stained cells from gill tissue or haemolymph are used, and in the case of the latter the cells can be extracted without killing the animal. This method is fast, accurate and can be used by anyone trained in the use of a microscope. However, the technique does depend on the animals being large enough so that samples of tissue, such as gill, can be easily obtained.

Flow cytometry is a technique used to estimate DNA content in cells. A fluorescent dye which bonds specifically to nucleic acids is added to a small sample of cells from fresh or frozen haemolymph, mantle, gill, siphon or foot tissue (Allen, 1983). The cells take up the dye in proportion to their DNA content and a flow cytometer measures the fluorescent intensity, and registers nuclei of different ploidies in different peaks (Fig. 10.10). Flow cytometry is the technique of choice in estimating triploidy because it is fast, accurate and can be used on a variety of tissues that can be sampled without killing the animal (Allen, 1983). However, one big drawback is that availability of the technique is restricted because of the high cost of flow cytometers. In micro-fluorometry nuclei are stained with DAPI, are then excited with UV light (365 nm), and the fluorescence intensity is measured with a photometer attached to a light microscope. The technique is simple and accurate and is a much cheaper alternative to flow cytometry.

Image analysis uses a microscope and a computer, both of which are linked to a camera. A programme analyses the photometric intensity of stained nuclei

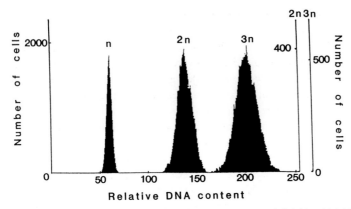

Fig. 10.10. Histograms of fluorescence intensity in spermatozoa, and diploid and triploid scallop (*Chlamys nobilis*) haemocytes analysed by flow cytometry. From Komaru *et al.* (1988) and reprinted with permission from Elsevier Science.

and gives their individual integrated optical density (IOD), which is positively correlated with DNA content. The technique is efficient and can be used at different stages of the life cycle. It is cheaper than flow cytometry but more expensive than microfluorometry. One drawback is that ploidy determination is highly dependent on homogeneous slide staining (Gérard *et al.*, 1994b).

Results from ploidy manipulations

Triploids

The biggest putative advantage of triploids is their sterility, but are triploids really sterile? Theoretically they should be (see above), but several studies have shown that triploids are not completely sterile. Indeed, the production of first generation tetraploids depends on obtaining eggs from triploids! Retarded gonad development and abnormal gametogenesis have been reported in *Mya arenaria*, *Mercenaria mercenaria*, *Crassostrea gigas*, *C. virginica*, *Saccostrea glomerata*, *Chlamys nobilis* and *Pinctada martensii* (Allen, 1987; Komaru & Wada, 1989; Jiang *et al.*, 1993; Cox *et al.*, 1996; Eversole *et al.*, 1997). In *Crassostrea gigas* the relative fecundity of triploid females is about 2% that of normal diploids (Guo & Allen, 1994a). Triploids of some species are actually fertile (Guo & Allen, 1994b), although survival of fertilised eggs to metamorphosis and settlement is extremely low (<0.01%). Gonad development is always more extensive in male than in female triploids, although this is not reflected in a slower growth rate in males.

Because triploids have a considerably reduced reproductive potential compared to diploids, they should exhibit faster growth rates once maturity has been reached. Results on several species support this. No difference in growth rate was observed between triploid and diploid *Crassostrea virginica* up to 9 months of age, but subsequently growth diverged so that 3-year-old triploids had 40% higher meat weight than diploid controls (Stanley *et al.*, 1984). An added benefit of faster growth in triploid *C. virginica* is that a larger proportion of oysters reach market size before being killed by the parasite *Perkinsus marinus* (Barber & Mann, 1991). Faster growth has also been reported for triploid *Crassostrea gigas* (Akashige & Fushimi, 1992), *Argopecten irradians* (Tabarini, 1984), *Pinctada martensii* (Jiang *et al.*, 1993), *Saccostrea glomerata* (Hand *et al.*, 1998; Smith *et al.*, 2000), *Mercenaria mercenaria* (Eversole *et al.*, 1997) and *Chlamys farreri* (Yang *et al.*, 2002).

Since triploids have three instead of two copies of each chromosome in the cell nucleus they are expected to be genetically more variable than diploids. This has been confirmed in juvenile *Mya arenaria* by electrophoresis; triploids were shown to be nearly twice as heterozygous as their diploid siblings (Mason *et al.*, 1988). However, despite higher heterozygosity these authors observed no difference in oxygen consumption or filtration rate between triploid and diploid clams. Triploids that are produced by blocking meiosis I are expected to be more heterozygous then meiosis II triploids (Fig. 10.8). However, for loci with high recombination frequencies, MII triploids are expected to be more heterozygous than MI triploids (Beaumont, 2000). To date, despite expectations that MI and MII triploids are equally genetically variable, significant growth rate differences have been observed between them. In *C. gigas* (Yamomoto *et al.*, 1988) mean shell length of meiosis I triploid larvae was significantly larger (213 μm) than either meiosis II (190 μm) or diploids (185 μm). Similar results have been observed for both larval and adult stages of other species (Stanley *et al.*, 1984; Beaumont & Kelly, 1989). Hawkins *et al.* (1994) have shown that in the oyster *Ostrea edulis* the faster growth of meiosis-I triploids resulted from reduced energy expenditure associated with lower concentrations of RNA per unit of total tissue protein, which infers reduced rates of protein turnover. The implication is that the alleged increased heterozygosity of meiosis I triploids enables them to sustain basal metabolism with lower expenditure of energy (see Chapter 6).

Triploid production of *C. gigas* has been enthusiastically adopted by commercial operations on the Pacific coast of North America and is set to continue with the newer method of triploid production using sperm from tetraploids. However, with the exception of France, there has been limited interest in commercial production outside of North America. In France, the government agency IFREMER has begun to make sperm from tetraploid *C. gigas* available to commercial hatcheries. It is expected that in the very near future about 50% of all spat supplied by hatcheries will be triploids (Nell *et al.*, 2002)

Tetraploids

Since 1994, when viable tetraploids were first produced, methods have been refined, and in *Crassostrea gigas* mean tetraploid production is 45%, an acceptable figure considering that tetraploid progeny are ultimately to be used as broodstock (Eudeline *et al.*, 2000). Using tetraploids crossed with diploids Guo *et al.* (1996) have now successfully produced 100% triploid offspring and these at 8 and 10 months of age were 13–51% larger than normal diploids, possibly due to polyploid gigantism (increased cell volume and a lack of cell number compensation). More recently, Wang *et al.* (2002) observed significantly greater flesh weight in *C. gigas* triploids produced from a diploid by tetraploid mating than in triploids produced from CB treatment.

Gynogenesis

Viable gynogenetic diploids have been successfully produced in a number of fish species (Váradi *et al.* 1999 and references) but attempts to produce viable gynogens in marine bivalves have not been very successful to date (references in Guo *et al.*, 1993; Fairbrother, 1994). The most cited reason for reduced survival of gynogens is damaged DNA fragments from sperm irradiation (Guo *et al.*, 1993; Guo & Allen, 1994c). It is likely however that the production of viable gynogens is attainable with technique improvement.

Transgenics

The injection of foreign DNA into animal ova has been used since the early 1980s to produce what are known as 'transgenics'. The first attempt at transgenic production was in the early 1980s when the rat growth hormone gene was injected into mice, producing a dramatic increase in growth rate (Palmiter *et al.*, 1982). These experiments paved the way for the production of transgenic farm animals and towards the end of the 1980s the technology was applied to several commercially important fish species (reviews by Iyengar *et al.*, 1996 and Devlin, 1997). The principal genes that have been used in fish transgenic studies to date are fish growth hormone gene, metallothionein (metal binding) and antifreeze genes, chosen primarily because of the potential commercial contribution of the produced transgenics to the aquaculture industry. Briefly, the gene with flanking sequences, such as promotor and tissue enhancer sequences, is obtained from a gene library. To provide a high copy number, many millions of copies of the acquired gene are inserted into a DNA vector, grown up in a suitable bacterial strain, and then harvested. About one million copies of the gene are introduced close to the nucleus of the fertilised egg. In bivalves the major transfection methods are microinjection, lipofection (liposome mediated) and electroporation, whereby eggs are subjected to electrical pulses that increase membrane porosity and facilitate the uptake of the DNA (Cadoret *et al.*, 2000). The next step is to ascertain whether the transgene has been integrated into the host's genome and, if so, whether it is expressed. Expression can be established either by assaying for the appropriate mRNA or, more commonly, by assaying directly for the translated protein product. Finally, it must be established whether the gene has been incorporated into the germ-line.

Most of the literature on fish transgenics has been concerned with the technical aspects of the method, with only a few published reports to date on the actual phenotypic effects of gene transfer. Almost all deal exclusively with the effect of growth hormone on fish growth, with all reporting an improvement in transgenic versus untreated fish in the range 15–150%. There are no equivalent results for bivalves although Cadoret *et al.* (2000) have recently isolated a portion of an oyster actin gene that has turned out to be a very efficient promoter. These results indicate that the introduction of novel genes into fish, and eventually shellfish, may make a major contribution to aquaculture. For example, transgenesis could be a fast and efficient method for producing disease resistant lines in bivalves. Several anti-microbial peptides have already been isolated, and attempts to identify and study the genes coding for these peptides, as well as their regulatory mechanism, is underway (Cadoret *et al.*, 2000). However, the cost of the technology is prohibitive for developing countries, where transgenics and other genetically modified organisms are perceived in terms of improved diet. Ethical considerations, non-acceptance by the consumer, and the lack of specific guidelines for experiments with transgenic aquatic organisms are just some of the problems facing the new technology (Hallerman & Kapuscinski, 1995; Hallerman *et al.*, 1999).

References

Akashige, S. & Fushimi, T. (1992) Growth, survival and glycogen content of triploid Pacific oyster *Crassostrea gigas* in the waters of Hiroshima, Japan. *Nippon Suisan Gakkaishi*, **58**, 1063–71.

Allen, S.K. Jr. (1983) Flow cytometry: assaying experimental polyploid fish and shellfish. *Aquaculture*, **33**, 317–28.

Allen, S.K. Jr. (1987) Gametogenesis in three species of triploid shellfish: *Mya arenaria, Crassostrea gigas* and *Crassostrea virginica*. In: *Selection, Hybridization and Genetic Engineering in Aquaculture* (ed. K. Tiews), pp. 208–17. Heenemann, Berlin.

Allen, S.K. Jr. (1988) Triploid oysters ensure year-round supply. *Oceanus*, **31**, 58–63.

Allen, S.K. Jr. (1998) Commercial applications of bivalve genetics; not a solo effort. *World Aquaculture*, **29**, 38–43.

Allen, S.K. Jr. & Bushek, D. (1992) Large scale production of triploid *Crassostrea virginica* (Gmelin) using 'stripped' gametes. *Aquaculture*, **103**, 241–51.

Allen, S.K. Jr. & Gaffney, P.M. (1993) Genetic confirmation of hybridisation between *Crassostrea gigas* (Thunberg) and *Crassostrea rivularis* (Gould). *Aquaculture*, **113**, 291–300.

Allen, S.K. Jr., Downing, S.L., Chaiton, J. & Beattie, J.H. (1986) Chemically and pressure-induced triploidy in the Pacific oyster *Crassostrea gigas* (abstract only). *Aquaculture*, **57**, 359–60.

Allen, S.K. Jr., Downing, S.L. & Chew, K.K. (1989) *Hatchery Manual for Producing Triploid Oysters*. University of Washington Press, Seattle.

Allen, S.K. Jr., Gaffney, P.M., Scarpa, J. & Bushek, D. (1993) Inviable hybrids of *Crassostrea virginica* (Gmelin) with *C. rivularis* (Gould) and *C. gigas* (Thunberg). *Aquaculture*, **113**, 269–89.

Arnold, W.S., Bert, T.M., Quitmyer, I.R. & Jones, D.S. (1998) Contemporaneous deposition of annual growth bands in *Mercenaria mercenaria* (Linnaeus), *Mercenaria campechiensis* (Gmelin), and their natural hybrid forms. *J. Exp. Mar. Biol. Ecol.*, **223**, 93–109.

Barber, B.J. & Mann, R. (1991) Sterile triploid *Crassostrea virginica* (Gmelin, 1791) grow faster than diploids but are equally susceptible to *Perkinsus marinus*. *J. Shellfish Res.*, **10**, 445–50.

Baron, J., Diter, A. & Bodoy, A. (1989) Triploidy induction in the black scallop (*Chlamys varia* L.) and its effect on larval growth and survival. *Aquaculture*, **77**, 103–11.

Bayne, B.L., Hedgecock, D., McGoldrick, D. & Rees, R. (1999) Feeding behaviour and metabolic efficiency contribute to growth heterosis in Pacific oysters (*Crassostrea gigas* Thunberg). *J. Exp. Mar. Biol. Ecol.*, **233**, 115–130.

Beaumont, A.R. (1982) Geographic variation in allele frequencies at three loci in *Chlamys opercularis* from Norway to the Brittany coast. *J. mar. biol. Ass. U.K.*, **62**, 243–61.

Beaumont, A.R. (1986) Genetic aspects of hatchery rearing of the scallop *Pecten maximus* (L.). *Aquaculture*, **57**, 99–110.

Beaumont, A.R. (1994) Application and relevance of genetics in aquaculture. In: *Genetics and Evolution of Aquatic Organisms* (ed. A.R. Beaumont), pp. 467–86. Chapman and Hall, London.

Beaumont, A.R. (2000) Genetic considerations in hatchery culture of bivalve shellfish. In: *Recent Advances in Marine Biotechnology: Aquaculture: Seaweeds and Invertebrates*, Vol. 4, Part A (eds M. Fingerman & R. Nagabhushanam), pp. 87–109. Science Publishers Inc., USA.

Beaumont, A.R. & Budd, M.D. (1983) Effects of self-fertilisation and other factors on the early development of the scallop, *Pecten maximus*. *Mar. Biol.*, **76**, 285–89.

Beaumont, A.R. & Contaris, M.H. (1988) Production of triploid embryos of *Tapes semidecussatus* by the use of cytochalasin B. *Aquaculture*, **73**, 37–42.

Beaumont, A.R. & Kelly, K.S. (1989) Production and growth of triploid *Mytilus edulis* larvae. *J. Exp. Mar. Biol. Ecol.*, **132**, 69–84.

Beaumont, A.R. & Fairbrother, J.E. (1991) Ploidy manipulation in molluscan shellfish: a review. *J. Shellfish Res.*, **10**, 1–18.

Benzie, J. (ed) (2002) *Genetics in Aquaculture* VII. Elsevier Science B.V., Amsterdam.

Blot, M., Thiriot-Quiévreux, C. & Soyer, J. (1989) Genetic differences and environments of mussel populations in Kerguelen Islands. *Polar Biol.*, **10**, 167–74.

Boudry, P., Heurtebise, S., Collet, B., Cornette, F. & Gérard, A. (1998) Differentiation between populations of the Portugese oyster, *Crassostrea angulata* (Lamark) and the Pacific oyster, *Crassostrea gigas* (Thunberg), revealed by mtDNA RFLP analysis. *J. Exp. Mar. Biol. Ecol.*, **226**, 279–91.

Boudry, P., Collet, B., Cornette, F., Hervouet, V. & Bonhomme, F. (2002) Microsatellite markers as a tool to study reproductive success in the Pacifc oyster, *Crassostrea gigas* (Thunberg), crossed under controlled hatchery condition. *Aquaculture*, **204**, 283–96.

Bowman, J.C. (1974) *An Introduction to Animal Breeding*. Edward Arnold Ltd., Southampton, UK.

Brown, B.L. & Paynter., K.T. (1991) Mitochondrial DNA analysis of native and selectively inbred Chesapeake Bay oysters, *Crassostrea virginica*. *Mar. Biol.*, **110**, 343–52.

Brown, W.M., George, M. Jr. & Wilson, A.C. (1979) Rapid evolution of animal mitochondrial DNA. *Proc. Natl. Acad, Sci.*, **76**, 1967–71.

Buroker, N.E., Chanley, P., Cranfield, H.J. & Dinamani, P. (1983) Systematic status of two oyster populations of the genus *Tiostrea* from New Zealand and Chile. *Mar. Biol.*, **77**, 191–200.

Cadoret, J.-P., Bachère, E., Roch, P., Mialhe, E. & Boulo, V. (2000) Genetic transformation of farmed marine bivalve molluscs. In: *Recent Advances in Marine Biotechnology: Aquaculture: Seaweeds and Invertebrates*, Vol. 4, Part A (eds M. Fingerman & R. Nagabhushanam), pp. 111–126. Science Publishers Inc., USA.

Carvahlo, G.R. & Pitcher, T.J. (eds) (1995) *Molecular Genetics in Fisheries*. Chapman and Hall, London.

Chaiton, J.A. & Allen, S.K. Jr. (1985) Early detection of triploidy in the larvae of Pacific oysters *Crassostrea gigas* by flow cytometry. *Aquaculture*, **48**, 35–43.

Chew, K.K. (1994) Tetraploid Pacific oysters offer promise to future production of triploids. *Aquaculture Magazine*, **20**, 69–74.

Chew, K.K. (2000) Update on triploid Pacific oysters. *Aquaculture Magazine*, **26**, 87–89.

Child, A.R. & Watkins, H.P. (1994) A simple method to identify triploid molluscan bivalves by the measurement of cell nucleus diameter. *Aquaculture*, **125**, 199–204.

Cox, E.S., Smith, M., Nell, J.A. & Maguire, G.B. (1996) Studies on triploid oysters in Australia. VI. Gonad development in diploid and triploid Sydney rock oysters *Saccostrea commercialis* (Iredale and Roughley). *J. Exp. Mar. Biol. Ecol.*, **197**, 101–20.

Crenshaw, J.W.J., Heffernan, P.B. & Walker, R.L. (1991) Heritability of growth rate in the southern bay scallop, *Argopecten iradians concentricus* (Say, 1822). *J. Shellfish Res.*, **10**, 55–63.

Davis, G.P. & Hetzel, D.J.S. (2000) Integrating molecular genetic technology with tradition: approaches for genetic improvement in aquaculture species. *Aquaculture Res.*, **31**, 3–10.

Desrosiers, R.R., Gérard, A., Peignon, J.M. *et al.* (1993) A novel method to produce triploids in bivalve molluscs by the use of 6-dimethylaminopurine. *J. Exp. Mar. Biol. Ecol.*, **170**, 29–43.

Devlin, R.H. (1997) Transgenic salmonids. In: *Transgenic Animals: Generation and Use* (ed. L.M. Houbedine), pp. 105–17. Harwood Academic Publishers, Amsterdam.

Dillon, R.T. (1992) Minimal hybridization between populations of the hard clam, *Mercenaria mercenaria* and *Mercenaria campechiensis* co-occurring in South Carolina. *Bull. Mar. Sci.*, **50**, 411–16.

Dillon, R.T. & Manzi, J.J. (1987) Hard clam, *Mercenaria mercenaria*, broodstocks: genetic drift and loss of rare alleles without reduction in heterozygosity. *Aquaculture*, **60**, 99–105.

Dillon, R.T. & Manzi, J.J. (1989) Genetics and shell morphology in a hybrid zone between the hard clams *Mercenaria mercenaria* and *M. campechiensis*. *Mar. Biol.*, **100**, 217–22.

Dowling, T.E., Moritz, C., Palmer, J.D. & Rieseberg, L.H. (1996) Nucleic acids III: Analysis of fragments and restriction sites. In: *Molecular Systematics*, 2nd edn. (eds D.M. Hillis, C. Moritz & B.K. Mable), pp. 249–320. Sinauer Associates Inc., Sunderland, Massachusetts, USA.

Downing, S.L. & Allen, S.K. Jr. (1987) Induced triploidy in the Pacific oyster, *Crassostrea gigas*: optimal treatments with cytochalasin B depend on temperature. *Aquaculture*, **61**, 1–15.

Doyle, R.W., Herbinger, C.M., Ball, M. & Gall, G.A.E. (eds) (1996) *Genetics in Aquaculture* V. Elsevier Science B.V., Amsterdam.

Dufy, C. & Diter, A. (1990) Polyploidy in the Manila clam *Ruditapes philippinarum*. I. Chemical induction and larval performance of triploids. *Aquat. Living Resour.*, **3**, 55–60.

English, L.J., Maguire, G.B. & Ward, R.D. (2000) Genetic variation of wild and hatchery populations of the Pacific oyster, *Crassostrea gigas* (Thunberg) in Australia. *Aquaculture*, **187**, 283–98.

Eudeline, B., Allen, S.K. Jr. & Guo, X. (2000) Optimization of tetraploid induction in Pacific oysters, *Crassostrea gigas*, using first polar body as a natural indicator. *Aquaculture*, **187**, 73–84.

Eversole, A.G., Kempton, C.J., Hadley, N.H. & Buzzi, W.R. (1997) Comparison of growth, survival, and reproductive success of diploid and triploid *Mercenaria mercenaria*. *J. Shellfish Res.*, **15**, 689–94.

Fairbrother, J.E. (1994) Viable gynogenetic diploid *Mytilus edulis* (L.) larvae produced by ultraviolet light irradiation and cytochalasin B shock. *Aquaculture*, **126**, 25–34.

Falconer, D.S. (1960) *Introduction to Quantitative Genetics*. Oliver & Boyd, Edinburgh.

Falconer, D.S. & Mackay, T.F.C. (1996) *Introduction to Quantitative Genetics*, 4nd edn. Longman Scientific and Technical, England.

Ferguson, M. (1995) The role of molecular genetic markers in the management of cultured fishes. In: *Molecular Genetics in Fisheries* (ed. G.R. Carvahlo & T.J. Pitcher), pp. 81–103. Chapman and Hall, London.

Ford, S.E. & Haskin, H.H. (1987) Infection and mortality patterns in strains of oysters *Crassostrea virginica* selected for resistance to the parasite *Haplosporidium nelsoni* (MSX). *J. Parasitol.*, **73**, 368–76.

Gaffney, P.M. (1990) Breeding designs for bivalve molluscs (abstract only). *Aquaculture*, **85**, 321.

Gaffney, P.M. & Allen, S.K. Jr. (1993) Hybridization among *Crassostrea species*: a review. *Aquaculture*, **116**, 1–13.

Gaffney, P.M., Davis, C.V. & Hawes, R.O. (1992) Assessment of drift and selection in hatchery populations of oysters (*Crassostrea virginica*). *Aquaculture*, **105**, 1–20.

Gaffney, P.M., Reece, K.S., Ribeiro, W. & Pierce, J. (2002) Development of molecular markers for constructing a genetic linkage map of the eastern oyster *Crassostrea virginica* (abstract only). *Aquaculture*, **204**, 205.

Gall, G.A.E. (1990) Basis for evaluating breeding plans. *Aquaculture*, **85**, 125–42.

Gall, G.A.E. & Busack, C.A. (eds) (1986) *Genetics in Aquaculture* II. Elsevier Science B.V., Amsterdam.

Gall, G.A.E. & Chen, H. (eds) (1993) *Genetics in Aquaculture* IV. Elsevier Science B.V., Amsterdam.

Gardner, J.P.A. & Skibinski, D.O.F. (1991) Mitochondrial DNA and allozyme covariation in a hybrid mussel population. *J. Exp. Mar. Biol. Ecol.*, **149**, 45–54.

Geller, J. B., Carlton, J.T. & Powers, D.A. (1994) PCR based detection of mtDNA haplotypes of native and invading mussels on the northeastern Pacific coast: latitudinal pattern of invasion. *Mar. Biol.*, **119**, 243–49.

Gendreau, S. & Grizel, H. (1990) Induced triploidy and tetraploidy in the European flat oyster, *Ostrea edulis*. *Aquaculture*, **90**, 229–38.

Gérard, A., Naciri, Y., Peignon, J.M., Ledu, C. & Phelipot, P. (1994a) Optimization of triploid induction by the use of 6-DMAP for the oyster *Crassostrea gigas* (Thunberg). *Aquaculture Fish. Manag.*, **25**, 709–19.

Gérard, A., Naciri, Y., Peignon, J.M. *et al.* (1994b) Image analysis: a new method for estimating triploidy in commercial bivalves. *Aquaculture Fish. Manag.*, **25**, 697–708.

Gjedrem, T. (1983) Genetic variation in quantitative traits and selective breeding in fish and shellfish. *Aquaculture*, **33**, 51–72.

Gjedrem, T. (ed) (1990) *Genetics in Aquaculture* III. Elsevier Science B.V., Amsterdam.

Gjedrem, T. (1992) Breeding plans for rainbow trout. *Aquaculture*, **100**, 73–83.

Gosling, E.M. (1982) Genetic variability in hatchery produced Pacific oysters (*Crassostrea gigas* Thunberg). *Aquaculture*, **26**, 273–87.

Gosling, E.M. (1992) Genetics. In: *The Mussel* Mytilus: *Ecology, Physiology, Genetics and Culture* (ed. E.M. Gosling), pp. 309–82. Elsevier Science Publishers B.V., Amsterdam.

Gosling, E.M. & Burnell, G.M. (1988) Evidence for selective mortality in *Chlamys varia* transplant experiments. *J. Mar. Biol. Ass. U.K.*, **68**, 251–58.

Gosling, E.M. & Nolan, A. (1989) Triploidy induction by thermal shock in the Manila clam, *Tapes semidecussatus*. *Aquaculture*, **78**, 223–28.

Guo, X. & Allen, S.K. Jr. (1994a) Viable tetraploids in the Pacific oyster (*Crassostrea gigas* Thunberg) produced by inhibiting polar body 1 in eggs from triploids. *Mol. Mar. Biol. Biotech.*, **3**, 42–50.

Guo, X. & Allen, S.K. Jr. (1994b) Reproductive potential and genetics of triploid Pacific oysters, *Crassostrea gigas* (Thunberg). *Biol. Bull.*, **187**, 309–18.

Guo, X. & Allen, S.K. Jr. (1994c) Sex determination and polyploid gigantism in the dwarf clam (*Mulinia lateralis* Say). *Genetics*, **138**, 1199–206.

Guo, X., Hershberger, W.K., Cooper, K. & Chew, K.K. (1993) Artificial gynogenesis with ultraviolet-irradiated sperm in the Pacific oyster, *Crassostrea gigas*. I. Induction and survival. *Aquaculture*, **113**, 201–14.

Guo, X., Debrosse, G.A., Allen, S.K. Jr. (1996) All-triploid Pacific oysters (*Crassostrea gigas* Thunberg) produced by mating tetraploids and diploids. *Aquaculture*, **142**, 149–61.

Hadley, N.H., Dillon, R.T. & Manzi, J.J. (1991) Realised heritability of growth rate in the hard clam *Mercenaria mercenaria*. *Aquaculture*, **93**, 109–19.

Hadrys, H., Balick, M. & Schierwater, B. (1992) Applications of random amplified polymorphic DNA (RAPD) in molecular ecology. *Mol. Ecol.*, **1**, 55–63.

Hallerman, E.M. & Kapuscinski, A.R. (1995) Incorporating risk assessment and risk management into public policies on genetically modified finfish and shellfish. *Aquaculture*, **137**, 9–17.

Hallerman, E., King, D. & Kapuscinski, A. (1999) A decision support software for safely conducting research with genetically modified fish and shellfish. *Aquaculture*, **173**, 309–18.

Hand, R.E., Nell, J.A. & Maguire, G.B. (1998) Studies on triploid oysters in Australia. X. Growth and mortality of diploid and triploid Sydney rock oysters *Saccostrea commercialis* (Iredale & Roughley). *J. Shellfish Res.*, **17**, 1115–27.

Haskin, H.H. & Ford, S.E. (1979) Development of resistance to *Minchinia nelsoni* (MSX) mortality in laboratory-reared and native oyster stocks in Delaware Bay. *Mar. Fish. Rev.*, **41**, 54–63.

Hawkins, A.J.S., Day, A.J., Gérard, A. *et al.* (1994) A genetic and metabolic basis for faster growth among triploids induced by blocking meiosis I but not meiosis II in the larviparous European flat oyster, *Ostrea edulis* L. *J. Exp. Mar. Biol. Ecol.*, **184**, 21–40.

He, M., Lin, Y., Shen, Q., Hu, J. & Jiang, W. (2000) Production of tetraploid pearl oyster (*Pinctada martensii* Dunker) by inhibiting the first polar body in eggs from triploids. *J. Shellfish Res.*, **19**, 147–51.

Hedgecock, D. & Sly, F. (1990) Genetic drift and effective sizes of hatchery-propagated stocks of the Pacific oyster, *Crassostrea gigas*. *Aquaculture*, **88**, 21–38.

Hershberger, W.K., Perdue, J.A. & Beattie, J.H. (1984) Genetic selection and systematic breeding in Pacific oyster culture. *Aquaculture*, **39**, 237–45.

Hillis, D.M., Moritz, C. & Mable, B. (eds) (1996) *Molecular Systematics*, 2nd edn. Sinauer Associates, Sunderland, Massachusetts, USA.

Hoelzel, A.R. (ed) (1992) *Molecular Genetic Analysis of Populations*. IRL Press at Oxford University Press, Oxford.

Høyheim, B., Kampenhaug, E., Lunner, S. *et al.* (2002) SALMAP: Constructing genetic maps of salmonid fish (abstract only). *Aquaculture*, **204**, 215.

Ibarra, A.M., Ramirez, J.L., Ruiz, C.A., Cruz, P. & Avila, S. (1999) Realised heritabilities and genetic correlation after dual selection for total weight and shell width in catarina scallop (*Argopecten ventricosus*). *Aquaculture*, **175**, 227–41.

Inoue, K., Waite, J., Matsuoka, M., Odo, S., Harayama, S. (1995) Interspecific variations in adhesive protein sequences of *Mytilus edulis*, *Mytilus galloprovincialis*, and *Mytilus trossulus*. *Biol. Bull.*, **189**, 370–75.

Iyengar, A., Müller, F. & Maclean, N. (1996) Regulation and expression of transgenes in fish – a review. *Transgenic Res.*, **5**, 1–19.

Jiang, W., Li, G., Xu, G, Lin, Y. & Qing, N. (1993) Growth of the induced triploid pearl oyster, *Pinctada martensii* (D). *Aquaculture*, **111**, 245–53.

Johannesson, K., Kautsky, N. & Tedengren, M. (1990) Genotypic and phenotypic differences between Baltic and North Sea populations of *Mytilus edulis* evaluated through reciprocal transplantations. II. Genetic variation. *Mar. Ecol. Prog. Ser.*, **59**, 211–19.

Jørstad, K.E., Naevdal, G., Paulsen, O.I. & Thorkildsen, S. (1994) Release and recapture of genetically tagged cod fry in a Norwegian fjord system. In: *Genetics and Evolution of Aquatic Organisms* (ed. A.R. Beaumont), pp. 519–28. Chapman and Hall, London.

Kapuscinski, A.R. & Jacobson, L.D. (1987) *Genetic Guidelines for Fisheries Management*. Minnesota Sea Grant, University of Minnesota, USA.

Kenchington, E., Naidu, K.S., Roddick, D.L., Cook, D.I. & Zouros, E. (1993) Use of biochemical genetic markers to discriminate between adductor muscles of the sea scallop (*Placopecten magellanicus*) and the Iceland scallop (*Chlamys islandica*). *Can. J. Fish. Aquat. Sci.*, **50**, 1222–28.

Kincaid, H.L. (1976a) Effects of inbreeding in rainbow trout populations. *Trans. Am. Fish. Soc.*, **105**, 273–80.

Kincaid, H.L. (1976b) Inbreeding in rainbow trout (*Salmo gairdneri*). *J. Fish. Res. Board Can.*, **33**, 2420–26.

Kincaid, H.L. (1977) Rotational line crossing: an approach to the reduction of inbreeding accumulation in trout brood stocks. *Prog. Fish Culturist*, **39**, 179–81.

Kincaid, H.L. (1983) Inbreeding in fish populations used in aquaculture. *Aquaculture*, **33**, 215–27.

Komaru, A. & Wada, K.T. (1989) Gametogenesis and growth of induced triploid scallops, *Chlamys nobilis*. *Nippon Suisan Gakkaishi*, **55**, 447–52.

Komaru, A., Uchimura, Y. & Wada, K.T. (1988) Detection of induced triploid scallop, *Chlamys nobilis*, by DNA microflurometry with DAPI staining. *Aquaculture*, **69**, 201–09.

Komaru, A., Matsuda, H., Yamakawa, T. & Wada, K.T. (1990) Meiosis and fertilization of the Japanese pearl oyster eggs at different temperatures observed with a fluorescent microscope. *Nippon Suisan Gakkaishi*, **56**, 425–30.

Kraeuter, J., Adamkewicz, L., Castagna, M., Wall, R. & Karney, R. (1984) Ridge number and shell colour in hybridised subspecies of the Atlantic bay scallop, *Argopecten irradians*. *Nautilus*, **98**, 17–20.

Lannan, J.E. (1972) Estimating heritability and predicting response to selection for the Pacific oyster, *Crassostrea gigas*. *Proc. Natl. Shellfish. Assoc.*, **62**, 62–66.

Lannan, J.E. (1980) Broodstock management of *Crassostrea gigas*: IV. Inbreeding and larval survival. *Aquaculture*, **21**, 353–56.

Li, Q., Osaka, M., Kashihara, M., Hirohashi, K. & Kijima, A. (2000) Effects of ultraviolet irradiation on genetical inactivation and morphological structure of sperm of the Japanese scallop, *Patinopecten yessoensis*. *Aquaculture*, **186**, 233–42.

Longwell, A.C. (1976) Review of genetic and related studies on commercial oysters and other pelecypod mollusks. *J. Fish. Res. Board Can.*, **33**, 1100–07.

Longwell, A.C. & Stiles, S. (1973) Gamete cross incompatibility and inbreeding in the commercial oyster, *Crassostrea virginica*. *Cytologia*, **38**, 521–33.

Lymbery, A.J. (2000) Genetic improvement in the Australian aquaculture industry. *Aquaculture Res.*, **31**, 145–49.

Mahon, G.A.T. (1983) Selection goals in oyster breeding. *Aquaculture*, **33**, 141–48.

Mallet, A.L. & Haley, L.E. (1983) Effects of inbreeding on larval and spat performance in the American oyster. *Aquaculture*, **33**, 229–35.

Mallet, A. L. & Haley, L.E. (1984). General and specific combining abilities of larval and juvenile growth and viability estimated from natural oyster populations. *Mar. Biol.*, **81**, 53–59.

Manzi, J.J., Hadley, N.H. & Dillon, R.T. (1991) Hard clam, *Mercenaria mercenaria*, broodstocks: growth of selected hatchery stocks and their reciprocal crosses. *Aquaculture*, **94**, 17–26.

Mason, K.M., Shumway, S.E., Allen, S.K. & Hidu, H. (1988) Induced triploidy in the soft-shelled clam *Mya arenaria*: energetic implications. *Mar. Biol.*, **98**, 519–28.

McAndrew, B. & Penman, D. (eds) (1999) *Genetics in Aquaculture* VI. Elsevier Science B.V., Amsterdam.

McGoldrick, D.J., Bayne, B.L., Innes, B., Ho Ha, J. & Ward, R.D. (2002) Updating our progress in the mapping of major genes in the Pacific oyster (*Crassostrea gigas*) (abstract only). *Aquaculture*, **204**, 229.

Meehan, B.W. (1985) Genetic comparison of *Macoma balthica* (Bivalvia, Telinidae) from the eastern and western North Atlantic Ocean. *Mar. Ecol. Prog. Ser.*, **22**, 69–76.

Meehan, B.W., Carlton, J.T. & Wenne, R. (1989) Genetic affinities of the bivalve *Macoma balthica* from the Pacific coast of North America: evidence for recent introduction and historical distribution. *Mar. Biol.*, **102**, 235–41.

Menzel, W. (1962) Seasonal growth of the northern and southern quahogs and their hybrids in Florida. *Proc. Natl. Shellfish Assoc.*, **53**, 111–18.

Menzel, W. (1968) Cytotaxonomy of species of clams (*Mercenaria*) and oysters (*Crassostrea*). *Symp. Mollusca, Mar. Biol. Ass. India*, Part 1, 75–84.

Menzel, W. (1989) The biology, fishery and culture of quahog clams, *Mercenaria*. In: *Clam Mariculture in North America* (eds J.J. Manzi & M. Castagna), pp. 201–42. Elsevier Science Publishing Company Inc., New York.

Murphy, R.W., Sites, J.W., Jr., Buth, D.G. & Haufler, C.H. (1996) Proteins: isozyme electrophoresis. In: *Molecular Systematics*, 2nd edn. (eds D.M. Hillis, C. Moritz & B.K. Mable), pp. 51–120. Sinauer Associates Inc., Sunderland, Massachusetts, USA.

Naciri-Graven, Y., Martin, A.G., Baud, J.P., Renault, T. & Gérard, A. (1998) Selecting the flat oyster *Ostrea edulis* (L.) for survival when infected with the parasite *Bonamia ostreae*. *J. Exp. Mar. Biol. Ecol.*, **224**, 91–107.

Nell, J.A. (2002) Farming triploid oysters. *Aquaculture*, **210**, 69–88.

Nell, J.A., Smith, I.R. & Maguire, G.B. (1994) Studies on triploid oysters in Australia. I. The farming potential of triploid oysters *Saccostrea commercialis* (Iredale and Roughley). *Aquaculture*, **126**, 243–55.

Nell, J.A., Sheridan, A.K. & Smith, I.R. (1996) Progress in a Sydney rock oyster, *Saccostrea commercialis* (Iredale and Roughley) breeding programme. *Aquaculture*, **144**, 295–302.

Nell, J.A., Smith, I.R. & Sheridan, A.K. (1999). Third generation evaluation of Sydney rock oyster *Saccostrea commercialis* (Iredale and Roughley) breeding lines. *Aquaculture*, **170**, 195–203.

Newkirk, G.F. (1978) Interaction of genotype and salinity in larvae of the oyster *Crassostrea virginica*. *Mar. Biol.*, **48**, 227–34.

Newkirk, G.F. (1983) Applied breeding of commercially important molluscs: a summary of discussion. *Aquaculture*, **33**, 415–22.

Newkirk, G.F. (1993) A discussion of genetic aspects of broodstock establishment and management. In: *Genetic Aspects of Conservation and Cultivation of Giant Clams* (ed. P. Munro), pp. 6–13. ICLARM Conf. Proc., 39, 47p.

Palmiter, R.D., Brinster, R.L., Hammer, R.E. *et al.* (1982) Dramatic growth of mice that develop from eggs microinjected with metallothionein-growth hormone fusion genes. *Nature* (Lond.), **300**, 611–15.

Park, L.K. & Moran, P. (1995) Developments in molecular genetic techniques in fisheries. In: *Molecular Genetics in Fisheries* (eds G.R. Carvahlo & T.J. Pitcher), pp. 1–28. Chapman and Hall, London.

Purdom, C.E. (1972) *Genetics and fish farming*. Lab. Leafl., MAFF Dir. Fish. Res., Lowestoft, 33, 17pp.

Purdom, C.E. (1987) Methodology on selection and intraspecific hybridisation in shellfish: a critical review. In: *Selection, Hybridization and Genetic Engineering in Aquaculture* (ed. K. Tiews), pp. 285–91. Heenemann, Berlin.

Rawson, P.D. & Hilbish, T.J. (1990) Heritability of juvenile growth rate for the hard clam *Mercenaria mercenaria*. *Mar. Biol.*, **105**, 429–36.

Scarpa, J., Komaru, A. & Wada, K.T. (1994) Gynogenetic induction in the mussel, *Mytilus galloprovincialis. Bull. Natl. Res. Inst. Aquaculture*, **23**, 33–41.

Shen, Y.P., Zhang, X.Y., He, H.P. & Ma, L.J. (1993) Triploidy induction by hydrostatic pressure in the pearl oyster, *Pinctada martensii* Dunker. *Aquaculture*, **110**, 221–27.

Smith, I.R., Nell, J.A. & Adlard, J. (2000) The effect of growing level and growing method on winter mortality, *Mikrocytos roughleyi*, in diploid and triploid Sydney rock oysters, *Saccostrea glomerata. Aquaculture*, **185**, 197–205.

Smith, P.J., Ozaki, H. & Fujio, Y. (1986) No evidence for reduced genetic variation in the accidentally introduced oyster *Crassostrea gigas* in New Zealand. *N. Z. J. Mar. Freshw. Res.*, **20**, 569–74.

Sobolewska, H., Beaumont, A.R. & Hamilton, A. (2001) Dinucleotide microsatellites isolated from the European flat oyster, *Ostrea edulis. Mol. Ecol. Notes*, **1**, 79–80.

Stanley, J.G., Allen, S.K. Jr. & Hidu, H. (1981) Polyploidy induced in the American oyster, *Crassostrea virginica*, with cytochalasin B. *Aquaculture*, **23**, 1–10.

Stanley, J.G., Hidu, H. & Allen, S.K. Jr. (1984) Growth of American oysters increased by polyploidy induced by blocking meiosis I but not meiosis II. *Aquaculture*, **37**, 147–55.

Suchanek, T.H., Geller, J.B., Kreiser, B.R. & Mitton, J.B. (1997) Zoogeographic distributions of the sibling species *Mytilus galloprovincialis* and *M. trossulus* (Bivalvia: Mytilidae) and their hybrids in the North Pacific. *Biol. Bull.*, **193**, 187–94.

Supan, J.E. (2000) Tetraploid eastern oysters: an arduous effort (abstract only). *J. Shellfish Res.*, **19**, 655.

Tabarini, C.L. (1984) Induced triploidy in the bay scallop, *Argopecten irradians*, and its effect on growth and gametogenesis. *Aquaculture*, **42**, 151–60.

Tave, D. (1993) *Genetics for Fish Hatchery Managers*, 2nd edn. Van Nostrand Reinhold, New York.

Theisen, B.F. (1978) Allozyme clines and evidence of strong selection in three loci in *Mytilus edulis* (Bivalvia) from Danish waters. *Ophelia*, **17**, 135–42.

Toro, J.E. & Newkirk, G.F. (1990) Divergent selection for growth rate in the European oyster *Ostrea edulis*: response to selection and estimation of genetic parameters. *Mar. Ecol. Prog. Ser.*, **62**, 219–27.

Toro, J.E., Sanhueza, M.A., Winter, J.E., Aguila, P. & Vergara, A.M. (1995) Selection responses and heritability estimates for growth in the Chilean oyster *Ostrea chilensis* (Philippi, 1845). *J. Shellfish Res.*, **14**, 87–92.

Toro, J.E., Aguila, P. & Vergara, A.M. (1996) Spatial variation in response to selection for live weight and shell length from data on individually tagged Chilean native oysters (*Ostrea chilensis* Philippi, 1845). *Aquaculture*, **146**, 27–36.

Váradi, L., Benkó, I., Varga, J. & Horváth, L. (1999) Induction of diploid gynogenesis using interspecific sperm and production of tetraploids in African catfish, *Clarias gariepinus* Burchell (1822). *Aquaculture*, **173**, 401–11.

Vrijenhoek, R.C., Ford, S.E. & Haskin, H.H. (1990) Maintenance of heterozygosity during selective breeding of oysters for resistance to MSX disease. *J. Hered.*, **81**, 418–23.

Wada, K.T. (1984) Breeding study of the pearl oyster *Pinctada fucata. Bull. Natl. Res. Inst. Aquaculture*, **6**, 79–157.

Wada, K.T. (1986) Genetic selection for shell traits in the Japanese pearl oyster *Pinctada fucata martensii. Aquaculture*, **57**, 171–76.

Wada, K.T. (2000) Genetic improvement of stocks of the pearl oyster. In: *Recent Advances in Marine Biotechnology: Aquaculture: Seaweeds and Invertebrates*, Vol. 4, Part A (eds M. Fingerman & R. Nagabhushanam), pp. 75–85. Science Publishers Inc., USA.

Wada, K.T. & Komaru, A. (1994) Effect of selection for shell coloration on growth rate and mortality in the Japanese pearl oyster, *Pinctada fucata martensii*. *Aquaculture*, **125**, 59–65.

Wang, Z., Guo, X. & Wang, R. (2002) Studies on heterozygosity and growth in triploid Pacific oyster (*Crassostrea gigas*) (abstract only). *Aquaculture*, **204**, 337–48.

Wilkins, N.P. (1981) The rationale and relevance of genetics in aquaculture: an overview. *Aquaculture*, **22**, 209–28.

Wilkins, N.P. & Gosling, E.M. (eds) (1983) *Genetics in Aquaculture* I. Elsevier Science B.V., Amsterdam.

Wirgin, I.I. & Waldman, J.R. (1994) What DNA can do for you. *Fisheries*, **19**, 16–27.

Wright, J.M. & Bentzen, P. (1995) Microsatellites: genetic markers for the future. In: *Molecular Genetics in Fisheries* (eds G.R. Carvahlo & T.J. Pitcher), pp. 117–19. Chapman and Hall, London.

Yang, H., Li, L., Wang, H. & Guo, X. (2002) Increased cell size as a cause for polyploid gigantism in triploid Zhikong scallop *Chlamys farreri* (abstract only). *Aquaculture*, **204**, 252.

Yamamoto, S. & Sugawara., Y. (1988) Induced triploidy in the mussel, *Mytilus edulis*, by temperature shock. *Aquaculture*, **72**, 21–29.

Yamamoto, S., Sugawara, Y., Nomura, T. & Oshino, A. (1988) Induced triploidy in the Pacific oyster *Crassostrea gigas*, and the performance of triploid larvae. *Tohoku J. Ag. Res.*, **39**, 47–59.

11 Diseases and Parasites

Introduction

Because of the considerable economic importance of bivalves there is a wealth of information available on diseases that affect them. To quote Kinne, who said in 1983: 'we now know more about the diseases of bivalves than of the diseases of all other marine invertebrate groups together'. Twenty years later this still applies. The major disease-causing agents of marine bivalves are: viruses, bacteria, fungi, protozoans, helminths and parasitic crustaceans. Because of the considerable amount of information available, only key bivalve pathogens within each group will be described and readers are referred to Lauckner (1983), Sparks (1985), Sindermann (1990), Bower et al. (1994), Ford & Tripp (1996) and Ford (2001) for additional information. In addition, the different defence mechanisms utilised by bivalves to combat disease are dealt with.

Viruses

Viruses are ultramicroscopic infectious agents (~20–350 nm long), which can only multiply inside living cells. Once infected, the host cell begins to degenerate and is soon destroyed, liberating new virus particles in the process. These in turn infect new cells. The first incidence of a viral disease in shellfish was reported by Farley et al. (1972) and since then about 20 more have been described. However, in most cases only a tentative link can be made between a particular virus and the pathology observed. In order to demonstrate that a virus is the causal agent it must first be isolated from a diseased host and passed via a cell-free extract to a susceptible host in which the same disease is later manifested. This is not an easy task because there is seldom conclusive evidence that the pathology is virus related; there are also problems in identification, isolation and characterisation of the virus; experimental transmission of the virus is virtually impossible due to a lack of standardised bivalve cell cultures; and there is often uncertainty as to whether the virus is a primary pathogen or secondary invader. Some examples of viral diseases in bivalves are presented in Table 11.1. The role of bivalves as carriers of enteroviruses, such as hepatitis A or polio, will be dealt with in the chapter on public health.

Bacteria

The aquatic environment harbours a rich bacterial flora and bivalves because of their efficient filter-feeding mechanism, can ingest many different kinds of microorganisms. While the majority of marine bacteria are not harmful, unless present in excessive numbers, some can be pathogenic to their host. In some cases bivalves can act as passive carriers of human pathogens, without themselves contracting a bacterial disease (see Chapter 12). Most of the informa-

Table 11.1. Examples of some disease-causing viruses in bivalves.

Species	Effects on host	Distribution	Reference
Mytilus edulis	Lesions in digestive gland and mantle.	Denmark	Rasmussen, 1986
Crassostrea virginica	Massive hypertrophy of gametes; lysis of infected cells.	Long Island, USA	Meyers, 1981
Crassostrea angulata	Gill erosion and necrosis; believed to be cause of *maladie des branchies* that was responsible for mass mortalities in France between 1966 and 1969.	France, Spain, Portugal	References in Sindermann, 1990
Crassostrea gigas larvae	Feeding stopped 3–4 days after spawning, followed by 60–100% mortalities.	New Zealand hatchery	Hine *et al.*, 1992

tion on bacterial diseases comes from studies on cultured larvae, which are more susceptible than adults to infection (Table 11.2). This is because larvae are reared in static systems at elevated temperatures and high densities. Information on bacterial infections and their treatment in hatchery-reared larvae and juveniles is in Chapter 9. Most bacterial diseases of bivalves are caused by species of *Vibrio* and *Pseudomonas* (Table 11.2), even though species in these genera are commonly found on the shell, in the pallial fluid, and digestive tract (Elston *et al.*, 1982; Paillard *et al.*, 1996). One of these diseases, brown ring disease in clams will now be described in some detail. This is one of the few well-documented bacterial diseases that affect adults as well as juveniles.

Brown ring disease

During the summer of 1987 mass mortality of adult and juvenile *Ruditapes philippinarum* were reported from culture beds in the north west of France. All of the moribund clams exhibited a brown deposit on the inside of the shell, between the pallial line and the shell margin (Fig. 11.1). The deposit forms a characteristic ring, hence the name brown ring disease or BRD (Paillard *et al.*, 1989). The etiological agent responsible for the disease is *Vibrio tapetis*. In the field the pathogen appears to be species specific, although in laboratory challenge tests it is possible to infect the closely related species *R. decussatus* (Maes & Paillard, 1992). Since 1987 the disease has been reported from other areas of France, and in areas of Spain, Portugal, Italy, the UK and Ireland where the Manila clam is cultivated (Paillard *et al.*, 1994; Figueras *et al.*, 1996; Allam *et al.*, 2000). BRD seems to be indigenous to Europe since there have been no reports of the disease from outside this region.

The typical sign of BRD is the brown organic deposit on the inside of the shell (Paillard & Maes, 1994; Paillard & Maes, 1995a,b) but as the disease progresses there is degeneration of the digestive diverticula manifested in an inability to store reserves, mainly glygogen, and distribute them to other tissues

Table 11.2. Examples of some pathogenic bacteria in bivalves.

Bacterium	Disease	Bivalve	Symptoms	Reference
Vibrio sp.	Vibriosis (bacillary necrosis)	*Crassostrea virginica* larvae	Up to 100% mortality	In Sindermann, 1990
New species of α-proteobacteria (CVSP)	Juvenile oyster disease (JOD)	*C. virginica* juveniles <25 mm shell length	Mortalities up to 90%	Davis & Barber, 1999; Boettcher et al., 2000
Vibrio anguillarum-related (VAR)		*Argopecten purpuratus* larvae	High mortalities	Riquelme et al., 1995
Vibrio sp.		*Tridacna gigas* larvae	Rapid and high mortalities	Sutton & Garrick, 1993
Vibrio tapetis	Brown ring disease (BRD)	*Ruditapes philippinarum*	Brown deposit on shell; degeneration of digestive gland followed by metabolic disorder and death	See text for references
Nocardia sp.	Pacific oyster nocardiosa (PON) or summer disease	*Crassostrea gigas*	Connective tissue degeneration folowed by high mortalities	References in Paillard et al., 1994
Pseudomonas sp.		*Ostrea edulis* larvae	High mortalities	Helm & Smith, 1971
Rickettsia		*Mytilus galloprovincialis*	Some gill malformation and lowered condition	Figueras et al., 1991
Cyanobacteria		*Perna perna*	Shell damage; lowered reproductive output; mortalities up to 50%	Kaehler & McQuaid, 1999

(Plana *et al.*, 1996). Growth, weight and condition indices are subsequently affected. In addition, accumulation of the brown deposit induces mantle recession and ultimately the two shell edges are no longer watertight. Various types of debris accumulate, including sand and algae, which in turn may promote colonisation by fungi, worms, and secondary infectious agents (Paillard *et al.*, 1994). Antibiotics have been used successfully on diseased clams (Noel *et al.*, 1992), and are also used to prevent infections of spat production in hatcheries (Noel *et al.*, 1996). The development of an immunological assay to identify and quantify *V. tapetis* means that it is now possible to develop epidemiological research on BRD, follow the kinetics of infection, and look for the vibrio in other bivalve species (Castro *et al.*, 1995; Noel *et al.*, 1996). However, this assay is not very sensitive and may soon be replaced by a DNA assay (C. Paillard, personal communication 2001).

Before leaving this section on bacterial diseases a few points are worth noting:

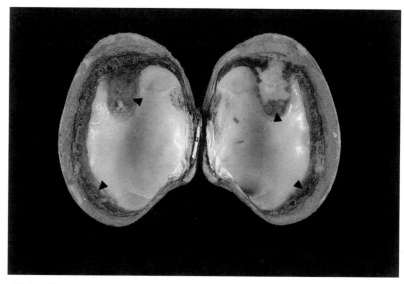

Fig. 11.1. Characteristic brown ring disease organic material (arrowed) symptoms on the inner-shell surface of the Manila clam, *Ruditapes philippinarum*. Photo courtesy of C. Paillard, Institut Universitaire Européen de la Mer, Université de Bretagne Occidentale, France.

- Bacteria that are pathogenic in larval rearing systems may or may not be pathogenic in the wild. Adults in culture are frequently unaffected by concentrations of bacteria that are pathogenic to larvae.
- Very often bacteria are not the primary causative agent in disease; stress factors can lower resistance, and thus increase the probability of infection.
- The study of pathogens like *Vibrio* and *Pseudomonas* is problematic in that their taxonomy is disorganised; they consist of many strains, which appear to be evolving rapidly; and their pathogenicity varies depending on the host species (Sindermann, 1990).
- Remedial measures such as the use of antibiotics in culture systems are not always successful; elimination of certain species of bacteria may favour the appearance of other bacteria, and reduce bivalve resistance to infection by pathogenic forms. Also, there is the danger that with persistent use of antibiotics resistant strains of bacteria will appear. Alternatives to systematic use of antibiotics are UV treatment of incoming waters, maintenance of bacterial-free algae cultures, and good hygiene practice (Prieur *et al.*, 1990).
- Results from a search for naturally occurring disease control agents look very promising. Strains of bacteria have been isolated that are able to promote the growth and survival of larvae by inhibiting the activity of other deleterious bacteria that flourish in hatchery cultures (see Chapter 9).

Fungi

Few fungi are pathogenic to bivalves. Those that are confine themselves to external parts such as shell or byssus. It is the larval stages that are most prone to disease, although it is not always clear whether the fungus is a primary pathogen or secondary invader. There is one disease, however, maladie du pied or shell disease, which has severe effects on the oyster *Ostrea edulis* in western Europe. The causative agent, *Ostracoblabe implexa*, lives in the shell (Alderman & Jones, 1971a). Initially, small white spots appear on the inner shell surfaces but as the fungus proliferates these develop into greenish rubbery warts of conchiolin, particularly in the region of the adductor muscle attachment. In the advanced stages of the disease the valves become grossly deformed, and shell closure may become impossible, leading to the eventual death of the oyster. In addition to *O. edulis*, another oyster species, *Crassostrea angulata*, is susceptible to the disease but only suffers light infections (Alderman & Jones, 1971b).

The disease was first reported in Holland in 1902, but was not serious in that only a small percentage of oysters were infected. However, with increased oyster production in the 1930s, and subsequent widespread importation and dispersal of oysters, the disease has spread through western Europe. Progress of the disease is usually slow but high water temperatures (above 19°C) favour the spread and virulence of the fungus (Lauckner, 1983). Incidences can be as high as 70%, and young oysters are more susceptible than older individuals; apparently, the latter may recover from an infection under good feeding conditions.

Protozoa

These single-celled organisms are the most common cause of disease in bivalves. Most belong to the phyla Acetospora and Apicomplexa, and infect mainly oyster and clam species. The enormous literature on these pathogens is a reflection of the extent of damage that they cause to commercial production. Undoubtedly, they are also significant disease-causing organisms in non-economic species. The most important protozoan pathogens are listed in Table 11.3; some of these are considered below.

Bonamia ostreae

Bonamia ostreae is a haplosporidian protozoan (Phylum Acetospora), which infects haemocytes of the European oyster, *Ostrea edulis* (Fig. 11.2). The disease, bonamiasis, was first reported from the coasts of France in 1979 (Comps *et al.*, 1980). Since then there have been significant and widespread mortalities in oyster growing regions, not just in Europe (references in Montes *et al.*, 1994), but further afield on the west coast of the United States (Elston *et al.*, 1986) and also in New Zealand (Dinamani *et al.*, 1987). The parasite has been detected in Maine on the east coast of the United States, but so far no mortalities have been reported, probably due to the low density of oysters in natural beds and the relatively cold winters (Friedman & Perkins, 1994; Zabaleta & Barber, 1996).

To date, no agreement has been reached regarding the life cycle of *B. ostreae* although there is a growing body of information on the morphology and fine structure of the parasite, and the histopathology and epidemiology of the

Table 11.3. Important protozoan pathogens of oysters; *indicates disease organisms dealt with in text. Information on non-asterisked pathogens from text of Sindermann (1990).

Pathogen	Disease name	Host	Locality	Pathology
*Bonamia ostreae**	Bonamiasis	*Ostrea edulis*	Atlantic coasts of western Europe	Yellow discolouration of tissue, extensive gill lesions, breakdown of connective tissue and loss of condition; mortalities up to 90%.
Marteilia refringens	Digestive gland (or Aber) disease	*O. edulis* but low prevalence in *Crassostrea gigas* and *Mytilus edulis*	Atlantic coasts of France and Spain	Pale digestive gland, severe emaciation, failure to close valves; maximum mortalities (≤90%) in May–August.
Marteilia sydneyi		*Saccostrea glomerata*	E. Australia	Necrosis of digestive gland, loss of condition, inhibition of gametogenesis; mortalities up to 80%.
*Haplosporidium nelsoni**	MSX	*Crassostrea virginica*	Maine to Florida, USA	Emaciation, reduced condition, and fecundity; maximum mortalities (up to 100%) in summer.
H. costale	Seaside disease	*C. virginica*	Virginia and Maryland, USA	Emaciation, discolouration of digestive gland, mantle recession; maximum mortalities (≤40%) in May–June.
*Perkinsus marinus**	'Dermo' disease	*C. virginica*	Gulf of Mexico northwards to Delaware Bay, Cape Cod, and Maine, USA	Severe emaciation, loss of condition; mortalities as high as 100% depending on temperature and salinity.

disease. Van Banning (1990) proposed a presumptive life cycle in which the first phase of infection occurred in the ovarian tissue of the oyster followed by a haemocytic phase. However, electron microscope studies have since shown that *B. ostreae* infects not just haemocytes but also gill epithelial cells, and there does not appear to be an initial infective stage in ovarian tissue (Montes *et al.*, 1994; Culloty & Mulcahy, 1996). The parasite proliferates in both epithelial cells and haemocytes, and because free parasite cells have been observed in the haemolymph it is believed that infected haemocytes eventually die and release their contents. The freed parasites possibly then infect other haemocytes and gill epithelial cells or alternatively, enter the water column where they subsequently infect other oysters (Montes *et al.*, 1994). Uninfected oysters introduced in contaminated areas contract the disease after about three

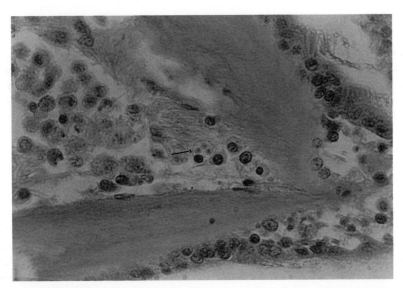

Fig. 11.2. Tissue cross-section from the oyster, *Ostrea edulis*, showing *Bonamia ostreae* protozoa within a haemocyte (arrowed). Haematoxylin and eosin stain (×100). Photo courtesy of the Marine Institute, Dublin, Ireland.

to five months (Tigé & Grizel, 1984; Elston *et al.*, 1987; Montes, 1991). It is worth mentioning that other species, e.g. *Mytilus edulis*, *M. galloprovincialis*, *Ruditapes philippinarum*, *R. decussatus* and *Crassostrea gigas*, cannot be infected either naturally or experimentally, and therefore do not appear to act as vectors or intermediate hosts for the parasite (Culloty *et al.*, 1999).

Yellow discolouration of oyster tissue, extensive gill lesions, breakdown of connective tissue and loss of condition characterise the disease (Sindermann, 1990; Rogan *et al.*, 1991). *Bonamia* can be detected in oysters of all sizes, but the percentage is close to zero in individuals <2 cm shell length; after that infection increases with increasing size reaching values higher than 30% (Cáceres-Martínez *et al.*, 1995). The parasite is present and transmissible throughout the year, and a continual cycle of infection and mortality occurs (Grizel *et al.*, 1988; Culloty & Mulcahy, 1996). To date it has not been possible to culture *B. ostreae*, but Miahle *et al.* (1988) have developed a method for the isolation and purification of parasitic cells from infected oysters. In addition, primary cultures of haemocytes can be prepared with relative ease. Therefore, it has been possible to carry out *in vitro* studies on interactions between *B. ostreae* and haemocytes. The parasite enters the cells by phagocytosis, and is then enclosed by a membrane to form a phagosome. The parasite proliferates within the phagosome and digests host haemocyte cytoplasm using hydrolytic enzymes contained in lipoid bodies within the parasite. In the host cell the parasite is neither injured nor destroyed because the particular haemocytes that are infected lack a lysosomal system (Hine & Wesney, 1994). The 50% infectious dose (ID), estimated from inoculations with from 10 to 10^6 *B. ostreae* per individual, was estimated to be 80 000 parasites per 3-year-old oyster. This 50% ID has been used to screen for parasite-resistant oysters and preliminary results have shown that the F_1 progeny of oysters that survived previous outbreaks of bonamiasis were more resistant to the disease (Hervio *et al.*, 1995). Until recently, detection of *Bonamia* has relied heavily on light microscopic examination of haemolymph smears from heart tissue, or stained sections or smears from heart, digestive gland and gill. These methods are time consuming and are not efficient in detecting low levels of infection. An immunoassay, which is fast and accurate, has been developed (Cochennec *et al.*, 1992), and also, more recently, a polymerase chain reaction (PCR) assay capable of detecting very small amounts of parasite ribosomal DNA in bulk DNA from oyster gill (Carnegie *et al.*, 2000). See Miahle *et al.* (1992) on development of DNA probes for diagnosis of infectious diseases.

Bonamiasis continues to spread, both within and between countries, through movement of infected oysters. Transmission of the disease within infected areas seems to depend on proximity of oyster stocks and local hydrography. Speed of the spread is mainly influenced by the density of oysters and their resistance to the parasite. Resistance seems to be significantly affected by dredging, handling, temperature and salinity fluctuations. In France the government agency IFREMER have produced several strains of *O. edulis* that are resistant to the parasite (Naciri-Graven *et al.*, 1998, 1999). It should be noted that their use of the term 'resistant' refers only to improved survival, and implies nothing about host-parasite interactions (Naciri-Graven *et al.*, 1999). Culloty *et al.* (2001) have reported an Irish strain that is sufficiently tolerant of the parasite so that oysters can be grown to market size. Various national control measures to minimise the consequences of bonamiasis are in place (Table 11.4) but despite these the disease continues to spread.

Haplosporidium nelsoni

Haplosporidium nelsoni is an acetosporan parasite that causes the disease MSX (multinucleate sphere X), an often fatal disease of *Crassostrea virginica*. Unknown prior to 1957, MSX killed 90–95% of the oysters in Delaware and

Table 11.4. UK guidelines to minimise the consequences of bonamiasis in relaid stocks. *procedure covered in Chapter 12. Modified from Hudson & Hill (1991).

Do not accept undocumented oysters for deposit.
Do not relay diseased oysters.
Only relay oysters for one growing season.
Do not return undersized oysters to lays.
Use oyster beds in rotation.
Reduce density of lays.
Destroy unwanted oysters on land, away from water.
Avoid putting oysters under unnecessary stress.
Do not depurate* in areas free of the disease.
Clean all fishing gear before returning to base.
Clean depuration sites regularly.
Clear infected lays quickly and immediately when infestation levels exceed 10%.
Oysters *en route* to market should only be deposited (if necessary) in tanks or ponds that can be readily cleaned after use.

lower Chesapeake Bays, United States, between 1957 and 1960 (Haskin *et al.*, 1966; Andrews & Wood, 1967). The parasite remains prevalent in these areas, but has also spread northwards as far as Maine, and southwards to Florida. Initial infections occur in the gill epithelium, where the parasite proliferates, eventually breaking through the basement membrane into the circulatory system and is spread to all organs. Infections confined to the gill epithelium are not lethal and have little measurable effect on the oyster (Ford *et al.*, 1999). However, the effects of systemic infection on the host are severe in that MSX reduces food intake and competes for energy reserves. Not surprisingly, parameters such as condition index and fecundity are affected. For example, oysters with systemic MSX infections had a condition index that was 31% lower, and a relative fecundity that was 81% lower than uninfected oysters (Barber *et al.*, 1988). It is still not known how MSX actually kills its host. Highly susceptible oysters die rapidly and in 'good condition', possibly killed by a toxin that disrupts some key metabolic process. More resistant oysters take longer to die, and when they do, they are emaciated, indicating that the metabolic burden of prolonged parasitism finally killed them (Ford & Tripp, 1996 and S. Ford, personal communication 2001).

The parasite is thought to enter the oyster as a uninucleate spore that divides to produce a multinucleate plasmodium, from 4–30 μm in diameter, containing up to 60 nuclei (Figs. 11.3 & 11.4). This stage is found throughout all tissues. Spore stages are rare in adults but are regularly formed in juvenile oysters with advanced infections. The reason may be that juvenile oysters with their higher energy reserves provide a superior internal environment for sporulation than do adult oysters (Ford *et al.*, 1999). Sporulation is confined to the epithelial lining of the digestive tubules and mature spores are shed from live or dying oysters. Transmission in the field does not depend on the presence of infected oysters and to date transmission has not been achieved experimentally. It is not known whether spores are directly infective to oysters, or whether they infect an alternative host.

In the field new infections are acquired from June through to October each year (Fig. 11.5A). Maximum mortality occurs in the summer, but declines as water temperatures decrease in the autumn (Fig. 11.5B). Infection levels do, however, remain high during the winter months, and parasite activity and oyster mortality resume when water temperature starts to increase the following spring (Barber *et al.*, 1991). Superimposed on this pattern there is a periodicity to infection with peaks spaced six to eight years apart; low disease activity appears to be associated with unusually low winter temperatures in the preceding year (Ford & Haskin, 1982). Salinity is another important controlling factor in the distribution of MSX. *Crassostrea virginica* is found in salinities from 5 to 30 psu but infections are extremely rare at salinities below 10 psu, and the disease does not become epizootic (spreading rapidly) at salinities below 15 psu. Results from laboratory and field studies indicate that this is most probably due to the physiological inability of the parasite to tolerate salinities below 10 psu (Sprague *et al.*, 1969; Haskin & Ford, 1982), rather than to enhanced effectiveness of host defence mechanisms (Ford & Haskin, 1988). In a set of inter-related papers Powell *et al.* (1999), Ford *et al.* (1999) and Paraso *et al.* (1999) have developed complex mathematical models, from

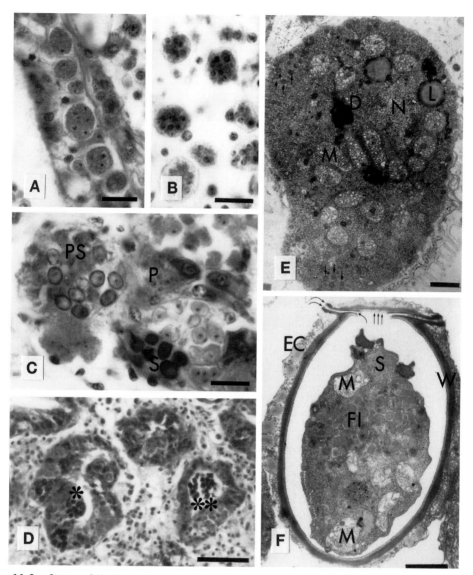

Fig. 11.3. Stages of *Haplosporidium nelsoni* found in *Crassostrea virginica*. (A) Plasmodia in gill epithelium (earliest recognised stage). (B) Plasmodia in haemolymph vessels. (C) Plasmodial [P], prespore [PS] and spore [S] stages. (D) Sporulating stages in epithelium * and lumen ** of digestive tubules. (E) Plasmodium showng mitochondria [M], nucleus [N], lipid body [L], digestive lamellae [D] and haploid spores (arrows). (F) Mature spore showing spherule [S], mitochondria [M], spore wall [W], operculum [straight arrows], juncture between operculum and lip of spore wall [curved arrows], formative inclusions [FI] thought to produce haplosporosomes, and epispore cytoplasm [EC]. Scale bars = 10 μm in A to C; 50 μm in D; 1 μm in E and F. E from Scro & Ford (1990); rest from Ford & Tripp (1996).

experimental and field data, to describe the transmission of the disease and host-parasite-environmental interactions.

The effects of the parasite on the oyster industry in the United States have been devastating. At least half of all oyster deaths in lower Delaware Bay from the early 1960s to the early 1980s were attributed to MSX (Ford & Haskin,

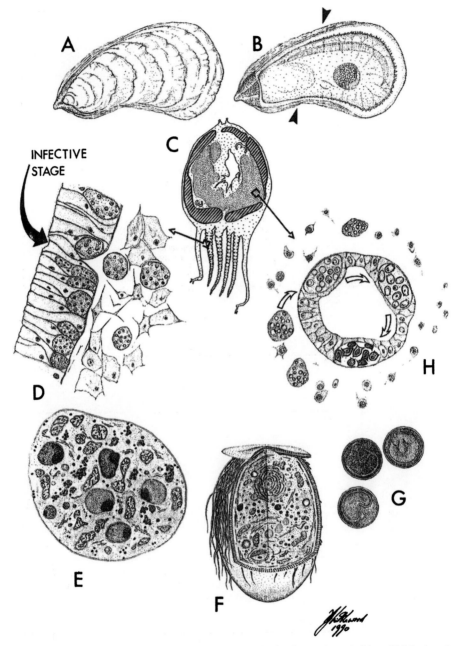

INFECTIVE STAGE

Fig. 11.4. Diagram of stages of *Haplosporidium nelsoni* seen in *Crassostrea virginica*: (A) intact oyster; (B) oyster with right valve removed showing plane [arrows] of section in (C); (D) gill epithelium (entry site) showing plasmodia between cells and in spaces beneath epithelium; (E) plasmodium; (F) spore; (G) haploid spores; (H) development of plasmodia into sporocysts in epithelium of digestive tubules [open arrows]. From Ford & Tripp (1996). Original drawing by Dr. Tim Littlewood.

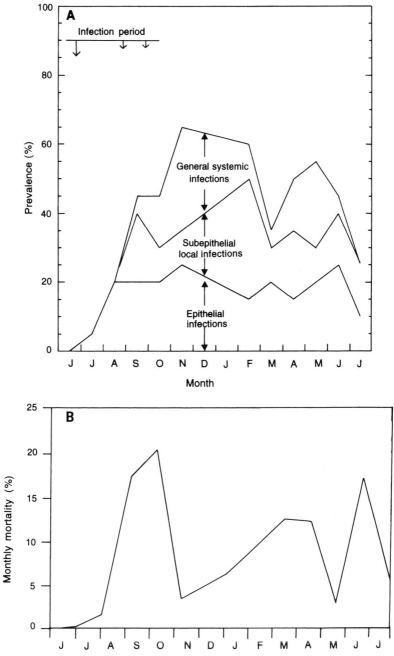

Fig. 11.5. (A) Annual *Haplosporidium nelsoni* infection in *Crassostrea virginica* in Delaware Bay, New Jersey. Filled arrows represent the percentage of oysters in each infection level (prevalence). Epithelial infections are the lightest infections and systemic infections are the heaviest. The infection period (open arrows) designates the time in which infections begin in uninfected oysters; the length of the arrows represent the relative intensity of infection. From Paraso *et al.* (1999) after Ford & Haskin (1982). (B) Monthly mortality of oysters subject to first year exposure to the parasite. From Ford *et al.* (1999).

1982). However, since about 1990 there has not been much mortality in the bay (S. Ford personal communication 2001). The parasite is still present because it continues to infect and kill susceptible oysters brought into the area. It is believed that the native Delaware oysters have, in a relatively short space of time, become highly resistant to the parasite. Resistant oysters are infected but are able to contain and localise infections, preventing them from becoming systemic (Ford & Haskin, 1982). In addition, infected resistant oysters are able to tolerate parasite numbers that would be fatal to susceptible oysters (Ford & Haskin, 1987). These findings have stimulated a programme of selective breeding to create highly resistant oyster strains (see Chapter 10). Strains of oysters now exist which have up to ten times the survival of unselected oysters. However, even highly selected oysters may eventually die with *H. nelsoni* infections (Ford & Haskin, 1987). Also, MSX-resistant oysters do not have enhanced resistance to *Perkinsus marinus* (Chintala & Fisher, 1989), another parasite of this species (see below). Disease resistant oysters might also be produced by gynogenesis. Lightly irradiated sperm from *Crassostrea gigas*, a species resistant to MSX and *P. marinus* infection, could be used to fertilise *C. virginica* eggs. If the paternal chromosome fragments contain elements that provide disease resistance, and these are successfully incorporated into the genome and expressed, the result could be a disease-resistant oyster (Gaffney & Bushek, 1996). Resistance to *H. nelsoni* may, according to Ford & Haskin (1987) involve a combination of failure to provide a completely suitable environment for the parasite (i.e., being insusceptible), which inhibits the ability to complete its life cycle, and a capacity to tolerate parasitism while continuing to carry out life processes while parasitised (Fig. 11.6). A range of para-

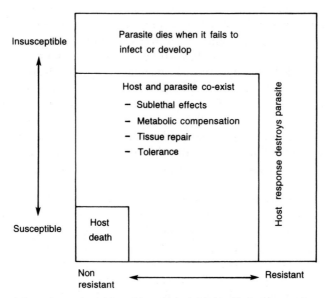

Fig. 11.6. Schematic representation of hypothetical relationship between resistance and susceptibility to *Haplosporidium nelsoni* infection in the oyster *Crassostrea virginica*. From Ford (1988). Reprinted with permission from the America Fisheries Society.

site burdens are found in unselected oysters indicating that individuals vary in their ability to provide a suitable environment for plasmodial growth (Ford, 1986).

There are now several DNA-based diagnostic assays available, initiated using parasites in infected oysters (Stokes & Burreson, 1995; Day *et al.*, 2000).

Perkinsus marinus

A third serious protozoan parasite, *Perkinsus marinus*, Phylum Apicomplexa, causes dermo disease in *Crassostrea virginica*. The pathogen, first described from the Gulf of Mexico (Mackin *et al.*, 1950), was found all along the south-east coast of the United States as far north as the Virginia portion of Chesapeake Bay during the late 1940s and early 1950s (see Ray, 1996, for a lively account of the history of the disease). In the mid 1980s it spread northwards within Chesapeake Bay, and between 1990 and 1992 it was found from Delaware Bay to Cape Cod, a distance of about 500 km (Ford, 1996). The parasite is now firmly established in oyster populations along the Connecticut shoreline and as far north as the state of Maine. Essentially, the parasite is present throughout the Gulf of Mexico and all along the east coast of the United States, and causes mortalities over most of that range. Typical prevalence (number of individuals infected) rates are in the range 50–100% (Wilson *et al.*, 1990; Karolus *et al.*, 2000).

The earliest recognised stage in oysters is unicellular trophozoites that proliferate by multiple fission intra- or extracellularly throughout the host tissue (Fig. 11.7). This process may occur in any tissue but is most often found in connective tissue, between epithelial cells of the gut, gill, digestive gland, and in lesions of the gut and gill epithelia (Perkins, 1996). When released into seawater the trophozoites enlarge and repeatedly divide to form hypnospores that contain numerous biflagellated zoospores. These are eventually discharged through a tube (Fig. 11.7) and the liberated zoospores swim to, or are pulled into, an oyster's mantle cavity, thus setting up new infections. Primary infection is most likely through the gill, mantle and gut epithelia. The parasite destroys the epithelium, lyses the basement membrane, and is distributed via haemocytes to all parts of the body. Diseased oysters exhibit severe emaciation and multiple abscesses. Reproduction is also affected in that diseased oysters produce fewer eggs, albeit with a similar lipid content to the eggs of healthy individuals. The retarded growth exhibited by diseased oysters suggests that energy for somatic growth is being diverted to reproduction (Kennedy *et al.*, 1995).

The primary mechanism of disease transmission from one oyster to another is through the water (Ray, 1954). However, there is some evidence that a mobile vector, the ectoparasitic gastropod, *Boonea impressa*, may play a role in spreading infection (White *et al.*, 1989). The slipper limpet (*Crepidula* spp.), one of the most important fouling competitors of oysters, may serve as a conduit between oysters and *B. impressa*, since juvenile snails may depend on the limpet as a primary food source (Powell *et al.*, 1987). Scavengers, e.g. crabs or fish, feeding on dead or dying oysters, may also serve as additional routes of transmission (Hoese, 1964).

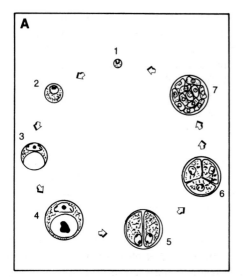 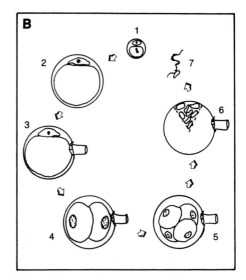

Fig. 11.7. Development cycle of *Perkinsus marinus* in the oyster *Crassostea virginica*. (A) Immature tropho-zoite (1) becomes a mature trophozoite (2–4), multiple fission (5–7) produces 8–16 immature trophozoites (range *c*. 4–64) which are liberated through a tear in the wall into the sea. (B) In seawater mature trophozoites (1–2) enlarge and a discharge tube and pore develop in the wall (3), multiple fission occurs producing numerous bifla-gellated zoospores (4–6) which are eventually discharged through the tube (6–7). The zoospores swim to, or are pulled into an oyster's mantle cavity, resulting in new infections being established in the gill, mantle and gut epithelia. The zoospores probably loose their flagella and become rounded to form immature trophozoites. From Perkins (1996).

Temperature appears to be the major large-scale controlling factor in the epizootiology of the disease, explaining about 40% of the variability in preva-lence and intensity (Burreson & Calvo, 1996). In Chesapeake and Delaware Bays infections and accompanying mortalities rise during the summer months, peak in early autumn, and decline dramatically over the late winter and early spring. However, some parasites remain and proliferate as temperatures rise in the spring (references in Ford & Tripp, 1996). The sudden appearance of the disease between 1990 and 1992 at sites from Delaware Bay to Cape Cod has been ascribed to above average winter, rather than summer temperatures (Ford, 1996; Cook *et al.*, 1998). Undoubtedly, long-term changes in climate could have a significant impact on *P. marinus* prevalence (Powell *et al.*, 1992). Salin-ity is another important controlling factor in the incidence of the disease, and within estuaries (i.e. on a local scale) is probably as important as tem-perature. The parasite appears to require salinities in excess of 9 psu to develop (Lauckner, 1983; Burreson & Calvo, 1996), but excessively high salinities also retard development. It would appear that the interaction of temperature and salinity are more important in regulating *P. marinus* epizootics than either factor acting alone (Calvo & Burreson, 1994; Chu 1996). This synergistic interplay could possibly explain the periodic nature of outbreaks that are so characteristic of this disease.

Prevalence also varies with oyster age. Death rates are low during the first year of life but increase with age and size of the oyster. Other factors such as food availability (Hofmann *et al.*, 1995) and oyster density (Andrews, 1965) have been cited as being important in the incidence of the disease. Several studies have reported a link between environmental contaminants and disease development (Wilson *et al.*, 1990; Chu & Hale, 1994; Anderson *et al.*, 1996; Fisher *et al.*, 1999). However, Ford & Tripp (1996) contend that while some contaminants may accelerate disease development there is no relationship between the distribution of the disease in nature and pollution. Most epizootics have occurred in perfectly clean waters (S. Ford personal communication 2001). Consequently, in oyster-growing areas culture schedules and planting densities have been adapted to reduce the impact of the disease. Low-density planting, early harvest at minimal commercial size, planting of large seed oysters in high salinity water late in the growing season, and utilisation of low-salinity growing areas are some of the measures that are currently employed.

There are indications that oysters exposed to *P. marinus* for several decades have developed some resistance to the parasite. Bushek & Allen (1996) examined the resistance of four genetically distinct oyster populations that had different natural histories of exposure to the pathogen. Oysters showed levels of resistance roughly corresponding to the duration of exposure (Fig. 11.8). In addition, they also found that distinct isolates of *P. marinus* varied in virulence. A number of selective breeding programmes are ongoing in Chesapeake and Delaware Bays, but progress in the development of disease-resistant lines of *C. virginica* is slow, although a search for potential biochemical markers for resistance may prove worthwhile. For example, in selectively bred 2-year-old oysters under field challenge, families with the highest survival rates had the highest protease inhibitory activity against *P. marinus* proteases (Oliver *et al.*, 2000). An alternative approach could involve hybridisation between *C. virginica* and *C. gigas* (which appears to be highly resistant to dermo and MSX), with subsequent backcrossing under selection for resistance (Gaffney & Bushek, 1996).

In the laboratory the parasite has been cultured successfully from tissue fragments of infected oysters. Oysters have been infected experimentally with these cultured cells and the parasite has been re-isolated from the infected oysters (La Peyre *et al.*, 1993; La Peyre & Faisal, 1995; La Peyre, 1996). However, results by Ford *et al.* (2002) demonstrate that cultured parasites have lower pathogenicity than wild-type parasites. Initially, diagnosis of *P. marinus* infections relied on examination of tissue sections but since the early 1950s the fluid thioglycollate assay (RFTG) of Ray (1952, 1966) has become the detection method of choice because it is faster, cheaper and more sensitive than histology. When infected oyster tissue is incubated in the RFTG medium *P. marinus* trophozoites are induced to differentiate into large thick-walled hynospores that are easily stained with Lugol's iodine. The assay is also used on haemolymph to detect circulating parasites (Gauthier & Fisher, 1990) and to isolate and quantify hypnospores from oyster tissue (Choi *et al.*, 1989). A more recent development is whole clam culture in RFTG medium, a quantitative technique that provides detection of low levels or early infections of *Perkinsus* spp. (Almeida *et al.*, 1999). Another detection assay is based on the production of a soluble formazan chromophore upon intracellular reduction

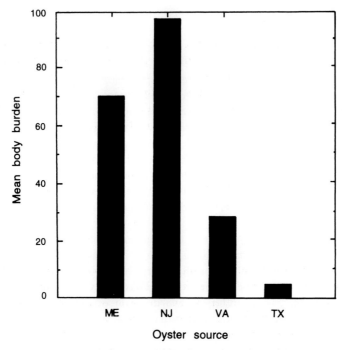

Fig. 11.8. Geometric mean infection intensity (cells per g wet tissue) of oyster (*Crassostrea virginica*) populations 94 days after inoculation with *in vitro* cultured *Perkinsus marinus*. Populations with a long history of exposure to the pathogen [Virginia (VA) and Texas (TX)] showed significantly ($p = 0.004$) lower infection intensity than populations with little or no history of exposure [Maine (ME) and New Jersey (NJ)]. From Gaffney & Bushek (1996), after Bushek (1994).

by the parasite at a rate proportional to parasite density (Dungan & Hamilton, 1995). The parasite can also be detected using monoclonal antibodies (Dungan & Roberson, 1993), or PCR amplified DNA (Marsh *et al.*, 1995; Yarnall *et al.*, 2000). The PCR assays are species specific and sensitive enough to detect a single *P. marinus* trophozoite in 30 g of oyster tissue (Robledo *et al.*, 1998).

Perkinsus marinus does not appear to be host-specific, as it has been reported in *Ostrea* species in the United States, although not in *Ostrea edulis* or *Saccostrea glomerata* (Sinderman, 1990). A different species, *Perkinsus atlanticus*, has been reported in clam species, *Ruditapes decussatus*, *R. philippinarum*, *Venerupis pullastra* and *V. aureus*, and is responsible for extensive mortalities in clam-growing areas in Spain (Navas *et al.*, 1992) and Portugal (Auzouxbordenave *et al.*, 1995). Optimal conditions for *P. atlanticus* development are temperatures of 24–28°C and salinities of 25–35 psu. A new species, *Perkinsus chesapeaki*, has been described from the clam, *Mya arenaria* (McLaughlin *et al.*, 2000) and another, *P. andrewsi*, has been isolated from the Baltic clam, *Macoma balthica*, and has also been detected in two other clams, *Macoma mitchelli* and *Mercenaria mercenaria*, and in *C. virginica* (Coss *et al.*, 2001). Yet another species, *P. qugwadi*, has been reported in cultured scallops (*Patinopecten yessoensis*) in British Columbia, Canada, with reported losses exceeding 90% (Bower *et al.*, 1999).

Porifera

Boring sponges (*Cliona* spp.) have been reported in the shells of mussels, oysters and scallops. The sponge excavates into the shell and in heavy infestations may even penetrate through to the inner shell. If perforation is extensive the shell is weakened and affords little protection from crabs and other predators. In addition, infestation may interfere with adductor muscle attachment, impede feeding and cause mortality. In the pearl oyster, *Pinctada maxima*, pearl production and quality may be seriously affected in heavy infestations. This is because oysters expend energy in depositing thickened nacre as a protection against the invader at the expense of depositing nacre around the inserted pearl nuclei (see Chapter 2). Growing bivalves in suspended culture is the simplest method to reduce shell damage (Bower, 2001).

Helminths

The main helminth parasites of bivalves are trematodes, cestodes and nematodes. The first two belong to the Phylum Platyhelmintha, while the latter are in the Phylum Nematoda. In terms of their disease-causing potential helminths are nowhere near as important as protozoan parasites. However, an enormous literature is available on this group, chiefly because they are visible to the naked eye, often have complex life cycles with several hosts, and in some cases may be passed on to humans through consumption of infected shellfish. Since trematodes, in particular the larval stages, are more important as bivalves pathogens than cestodes and nematodes, most of the following section will be devoted to them. For more details see Sinderman (1990), and the extensive review by Lauckner (1983) on helminth parasites.

Trematodes

Trematodes may use bivalves as the first host (sporocyst, redial and cercarial stages) or as the secondary host (metacercarial stage), or the bivalve may be the host for all stages of the life cycle (Fig. 11.9). An example of the latter is *Proctoeces maculatus*, which parasitises mussels, such as *Mytilus edulis* and *M. galloprovincialis*. The parasite has been reported in up to 46% of these species on both sides of the North Atlantic Ocean, and from the Mediterranean and Black Seas (references in Bower, 1992). A more recent report (Calvo-Ugarteburu & McQuaid, 1998) shows that *Perna perna* in South Africa has infection rates as high as 62%, but that introduced *M. galloprovincialis* is free of the parasite. This may give *M. galloprovincialis* a competitive advantage over *P. perna* and may explain its rapid, invasive spread from its introduction on the west coast in the early 1970s onto the south and east coasts of South Africa in the 1990s. Cercaria infect mussels and numerous sporocysts, as many as 28 000 per mussel develop in the vascular tissue of the mantle, giving it an orange hue (Machkevski, 1985). The adult stages of the life cycle are completed in the kidney. The fact that all stages of the life cycle are completed in mussels means that the parasite is easily spread through movement of infected mussels (Bower, 1992). Heavy infection can seriously affect gametogenesis,

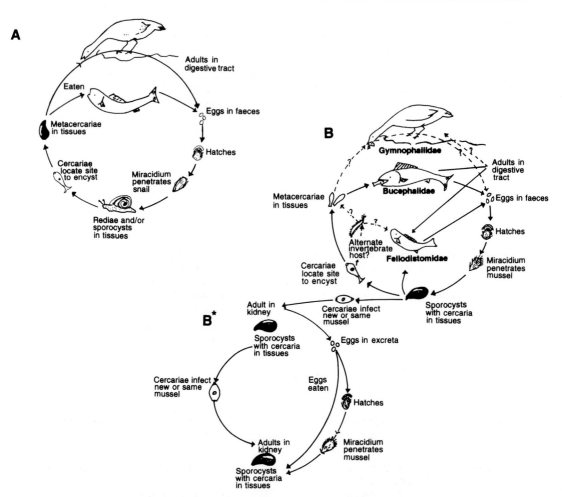

Fig. 11.9. Generalised life cycles of trematoda that infect mussels (*Mytilus* sp.). (A) Represents the life cycle of the majority of trematode species in the marine environment, where the metacercariae encyst in, or on, mussels as well as other animals and/or plants. (B) Represents the life cycle for members of the family Fellodistomidae, Bucephalidae, and possibly Gymnophallidae as indicated. (B*) Represents the abbreviated development of *Proctoeces maculatus*, which is capable of completing its entire life cycle in mussels. However, like other Fellodistomidae, adult forms also occur in the intestinal tract of fishes (usually tropical mollusc-eating fishes of the family Sparidae and Labridae) as illustrated in (B). From Bower (1992). Reprinted with permission from Elsevier Science.

leading to castration and death. In some regions, e.g. Portugal, castrated mussels do not die despite heavy infections, which indicates possible host adaptation to the parasite (Teia dos Santos & Coimbra, 1995).

Other species of trematode use mussels as secondary or final hosts but unlike *P. maculatus*, these are nonpathogenic although infection can cause organ deformation, reduced byssal production and induction of pearl formation. Pearls are formed because of an infection, probably by *Gymnophallus bursicola*. The fluke enters the mussel as a small larva and lodges in the mantle epithelium. The mussels encapsulate the fluke with layers of nacreous shell material, as a

Table 11.5. Some trematode parasites of bivalves.

Parasite	Host species	Locality	Pathology	References
Bucephalus cuculus	*Crassostrea virginica* and *Mugil* sp. of fish	Atlantic coasts of N America	Infection of gonad, digestive gland and eventually all organs; castration; up to 30% prevalence.	Hopkins, 1957
B. longicornutus	*Ostrea lutaria* and various fish species	New Zealand	Responsble for a decline in oyster numbers.	Millar, 1963; Howell, 1967
B. varicus	*Pinctada martensii* and *Caranx* species of fish	Japan	Loss of condition leading to death; low quality pearls in surviving oysters.	Sakaguchi, 1967 & 1968
Gymnophalloides tokiensis	*Crassostrea gigas* and marine birds	Japan	In mantle and gill; responsible for retardation in growth and inhibition of reproduction.	Hoshina & Ogino, 1951
Bucephalus sp.	*Pecten fumatus*	Australia	Castration; prevalence up to 60% in scallops >80 mm shell height.	Heasman *et al.*, 1996
Bucephalus sp.	*Pecten alba*	Australia	Castration; loss of energy reserves.	Sanders & Lester, 1981
Himasthla quissetensis	*Mya arenaria*, *Argopecten irradians* and herring gull *Larus argentatus*	Maine, USA	Infects gills and palps; up to 100% prevalence.	Uzmann, 1951; Getchell, 1991
H. elongatus	*Cerastoderma edule*, *Mytilus edulis*, gastropod *Littorina littorea*	Europe	Infects the foot; up to 100% prevalence; seriously affects burrowing in *C. edule* and byssal production in *M. edulis*, leading to reduced survival.	Lauckner, 1983
Postmonorchis donacis	*Donax gouldi* and surf perch fish species	California, USA	Sterilisation; trematode may be responsible for population crashes.	References in Sindermann, 1990

protectant, and a pearl is formed. Heavy infestation does not seem to harm the mussel, but it seriously reduces the commercial value of a mussel population. This problem can be eliminated if mussels are grown to harvest over a shorter period of time, e.g. on ropes; this means that the mussels are marketed before the pearls reach a detectable size (Lutz, 1980). Information on other bivalve trematode parasites is presented in Table 11.5. Several points are worth noting from this table: *Bucephalus* species are the most common parasites of commercial bivalves; the adult stage of the trematode life cycle is completed in fish or bird species; the main symptoms of infection are tissue destruction and castration; and, while mortalities are generally low, heavy infestation in some instances may be responsible for population crashes.

Cestodes

Marine bivalves act as hosts for the pre-adult stages of the tapeworm, while the adults are parasites in the intestine of elasmobranch fish (sharks, skates and ray). Infection in bivalves is, therefore, most common in tropical and subtropical waters, where elasmobranchs constitute an important proportion of the vertebrate fauna (Lauckner, 1983). Larval cestodes of the genus *Tylocephalum* occur, very often in great numbers, in a variety of marine bivalves, especially oysters (*Crassostrea*, *Pinctada*). Bivalves are initially infected when they ingest free-swimming coracidia or cestode eggs excreted by an elasmobranch. Once inside the host they develop and encyst as metacestodes in the tissue of the digestive system. The host's response is to secrete a thick fibrous capsule around each metacestode. Infestation can be heavy with up to 200 metacestodes reported from a single *Pinctada margaritifera* (Lauckner, 1983). Despite heavy infestation larval cestodes elicit few ill effects on their host. For example, among 60 heavily infected *Crassostrea virginica* none was weak or moribund, or exhibited any significant loss of body volume and weight (Cake & Menzel, 1980). Other cestode parasites include species of *Echeneibothrium*, *Parachristianella*, *Acanthobothrium* and *Rhinebothrium* (Lauckner, 1983; Sindermann, 1990). There is no evidence that any of these can infest humans.

Nematodes

Nematodes are uncommon parasites of marine bivalves, but two are worth mentioning: *Echinocephalus sinensis* in oysters, and *Sulcascaris sulcata* in scallops and clams. The larval stages of both species are found in bivalves, while the adult stages are found in elasmobranch and turtle species. Coiled larvae of *Echinocephalus sinensis* inhabit the gonoduct lumen of *Crassostrea gigas* and cause tissue damage, possibly due to the movement of the spiny anterior end of the worm. It is not clear whether *E. sinensis* disrupts reproduction, or whether there is higher mortality among heavily infected oysters. What is particularly interesting about this parasite is that it is a potential health hazard to humans as experiments have shown that the larvae are capable of infecting mammals (Ko, 1976). *Sulcascaris sulcata* is a parasite of the clam, *Spisula solidissima*. It inhabits all tissues and sometimes gives a brown colour to the meats, due to the presence of another parasite, a haplosporidian *Urosporidium spisuli*, infecting the worms. In the 1970s an epizootic of *U. spisuli* on *Sulcascaris sulcata* occurred along the mid-Atlantic coast of the United States causing loss of revenue due to closure of the clam fishery. However, this has not re-occurred, although *Sulcascaris sulcata* continues to be a significant parasite of *Spisula solidissima* (Sindermann, 1990). Neither the nematode nor its protozoan parasite is regarded as hazardous to humans (Lauckner, 1983).

Sulcascaris sulcata also parasitises several scallop species (Getchell, 1991) but is restricted to the adductor muscle. High infection levels – values of 64% have been reported in *Amusium balotti* – can reduce the commercial value of scallops.

Annelids

The polychaete annelid *Polydora* burrows into the shells of bivalves and excavates U-shaped tunnels that subsequently become filled with compacted mud, causing 'mudblisters' on the shell. Oyster shells containing them are unsuitable for the lucrative half-shell market, as mudblisters are unsightly and, if punctured, can release sediments, faecal deposits and anaerobic metabolites such as hydrogen sulphide (Handley & Bergquist, 1997). The burrows also weaken the shell, thereby increasing susceptibility to predation, especially in thin-shelled species such as mussels, although crabs have been shown to show preference for non-infested mussels (Ambariyanto & Seed, 1991). In oysters this weakening can cause problems during shucking, packing and transport (Korringa, 1951a). Other effects of the parasite include atrophy and detachment of muscle when burrowing occurs in the region of the adductor muscle, loss of condition, retarded growth and mortality (Lauckner, 1983). Heavy infestations, up to 300 worms per host, exacerbate these effects. Mortalities as high as 84% in the scallop *Patinopecten yessoensis* were attributed to heavy infestations by *Polydora websteri* (Bower *et al.*, 1992).

Various methods have been tested to try to eradicate the parasite in bivalve-growing areas. Infestations are responsive to factors such as temperature, salinity, host density and intertidal level of cultivation. Handley & Bergquist (1997) have found that increasing the aerial exposure of oyster stocks (*Crassostrea gigas*) significantly decreased the incidence of new infections. Nel *et al.* (1996) exposed infested oysters of the same species to freshwater for 12 hours, or to heated (70°C) seawater for 40 seconds. Oysters were then placed in the field for a two-month recovery period. *Polydora* infestation was significantly reduced by both treatments, with heat treatment yielding the lowest average infestation (~1 worm per oyster) compared to the untreated control (~3 per oyster).

Crustaceans

Compared with bacterial or protozoan parasites, crustaceans are only mildly pathogenic to bivalves. The best-documented disease organism is the copepod *Mytilicola intestinalis* that inhabits the intestinal tract of *Mytilus edulis* and *M. galloprovincialis*. Another copepod, *M. orientalis*, originally confined to Japan, has spread to mussels (*M. trossulus*) on the Pacific coast of North America, and has been introduced into France with imported oysters (*Crassostrea gigas*). It is also found in *M. edulis* and *M. galloprovincialis*, where dual infections with *M. intestinalis* occur (Bower & Figueras, 1989; Goater & Weber, 1996; see also Lauckner, 1983 for details on the spread of *Mytilicola*). The copepod is red coloured and large; females reach a size of 7–8 mm and males 3–4 mm. Mature females carry paired egg sacs (Fig. 11.10). The eggs hatch in the host and are

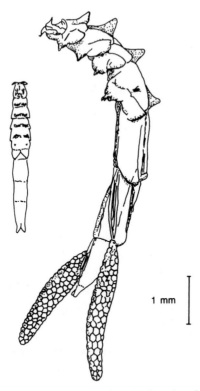

1 mm

Fig. 11.10. The copepod *Mytilicola intestinalis* from the intestine of the mussel *Mytilus edulis.* Adult male (left, ventral aspect) and female with egg-sacs (right, dorso-lateral view). From Lauckner (1983), after Hockley (1951).

expelled into the water column. The infective stage is a free-swimming larva that enters the mussel by the inhalant siphon. Infestation is a passive process depending on the chance encounter by the larvae with the host's field of filtration, and also on the strength of the inhalant current (Gee & Davey, 1986). Not surprisingly, infestation success is dependent on host size and density (Fig. 11.11). In addition, intensity of infestation is negatively correlated with exposure, i.e. mussels higher up the shore are less likely to be infected than those lower down (Fig. 11.12). Prevalence as high as 100% on a local scale is common, and as many as 90 copepods have been reported from the intestines of a single mussel (Sindermann, 1990), although numbers are usually in the range 5–20.

Mytilicola intestinalis has been blamed for widespread mortality and loss of condition in European populations of mussels. For example, the crash of the Dutch and German mussel fisheries in 1949 and 1950 was attributed to the species (Korringa, 1951b). For the next 30 years or so conflicting reports on the pathogenicity of the copepod appeared. Some of these reported serious effects on the health, growth and condition of parasitised mussels (Cole & Savage, 1951; Bayne *et al.*, 1978; Theisen, 1987; Robledo *et al.*, 1995),

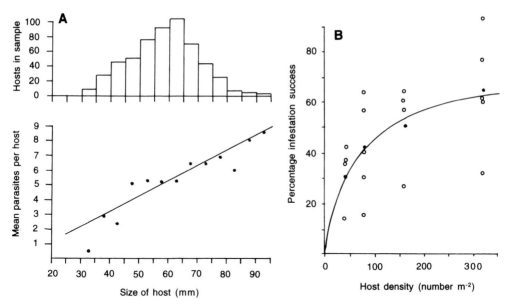

Fig. 11.11. (A) The relationship between mean number of the copepod *Mytilicola intestinalis* and size of host *Mytilus edulis* from sheltered estuarine sites. The histogram shows the number of mussels in each 5 mm size group. From Davey & Gee (1976). (B) The relationship between mussel density and infestation success of *M. intestinalis*. Open circles are the mean for each group of ten mussels; closed circles the mean for each group of 50 mussels. From Gee & Davey (1986).

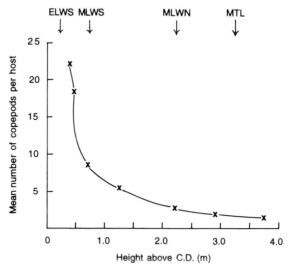

Fig. 11.12. Mean numbers of copepods *Mytilicola intestinalis* per host *Mytilus edulis* in relation to height above chart datum (C.D.) at Lynher River, south west England. ELWS: extreme low water spring tide; MLWS: mean low water spring tide; MLWN: mean low water neap tide; MTL: mean tide level. Modified from Davey & Gee (1976).

especially when parasite burdens exceeded 25 individuals per host (Gee *et al.*, 1977). However, in no instance was the copepod deemed to be responsible for the death of its host. In contrast, others have reported no loss of condition in parasitised mussels (Campbell, 1970; Figueras *et al.*, 1991; Gilek *et al.*, 1992), even when parasite numbers exceeded 20 individuals per host (Dethlefsen, 1975). It now seems likely that the *Mytilicola–Mytilus* relationship is commensal rather than parasitic (Davey, 1989), with the copepod feeding on host gut contents rather than on host tissue, as previously stated by Guisti (1967). Reported negative effects of the copepod on mussels may be explained by environmental variables, host gametogenic stage, and even the presence of microscopic pathogens. Davey & Gee (1988) have suggested that *M. intestinalis* may be responsible for introducing more virulent pathogens, or that synergistic effects operate to enhance the response of mussels infected by copepods to such pathogens.

Another copepod, *Pseudomyicola spinosus*, is found in the mantle cavity and gills of 54 bivalve species around the world, but also inhabits the intestine of many of these species. Movement of the copepod both within and through the walls of the intestinal tract causes tissue damage in the host, and may lead to loss of condition in heavily-infected individuals (Cáceres-Martínez & Vásquez-Yeomans, 1997). Another crustacean, the pea crab, *Pinnotheres* spp., inhabits the mantle cavity of many bivalve species, but its status as a parasite or commensal is unresolved. The crab feeds on material collected by the gills of the host, and causes gill erosion, weight loss through reduced filtration, loss of condition and impaired gametogenesis (Pregenzer, 1981 and references therein; Lauckner, 1983; Obeirn & Walker, 1999). Increased mortalities have not been reported in infected hosts. Indeed, pea crabs may actually protect their hosts from predation. Campbell (1993) has found that in selection experiments starfish prefer to feed on uninfested mussels, although pea crabs usually abandon their host during attacks.

Neoplasia

To conclude this section on disease and parasites, haemocytic neoplasia, a complex of proliferative haemolymph disorders, will now be considered. The condition is characterised by proliferative growth of abnormal haemocytes to a terminal stage where these abnormal cells are present in overwhelming numbers, and possibly occlude most vascular spaces (Bower, 1992)[Fig. 11.13]. In most bivalves prevalence is low (<5%) and to date, the link between haemocytic neoplasia and reported mortality in oysters and mussels is purely inferential. In contrast, prevalence of neoplasia is high (up to 90% in some years) in soft-shell clams, *Mya arenaria*, from Chesapeake Bay, United States, and mortality correlates with advanced stages of the disease (Brousseau & Baglivo, 1991; Farley *et al.*, 1991). Transmissibility of the disease has been demonstrated; 95% of injected clams developed neoplasms and advanced infections were lethal within four months (Sindermann, 1990). An infectious agent, possibly a virus, may be responsible (Oprandy *et al.*, 1981). There is also some evidence to suggest that activation of oncogenes, in response to environmental pollutants, can also induce neoplasms (Rhodes & Vanbeneden, 1996).

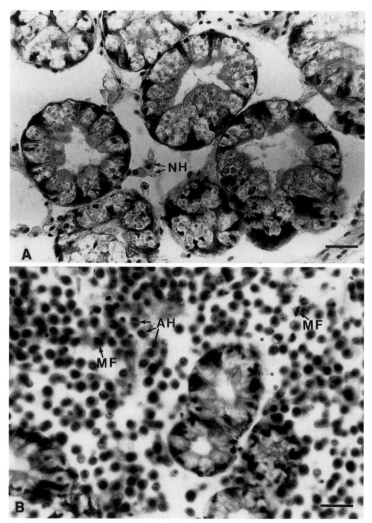

Fig. 11.13. Histological sections of the digestive gland of *Mytilus trossulus* illustrating: (A) a few normal haemocytes (NH) in the spaces between the tubules of a normal mussel; (B) numerous neoplastic haemocytes, some with a large nucleus containing two or more nucleoli and a small amount of peripheral cytoplasm (AH), and some undergoing mitosis (MF) in the distended spaces between the digestive gland tubules. Haematoxylin and eosin stain. Scale bar = 50 μm. Photos courtesy of S. Bower, Pacific Biological Station, Nanaimo, British Columbia, Canada.

Defence mechanisms

All metazoan animals are protected against invading organisms or foreign substances by an internal defence system. The system includes two forms of immunity, acquired or specific immunity, and innate or natural immunity (Livingstone *et al.*, 2000). Natural immunity allows the animal to combat foreign components without any previous contact with them. Acquired immunity provides rapid and selective protection against a specific foreign body (antigen), but requires previous exposure to the antigen. A second contact with the same antigen produces an enhanced response in the form of specific anti-bodies, produced as a result of memory formation (Prieur *et al.*, 1990). This type of immunity is only present in vertebrates; but both types, natural and acquired, rely on cellular (whole cell) and humoral (cell product) mechanisms. In bivalves phagocytosis and haemocytic infiltration are the primary cellular responses, while a multitude of haemolymph factors such as lysins, agglutinins and enzymes – either normally present or induced – constitutes the humoral defence mechanism (Sindermann, 1990). Haemocytes are the first line of defence against foreign substances. Although there is no common system of haemocyte classification, detailed morphological studies, along with inves-tigations based on molecular characterisation of haemocyte cell surfaces, e.g. lectin-binding and antigenic properties, have gone some way in helping to identify sub-populations of these cells in bivalves (Bachère *et al.*, 1995; Cheng, 1996; Dyrynda *et al.*, 1997).

Cellular defence mechanisms

There are four types of defence mechanisms utilised by bivalves: haemo-cytosis, phagocytosis, encapsulation and nacresation. Haemocytosis is a first response to infection and involves a measurable increase in the number of circulating haemocytes, which then infiltrate infected or injured tissues. Phagocytosis is the next step in the defence process and can be divided into a number of stages: chemotaxis, recognition, adhesion, endocytosis and destruction. Chemotaxis, which is poorly understood in bivalves, involves the movement of haemocytes towards the target, probably through chemo-attractant substances secreted by the target (Prieur *et al.*, 1990; Fawcett & Tripp, 1994). After non-self recognition, adhesion between haemocytes and foreign materials is principally through cell surface receptors. At the site of adhesion the foreign material is taken into the cell by endocytosis (Fig. 11.14) and is enclosed in a vesicle called a primary phagosome. This fuses with a lysosome to form a phagolysosome. The fate of ingested particles varies. Digestible material is degraded by enzymes within the phagolysosome and nutrients then diffuse into the cytoplasm, thus providing the host with a supplementary source of nourishment. Indigestible material is either stored within haemo-cytes or else physically removed through migration of haemocytes out of the host (references in Sindermann, 1990). In some cases disease organisms remain alive within the phagolysosome, continue to develop and eventually kill the host (see below). Phagocytosis is accompanied by a diversity of killing methods and these will be covered later in the section on humoral defence mechanisms.

When the invading organism or particle is too large to be phagocytosed then encapsulation by haemocytes is the defence method of choice. An enve-lope of concentric layers of cells is laid down around the invader, which is usually a larval cestode or trematode, or groups of *Perkinsus*. Disintegration of the parasite, followed by resorption of cellular debris, may follow. Another defence mechanism, and one which is distinct from encapsulation, is nacresa-tion whereby nacre is laid down around parasites or foreign material that invade the space between mantle and shell. An example of nacresation is pearl

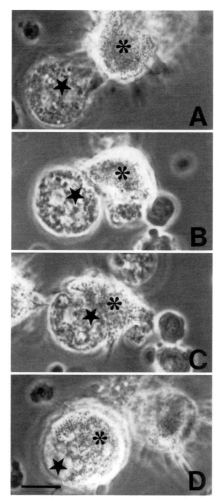

Fig. 11.14. *In vitro* encounters between haemocytes from the mussel *Geukensia demissa* and *Haplosporidium nelsoni* plasmodia. (A) Granular haemocyte (asterisk) contacts plasmodium (star) with extended filipodia. (B) Granular haemocyte (asterisk) adheres to plasmodium (star). (C) Granular haemocyte (asterisk) extends cytoplasm around plasmodium (star). (D) Haemocyte (asterisk) has almost completely engulfed plasmodium (star). Photomicrographs were taken of 2 encounters; the time between initial contact and complete engulfment ranged from 2 to 4 min. Scale bar = 10 μm. Photomicrographs courtesy of S. Ford, Institute of Marine and Coastal Sciences, Rutgers University, New Jersey, USA.

formation by *Mytilus* in response to invasion by the trematode parasite *Gymnophallus bursicola* (see above and Chapter 2).

Humoral defence mechanisms

Haemocytes or other cells secrete a variety of biologically active molecules into the haemolymph, and these are believed to play an important role in defence. During phagocytosis hydrolytic enzymes, e.g. lysozyme, are secreted into the haemolymph where they have been shown to exert microbiocidal effects on bacterial membranes (reviewed in Cheng, 1996). Lectins are also synthesised by haemocytes and occur on the cell surface as protein receptors for foreign antigens, e.g. bacteria. They probably play a role in eliminating bacteria by promoting their immobilisation, binding and eventual destruction by phagocytosis. In response to pathogens and various pollutants haemocytes release a variety of reactive oxygen intermediates (ROI). Although ROIs are important in vertebrate phagocytic activity their exact role in bivalve defence is not at all clear (Pipe & Coles, 1995; Roch, 1999 for review). In the past few years antibacterial and antifungal proteins, akin to arthropod defensin, have been isolated from the haemolymph of *Mytilus* that had been injected with bacteria (Charlet *et al.*, 1996; Hubert *et al.*, 1996; Roch, 1999).

It is now widely accepted that pollution can suppress the immune system in bivalves, leading to enhanced disease susceptibility (Dyrynda *et al.*, 1998, 2000; and see also Chapter 7). A range of assays based on haemocyte counts, phagocytosis, degradative enzyme levels and release of reactive oxygen metabolites, have been developed to measure the competence of the immune system (Volety *et al.*, 1999). Livingstone *et al.* (2000) have suggested that such assays could also be used as biomarkers of environmental perturbation.

Host-parasite interactions

Several of the major diseases covered above will now be discussed with reference to host–pathogen interactions on two levels: host defence reactions initiated by the host against the parasite, and mechanisms employed by the parasite to evade the host's defence system. Attention will focus primarily on protozoan infections since these are well documented, mostly because of recent advances in pathogen purification and haemocyte culture methods (Bachère *et al.*, 1995).

In some *Perkinsus* sp. infections haemocytes migrate from the haemolymph to the connective tissue of parasitised organs and synthesise a capsule that encloses, but does not kill, the parasite. Encapsulation effectively removes *Perkinsus* from the haemolymph, the main distribution pathway of this parasite. In addition, haemocytes secrete a small protein into the capsule, which has a high capacity for agglutination (Montes *et al.*, 1995a). However, the parasite mounts its own defence in that the parasite actually provokes haemocyte infiltration into parasitised tissues (Montes *et al.*, 1995b). It then promotes capsule formation, and is responsible for haemocyte shrinkage and the eventual death of these cells around the capsule. In addition, it appears that *P. marinus* is able to inhibit agglutination and suppress ROI release from haemocytes, thus evading other components of the host defence system (Cheng & Dougherty, 1995; Volety & Chu, 1995; Anderson, 1999). Faisal *et al.* (1999) have shown that *P. marinus* also secretes extracellular proteases that enhance its propagation and compromise host defences. However, *C. virginica* possesses several inhibitors of these proteases. Interestingly, *C. gigas*, which is resistant to *P. marinus*, possesses protease inhibitors with significantly higher specific activities than those in *C. virginica*. The recent development of *in vitro* culture of *Perkinsus* spp. cells provides an excellent opportunity to study host-parasite interaction. However, there are important differences between cultured and natural parasites in that higher numbers of cultured cells are required to initiate infection, and different isolates vary in their ability to infect. In addition, cultured parasites are digested by haemocytes a few hours after injection into oysters while, as already mentioned above, this is not the case for natural parasites (Bushek *et al.*, 1997; Volety & Fisher, 2000).

Haplosporidium nelsoni infections (MSX) in *Crassostrea virginica* provoke haemocyte infiltration into tissues where parasites are located. This reaction is most likely due to the considerable damage associated with MSX infections rather than to non-self recognition, since haemocytes rarely phagocytose parasites (Ford, 1988). Time-lapse pictures reveal that haemocytes move towards parasites, make contact, and may even begin to phagocytose them, but then withdraw and move away. On contact the parasite may emit a noxious substance with repellent properties, or the parasite may possess cell surface components not recognised as foreign by haemocytes (Ford *et al.*, 1993). Differences in serum protein and hydrolytic enzyme concentrations in the haemolymph of infected and non-infected oysters have been reported, but these may be symptoms of disease rather than humoral defences against the parasite (Ford, 1988; references in Sindermann, 1990).

The protozoan *Bonamia ostreae* is an intra-haemocytic parasite of the European oyster, *Ostrea edulis*, but the Pacific oyster, *Crassostrea gigas*, is refractory to it. *B. ostreae* is phagocytosed by haemocytes of both oyster species but the parasite is able to bypass the cytotoxic system of the haemocytes (Bachère *et al.*, 1995). Why one species succumbs to infection and the other does not has yet to be determined.

Vibrio tapetis that infects the clam *Ruditapes philippinarum* and causes brown ring disease (BRD) is a good experimental model to study host-pathogen interactions. Experimental challenge of *R. philippinarum* with the vibrio induces elevation of haemocyte numbers in the haemolymph after 24–72 hours. Haemocytes migrate towards the bacteria and phagocytose them, although the process is apparently non-specific (Lopez-Cortes *et al.*, 1999). Two weeks post-infection there is a decline in circulating haemocyte numbers that is accompanied by an increasing prevalence of BRD in the challenged clams (Fig. 11.15). Simultaneous with this decline is an increase in the numbers of haemocytes in the extrapallial fluid, probably due to mobilisation of haemocytes to the site of infection (Allam *et al.*, 2000). This increase is also accompanied by an increase in the activity of lysozyme, a hydrolytic enzyme that probably plays a defensive role against the pathogen. As infection proceeds there is an increasing number of dead haemocytes most likely due to the haemolytic and cytotoxic properties of the parasite. The host also employs an external defence mechanism by embedding bacteria within a brown organic deposit on the inside of the shell, a process similar to, but more complex than, nacresation already described. It should be noted, however, that while BRD can be fatal in the field, death rarely occurs in experimental infections (Allam *et al.*, 2000).

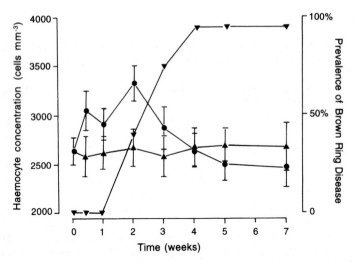

Fig. 11.15. Significant (p < 0.05) and irregular increase in haemocyte concentration of the clam *Ruditapes philippinarum* within 2 weeks after challenge with a bacterial suspension (5×10^7) of *Vibrio tapetis* (●). Control (▲) animals were injected with filtered seawater. The decrease in circulating haemocyte levels after 2 weeks was accompanied by an increasing prevalence of BRD in the challenged clams (▼). Haemolymph samples were taken from individual clams (N = 20 per sample). Vertical bars are standard errors of the mean. From Paillard *et al.* (1994). Reprinted with permission from Elsevier Science.

Further research

The control of infectious diseases is one of the main challenges facing those involved in the mollusc aquaculture industry. Miahle *et al.* (1995), Bachère *et al.* (1995) and Roch (1999) have proposed that major research efforts should be focused on:

- Development of quick, sensitive and specific diagnostic techniques e.g. immuno- and DNA-based diagnostics, which are especially useful in virus detection.
- Development of *in vitro* culture of mollusc cells for the study of host–pathogen interactions.
- Immunological research on host–pathogen interactions through recently developed protocols for purifying specific disease organisms.
- Selection of pathogen-resistant strains of bivalves either by classical genetics, or by finding marker genes that are linked to genes for resistance (see Chapter 10).
- Identification of inducible antibacterial peptides. Genes for these proteins could be cloned and characterized, and could be candidates for selecting resistant strains.

In addition, basic information on the biology, ecology and transmission of pathogens must be accumulated. Ultimately, progress in combating disease is dependent on increased international cooperation between pathologists, immunologists and geneticists, as well as the bivalve industry and its regulators.

References

Alderman, D.J. & Jones, E.B.G. (1971a) Physiological requirements of two marine phycomycetes, *Althornia crouchii* and *Ostracoblabe implexa*. *Trans. British Mycol. Soc.*, **57**, 213–25.

Alderman, D.J. & Jones, E.B.G. (1971b) *Shell disease of oysters. MAFF Fish. Invest. Ser. 2*, **26**, 1–19. HMSO, London.

Allam, B., Paillard, C., Howard, A. & Le Pennec, M. (2000) Isolation of the pathogen *Vibrio tapetis* and defence parameters in brown ring diseased Manila clams *Ruditapes philippinarum* cultivated in England. *Dis. Aquat. Org.*, **41**, 105–13.

Almeida, M., Berthe, F., Thébault, A. & Dinis, M.T. (1999) Whole clam culture as a quantitative diagnostic procedure of *Perkinsus atlanticus* (Apicomplexa, Perkinsea) in clams *Ruditapes decussatus*. *Aquaculture*, **177**, 325–32.

Ambariyanto & Seed, R. (1991) The infestation of *Mytilus edulis* Linnaeus by *Polydora ciliata* (Johnson) in the Conwy Estuary, North Wales. *J. Moll. Stud.*, **57**, 413–24.

Anderson, R.S., Unger, M. & Burreson, E.M. (1996) Enhancement of Perkinsus marinus disease progression in TBT-exposed oysters (*Crassostrea virginica*). *Mar. Environ. Res.*, **42**, 177–80.

Anderson, R.S. (1999) Lack of hemocyte chemiluminescence stimulation by *Perkinsus marinas* in eastern oysters *Crassostrea virginica* with dermo disease. *J. Aquat. Anim. Health*, **11**, 179–82.

Andrews, J.D. (1965) Infection experiments in nature with *Dermocystidium marinum* in Chesapeake Bay. *Chesapeake Sci.*, **6**, 60–67.

Andrews, J.D. & Wood, J.L. (1967) Oyster mortality studies in Virginia. VI. History and distribution of *Minchinia nelsoni*, a pathogen of oysters, in Virginia. *Chesapeake Sci.*, **8**, 1–13.

Auzouxbordenave, S., Vigario, A., Ruano, F., Domartcoulon, I. & Doumenc, D. (1995) In-vitro sporulation of the clam pathogen *Perkinsus atlanticus* (Apicomplexa, Perkinsea) under various environmental conditions. *J. Shellfish Res.*, **14**, 469–75.

Bachère, E., Mialhe, E., Noel, D., Boulo, V., Morvan, A. & Rodriguez, J. (1995) Knowledge and research prospects in marine mollusc and crustacean immunology. *Aquaculture*, **132**, 17–32.

Barber, B.J., Ford, S.E. & Haskin, H.H. (1988) Effects of the parasite MSX (*Haplosporidium nelsoni*) on oyster (*Crassostrea virginica*) energy metabolism. II. Tissue biochemical composition. *Comp. Biochem. Physiol.*, **91**A, 603–608.

Barber, B.J., Ford, S.E. & Littlewood, D.T.J. (1991) A physiological comparison of resistant and susceptible oysters *Crassostrea virginica* (Gmelin) exposed to the endoparasite *Haplosporidium nelsoni* (Haskin, Stauber & Mackin). *J. Exp. Mar. Biol. Ecol.*, **146**, 101–12.

Bayne, B.L., Gee, J.M., Davey, J.T. & Scullard, C. (1978) Physiological responses of *Mytilus edulis* L. to parasitic infestation by *Mytilicola intestinalis*. *J. Cons. int. explor. Mer*, **38**, 12–17.

Boettcher, K.J., Barber, B.J. & Singer, J.T. (2000) Additional evidence that juvenile oyster disease is caused by a member of the Roseobacter group and colonization of nonaffected animals by *Stappia stellulata*-like strains. *Appl. Environ. Microbiol.*, **66**, 3924–30.

Bower, S.M. (1992) Diseases and parasites of mussels. In: *The Mussel* Mytilus: *Ecology, Physiology, Genetics and Culture* (ed. E.M. Gosling), pp. 543–63. Elsevier Science Publishers B.V., Amsterdam.

Bower, S.M. (2001) *Synopsis of infectious diseases and parasites of commercially exploited shellfish: shell-burrowing sponges of oysters*. URL: http://www-sci.pac.dfo-mpo.gc.ca/sealane/aquac/pages/sbsoy.htm.

Bower, S.E. & Figueras, A. (1989) Infectious diseases of mussels, especially pertaining to mussel transplantation. *World Aquacul. Rev.*, **20**, 89–93.

Bower, S.M., Blackbourn, J., Meyer, G.R. & Nishimura, D.J.H. (1992) Diseases of cultured Japanese scallops (*Patinopecten yessoensis*) in British Columbia, Canada. *Aquaculture*, **107**, 201–10.

Bower, S.M., McGladdery, S.E. & Price, I.M. (1994) Synopsis of infectious disease and parasites of commercially exploited shellfish. *Ann. Rev. Fish Dis.*, **4**, 1–200.

Bower, S.M., Blackbourn, J., Meyer, G.R. & Welch, D.W. (1999) Effect of *Perkinsus qugwadi* on various species and strains of scallops. *Dis. Aquat. Org.*, **36**, 143–51.

Brousseau, D.J. & Baglivo, J.A. (1991) Field and laboratory comparisons of mortality in normal and neoplastic *Mya arenaria*. *J. Invertebr. Pathol.*, **57**, 59–65.

Burreson, E.M. & Calvo, L. (1996) Epizootiology of *Perkinsus marinus* disease of oysters in Chesapeake Bay, with emphasis on data since 1985. *J. Shellfish Res*, **15**, 17–34.

Bushek, D. (1994) *Dermo disease in American oysters; genetics of host-parasite interactions.* Ph.D Thesis, Rutgers University, New Jersey, USA.

Bushek, D. & Allen, S. (1996) Host-parasite interactions among broadly distributed populations of the eastern oyster *Crassostrea virginica* and the protozoan *Perkinsus marinus*. *Mar. Ecol. Prog. Ser.*, **139**, 127–41.

Bushek, D., Allen, S.K. Jr., Alcox, K.A., Gustafson, R.G. & Ford, S.E. (1997) Response of *Crassostrea virginica* to *in vitro* cultured *Perkinsus marinus*: preliminary comparisons of three inoculation methods. *J. Shellfish Res.*, **16**, 479–85.

Cáceres-Martínez, J. & Vasquez-Yeomans, Y. (1997) Presence and histopathological effects of the copepod *Pseudomyicola spinosus* in *Mytilus galloprovincialis* and *Mytilus californianus*. *J. Invertebr. Pathol.*, **70**, 150–55.

Cáceres-Martínez, J., Robledo, J.A.F. & Figueras, A. (1995) Presence of *Bonamia* and its relation to age, growth rates and gonadal development of the flat oyster, *Ostrea edulis*, in the Ria de Vigo, Galicia (NW Spain). *Aquaculture*, **130**, 15–23.

Cake, E.W. & Menzel, R.W. (1980) Infections of *Tylocephalum* metacestodes in commercial oysters and three predaceous gastropods of the eastern Gulf of Mexico. *Proc. Natl. Shellfish Ass.*, **70**, 94–104.

Calvo, L.M.R. & Burreson, E. (1994) Characterization of overwintering infections of *Perkinsus marinus* (Apicomplexa) in Chesapeake Bay oysters. *J. Shellfish Res.*, **13**, 123–30.

Calvo-Ugarteburu, G. & McQuaid, C.D. (1998) Parasitism and introduced species: epidemiology of trematodes in the intertidal mussels *Perna perna* and *Mytilus galloprovincialis*. *J. Ecol. Mar. Biol. Ecol.*, **220**, 47–65.

Campbell, D.B. (1993) The effect of pea crabs on predation of host mussels by sea stars. *Mar. Behav. Physiol.*, **24**, 93–99.

Campbell, S.A. (1970) The occurrence and effects of *Mytilicola intestinalis* in *Mytilus edulis*. *Mar. Biol.*, **5**, 89–95.

Carnegie, R.B., Barber, B.J., Culloty, S.C., Figueras, A.J. & Distel, D.L. (2000) Development of a PCR assay for detection of the oyster pathogen *Bonamia ostreae* and support for its inclusion in the Haplosporidia. *Dis. Aquat. Org.*, **42**, 199–206.

Castro, D., Luque, A., Santamara, J.A., Maes, P., Martínez-Manzanares & Borrego, J.J. (1995) Development of immunological techniques for the detection of the potential causative agent of the brown ring disease. *Aquaculture*, **132**, 97–104.

Charlet, M., Cherysh, S., Philippe, H., Hetru, C., Hoffmann, J.A. & Bulet, P. (1996) Innate immunity-isolation of several cysteine-rich antimicrobial peptides from the blood of a mollusc, *Mytilus edulis*. *J. Biol. Chem.*, **271**, 21808–13.

Cheng, T. (1996) Haemocytes: form and function. In: *The Eastern Oyster* Crassostrea virginica (eds V.S. Kennedy, R.I.E. Newell & A.F. Eble), pp. 299–333. Maryland Sea Grant, College Park, Maryland.

Cheng, T.C. & Dougherty, W. (1995) Partial inhibition of hemocyte agglutination by lathyrus-odoratus lectin in *Crassostrea virginica* infected with *Perkinsus marinus*. *Mem. Inst. Oswado Cruz*, **90**, 407–10.

Chintala, M.M. & Fisher, W.S. (1989) Comparison of oyster defense mechanisms for MSX-resistant and susceptible stocks held in Chesapeake Bay. *J. Shellfish Res.*, **8**, 467–68.

Choi, K.-S., Wilson, E.A., Lewis, D.H., Powell, E.N. & Ray, S.M. (1989) The energetic cost of *Perkinsus marinus* parasitism in oysters: quantification of the thioglycollate method. *J. Shellfish Res.*, **8**, 125–36.

Chu, F.L.E. (1996) Laboratory investigations of susceptibility, infectivity, and transmission of *Perkinsus marinus* in oysters. *J. Shellfish Res.*, **15**, 57–66.

Chu, F.L.E. & Hale, R.C. (1994) Relationship between pollution and susceptibility to infectious disease in the eastern oyster, *Crassostrea virginica. Mar. Environ. Res.*, **38**, 243–56.

Cochennec, N., Hervio, D., Panatier, B. *et al.* (1992) A direct monoclonal-antibody sandwich immunoassay for detection of *Bonamia ostreae* (Acetospora) in hemolymph samples of the flat oyster *Ostrea edulis* (Mollusca, Bivalvia). *Dis. Aquat. Org.*, **12**, 129–34.

Cole, H.A. & Savage, R.E. (1951) The effect of the parasitic copepod, *Mytilicola intestinalis* (Steuer) upon the condition of mussels. *Parasitology*, **41**, 156–61.

Comps, M., Tigé, G. & Grizel, H. (1980) Études ultrastructurales sur un protiste parasite de l'huitre plate *Ostrea edulis. C. R. Acad. Sci. Paris*, **290**, 383–84.

Cook, T., Folli, M., Klinck, J., Ford, S. & Miller, J. (1998) The relationship between increasing sea-surface temperature and the northward spread of *Perkinsus marinus* (Dermo) disease epizootics in oysters. *Est. Coast. Shelf Sci.*, **46**, 587–97.

Coss, C., Robledo, J.A.F., Ruiz, G.M. & Vasta, G.R. (2001) Description of *Perkinsus andrewsi* n. sp. isolated from the Baltic clam by characterization of the ribosomal RNA locus, and development of a species-specific PCR-based diagnostic assay. *J. Euk. Microbiol.*, **48**, 52–61.

Culloty, S.C. & Mulcahy, M.F. (1996) Season-, age-, and sex-related variation in the prevalence of bonamiasis in flat oysters (*Ostrea edulis* L.) on the south coast of Ireland. *Aquaculture*, **144**, 53–63.

Culloty, S.C., Novoa, B. & Pernas, M. (1999) Susceptibility of a number of bivalve species to the protozoan parasite *Bonamia ostreae* and their ability to act as vectors for this parasite. *Dis. Aquat. Org.*, **37**, 73–80.

Culloty, S.C., Cronin, M.A. & Mulcahy, M.F. (2001) An investigation into the relative resistance of Irish flat oysters *Ostrea edulis* L. to the parasite *Bonamia ostreae* (Pichot *et al.*, 1980). *Aquaculture*, **199**, 229–44.

Davey, J.T. (1989) *Mytilicola intestinalis* (Copepoda: Cyclopoida): a ten year survey of infested mussels in a Cornish estuary, 1978–1988. *J. mar. biol. Ass. U.K.*, **69**, 823–36.

Davey, J.T. & Gee, J.M. (1976) The occurrence of *Mytilicola intestinalis* Steuer, an intestinal copepod parasite of *Mytilus*, in the south-west of England. *J. mar. biol. Ass. U.K.*, **56**, 85–94.

Davey, J.T. & Gee, J.M. (1988) *Mytilicola intestinalis*, a copepod parasite of blue mussels. *Am. Fish. Soc. Spec. Pub.*, **18**, 64–73.

Davis, C.V. & Barber, B.J. (1999) Growth and survival of selected lines of eastern oysters, *Crassostrea virginica* (Gmelin 1791) affected by juvenile oyster disease. *Aquaculture*, **178**, 253–71.

Day, J.M., Franklin, D.E. & Brown, B.L. (2000) Use of competitive PCR to detect and quantify *Haplosporidium nelsoni* infection (MSX disease) in the eastern oyster (*Crassostrea virginica*). *Mar. Biotechnol.*, **2**, 456–65.

Dethlefsen, V. (1975) The influence of *Mytilicola intestinalis* Steuer on the meat content of the mussel *Mytilus edulis*. *Aquaculture*, **6**, 83–97.

Dinamani, P. Hine, P.M. & Jones, J.B. (1987) Occurrence and characteristics of the haemocyte parasite *Bonamia* sp. in the New Zealand dredge oyster *Tiostrea lutaria*. *Dis. Aquat. Org.*, **3**, 37–44.

Dungan, C.F. & Roberson, B.S. (1993) Binding specificities of monoclonal and poly-clonal antibodies to the protozoan oyster pathogen *Perkinsus marinus*. *Dis. Aquat. Org.*, **15**, 9–22.

Dungan, C.F. & Hamilton, R.M. (1995) Use of a tetrazolium-based cell-proliferation assay to measure effects of in-vitro conditions on *Perkinsus marinus* (Apicomplexa) proliferation. *J. Euk. Microbiol.*, **42**, 379–88.

Dyrynda, E.A., Pipe, R.K. & Ratcliffe, N.A. (1997) Sub-populations of haemocytes in the adult and developing marine mussel, *Mytilus edulis*, identified by use of mono-clonal antibodies. *Cell Tissue Res.*, **289**, 527–36.

Dyrynda, E.A., Pipe, R.K., Burt, G.R. & Ratcliffe, N.A. (1998) Modulations in the immune defences of mussels (*Mytilus edulis*) from contaminated sites in the UK. *Aquat. Toxicol.*, **42**, 169–85.

Dyrynda, E.A., Law, R.J., Dyrynda, P.E.J., Kelly, C.A., Pipe, R.K. & Ratcliffe, N.A. (2000) Changes in immune parameters of natural mussel *Mytilus edulis* populations following a major oil spill ('Sea Empress', Wales, UK). *Mar. Ecol. Prog. Ser.*, **206**, 155–70.

Elston, R.A., Elliot, E.L. & Colwell, R.R. (1982) Conchiolin infection and surface coating *Vibrio*: shell fragility, growth depression and mortalities in cultured oysters and clams, *Crassostrea virginica*, *Ostrea edulis* and *Mercenaria mercenaria*. *J. Fish Dis.*, **5**, 265–84.

Elston, R.A., Farley, C.A. & Kent, M.L. (1986) Occurrence and significance of bonamiasis in European flat oysters *Ostrea edulis* in North America. *Dis. Aquat. Org.*, **2**, 49–54.

Elston, R.A., Kent, M.L. & Wilkinson, M.T. (1987) Resistance of *Ostrea edulis* to *Bonamia ostreae* infection. *Aquaculture*, **64**, 237–42.

Faisal, M., Oliver, J.L. & Kaattari, S.L. (1999) Potential role of protease-antiprotease interactions in *Perkinsus marinus* infection in *Crassostrea* spp. *Bull. Eur. Assoc. Fish Pathol.*, **19**, 269–76.

Farley, C.A., Banfield, W.G., Kasnic, G., Jr. & Foster, W.S. (1972) Oyster herpes-type virus. *Science*, **178**, 759–60.

Farley, C.A., Plutschak, D.L. & Scott, R.F. (1991) Epizootiology and distribution of transmissible sarcoma in Maryland softshell clams, *Mya arenaria*, 1984–1988. *Environ. Health Perspect.*, **90**, 35–41.

Fawcett, L.B. & Tripp, M. (1994) Chemotaxis of *Mercenaria mercenaria* hemocytes to bacteria in-vitro. *J. Invertebr. Pathol.*, **63**, 275–84.

Figueras, A., Jardon, C.F. & Caldas, J.R. (1991) Diseases and parasites of rafted mussels (*Mytilus galloprovincialis* Lmk): preliminary results. *Aquaculture*, **99**, 17–33.

Figueras, A., Robledo, J. & Novoa, B. (1996) Brown ring disease and parasites in clams (*Ruditapes decussatus* and *R. philippinarum*) from Spain and Portugal. *J. Shellfish Res.*, **15**, 363–68.

Fisher, W.S., Oliver, L.M., Walker, W.W., Manning, C.S. & Lytle, T.F. (1999) Decreased resistance of eastern oysters (*Crassostrea virginica*) to a protozoan pathogen (*Perkinsus marinus*) after sublethal exposure to tributyltin oxide. *Mar. Environ. Res.*, **47**, 185–201.

Ford, S.E. (1986) Comparison of haemolymph proteins between resistant and suscep-tible oysters, *Crassostrea virginica*, exposed to the parasite *Haplosporidium nelsoni* (MSX). *J. Invertebr. Pathol.*, **47**, 283–94.

Ford, S.E. (1988) Host-parasite interactions in eastern oysters selected for resistance to *Haplosporidium nelsoni* (MSX) disease: survival mechanism against a natural pathogen. *Am. Fish. Soc. Spec. Pub.*, **18**, 206–24.

Ford, S.E. (1996) Range extension by the oyster parasite *Perkinsus marinus* into the northeastern United States: response to climate change. *J. Shellfish Res*, **15**, 45–56.

Ford, S.E. (2001) Pests, parasites, diseases, and defense mechanisms. In: *The Biology of the Hard Clam, Mercenaria mercenaria (Linné)* (eds J.N. Kraeuter & M. Castagna), pp. 591–628. Elsevier Science Publishers B.V., Amsterdam.

Ford, S.E. & Haskin, H.H. (1982) History and epizootiology of *Haplosporidium nelsoni* (MSX), an oyster pathogen, in Delaware Bay, 1957–1980. *J. Invertebr. Pathol.*, **40**, 118–41.

Ford, S.E. & Haskin, H.H. (1987) Infection and mortality patterns in strains of oysters *Crassostrea virginica* selected for resistance to the parasite *Haplosporidium nelsoni* (MSX). *J. Parasitol.*, **73**, 368–76.

Ford, S.E. & Haskin, H.H. (1988) Comparison of in vitro salinity tolerance of the oyster parasite, *Haplosporidium nelsoni* (MSX) and hemocytes from the host, *Crassostrea virginica*. *Comp. Biochem. Physiol.*, **90**A, 183–87.

Ford, S.E. & Tripp, M.R. (1996) Diseases and defense mechanisms. In: *The Eastern Oyster* Crassostrea virginica (eds V.S. Kennedy, R.I.E. Newell & A.F. Eble), pp. 581–660. Maryland Sea Grant, College Park, Maryland.

Ford, S.E., Ashton-Alcox, K.A. & Kanaley, S.A. (1993) In vitro interactions between bivalve hemocytes and the oyster pathogen *Haplosporidium nelsoni* (MSX). *J. Parasitol.*, **79**, 255–65.

Ford, S.E., Powell, E.N., Klinck, J.M. & Hofman, E.E. (1999) Modeling the MSX parasite in eastern oyster (*Crassostrea virginica*) populations. I. model development, implementation and verification. *J. Shellfish Res.*, **18**, 475–500.

Ford, S.E., Chintala, M.M. & Bushek, D. (2002) Comparison of *in vitro* cultured and wild-type *Perkinsus marinus* II. Pathogen virulence. *Dis. Aquat. Org.*, **51**, 187–201.

Friedman, C.S. & Perkins, F.O. (1994) Range extension of *Bonamia ostreae* to Maine, USA. *J. Invertebr. Pathol.*, **64**, 179–81.

Gaffney, P.M. & Bushek, D. (1996) Genetic aspects of disease resistance in oysters. *J. Shellfish Res.*, **15**, 135–140.

Gauthier, J.D. & Fisher, W.S. (1990) Haemolymph assay for diagnosis of *Perkinsus marinus* in oysters *Crassostrea virginica* (Gmelin 1791). *J. Shellfish Res*, **9**, 367–72.

Gee, J.M. & Davey, J.T. (1986) Experimental studies on the infestation of *Mytilus edulis* (L.) by *Mytilicola intestinalis* Steuer (Copepoda, Cyclopoida). *J. cons. int. Explor. Mer*, **42**, 265–71.

Gee, J.M., Maddock, L. & Davey, J.T. (1977) The relationship between infestation by *Mytilicola intestinalis*, Steuer (Copepoda: Cyclopoidea) and the condition index of *Mytilus edulis* in southwest England. *J. cons. int. Explor. Mer*, **37**, 300–308.

Getchell, R.G. (1991) Diseases and parasites of scallops. In: *Scallops: Biology, Ecology and Aquaculture* (ed. S.E. Shumway), pp. 471–94. Elsevier Science Publishers B.V., Amsterdam.

Gilek, M., Tedengren, M. & Kautsky, N. (1992) Physiological performance and general histology of the blue mussel, *Mytilus edulis* L., from the Baltic and North Seas. *Neth. J. Sea Res.*, **30**, 11–21.

Giusti, F. (1967) The action of *Mytilicola intestinalis* Steuer on *Mytilus galloprovincialis* Lmk. of the Tuscan Coast. *Parasitology*, **28**, 17–26.

Goater, C.P. & Weber, A.E. (1996) Factors affecting the distribution and abundance of *Mytilicola orientalis* (Copepoda) in the mussel, *Mytilus trossulus*, in Barkley Sound, B.C. *J. Shellfish Res.*, **15**, 681–84.

Grizel, H., Miahle, E., Chagot, D., Boulo, V. & Bachère, E. (1988) Bonamiasis: a model study of diseases in marine molluscs. *Am. Fish. Soc. Publ.*, **18**, 1–4.

Handley, S.J. & Bergquist, P.R. (1997) Spionid polychaete infestations of intertidal Pacific oysters *Crassostrea gigas* (Thunberg), Mahurangi Harbour, northern New Zealand. *Aquaculture*, **153**, 191–205.

Haskin, H.H. & Ford, S.E. (1982) *Haplosporidium nelsoni* (MSX) on Delaware Bay seed oyster beds: a host-parasite relationship along a salinity gradient. *J. Invertebr. Pathol.*, **40**, 388–405.

Haskin, H.H., Stauber, L.A. & Mackin, J.G. (1966) *Minchinia nelsoni* n.sp. (Haplosporidia, Haplosporidiidae): causative agent of the Delaware Bay oyster epizootic. *Science*, **153**, 1414–16.

Heasman, M.P., O Connor, W. & Frazer, A.W.J. (1996) Digenean (Bucephalidae) infections in commercial scallops, *Pecten fumatus* Reeve, and doughboy scallops, *Chlamys (Mimachlamys) asperrima* (Lamarck), in Jervis Bay, New South Wales. *J. Fish Dis.*, **19**, 333–39.

Helm, M.M. & Smith, F.M. (1971) Observations on a bacterial disease of the European flat oyster *Ostrea edulis* L. *ICES, CM* 1971/K:**10**, 7p.

Hervio, D., Bachère, E., Boulo, V. *et al.* (1995) Establishment of an experimental infection protocol for the flat oyster, *Ostrea edulis*, with the intrahaemocytic protozoan parasite, *Bonamia ostreae* – application in the selection of parasite resistant oysters. *Aquaculture*, **132**, 183–94.

Hine, P.M. & Wesney, B (1994) Interaction of phagocytosed *Bonamia* sp (Haplosporidia) with hemocytes of oysters *Tiostrea chilensis*. *Dis. Aquat. Org.*, **20**, 219–29.

Hine, P.M., Wesney, B. & Hay, B.E. (1992) Herpes virus associated with mortalities among hatchery-reared larval Pacific oysters *Crassostrea gigas*. *Dis. Aquat. Org.*, **12**, 135–42.

Hockley, A.R. (1951) On the biology of *Mytilicoa intestinalis* (Steuer). *J. mar. biol. Ass. U.K.*, **30**, 223–32.

Hoese, H.D. (1964) Studies on oyster scavengers and their relation to the fungus *Dermocystidium marinum*. *Proc. Natl. Shellfish Ass.*, **53**, 161–74.

Hofmann, E.E., Powell, E., Klinck, J.M. & Saunders, G. (1995) Modelling diseased oyster populations. 1. Modelling *Perkinsus marinus* infections in oysters. *J. Shellfish Res.*, **14**, 121–51.

Hopkins, S.H. (1957) Our present knowledge of the oyster parasite 'Bucephalus'. *Proc. Natl. Shellfish Assoc.*, **47**, 58–61.

Hoshina, T. & Ogino, C. (1951) Studien uber Gymnophalloides tokiensis Fujita, 1925. I. Uber die Einwirkung der larvalen Trematoda auf die chemische Komponente und das Wachstum von *Ostrea gigas* Thunberg. *J. Tokyo Univ. Fish.*, **38**, 335–50.

Howell, M. (1967) The trematode, *Bucephalus longicornutus* (Manter, 1954) in the New Zealand mud-oyster, *Ostrea lutaria*. *Trans. R. Soc. New Zeal.*, **8**, 221–37.

Hubert, F., Noel, T. & Roch, P. (1996) A member of the arthropod defensin family from edible Mediterranean mussels (*Mytilus galloprovincialis*). *Eur. J. Biochem.*, **240**, 302–306.

Hudson, E.B. & Hill, B.J. (1991). Impact and spread of bonamiasis in the UK. *Aquaculture*, **93**, 279–85.

Kaehler, S. & McQuaid, CD (1999) Lethal and sub-lethal effects of phototrophic endoliths attacking the shell of the intertidal mussel *Perna perna*. *Mar. Biol.*, **135**, 497–503.

Karolus, J., Sunila, I., Spear, S. & Volk, J. (2000) Prevalence of *Perkinsus marinus* (Dermo) in *Crassostrea virginica* along the Connecticut shoreline. *Aquaculture*, **183**, 215–21.

Kennedy, V.S., Newell, R., Krantz, G.E. & Otto, S. (1995) Reproductive capacity of the eastern oyster *Crassostrea virginica* infected with the parasite *Perkinsus marinus*. *Dis. Aquat. Org.*, **23**, 135–44.

Ko, R.C. (1976) Experimental infection of mammals with larval *Echinocephalus sinensis* (Nematoda: Gnathostomatidae) from oysters (*Crassostrea gigas*). *Can. J. Zool.*, **54**, 597–609.

Korringa, P. (1951a) The shell of *Ostrea edulis* as a habitat. *Arch. Neer. Zool.*, **10**, 32–136.

Korringa, P. (1951b) Le *Mytilicola intestinalis* Steuer (Copepoda parasitica) menace l'insustrie mouliére en Zelande. *Revue Trav. Off. (scient. tech.) Pêch. Marit.*, **17**, 9–13.

La Peyre, J.F. (1996) Propagation and *in vitro* studies of *Perkinsus marinus*. *J. Shellfish Res.*, **15**, 89–101.

La Peyre, J.F. & Faisal, M. (1995) Improved method for the initiation of continuous cultures of the oyster pathogen *Perkinsus marinus* (Apicomplexa). *Trans. Am. Fish. Soc.*, **124**, 144–46.

La Peyre, J.F., Faisal, M. & Burreson, E.M. (1993) *In vitro* propagation of the protozoan *Perkinsus marinus*, a pathogen of the eastern oyster *Crassostrea virginica*. *J. Euk. Microbiol.*, **40**, 304–10.

Lauckner, G. (1983) Diseases of Mollusca: Bivalvia. In: *Diseases of Marine Animals*, Vol. 2 (ed. O. Kinne), pp. 477–961. Biologishe Anstalt Helgoland, Hamburg.

Livingstone, D.R., Chipman, J.K., Lowe, D.M. *et al.* (2000) Development of biomarkers to detect the effects of organic pollution on aquatic invertebrates: recent molecular, genotoxic, cellular and immunological studies on the common mussel (*Mytilus edulis* L.) and other mytilids. *Int. J. Environ. Pollut.*, **13**, 56–91.

Lopez-Cortes, L., Castro, D., Navas, J.I. & Borrego, J.J. (1999) Phagocytic and chemotactic responses of Manila and carpet shell clam haemocytes against *Vibrio tapetis*, the causative agent of brown ring disease. *Fish Shellfish Immunol.*, **9**, 543–55.

Lutz, R.A. (1980) Pearl incidences: mussel culture and harvest implications. In: *Mussel Culture and Harvest: a North American Perspective* (ed. R.A. Lutz), pp. 193–222. Elsevier Science Publishing Co. Inc., New York.

Machkevski, V.K. (1985) Some aspects of the biology of the trematode, *Proctoeces maculatus*, in connection with the development of mussel farms on the Black Sea. In: *Parasitology and Pathology of Marine Organisms of the World Oceans* (ed. J.W. Hargis), pp. 109–10. US Department of Commerce.

Mackin, J.G., Owen, H.M. & Collier, A. (1950) Preliminary note on the occurrence of a new protistan parasite, *Dermocystidium marinum* n. sp., in *Crassostrea virginica* (Gmelin). *Science*, **111**, 328–29.

Maes, P. & Paillard, C. (1992) Effet de *Vibrio* P1, pathogene de *Ruditapes philippinarum* sur d'autres especes de bivalves. *Haliotis*, **14**, 141–48.

Marsh, A.G., Gauthier, J. & Vasta, G.R. (1995) A semiquantitative PCR assay for assessing *Perkinsus marinus* infections in the eastern oyster, *Crassostrea virginica*. *J. Parasitol.*, **81**, 577–83.

McLaughlin, S.M., Tall, B.D., Shaheen, A., Elsayed, E.E. & Faisal, M. (2000) Zoosporulation of a new *Perkinsus* species isolated from the gills of the softshell clam *Mya arenaria*. *Parasite*, **7**, 115–22.

Meyers, T.R. (1981) Endemic diseases of cultured shellfish of Long Island, New York: adult and juvenile American oysters (*Crassostrea virginica*) and hard clams (*Mercenaria mercenaria*). *Aquaculture*, **22**, 305–30.

Miahle, E., Bachère, E., Chagot, D. & Grizel, H. (1988). Isolation and purification of the protozoan *Bonamia ostreae* (Pichot *et al.*, 1980) a parasite affecting the flat oyster *Ostrea edulis* L. *Aquaculture*, 71, 293–99.

Mialhe, E., Boulo, V., Bachère, E. *et al.* (1992) Development of new methodologies for diagnosis of infectious diseases in mollusc and shrimp aquaculture. *Aquaculture*, **107**, 155–64.

Miahle, E., Bachère, E., Boulo, V. & Cadoret, J.P. (1995) Strategy for research and international co-operation in marine invertebrate pathology, immunology and genetics. *Aquaculture*, **132**, 33–41.

Millar, R.H. (1963) Oysters killed by trematode parasites. *Nature* (Lond.), **197**, 616.

Montes, J. (1991) Lag time for the infestation of flat oyster (*Ostrea edulis* L.) by *Bonamia ostreae* in estuaries of Galicia (NW Spain). *Aquaculture*, **93**, 235–39.

Montes, J., Anadon, R. & Azevedo, C. (1994) A possible life-cycle for *Bonamia ostreae* on the basis of electron-microscopy studies. *J. Invertebr. Pathol.*, **63**, 1–6.

Montes, J.F., Durfort, M. & García-Valero, J. (1995a) Characterization and localization of a Mr 225 kDa polypeptide specifically involved in the defence mechanism of the clam *Tapes semidecussatus*. *Cell Tissue Res.*, **280**, 27–37.

Montes, J.F., Durfort, M. & García–Valero, J. (1995b) Cellular defence mechanism of the clam *Tapes semidecussatus* against infection by the protozoan *Perkinsus* sp. *Cell Tissue Res.*, **279**, 529–38.

Naciri-Graven, Y., Martin, A.G., Baud, J.P., Renault, T. & Gérard, A. (1998) Selecting the flat oyster *Ostrea edulis* (L.) for survival when infected with the parasite *Bonamia ostreae*. *J. Exp. Mar. Biol. Ecol.*, **224**, 91–107.

Naciri-Graven, Y., Haure, J., Gérard, A. & Baud, J.P. (1999) Comparative growth of *Bonamia ostreae* resistant and wild flat oyster *Ostrea edulis* in an intensive system. II. Second year of the experiment. *Aquaculture*, **171**, 195–208.

Navas, J.I., Castillo, M.C., Vera, P. & Ruizrico, M. (1992) Principal parasites observed in clams, *Ruditapes decussatus* (L.), *Ruditapes philippinarum* (Reeve), *Venerupis pullastra* (Montagu) and *Venerupis aureus* (Gmelin), from the Huelva coast (SW Spain). *Aquaculture*, **107**, 193–99.

Nel, R., Coetzee, P.S. & Vanniekerk, G. (1996) The evaluation of 2 treatments to reduce mud worm (*Polydora hoplura claporede*) infestation in commercially reared oysters (*Crassostrea gigas* Thunberg). *Aquaculture*, **141**, 31–39.

Noel, T., Aubree, E., Blateau, D., Mialhe, E. & Grizel, H. (1992) Treatments against the vibrio P1, suspected to be responsible for mortalities in *Tapes philippinarum*. *Aquaculture*, **107**, 171–74.

Noel, T., Nicolas, J-L., Boulo, V., Mialhe, E. & Roch, P. (1996) Development of a colony-blot ELISA assay using monoclonal antibodies to identify vibrio P1 responsible for 'brown ring disease' in the clam *Tapes philippinarum*. *Aquaculture*, **146**, 171–78.

Obeirn, F.X. & Walker, R.L. (1999) Pea crab, *Pinnotheres ostreum*, in the eastern oyster, *Crassostrea virginica* (Gmelin 1791): prevalence and apparent adverse effects on oyster gonad development. *Veliger*, **42**, 17–20.

Oliver, J.L., Gaffney, P.M., Allen, S.K., Faisal, M. & Kaattari, S.L. (2000) Protease inhibitory activity in selectively bred families of eastern oysters. *J. Aquat. Anim. Health*, **12**, 136–45.

Oprandy, J.J., Chang, P.W., Pronovost, A.D., Cooper, K.R., Brown, R.S. & Yates, V.J. (1981) Isolation of a viral agent causing hemotopoietic neoplasia in the soft-shell clam *Mya arenaria*. *J. Invertebr. Pathol.*, **38**, 45–51.

Paillard, C. & Maes, P. (1994) Brown ring disease in the Manila clam *Ruditapes philippinarum*: establishment of a classification system. *Dis. Aquat. Org.*, **19**, 137–46.

Paillard, C. & Maes, P. (1995a) The brown ring disease in the Manila clam, *Ruditapes philippinarum*. I. Ultrastructural alterations of the periostracal lamina. *J. Invertebr. Pathol.*, **65**, 91–100.

Paillard, C. & Maes, P. (1995b) The brown ring disease in the Manila clam, *Ruditapes philippinarum*. II. Microscopic study of the brown ring syndrome. *J. Invertebr. Pathol.*, 65, 101–10.

Paillard, C., Percelay, L., Le Pennec, M. & Le Picard, D. (1989) Origine pathogene de 'l'anneau brun' chez *Tapes philippinarum* (Mollusque, Bivalve). *C.R. Acad. Sci. Paris*, **309**, 235–41.

Paillard, C., Maes, P. & Oubella, R. (1994) Brown ring disease in clams. *Annu. Rev. Fish Dis.*, **4**, 219–40.

Paillard, C., Ashton-Alcox, K.A. & Ford, S.E. (1996) Changes in bacterial densities and haemocyte parameters in oysters affected by Juvenile Oyster Disease. *Aquat. Living Resour.*, **9**, 145–58.

Paraso, M.C., Ford, S.E., Powell, E.N., Hofman, E.E. & Klinck, J.M. (1999) Modeling the MSX parasite in eastern oyster (*Crassostrea virginica*) populations. II. salinity effects. *J. Shellfish Res.*, **18**, 501–16.

Perkins, F.O. (1996) The structure of *Perkinsus marinus* (Mackin, Owen and Collier, 1950) Levine, 1978 with comments on taxonomy and phylogeny of *Perkinsus* spp. *J. Shellfish Res.*, **15**, 67–87.

Pipe, R.K. & Coles, J.A. (1995) Environmental contaminants influencing immune function in marine bivalve molluscs. *Fish Shellfish Immunol.*, **5**, 581–95.

Plana, S., Sinquin, G., Maes, P., Paillard, C. & Le Pennec, M. (1996) Variations in biochemical composition of juvenile *Ruditapes philippinarum* infected by a *Vibrio* sp. *Dis. Aquat. Org.*, **24**, 205–13.

Powell, E.N., White, M.E., Wilson, E.A. & Ray, S.M. (1987) Changes in host preferences with age in the ectoparasitic pyramidellid snail *Boonea impressa* (Say). *J. Moll. Stud.*, **53**, 285–86.

Powell, E.N., Gauthier, J.D., Wilson, E.A., Nelson, A., Fay, R.R. & Brooks, J.M. (1992) Oyster disease and climatic change. Are yearly changes in *Perkinsus marinus* parasitism in oysters (*Crassostrea virginica*) controlled by climatic cycles in the Gulf of Mexico? *P.S.Z.N.I.: Marine Ecology*, **13**, 243–70.

Powell, E.N., Klinck, J.M., Ford, S.E., Hofman, E.E. & Jordan, S.J. (1999) Modeling the MSX parasite in eastern oyster (*Crassostrea virginica*) populations. III. Regional application and the problem of transmission. *J. Shellfish Res.*, **18**, 517–37.

Pregenzer, C.L. (1981) The effects of *Pinnotheres hickmani* on the meat yield (condition) of *Mytilus edulis* measured in several ways. *Veliger*, **23**, 250–53.

Prieur, D., Nicolas, J.L., Plusquellec, A. & Vigneulle, M. (1990) Interactions between bivalve mollusks and bacteria in the marine environment. *Oceanogr. Mar. Ecol.*, **28**, 277–352.

Rasmussen, L.P.D. (1986) Virus-associated granulocytomas in the marine mussel *Mytilus edulis* from three sites in Denmark. *J. Invertebr. Pathol.*, **48**, 117–23.

Ray, S.M. (1952) A culture technique for the diagnosis of infection with *Dermocystidium marinum* Mackin, Owen and Collier in oysters. *Science*, **116**, 360–61.

Ray, S.M. (1954) Experimental studies on the transmission and pathogenicity of *Dermocystidium marinum*, a fungus parasite of oysters. *J. Parasitol.*, **40**, 235.

Ray, S.M. (1966) A review of the culture method for detecting *Dermocystidium marinum*, with suggested modifications and precautions. *Proc. Natl. Shellfish Ass.*, **54**, 55–69.

Ray, S.M. (1996) Historical perspective on *Perkinsus marinus* disease of oysters in the Gulf of Mexico. *J. Shellfish Res.*, **15**, 9–11.

Rhodes, L.D. & Vanbeneden, R. (1996) Application of differential display Polymerase Chain-Reaction to the study of neoplasms of feral marine bivalves. *Mar. Environ. Res.*, **42**, 81–85.

Riquelme, C., Hayashida, G., Toranzo-AE., Vilches, J. & Chavez, P. (1995) Pathogenicity studies on a *Vibrio anguillarum*-related (var) strain causing an epizootic in *Argopecten purpuratus* larvae cultured in Chile. *Dis. Aquat. Org.*, **22**, 135–41.

Robledo, J.A.F., Santarém, M.M. & Figueras, A. (1995) Seasonal variations in the biochemical composition of the serum of *Mytilus galloprovincialis* Lmk. and its relationship to the reproductive cycle and parasitic load. *Aquaculture*, **133**, 311–22.

Robledo, J.A.F., Gauthier, J.D., Coss, C.A., Wright, A.C. & Vasta, G.R. (1998) Species-specificity and sensitivity of a PCR-based assay for *Perkinsus marinus* in the eastern oyster, *Crassostrea virginica*: a comparison with the fluid thioglycollate assay. *J. Parasitol.*, **84**, 1237–44.

Roch, P. (1999) Defense mechanisms and disease prevention in farmed marine invertebrates. *Aquaculture*, **172**, 125–45.

Rogan, E., Culloty, S.C., Cross, T.F. & Mulcahy, M.F. (1991) The detection of *Bonamia ostreae* (Pichot *et al.* 1980) in frozen oysters (*Ostrea edulis* L.) and the effect of the parasite on condition. *Aquaculture*, **97**, 311–15.

Sakaguchi, S. (1967) Studies on a trematode parasite of the pearl oyster, *Pinctada martensii*. IV. Artificial infection with the encysted metacercaria to the final host. *Bull. Natl. Pearl Res. Lab.*, **12**, 1445–54.

Sakaguchi, S. (1968) Studies on the life-history of the trematode parasitic in the pearl oyster, *Pinctada fucata*, and on the hindrance for pearl culture. *Bull. Natl. Pearl Res. Lab.*, **13**, 1635–88.

Sanders, M.J. & Lester, R.J. (1981) Further observations on a bucephalid trematode infection in scallops (*Pecten alba*) in Port Phillip Bay, Victoria. *Aust. J. Mar. Freshw. Res.*, **32**, 475–78.

Scro, R.A. & Ford, S.E. (1990) An electron miscroscope study of disease progression in the oyster *Crassostrea virginica* infected with the protozoan parasite *Haplosporidium nelsoni* (MSX). In: *Pathology in Marine Science* (ed. F.O. Perkins), pp. 229–54. Academic Press, Orlando, Florida.

Sindermann, C. (1990) *Principal Diseases of Marine Fish and Shellfish*, 2nd edn. Vol. II. Academic Press, San Diego, California.

Sparks, A.K. (1985) *Synopsis of Invertebrate Pathology Exclusive of Insects*. Elsevier Science Publishers B.V., Amsterdam.

Sprague, V., Dummington, E.A. & Drobeck, E. (1969) Decrease in incidence of *Minchinia nelsoni* in oysters accompanying reduction of salinity in the laboratory. *Proc. Natl. Shellfish Ass.*, **59**, 23–26.

Stokes, N.A. & Burreson, E. (1995) A sensitive and specific DNA probe for the oyster pathogen *Haplosporidium nelsoni*. *J. Euk. Microbiol.*, **42**, 350–57.

Sutton, D.C. & Garrick, R. (1993) Bacterial disease of cultured giant clam *Tridacna gigas* larvae. *Dis. Aquat. Org.*, **16**, 47–53.

Teia dos Santos, A.M.T. & Coimbra, J. (1995) Growth and production of raft-cultured *Mytilus edulis* L. in Ria de Aveiro; gonad symbiotic infestation. *Aquaculture*, **132**, 195–211.

Theisen, B.F. (1987) *Mytilicola intestinalis* Steuer and the condition of its host *Mytilus edulis*. *Ophelia*, **27**, 77–86.

Tigé, G. & Grizel, H. (1984) Essai de contamination d'*Ostrea edulis* Linné par *Bonamia ostreae* (Pichot *et al.*, 1980) en Rivière de Crach (Morbihan). *Rev. Trav. Inst. Pêch. Marit.*, **46**, 307–14

Uzmann, J.R. (1951) Record of the larval trematode *Himasthla quissetensis* (Miller and Northup, 1926) Stunkard, 1934 in the clam, *Mya arenaria*. *J. Parasitol.*, **37**, 327–28.

Van Banning, P. (1990) The life cycle of the oyster pathogen *Bonamia ostreae* with a presumptive phase in the ovarian tissue of the European flat oyster, *Ostrea edulis*. *Aquaculture*, **84**, 189–92.

Volety, A.K. & Chu, F. (1995) Suppression of chemiluminescence of eastern oyster (*Crassostrea virginica*) hemocytes by the protozoan parasite *Perkinsus marinus*. *Dev. Comp. Immunol.*, **19**, 135–42.

Volety, A.K. & Fisher, W.S. (2000) In vitro killing of *Perkinsus marinus* by hemocytes of oysters *Crassostrea virginica*. *J. Shellfish Res.*, **19**, 827–34.

Volety, A.K., Oliver, L.M., Genthner, F.J. & Fisher, W.S. (1999) A rapid tetrazolium dye reduction assay to assess the bactericidal activity of oyster (*Crassostrea virginica*) hemocytes against *Vibrio parahaemolyticus*. *Aquaculture*, **172**, 205–22.

White, M.E., Powell, E.N., Wilson, E.A. & Ray, S.M. (1989) The spatial distribution of *Perkinsus marinus*, a protozoan parasite, in relation to its oyster host (*Crassostrea virginica*) and an ectoparasitic gastropod (*Boonea impressa*). *J. mar. biol. Ass. U.K.*, **69**, 703–17.

Wilson, E.A., Powell, E.N., Craig, M.A., Wade, T.L. and Brooks, J.M. (1990) The distribution of *Perkinsus marinus* in Gulf Coast oysters: its relationship with temperature, reproduction, and pollutant body burden. *Int. Revue Ges. Hydrobiol.*, **75**, 533–50.

Yarnall, H.A., Reece, K.S., Stokes, N.A. & Burreson, E.M. (2000) A quantitative competitive polymerase chain reaction assay for the oyster pathogen *Perkinsus marinus*. *J. Parisitol.*, **86**, 827–37.

Zabaleta, A.I. & Barber, B. (1996) Prevalence, intensity, and detection of *Bonamia ostreae* in *Ostrea edulis* L in the Damariscotta River area, Maine. *J. Shellfish Res.*, **15**, 395–400.

12 Public Health

Introduction

By virtue of their feeding habit, bivalves concentrate and accumulate material from the environment. This, coupled with the tradition of consuming bivalves raw or partially cooked, means that they act as potential vectors for human infection from water-borne agents such as bacteria, viruses, algal toxins and heavy metals, and thus are high-risk products. Generally speaking, the bivalves are not themselves affected by the microorganisms or toxins, merely serving to concentrate and passively transport the ethiological agent. In this section on public health the role of bivalves in transmission of disease to humans will be dealt with under the following headings: bacterial infections, viral infections, biotoxins, and industrial pollutants. Methods of bacterial, viral and toxin detection, decontamination procedures, and monitoring measures will also be covered.

Bacterial infections

Bivalves living in polluted waters are often subject to contamination from domestic sewage and land run-off that typically contain pathogenic bacteria such as *Salmonella* spp., *Shigella* spp., *Campylobacter* spp., *Aeromonas* spp. and non-pathogenic *Escherichia coli* (Table 12.1). Before the 1950s the most common illness associated with the consumption of raw bivalves in the United States was typhoid fever caused by *Salmonella typhi* (Rippey, 1994). With improved sewage treatment and water quality standards in bivalve-growing areas, the incidence of typhoid has declined, and now gastroenteritis is the most common illness associated with the consumption of raw, contaminated bivalves.

Estuarine bacteria of the genus *Vibrio* are also potentially pathogenic to humans. The most important of these are *Vibrio parahaemolyticus* and *V. vulnificus*. Their presence is not associated with faecal contamination from human or animal source, and they are not detected by standard monitoring methods, or eliminated from shellfish by standard depuration processes (see below; and Shumway, 1992). *Vibrio parahaemolyticus* is a normal constituent of the inshore marine flora but its abundance is increased through organic enrichment of coastal and estuarine areas, where bivalve-growing areas are concentrated. The illness caused by this bacterium is usually confined to gastroenteritis, which is sometimes severe and of long duration. A more important disease organism is *V. vulnificus*, which can cause serious illness and death in persons with pre-existing liver disease, diabetes, or compromised immune systems. In the United States infections (septicaemia, wound infection, gastroenteritis) occur in the warmer months from May to October and are associated with raw oyster consumption (Hlady & Mullen, 1993). In the Gulf of Mexico states there is an annual incidence for infections of at least 0.6 per million population, and a case-fatality rate of about 20% (Levine & Griffin, 1993).

Table 12.1. Bivalve-borne disease agents occurring in, and transmitted by, sewage and or waste water (1898–1990) in the United States. Modified from Rippey (1994).

Agent	Number of cases	Number of incidents	Number of outbreaks
Unknown	7978	277	256
Salmonella typhi	3270	93	78
Hepatitis A virus	1798	51	42
Norwalk virus	311	7	7
Salmonella spp.	130	8	3
Shigella spp.	111	9	4
Campylobacter spp.	27	12	1
Aeromonas spp.	7	1	1
Escherichia coli	2	1	1

Rates of uptake and elimination of bacteria are species and temperature dependent. For example, *Mytilus edulis* accumulated coliforms at a higher rate than four other bivalve species, and also eliminated these bacteria more effectively (Bernard, 1989). Rates of accumulation and elimination were higher at 17°C than at 12°C. In addition, some bacterial types accumulate at lower rates than others. For example, *Escherichia coli* accumulates at a lower rate than other bacterial types in *M. edulis* (Webber, 1982) and in *Mercenaria mercenaria* (Burkhardt *et al.*, 1992). This is interesting in view of the fact that *E. coli* in bivalves is used as an indicator of other pathogens, including viruses (see below). All bacterial types are destroyed by heat, thus proper cooking is sufficient to eliminate them from contaminated bivalves. However, some bacteria, e.g. *V. cholerae* can survive boiling for up to 8 min and steaming for up to 25 min (Blake *et al.*, 1980). Therefore, the commercial practice of heat-shocking oysters in boiling water to facilitate opening is not sufficient to ensure safety (Huss, 1994).

Bacterial assays

As *Vibrio* spp. are not detected by the standard methods used in monitoring for bacterial contamination, alternative methods have been developed to detect these organisms in bivalves. The most sensitive method is the PCR technique that can rapidly detect and amplify DNA sequences from specific *Vibrio* genes. The method is currently employed to detect different strains of *V. cholerae* (Depaola & Hwang, 1995), *V. parahaemolyticus* (McCarthy *et al.*, 2000) and *V. vulnificus* (Kim & Jeong, 2001). PCR assays have also been used to detect non-*Vibrio* bacteria, e.g. *E. coli* (Gonzalez *et al.*, 1999), *Salmonella* and *Shigella* species (Jones *et al.*, 1993; Vantarakis *et al.*, 2000).

Viral infections

Bacterial agents of bivalve-associated human disease represent a small proportion of disease outbreaks (~4%) compared to viral agents (~21%), although it must be pointed out that in the vast majority of illness reports (~75%) no agent is identified (Table 12.1. At present there are more than 100 known enteric viruses that are excreted in human faeces and find their way into domestic sewage (Huss, 1994). Those associated with the consumption of bivalves from waters contaminated with human or animal faeces are hepatitis A, Snow Mountain Agent, small round viruses (SRSV) and polioviruses. Hepatitis A is a serious illness, causing debilitating chronic infections and sometimes death. Worldwide outbreaks of bivalve-borne hepatitis are reported frequently. In China in 1988 there were 292000 cases of the disease (9 deaths) that were associated with uncooked contaminated clams (Halliday et al., 1991). Snow Mountain Agent and SRSVs are responsible for outbreaks of gastroenteritis associated with swimming in water contaminated with human sewage, faecal contamination of food or drinking water, and consumption of uncooked or partially cooked bivalves from contaminated waters. Polio is no longer endemic in countries with national vaccination programmes. However, epidemics may still occur in underdeveloped countries through contamination of water supplies or food with human sewage. Symptoms of poliomyelitis include malaise, sore throat, headache and vomiting, but in 1% of cases there is nervous system damage with permanent paralysis (Kilgen & Cole, 1991).

Viruses can survive and remain infectious in seawater for prolonged periods. Once inside bivalves they seem to survive longer than coliform bacteria, and in several studies enteric viruses have been isolated from bivalves otherwise having a satisfactory coliform index (Gerba & Goyal, 1978). More worrying is the finding that in marine waters there is often no correlation between the presence of enteric viruses and indicator bacteria, i.e. so called 'clean' waters often harbour enteroviruses (Croci et al., 2000). Also, enteric viruses are resistant to acid pH, digestive enzymes and bile salts in the human gut, common disinfectants, and even steaming and frying (Huss, 1994). Heat treatment is the only sure way to eliminate viruses. It is recommended that internal temperature of bivalves must be maintained at 90°C for 1.5 min before they are consumed (Huss et al., 2000).

Viral assays

The detection of viruses is a complicated analytical procedure and at present there are no limit values or guideline levels available for viruses in water quality monitoring programmes. The most common methods for virus detection employ reverse transcriptase (RT)-PCR, a modification of the PCR method described above. The technique has been used routinely to screen for enteroviruses (Gualillo et al., 1999; Casas & Sunen, 2001), and to detect specific viruses such as hepatitis A, Norwalk, 'Norwalk-like' viruses (SRSVs) (Sunen & Sobsey, 1999; Schwab et al., 2001), herpes-like virus (Renault et al., 2000) and polio virus (Jaykus et al., 1996).

Biotoxins

Marine biotoxins, produced by dinoflagelates and diatoms, are a naturally occurring phenomenon and not associated with sewage contamination of coastal waters. The continuing increase in the number of toxic species, coupled with the increased occurrences of blooms of these species in coastal waters, pose a constant threat to public health worldwide (Fig. 12.1). Filter-feeding bivalves accumulate the toxic cells in their tissues and are, therefore, the main vectors of toxins that cause paralytic (PSP), amnesic (ASP), diarrhetic (DSP), and neurotoxic (NSP) shellfish poisoning in humans (Table 12.2). Cooking does not inactivate these toxins, and there is no known antidote. It is generally believed that the toxins do not harm the bivalves themselves, but more detailed investigation may show this to be untrue (Burkholder, 1999). Of all bivalves consumed mussels pose the greatest threat because they accumulate toxins more rapidly, and to a greater degree, than other commercially important bivalves. This, together with their widespread geographic distribution, has led to their prominent use as indicator organisms in toxic surveillance programmes, and as test organisms in toxic research. Other marine organisms such as gastropods, crustaceans and fish that prey on bivalves, and are harvested for consumption, should also be monitored.

Paralytic shellfish poisoning (PSP) is caused by eating bivalves that contain saxitoxin and its derivatives. The toxins, a group of about 20, are produced by dinoflagellates of the genera, *Alexandrium* (previously *Gonyaulax*), *Gymnodinium* and *Pyrodinium*. Even when blooms of these species are not apparent bivalves may ingest resting cysts, which are 10–1000 times more toxic than motile dinoflagellates (Sinderman, 1990). The toxins produced by these organisms are the most common and widespread of the shellfish toxins (Fig. 12.1), and are among the most potent neurotoxins known. They act by blocking the passage of sodium ions through cell membranes, thus inhibiting nerve impulse transmission. If a lethal dose is consumed (0.3–12 mg), and stomach pumping or artificial respiration is not administered, death occurs from respiratory failure within 24 hours. The safety limit for PSP toxins is 80 μg per 100 g of bivalve tissue (Table 12.3). PSP toxins have a predictable rate of decay within bivalve tissues. Application of decay constants for PSP (and other toxins) allow fairly accurate predictions of the amount of time that harvest of bivalves must be suspended provided that the starting toxicity level is known. From this the length of time before affected area(s) can be reopened for harvest can be deduced.

Amnesic shellfish poisoning (ASP) is caused when bivalves are eaten that contain domoic acid and/or its analogues. The toxins are secreted by diatoms *Pseudonitzschia* spp. Domoic acid (DA) is an analogue of glutamic acid, a neurotransmitter in the brain. Within 24 hours of consuming contaminated bivalves, usually mussels, there is nausea, vomiting and diarrhoea with the following neurologic perturbations within 48 hours: confusion, memory loss, disorientation, and even seizures, coma and death (Table 12.2). The severity of symptoms depends on the amount of DA ingested; generally the more serious symptoms are caused by amounts in excess of 4 mg per kg body weight. A value of 20 μg DA g^{-1} has been established as the safety limit in shellfish (Table 12.3). A recent report has shown that domoic acid is quickly degraded in *Mytilus edulis* and *Mya arenaria* by indigenous bacteria, in contrast to *Placopecten magellanicus*, which does not appear to possess such bacteria (Stewart *et al.*, 1998).

Diarrhetic (DSP) and neurotoxic shellfish poisoning (NSP) are much less

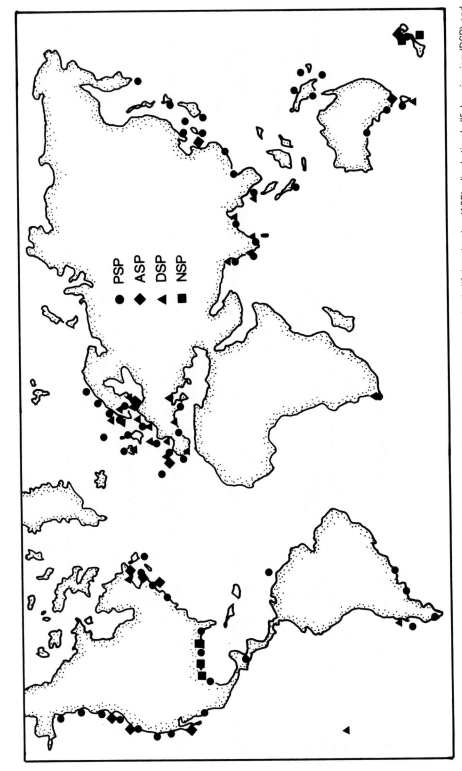

Fig. 12.1. Current global distribution of the toxins of paralytic shellfish poisoning (PSP), amnesic shellfish poisoning (ASP), diarrhetic shellfish poisoning (DSP) and neurotoxic shellfish poisoning (NSP) in phytoplankton and seafood. Modified from Leftley & Hannah (1998) and Whittle & Gallacher (2000). Reprinted with permission from Oxford University Press.

Table 12.2. Types of shellfish poisoning (Sindermann, 1990; Yasumoto *et al.*, 1995; Ishida *et al.*, 1996; Whittle & Gallacher, 2000).

Illness	Cause	Toxin	Symptoms/onset/ treatment	Mode of action	Notes
Paralytic shellfish poisoning (PSP)	Dinoflagellate species of *Alexandrium*, *Gymnodinium* and *Pyrodinium*	Saxitoxin and derivatives	Numbness of body parts, visual disturbances, paralysis, death: 30 min to 3–4 h: Artificial respiration.	Block sodium channels in cell membranes	Probably occurred as far back as 18th century; much more frequent outbreaks since the 1970s.
Amnesic shellfish poisoning (ASP)	Mainly diatoms *Pseudo-nitzschia* spp.	Domoic acid and analogues	Nausea, vomiting, disorientation, nervous system disfunction, coma, death: Gastrointestinal <24 h, neurological <48 h.	Agonists for glutamate, a brain neurotransmitter	First outbreak in New Brunswick, Canada in 1987.
Diarrhetic shellfish poisoning (DSP)	*Dinophysis* and *Prorocentrum* dinoflagellate species	Okadaic acid and derivatives	Cramps, severe diarrhoea, vomiting: 30 min to 3–12 h.	Inhibit protein phosphatases	First report in Europe in 1961.
Neurotoxic shellfish poisoning (NSP)	Dinoflagellate *Ptychodiscus brevis*	Brevetoxins	Dizziness, headaches, diarrhoea, muscle and joint pain, double vision, difficulty breathing: Within 3–6 h.	Open sodium channels in cell membranes	

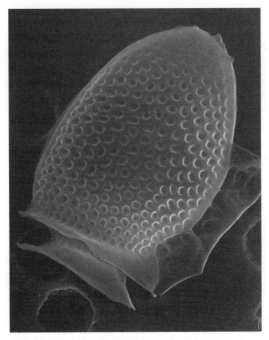

Fig. 12.2. Scanning electron micrograph of the dinoflagellate *Dinophysis acuminata*; maximum dimension approximately 50 μm. © C. Cusack.

Table 12.3. United States and European Union (EU) allowable levels for poisonous or deleterious substances in fish and shellfish. S: shellfish; F: fish; ppm: parts per million; DDT: dichlorodiphenyltrichloroethane; DDE: dichlorodiphenyldichloroethylene; DDD: dichloro-diphenyldichloroethane; PCB: polychlorinated biphenyls; PSP: paralytic shellfish poison; DSP: diarrhetic shellfish poison; NSP: neurotoxic shellfish poison; ASP: amnesic shellfish poison; *WHO (1989) specify 0–60 μg per 100 g; **value for Canada (Todd, 1993). Adapted from Shumway (1992) and Huss (1994).

	Allowable level	
Deleterious substance	US	EU
Dieldrin	0.30 ppm (F, S)	0.1 ppm (F)
DDT, and metabolites DDE and DDD	5.00 ppm (F)	2.0 ppm (F)
PCB	20 ppm (F, S)	2.0 ppm (F)
Mercury	1.0 ppm (F, S)	0.5 ppm (F)
Lead	?	2.0 ppm (F)
PSP	80 μg per 100 g meat (S)	80 μg per 100 g meat (S)
DSP	Not specified*	Not specified
NSP brevetoxins	No detectable amount	No detectable amount
ASP domoic acid	20 μg g^{-1} meat** (S)	20 μg g^{-1} meat (S)

serious than PSP or ASP. DSP is caused primarily by okadaic acid (OA) and its derivatives, the dinophysistoxins (DTX), produced by dinoflagellate species of *Dinophysis* (Fig. 12.2) and *Prorocentrum*. The main symptoms of DSP are cramps, severe diarrhoea and vomiting and patients usually recover within 3–4 days. However, recent findings that DSP toxins have tumour-promoting and immuno-suppressing activity in animal cells, is a cause for concern. The safety limit for OA in European Union (EU) countries is 0–0.60 g per 100 g (Table 12.3). NSP has only been reported from parts of North America and New Zealand (Fig. 12.1). It is caused by a group of toxins, the brevitoxins, secreted by the dinoflagellate *Ptychodiscus brevis* (formerly *Gymnodinium breve*). The main symptoms of NSP are headaches, dizziness, muscle and joint pain and difficulty in breathing. Full recovery generally occurs within 48 h. To date no safety limit has been set for brevitoxins.

Since the early 1990s a number of new shellfish toxins have been identified (Quilliam, 1999). One of these, azaspiracid, produces similar symptoms to DSP, but so far its source and mode of action are unknown (Satake *et al.*, 1998). It is not known why there has been an increase in the frequency of algal blooms in recent years. The increase may be due to a combination of natural and man-made factors. Frontal zones, or discontinuities between water masses, are the locations most likely to generate blooms. These fronts may result from tide-or wind-generated convergence and/or density discontinuities (Yentsch & Incze, 1980). Factors such as eutrophication of coastal waters and the accidental spread of algal species through ballast water to new locations may also be important (Todd, 1993). Monitoring programmes for biotoxins are well established in countries such as the United States, Canada, Japan, Tasmania, the EU and most other maritime European countries but to date, there is little, if any, monitoring carried out in developing countries (Shumway, 1995).

Biotoxin assays

The mouse bioassay is the official procedure for measuring PSP and DSP toxins. Liquid extracted from putative toxic bivalves is injected into mice; the time to death of mice of standard weight (18–22 g) is related to toxin level (Hollingworth & Wekell, 1990). However, the test has several drawbacks in that it is costly, inconvenient, has limited sensitivity and is regarded as unethical by defenders of animal welfare. Therefore, alternative methods are increasingly being developed, but none are validated yet for monitoring purposes (Whittle & Gallacher, 2000). One such alternative is cell-based bioassays that measure the effects of toxins on cultured animal or bacterial cells (Flanagan *et al.*, 2001). These assays are quick, inexpensive, and are highly sensitive. For example, a cytotoxicity assay for okadaic acid, the major DSP toxin, can detect 0.01 mg of toxin per gram of mussel tissue (Croci *et al.*, 2001). Methods based on perturbations to cellular activity, such as enzyme inhibition, have also been developed. One such assay, the protein phosphatase inhibition assay, detects okadaic acid down to $0.1 \mu g g^{-1}$ tissue, which is about twice as sensitive as the mouse bioassay test (Mountfort *et al.*, 2001).

As already indicated, DSP and PSP encompass a multiplicity of toxins, which makes detection a formidable challenge. High performance liquid chromatography (HPLC) with fluorometric detection, liquid chromatography in combination with mass spectroscopy (LC/MS), or ionspray (ISP) in combination with LC/MS, are used to detect PSP saxitoxins and DSP okadaic acid and its analogues dinophysistoxin (DPX)-1 through -3 (Draisci *et al.*, 1999; Holmes *et al.*, 1999; Gonzalez *et al.*, 2000; Suzuki & Yasumuto, 2000). Immunoassay methods are also used to detect PSP and DSP toxins (Huang *et al.*, 1996; Vale & Sampayo, 1999), and there is now an easy-to-use kit available for PSP detection (Jellett *et al.*, 1998).

Industrial pollutants

Bivalves accumulate high levels of pesticides, heavy metals, and hydrocarbons from contaminated water and there is ample evidence that these have a severe effect on their physiology and immune system (Widdows & Donkin, 1992; Livingstone & Pipe, 1992; Livingstone *et al.*, 2000). It is for this reason that bivalves, particularly mussels, are used as sentinel organisms in environmental monitoring programmes. Apart from the horrific Minamata disease episode in Japan (1953–61), caused by eating mercury contaminated fish and shellfish, there is little evidence for an established link between pollutant-contaminated bivalves and human disease. Neither is there evidence for a direct association between carcinogens, e.g. heavy metals, petroleum derivatives, in bivalve tissue and cancer in humans. Despite this, there is now enough information on the carcinogenic, mutagenic, and other long-term toxic effects of many industrial chemicals to justify further study of these contaminants in seafood.

Decontamination procedures

The capacity of bivalves to concentrate and accumulate bacteria, viruses and biotoxins and pollutants means that special decontamination procedures are often necessary before bivalves can be harvested and marketed. These methods could be seen as 'corrective' measures to compensate for the losses incurred through the application of increasing numbers of water regulations (see below). There are two methods used to purify bivalves contaminated with pathogenic viruses or bacteria. One method is relaying, where bivalves are moved from contaminated to approved waters for relatively long periods of time, usually 30 days or more (when they are metabolically active), to allow them to purge themselves of contaminants before harvesting. The major advantage of this method is that relayed bivalves are only required to meet open area bacterial standards (Shumway, 1992). However, the method is feasible only when: costs of relaying are low, retail value of the species is high, relayed stocks can be monitored adequately to prevent unauthorised harvesting, losses due to replanting are low, and reharvesting efficiency is high (Canzonier, 1988). The other method is depuration (controlled purification), where contaminated bivalves are placed for ≤48 hours in sterilised seawater to allow natural cleansing. Ultraviolet (UV) light, ozone or chlorine are used to sterilise the seawater. When chlorine is used the seawater must be dechlorinated before it can be used to depurate bivalves. Ozone, on the other hand, does not leave harmful residues in seawater. In the United States UV treatment is the method of choice, while chlorine and ozone are the preferred treatments in Europe (Shumway, 1992). Irradiation could prove to be an alternative method to depuration; Mallett *et al.* (1991) found that treatment of *Mercenaria mercenaria* with Cobalt[60] did not harm the species and effectively eliminated *Vibrio cholerae* and *V. parahaemolyticus.*

The efficiency of depuration depends on the nature of the contaminant, the species being depurated, the length of the depuration period, and the quality of the depuration plant itself. Complete elimination of *E. coli* and *Salmonella* normally occurs within 48–72 hours, although this period may not be sufficient for some bacterial types, e.g. *Streptococcus faecalis* (Plusquellec *et al.*, 1990) and some *Vibrio* species (Marino *et al.*, 1999). Harrisyoung *et al.* (1995) have found that *Vibrio vulnificus* numbers actually increase during UV depuration; it seems that this species employs encapsulation as a way of resisting ingestion and degradation by haemocytes during the purification process. Unlike most bacteria, viruses are released slowly and at different rates depending on the virus. For example, in mussels elimination of hepatitis A virus is slower than elimination of polio virus, taking about seven days for complete elimination (Enriquez *et al.*, 1992). This means that a much longer time-frame is necessary to ensure effective depuration of viruses, and this is not always feasible in a commercial plant. In the mussel *Perna canaliculus* no significant reduction in viral numbers had occurred after eight days of depuration, and there was no significant correlation between viral and faecal coliform numbers (Lewis *et al.*, 1986). Therefore, faecal coliform number, which is the index used to evaluate the efficacy of the depuration process, is an unreliable indicator of viral contamination (Chung *et al.*, 1998). Water temperature also seems to be an important factor in viral (and also bacterial) removal during depuration. During the winter commercial depuration plants in the UK maintain seawater temperatures above 5–8°C, depending on the bivalves species. Doré *et al.* (1998) have found that increasing the water temperature to 18°C significantly increased the efficiency of depuration. At 18°C levels of male-specific RNA bacteriophage (viral indicator) were reduced to just 2% of initial contamination levels compared with 40% of initial contamination levels for oysters depurated at 9°C (Fig. 12.3A). However, even at the higher temperature it was still not possible to eliminate all viruses, even after seven days of depuration. It should be noted that the current minimum requirements, in both the UK and other countries, are for depuration periods of the order of 48 hours. The speed of elimination was also dependent on initial viral load. At light contamination it was possible to reduce RNA bacteriophage to below detection level (<30 plaque forming units (pfu) per 100 g bivalve) after only two days of depuration at 18°C, whereas depuration at 9°C did not completely eliminate low levels even after seven days (Fig.12.3B).

Bivalves that have accumulated biotoxins, are normally left to purge themselves in their natural habitat. Time taken to reach acceptable levels varies between species and ranges between one week to three months (Shumway, 1992). For example, it took 40–50 days for domoic acid to decline to negligible levels in mussels from Prince Edwards Island, NE Canada (Todd, 1993). Elimination rates of domoic acid are faster in small (45–55 mm) than in large mussels (60–70 mm), and more rapid at 11°C than at 6°C, but factors such as salinity, or whether mussels are fed or starved, have no effect on elimination rates (Novaczek *et al.*, 1992).

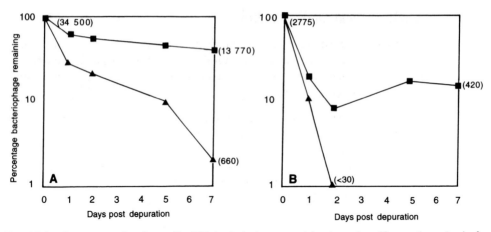

Fig. 12.3. Percentage of male-specific RNA bacteriophage remaining in oysters (*Crassostrea gigas*) after depuration at $9 \pm 1°C$ (■) and $18 \pm 1°C$ (▲) in 2 depuration cycles, A and B. Values in brackets are bacteriophage titres pfu per 100 g oyster for initial contamination and post-depuration levels. From Doré *et al.* (1998).

Table 12.4. Relationship between exposure time and depuration half-life for various hydrocarbons and polychlorobiphenyls (PCB) in bivalves. [1]Polynuclear aromatic hydrocarbons (PAH); [2]Number of chlorine atoms as subscript. Adapted from Livingstone & Pipe (1992).

Pollutant	Species	Exposure time (days)	Depuration half-life (days)	Reference
Hydrocarbon				
Alkyl-naphthalene	*Mytilus edulis*	2	0.9	Farrington *et al.*, 1982
Phenanthrene	*Modiolus modiolus*	2	~1–2	Palmork & Solbakken, 1981
4-ring PAH[1]	*M. edulis*	40	14–30	Pruell *et al.*, 1986
2–4 ring PAH	*Macoma balthica*	180	>60	Clement *et al.*, 1980
PCBs[2]				
Cl_3	*Cerastoderma edule*	10	~7	Langston, 1978
Cl_4			7–14	
Cl_5			21	
Cl_6			>21	
Cl_2	*Perna viridis*	17	0.5–2.5	Tanabe *et al.*, 1987
Cl_3			0.5–6.5	
Cl_5			4.9–8.3	

In the case of industrial contamination the uptake and clearance of these pollutants is a passive process also. However, it would appear that elimination rate is markedly affected by the duration of exposure to the chemical (Table 12.4). Elimination is rapid and complete following short–term exposure, but long–term exposure results in slower and often incomplete clearance of the chemical (Livingstone & Pipe, 1992). Elimination rate also depends on the molecular weight and water solubility of the chemical (Table 12.4), as well as on environmental factors such as temperature. All of these factors should be taken into account when designing programmes to monitor industrial pollutants.

Monitoring and quality control

Every country involved in bivalve production should have an effective sanitation programme that oversees the production, harvesting and marketing of species for human consumption. The basic requirements of such a programme should include (Canzonier, 1988):

- An infrastructure, responsible for the monitoring, culture and harvesting activities; also, adequate surveillance of the chain of supply from point of production to point of retail.
- An administrative system for coordinating the activities of the various public agencies responsible for running the programme.
- Appropriate legislation so that the agencies can prosecute those who breach regulations.

To illustrate how such a programme operates the National Shellfish Sanitation Program (NSSP) in the United States will be taken as an example. Set up in 1925, in response to outbreaks of bivalves-borne typhoid fever and other diseases, the NSSP operates under a set of guidelines that have been drawn up by state agencies, technical experts and representatives from the bivalve industry itself. Co-ordination, supervision and evaluation of the programme are the responsibility of the Food and Drug Administration (FDA). The guidelines are updated on a regular basis and are published in a *Manual of Operations* (NSSP 1989a,b) that is readily accessible to interested parties. The manual gives detailed instructions on how a monitoring programme should be run anywhere in the United States, describes the interrelation of individual state programmes, and cites the criteria to be applied by the FDA in evaluation of the programmes. Similar programmes operate in Canada, Japan, Australia, New Zealand, and several European countries.

The most logical and least onerous approach in applying the principles of bivalve sanitation is prevention of contamination at source, i.e. production and harvesting in clean waters. This requires knowledge of local geography, prevailing water currents, and the local discharge and treatment of sewage. In addition, monitoring of water quality is essential. A number of bacterial groups are used to indicate the sanitary quality of coastal waters. The most common indicators are the coliforms, specifically *Escherichia coli*. The presence of *E. coli* in water is proof that faecal contamination, of animal or human origin, has occurred, and is a definite indication that pathogens may be present. However, it should be noted that the ratio of coliforms to pathogens is not constant. The presence of coliforms is detected by plating a water sample on an agar-bile salt medium and analysing for gas production within 48 h at 37°C. *Escherichia coli* is distinguished by further tests. Two distinct analytical procedures are used routinely (Flanagan, 1992). The first is a multiple-tube method in which several replicates of each of three different dilutions of sample are incubated in test tubes containing the appropriate medium. After incubation the number of tubes in which gas has been produced is noted and the most probable number (MPN) of organisms in 100 ml of sample is obtained from probability tables. This test is a qualitative test, i.e. it indicates the presence of coliforms, but does not give the actual numbers of coliforms in 'positive' tubes.

Table 12.5. EU requirements (EU Council Directive 91/492/EEC) for live bivalves intended for immediate human consumption (EEC, 1991; Huss, 1994)

They must possess visual characteristics associated with freshness and viability, including shells free of dirt, an adequate response to percussion, and normal amounts of intravalvular liquid.

They must contain less than 300 faecal coliforms or less than 230 *E.coli* per 100 g of flesh and intravalvular liquid based on a 5-tube 3-dilution MPN test (see text), or any other bacteriological procedure of equivalent currency.

They must not contain *Salmonella* in 25 g of flesh.

They must not contain toxic or objectionable compounds occurring naturally or added to the environment.

The upper limit as regards radionucleide contents must not exceed the limits for foodstuffs as laid down by the EU.

The total paralytic shellfish poison (PSP) content in the edible part of molluscs must not exceed 80 µg per 100 g meat.

The customary biological testing methods must not give a positive result to the presence of diarrhetic shellfish poison (DSP) in the edible parts of the molluscs.

In the absence of routine virus testing procedures and the establishment of virological standards, health checks must be based on faecal bacteria counts.

The second procedure, membrane filtration, is carried out by passing samples through sterilised filter membranes. Micro-organisms present are retained on the membranes, which are then transferred to a suitable medium for culturing. The numbers of colonies are counted to give presumptive *E. coli* and total coliform counts.

The current standard for approved bivalves waters in the United States is 14 MPN faecal coliforms per 100 ml of water, and no more than 10% of the samples should exceed 43 MPN faecal coliforms per 100 ml (FDA, 1989). The standard for EU countries is 300 faecal coliforms per 100 ml of water, the same value as that recommended for bivalves flesh and intravalvular fluid (Table 12.5). It should be noted, however, that the resistance of *E. coli* and other faecal coliforms to adverse physical and chemical conditions is low, and this reduces the usefulness of the group as an indicator of water quality. At present there are no defined limits for viral contamination of harvesting areas, no appropriate treatment regimes for effective viral removal during processing, and no safe end-product standards for products sold to the consumer. Doré *et al.* (1998) have, therefore, suggested that bivalves containing levels of male-specific RNA bacteriophage (an indicator of viral contamination) below 100 plaque forming units (pfu) per 100 g bivalves are unlikely to be contaminated with viruses causing gastroenteritis, and they suggest that this might constitute a possible end-product standard for oysters. A tentative management strategy for the production of virologically assured oysters is outlined in Fig. 12.4.

Authorities in EU countries are required to monitor bivalve-growing areas regularly for toxic algae blooms, and bivalve tissue for the presence of toxins.

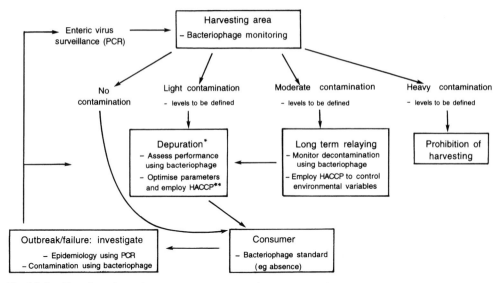

Fig. 12.4. Tentative scheme for management strategy for production of virologically assured oysters. * Relaying in uncontaminated waters may also be suitable under favourable environmental conditions; ** HACCP (Hazard Analysis Critical Control Point) system covered below. From Doré *et al.* (1998).

Areas are closed when toxins exceed the safety limits (Table 12.3) but in the case of heavy metals such as lead and mercury, EU requirements are rather vague. The directive governing bivalve water states that the concentration of the metal in bivalve water or flesh must not exceed a level which gives rise to harmful effects on the bivalves and their larvae. The synergistic effect of the metal and other specified metals must also be taken into consideration (EEC, 1979). The recommendation is similar for pesticides and PCBs (Table 12.3). For hydrocarbons, both dissolved and emulsified, there are no limit values quoted but the same directive stipulates that hydrocarbons must not be present in the bivalve water in such quantities as to produce a visible film on the surface of the water and/or a deposit on the bivalves, or have a harmful effect on the bivalves.

The HAACP system

To ensure that a high-quality product reaches the consumer several quality assurance schemes have been developed such as certification under an international accepted standard (ISO 9000 series), total quality management (TMQ), good manufacturing practice (GMP), good hygienic practice (GHP) and hazard analysis critical control point (HACCP). The United States, Canada and European Union recognise the importance of a formal procedure for food quality assurance, and have proposed a mandatory HACCP-based seafood regulation (FDA, 1994; White & Noseworthy, 1992; EEC, 1993). The HACCP system is based on the recognition that microbial hazards exist at various points

Table 12.6. Hazards and preventative measures during processing and distribution of chilled live molluscs. [1]Contamination with pathogenic bacteria, virus, biotoxins, parasites and chemicals. [2]Time × temperature control. CCP-1 will ensure full control of a hazard; CCP-2 will minimise but not assure full control. Adapted from Huss (1994).

Product flow	Hazard	Preventive measure	Degree of control
Live molluscs	Contamination[1]	Monitoring of environment	CCP-2
Catching	None		
Chilling	Growth of bacteria	T × t control[2]	CCP-1
Transport	Growth of bacteria	T × t control	CCP-1
Reception at plant			
Shucking	None		
Packaging	None		
All processing steps	Growth of bacteria	T × t control	CCP-1
	Contamination	Factory hygiene	CCP-2
		Water quality	CCP-1
		Sanitation	CCP-2
Chilling	Growth of bacteria	T × t control	CCP-1
Distribution	Growth of bacteria	T × t control	CCP-1

in food production, and that measures can be taken to control these hazards. The main elements of the system are:

- identification and assessment of potential hazards associated with growing, harvesting, processing, marketing, preparation and use of a raw material or food product;
- determination of critical control points (CCP) in order to eliminate or minimise hazards; and
- establishment of systems to monitor CCPs and take corrective action when a particular CCP is not under control (see Garrett & Hudak–Roos, 1991, Huss, 1994 and Dillon & Griffith, 1996 for details).

The US National Advisory Committee on Microbiological Criteria for Foods has defined a hazard as a biological, chemical or physical property that may cause a food to be unsafe for consumption. Hazards must be such that their elimination or reduction is essential to the production of safe food (NACMCF, 1992). Examples of hazards and preventative measures during processing and distribution of molluscs are given in Table 12.6. Further information, specific to the United States, is in Doré (1991) and Pierson & Corlett (1992); and in Huss (1994) and in Mortimore & Wallace (1994) for the EU.

Although it is now over 30 years since the HACCP concept was developed (APHA, 1971) the system has not been universally adopted by the fish or bivalve industry. Huss (1994) outlines some of the reasons for this, which include: a general misunderstanding of the concept through lack of training, difficulties in defining hazards and in applying the system at all stages of food production, and lack of mutual trust between regulatory bodies and the industry. These problems will only be overcome through greater understanding and communication between growers, processors, regulatory agencies, the scientific community and the general public.

References

APHA (American Public Health Association) (1971) *Proceedings of the 1971 National Conference on Food Protection.* US Department of Health, Education and Welfare, Public Health Service, Food and Drug Administration, Washington, D.C.

Bernard, F.R. (1989) Uptake and elimination of coliform bacteria by four marine bivalve mollusks. *Can. J. Fish. Aquat. Sci.*, **46**, 1592–9.

Blake, P.A., Allegra, D.T. & Snyder, J.D. (1980) Cholera – a possible epidemic focus in the U.S. *N. Eng. J. Med.*, **302**, 305–9.

Burkhardt, W., Watkins, W.D. & Rippey, S.R. (1992) Seasonal effects on accumulation of microbial indicator organisms by *Mercenaria mercenaria*. *Appl. Environ. Microbiol.*, **58**, 826–31.

Burkholder, J.M. (1999) The lurking perils of *Pfiesteria*. *Sc. Am.*, **28**, 28–35.

Canzonier, W.J. (1988) Public health component of bivalve shellfish production and marketing. *J. Shellfish Res.*, **7**, 261–6.

Casas, N. & Sunen, E. (2001) Detection of enterovirus and hepatitis A virus RNA in mussels (*Mytilus* spp.) by reverse transcriptase-polymerase chain reaction. *J. Appl. Microbiol.*, **90**, 89–95.

Chung, H., Jaykus, L.A., Lovelace, G. & Sobsey, M.D. (1998) Bacteriophages and bacteria as indicators of enteric viruses in oysters and their harvest waters. *Water Sci. Technol.*, **38**, 37–44.

Clement, L.E., Stekoll, M.S. & Shaw, D.G. (1980) Accumulation, fractionation and release of oil by the intertidal clam *Macoma balthica*. *Mar. Biol.*, **57**, 41–50.

Croci, L., De Medici, D., Scalfaro, C. *et al.* (2000) Determination of enteroviruses, hepatitis A virus, bacteriophages and *Escherichia coli* in Adriatic Sea mussels. *J. Appl. Microbiol.*, **88**, 293–8.

Croci, L., Stacchini, A., Cozzi, A. *et al.* (2001) Evaluation of rapid methods for the determination of okadaic acid in mussels. *J. Appl. Microbiol.*, **90**, 73–7.

Depaola, A. & Hwang, G. (1995) Effect of dilution, incubation-time, and temperature of enrichment on cultural and PCR detection of *Vibrio cholerae* obtained from the oyster *Crassostrea virginica*. *Mol. Cell Probes*, **9**, 75–81.

Dillon, M. & Griffith, C. (1996) *How to HACCP*, 2nd edn. M.D. Associates, Grimsby, UK.

Doré, I. (1991) *Shellfish: a Guide to Oysters, Mussels, Scallops, Clams and Similar Products for the Commercial User.* Van Nostrand Reinhold, New York.

Doré, W.J., Henshilwood, K. & Lees, D.N. (1998) The development of management strategies for control of virological quality in oysters. *Water Sci. Technol.*, **38**, 29–35.

Draisci, R., Palleschi, L., Giannetti, L. *et al.* (1999) New approach to the direct detection of known and new diarrhetic shellfish toxins in mussels and phytoplankton by liquid chromatography-mass spectrometry. *J. Chromatogr.* A, **847**, 213–21.

EEC (1979) Council Directive 79/923/EEC: Quality of shellfish. *Off. J. Euro. Comm.* **L 281/1**, 30.10.79.

EEC (1991) Council Directive 91/492/EEC: Laying down the health conditions for the production and the placing on the market of live bivalve molluscs. *Off. J. Euro. Comm.*, **L 268/1**, 24.09.91.

EEC (1993) Council Directive 93/43/EEC on the hygiene of foodstuffs. *Off. J. Euro. Comm.*, **L 175/I**, 19.07.93.

Enriquez, R., Frosner, G.G., Hochsteinmintzel, V., Riedemann, S. & Reinhardt, G. (1992) Accumulation and persistence of hepatitis-A virus in mussels. *J. Med. Virol.*, **37**, 174–9.

Farrington, J.W., Davis, R.C., Frew, N.M. & Rabin, K.S. (1982) No. 2 fuel oil compounds in *Mytilus edulis*: retention and release after an oil spill. *Mar. Biol.*, **66**, 15–26.

FDA (Food and Drugs Administration) (1989) *National Shellfish Sanitation Program. Manual of Operations*. Center for Food Safety and Applied Nutrition, Division of Cooperative Programs, Shellfish Sanitation Branch, Washington D.C.

FDA (Food and Drugs Administration) (1994) Proposal to establish procedures for the safe processing and importing of fish and fishery products. *Federal Register*, **59**, 4142–214.

Flanagan, P.J. (1992) *Parameters of Water Quality: Interpretation and Standards*, 2nd edn. Environmental Research Unit, Dublin, Ireland.

Flanagan, A.F., Callanan, K.R., Donlon, J., Palmer, R., Forde, A. & Kane, M. (2001) A cytotoxicity assay for the detection and differentiation of two families of shellfish toxins. *Toxicon*, **39**, 1021–7.

Garrett, E.S. & Hudak-Roos, M. (1991) U.S. seafood inspection and HACCP. In: *Microbiology of Marine Food Products* (eds D.R. Ward & C.R. Hackney), pp. 111–31. Van Nostrand Reinhold, New York.

Gerba, C.P. & Goyal, S.M. (1978) Detection and occurrence of enteric viruses in shellfish: a review. *J. Food. Prot.*, **41**, 743–54.

Gonzalez, I., Garcia, T., Fernandez, A., Sanz, B., Hernandez, P.E. & Martin, R. (1999) Rapid enumeration of *Escherichia coli* in oysters by a quantitative PCR-ELISA. *J. Appl. Microbiol.*, **86**, 231–6.

Gonzalez, J.C., Leira, F., Vieytes, M.R., Vieites, J.M., Botana, A.M. & Botana, L.M. (2000) Development and validation of a high-performance liquid chromatographic method using fluorimetric detection for the determination of the diarrhetic shellfish poisoning toxin okadaic acid without chlorinated solvents. *J. Chromatogr. A*, **876**, 117–25.

Gualillo, O., Biscardi, D., Di Carlo, R. & De Fusco, R. (1999) Simple method of detecting enteroviruses in contaminated molluscs and sewage by using polymerase chain reaction coupled with a colorimetric microwell detection assay. *Sci. Tot. Environ.*, **244**, 285–9.

Halliday, M.L., Kang, L.Y., Zhou, T.K. *et al.* (1991) An epidemic of hepatitis A attributable to the ingestion of raw clams in Shanghai, China. *J. Infect. Dis.*, **164**, 852–9.

Harrisyoung, L., Tamplin, M., Mason, J.W., Aldrich, H.C. & Jackson, J.K. (1995) Viability of *Vibrio vulnificus* in association with hemocytes of the American oyster (*Crassostrea virginica*). *Appl. Environ. Microbiol.*, **61**, 52–7.

Hlady, W.G. & Mullen, R.C. (1993) *Vibrio vulnificus* infections associated with raw oyster consumption: Florida, 1981–1992. *Arch. Dermatol.*, **129**, 957–8.

Hollingworth, P. & Wekell, M.M. (1990) Fish and other marine products. In: *Official Methods of Analysis*, 15th edn. (ed. K. Helrich), pp. 881–2. Association of Official Analytical Chemists, Arlington, Virginia.

Holmes, M.J., Teo, S.L.M., Lee, F.C. & Khoo, H.W. (1999) Persistent low concentrations of diarrhetic shellfish toxins in green mussels *Perna viridis* from the Johor Strait, Singapore: first record of diarrhetic shellfish toxins from South-East Asia. *Mar. Ecol. Prog. Ser.*, **181**, 257–68.

Huang, X., Hsu, K. & Chu, F.S. (1996) Direct competitive enzyme-linked-immuno-sorbent-assay for saxitoxin and neosaxitoxin. *J. Agricul. Food Chem.*, **44**, 1029–35.

Huss, H.H. (1994) *Assurance of seafood quality*. FAO Fisheries Technical Paper 334, Food and Agricultural Organization of the United Nations, Rome.

Huss, H.H., Reilly, A. & Karim Ben Embarek, P. (2000) Prevention and control of hazards in seafood. *Food Control*, **11**, 149–56.

Ishida, H., Muramatsu, N., Nukaya, H., Kosuge, T. & Tsuji, K. (1996) Study on neurotoxic shellfish poisoning involving the oyster, *Crassostrea gigas*, in New Zealand. *Toxicon*, **34**, 1050–3.

Jaykus, L.A., Deleon, R. & Sobsey, M.D. (1996) A virion concentration method for detection of human enteric viruses in oysters by PCR and oligoprobe hybridization. *Appl. Environ. Microbiol.*, **62**, 2074–80.

Jellett, J.F., Doucette, L.I. & Belland, E.R. (1998) The MIST™ shippable cell bioassay kits for PSP: an alternative to the mouse bioassay. *J. Shellfish Res.*, **17**, 1653–5.

Jones, D.D., Law, R. & Bej, A.K. (1993) Detection of *Salmonella* spp in oysters using Polymerase Chain Reactions (PCR) and gene probes. *J. Food Sci.*, **58**, 1191–7.

Kilgen, M.B. & Cole, M.T. (1991) Viruses in seafood. In: *Microbiology of Marine Food Products* (eds D.R. Ward & C.R. Hackney), pp. 197–209. Van Nostrand Reinhold, New York.

Kim, M.S. & Jeong, H.D. (2001) Development of 16S rRNA targeted PCR methods for the detection and differentiation of *Vibrio vulnificus* in marine environments. *Aquaculture*, **193**, 199–211.

Langston, W.J. (1978) Persistence of polychlorinated biphenyls in marine bivalves. *Mar. Biol.*, **46**, 35–40.

Leftley, J.W. & Hannah, F. (1998) Phycotoxins in seafood. In: *Natural Toxicants in Food* (ed. D.H. Watson), pp. 182–224. Sheffield Academic, Sheffield.

Levine, W.C. & Griffin, P.M. (1993) *Vibrio* infections on the Gulf Coast: the results of a first year of regional surveillance. *J. Infect. Dis.*, **167**, 479–83.

Lewis, G., Loutit, M.W. & Austin, F.J. (1986) Enteroviruses in mussels and marine sediments and depuration of naturally accumulated viruses by green lipped mussels (*Perna canaliculus*). *N. Z. J. Mar. Freshw. Res.*, **20**, 431–7.

Livingstone, D.R. & Pipe, R.K. (1992) Mussels and environmental contaminants: molecular and cellular aspects. In: *The Mussel* Mytilus: *Ecology, Physiology, Genetics and Culture* (ed. E.M. Gosling), pp. 425–64. Elsevier Science Publishers B.V., Amsterdam.

Livingstone, D.R., Chipman, J.K., Lowe, D.M. *et al.* (2000) Development of biomarkers to detect the effects of organic pollution on aquatic invertebrates: recent molecular, genotoxic, cellular and immunological studies on the common mussel (*Mytilus edulis* L.) and other mytilids. *Int. J. Environ. Pollut.*, **13**, 56–91.

Mallett, J.C., Beghian, L.E., Metcalf, T.G. & Kaylor, J.D. (1991) Potential of irradiation technology for improved shellfish sanitation. *J. Food Safety*, **11**, 231–45.

Marino, A., Crisafi, G., Maugeri, T.L., Nostro, A. & Alonzo, V. (1999) Uptake and retention of *Vibrio cholerae* non-01, *Salmonella typhi*, *Escherichia coli* and *Vibrio harvey* by mussels in seawater. *Microbiologica*, **22**, 129–38.

McCarthy, S.A., DePaola, A., Kaysner, C.A., Hill, W.E. & Cook, D.W. (2000) Evaluation of nonisotopic DNA hybridization methods for detection of the TDH gene of *Vibrio parahaemolyticus*. *J. Food Protect.*, **63**, 1660–4.

Mortimore, S. & Wallace, C. (1994) *HACCP: A Practical Approach*. Chapman & Hall, London.

Mountfort, D.O., Suzuki, T. & Truman, P. (2001) Protein phosphatase inhibition assay adapted for determination of total DSP in contaminated mussels. *Toxicon*, **39**, 383–90.

NACMCF (US National Advisory Committee on Microbial Criteria for Foods) (1992) Hazard Analysis Critical Control Point System. *Int. J. Food Microbiol.*, **16**, 1–23.

Novaczek, I., Madhyastha, M.S., Ablett, R.F. & Donald, A. (1992) Depuration of demoic acid from live blue mussels. *Can. J. Fish. Aquat. Sci.*, **49**, 312–18.

NSSP (National Shellfish Sanitation Program) (1989a) *Manual of Operations, Part 1. Sanitation of Shellfish, Growing Areas*. 1989 Revision. Food and Drug Administration, Shellfish Sanitation Branch, North Kingstown, Rhode Island, 115p.

NSSP (National Shellfish Sanitation Program) (1989b) *Manual of Operations, Part 2. Sanitation of the Harvesting, Processing and Distribution of Shellfish.* Food and Drug Administration, Shellfish Sanitation Branch, North Kingstown, Rhode Island, 166p.

Palmork, K.H. & Solbakken, J.E. (1981) Distribution and elimination of [9–14] phenanthrene in the horse mussel (*Modiolus modiolus*). *Environ. Contam. Toxicol.*, **26**, 196–201.

Pierson, M.D. & Corlett Jr., D.A. (1992) *HACCP: Principles and Applications.* Chapman & Hall, New York.

Plusquellec, A., Beucher, M., Prieur, D. & Le Gal, Y. (1990) Contamination of the mussel, *Mytilus edulis* Linnaeus, 1758 by enteric bacteria. *J. Shellfish Res.*, **9**, 95–101.

Pruell, R.J., Lake, J.L., Davis, W.R. & Quinn, J.G. (1986) Uptake and depuration of organic contaminants by blue mussels (*Mytilus edulis*) exposed to environmentally contaminated sediment. *Mar. Biol.*, **91**, 497–507.

Quilliam, M.A. (1999) Committee on natural toxins: phycotoxins. General referee reports. *J. AOAC Int.*, **82**, 773–81.

Renault, T., Le Deuff, R.M., Lipart, C. & Delsert, C. (2000) Development of a PCR procedure for the detection of a herpes-like virus infecting oysters in France. *J. Virol. Methods*, **88**, 41–50.

Rippey, S.R. (1994) Infectious diseases associated with molluscan shellfish consumption. *Clin. Microbiol. Rev.*, **7**, 419–25.

Satake, M., Ofuji, K., Naoki, H. *et al.* (1998) Azaspiracid, a new marine toxin having unique spiro ring assemblies isolated from Irish mussels *Mytilus edulis. J. Am. Chem. Soc.*, **120**, 9967–8.

Schwab, K.J., Neill, F.H., Le Guyader, F., Estes, M.K. & Atmar, R.L. (2001) Development of a reverse transcription-PCR-DNA enzyme immunoassay for detection of 'Norwalk-like' viruses and hepatitis A virus in stool and shellfish. *Appl. Environ. Microbiol.*, **67**, 742–9.

Shumway, S.E. (1992) Mussels and public health. In: *The Mussel* Mytilus: *Ecology, Physiology, Genetics and Culture* (ed. E.M. Gosling), pp. 511–42. Elsevier Science Publishers B.V., Amsterdam.

Shumway, S.E. (1995) Phycotoxin-related shellfish poisoning: bivalve molluscs are not the only vectors. *Rev. Fish. Sci.*, **3**, 1–31.

Sindermann, C. (1990) *Principal Diseases of Marine Fish and Shellfish*, 2nd edn. Vol. II. Academic Press, San Diego, California.

Stewart, J.E., Marks, L.J., Gilgan, M.W., Pfeiffer, E. & Zwicker, B.M. (1998) Microbial utilization of the neurotoxin domoic acid: blue mussels (*Mytilis edulis*) and soft shell clams (*Mya arenaria*) as sources of the microorganisms. *Can. J. Microbiol.*, **44**, 456–64.

Sunen, E. & Sobsey, M.D. (1999) Recovery and detection of enterovirus, hepatitis A virus and Norwalk virus in hardshell clams (*Mercenaria mercenaria*) by RT-PCR methods. *J. Virol. Methods*, **77**, 179–87.

Suzuki, T. & Yasumoto, T. (2000) Liquid chromatography-electrospray ionization mass spectrometry of the diarrhetic shellfish-poisoning toxins okadaic acid, dinophysistoxin-1 and pectenotoxin-6 in bivalves. *J. Chromatogr.* A, **874**, 199–206.

Tanabe, S., Tatsukawa, R. & Phillips, D.J.H. (1987) Mussels as bioindicators of PCB pollution: a case study on uptake and release of PCB isomers and congeners in green-lipped mussels (*Perna viridis*) in Hong Kong waters. *Environ. Pollut.*, **47**, 4–62.

Todd, E.C.D. (1993) Domoic acid and amnesic shellfish poisoning – a review. *J. Food Prot.*, **56**, 69–83.

Vale, P. & Sampayo, M.A.D. (1999) Comparison between HPLC and a commercial immunoassay kit for detection of okadaic acid and esters in Portuguese bivalves. *Toxicon*, **37**, 1565–77.

Vantarakis, A., Komninou, G., Venieri, D. & Papapetropoulou, M. (2000) Development of a multiplex PCR detection of *Salmonella* spp. and *Shigella* spp. in mussels. *Letts. Appl. Microbiol.*, **31**, 105–9.

Webber, D.L. (1982) *The accummulation of faecal indicator bacteria by the mussel*, Mytilus edulis. PhD. Thesis, University of Wales, UK.

White, D.R.L. & Noseworthy, J.E.P. (1992) The Canadian quality management programme. In: *Quality Assurance in the Fish Industry* (ed. H.H. Huss), pp. 509–13. Elsevier Science Publishers B.V., Amsterdam.

Whittle, K. & Gallacher, S. (2000) Marine toxins. *British Med. Bull.*, **56**, 236–53.

WHO (World Health Organization) (1989) *Report of WHO Consultation on Public Health Aspects of Seafood-Bourne Diseases*. WHO/CDS/VPH/90.86.

Widdows, J. & Donkin, P. (1992) Mussels and environmental contaminants: bio-accummmulation and physiological aspects. In: *The Mussel* Mytilus: *Ecology, Physiology, Genetics and Culture* (ed. E.M. Gosling), pp. 383–424. Elsevier Science Publishers B.V., Amsterdam.

Yasumoto, T., Fukui, M., Sasaki, K. & Sugiyama, K. (1995) Determination of marine toxins in foods. *J. AOAC Int.*, **78**, 574–82.

Yentsch, C.M. & Incze, L.S. (1980) Accumulation of algal biotoxins in mussels. In: *Mussel Culture and Harvest: a North American Perspective* (ed. R.A. Lutz), pp. 223–46. Elsevier Science Publishing Co. Inc., New York.

Subject Index

Species Index